FUNDAMENTAL APPROACH TO DISCRETE MATHEMATICS

FUNDAMENTAL APPROACH TO DISCRETE MATHEMATICS

D.P. Acharjya
Assistant Professor
School of Computing Sciences
VIT University, Vellore

Sreekumar
Associate Professor
Rourkela Institute of Management Studies
Rourkela

PUBLISHING FOR ONE WORLD

NEW AGE INTERNATIONAL (P) LIMITED, PUBLISHERS
New Delhi • Bangalore • Chennai • Cochin • Guwahati • Hyderabad
Jalandhar • Kolkata • Lucknow • Mumbai • Ranchi
Visit us at **www.newagepublishers.com**

Published by New Age International (P) Ltd., Publishers
First Edition: 2005
Second Edition: 2009
Reprint: 2009

Branches:

- 36, Malikarjuna Temple Street, Opp. ICWA, Basavanagudi, **Bangalore**. ✆ (080) 26677815
- 26, Damodaran Street, T. Nagar, **Chennai**. ✆ (044) 24353401
- CE–39/1016, Carrier Station Road, Near Paulson Park Hotel, Ernakulam South, Cochin ✆ (0484) 4051303
- Hemsen Complex, Mohd. Shah Road, Paltan Bazar, Near Starline Hotel, **Guwahati**. ✆ (0361) 2543669
- No. 105, 1st Floor, Madhiray Kaveri Tower, 3-2-19, Azam Jahi Road, Nimboliadda, **Hyderabad**. ✆ (040) 24652456
- RDB Chambers (Formerly Lotus Cinema) 106A, Ist Floor, S.N. Banerjee Road, **Kolkata**. ✆ (033) 22275247
- 16-A, Jopling Road, **Lucknow**. ✆ (0522) 2209578
- 142C, Victor House, Ground Floor, N.M. Joshi Marg, Lower Parel, **Mumbai**. ✆ (022) 24927869
- 22, Golden House, Daryaganj, **New Delhi**. ✆ (011) 23262370, 23262368

ISBN: 978-81-224-2607-6

Rs. 295.00

C-09-06-3742

Printed in India at Glorious Printers, Delhi.
Typeset at in-house.

PUBLISHING FOR ONE WORLD
NEW AGE INTERNATIONAL (P) LIMITED, PUBLISHERS
4835/24, Ansari Road, Daryaganj, New Delhi-110002
Visit us at **www.newagepublishers.com**

Dedicated to our Beloved Parents

D.P. Acharjya / Sreekumar

Preface to the Second Edition

It is our great pleasure to put forth the second edition of our book "Fundamental Approach to Discrete Mathematics". This edition is an outcome of the numerous feedback received from students and teachers who had welcomed the first edition. The major change in the second edition of this book is the addition of a new chapter on Generating Function and Recurrence Relation, Combinatorics, and Fuzzy Set Theory. These chapters have been introduced because we think these areas are very interesting and application oriented. The typographical errors and omissions, which were there in the first edition, are taken care of in this edition.

We continued our effort to keep the book student-friendly. By a problem solving approach, we mean that students learn the material through problem-type illustrative examples that provide the motivation behind the concepts and its relation to the theorems and definitions. At the same time, students must discover a solution for the non-trivial aspect of the solution. In such an approach, homework exercises contribute to a major part of the learning process. We have kept the same approach for the newly introduced chapters. We trust and hope that the new edition of the book will help to further illustrate the relevance of the discrete mathematics.

While writing we have referred to several books and journals, we take this opportunity to thank all those authors and publishers. Besides those thanked in the preface of the first edition, we are also thankful to S. Dehuri, B.D. Sahoo and A. Mitra for their constant motivation and contributions in this edition. We are extremely thankful to the editorial and production team of New Age International (P) Ltd. for encouraging us for the second edition and extending their cooperation and help in timely completion of this book.

Constructive criticism and suggestions for further improvement of the book is warmly welcomed. Feel free to mail us in **dpacharjya@gmail.com; dpacharjya@vit.ac.in; sreekumar42003@yahoo.com**

D.P. Acharjya

Sreekumar

Preface to the First Edition

Discrete Mathematics, the study of finite systems, remains at the heart of any contemporary study of computer science, which is a need of every student to attain mathematical maturity and ability to deal with abstraction, algorithms and graphs. Our intention in writing this book is to offer fundamental concepts and methods of discrete mathematics in a precise and clear manner. The objective of writing this book is to provide the students of computer science and information technology the fundamental mathematical basis required to achieve indepth knowledge in the field of computer science. It will also help the students, who have interest in mathematics, to keep insight into mathematical techniques and their importance in real life applications.

This book is intended for one semester introductory course in discrete mathematics. The book is also useful for the students of BE (Computer Science/IT), B.Tech. (Computer Science/IT), MCA and M.Sc.(Computer Science). The material in this book includes fundamental concepts, figures, tables, exercises and solved examples to guide the reader in mastering introductory discrete mathematics.

A discrete mathematics course has many objectives that students should learn the essentials of mathematics and how to think mathematically. To achieve these objectives, we emphasized on mathematical reasoning and problem solving techniques in this book. Each chapter begins with a clear statement of definitions, principles and theorems with illustrative and other descriptive materials. This is followed by sets of solved examples and exercises. The solved examples help to illustrate and amplify the material. This has been done to make the book more flexible and to stimulate further interest in topics. Once basic mathematical concepts have been developed then more complex material and applications are presented.

The mathematical topics to be discussed are mathematical logic, set theory, binary relation, function, algebraic structure such as group theory and ring theory, boolean algebra, graph theory and introduction to lattices. Although many excellent books exist in this area, we introduce this topic still keeping in mind that the reader will use them in practical applications related to computer science and information technology. It is hoped that the theoretical concepts present in this book will permit a student to understand most of the fundamental concepts. The text is designed that the students who do not have a strong background in discrete mathematics will find it is very useful to begin with and the students with an exposure to discrete mathematics will also find the book very useful as some of exercises given are thought provoking and help them for application building.

We have the unique opportunity to express our deepest sense of gratitude to Prof. S. Nanda, NIT, Rourkela; Prof. B.K. Tripathy, Berhampur University, Prof. G.N. Patel, Sambalpur University and Dr. Md. N. Khan, IGIT, Sarang for their effective guidance, sincere advise and valuable suggestions during the project work and thereby inspired us to take up an interesting and challenging project like this. We acknowledge to Prof. Sourya Pattnaik, Director, Rourkela Institute of Management Studies (RIMS), Rourkela who motivated and guided us in this project. We would like to acknowledge the contribution of many people, who have helped to bring this project successful.

In certainty, no technical book can be claimed as the product of its authors alone. We are pleased to acknowledge here the contributions of several colleagues who have had a major influence in this book and the course from which it arose. We shall be grateful to the readers for pointing out errors and omissions that, in spite of all care, might have crept in. We shall be delighted if this book is accepted and appreciated by the scholars of today.

You can **e-mail your comments to debi_69@rediffmail.com; debi_rims@yahoo.co.in or sreekumar42003@yahoo.com .**

At last but not the least we express our heartfelt thanks to M/s New Age International (P) Ltd, Publishers, New Delhi, for the cooperation and publication with high accuracy.

D.P. Acharjya
Sreekumar

Contents

Preface — *v*

List of Symbols — *xvii*

1. Mathematical Logic — **1–17**

- 1.0 Introduction — 1
- 1.1 Statement (Proposition) — 1
- 1.2 Logical Connectives — 2
- 1.3 Conditional — 3
- 1.4 Bi-Conditional — 4
- 1.5 Converse — 5
- 1.6 Inverse — 5
- 1.7 Contra Positive — 5
- 1.8 Exclusive OR — 6
- 1.9 NAND — 6
- 1.10 NOR — 6
- 1.11 Tautology — 7
- 1.12 Contradiction — 7
- 1.13 Satisfiable — 7
- 1.14 Duality Law — 8
- 1.15 Algebra of Propositions — 8
- 1.16 Mathematical Induction — 8
- *Solved Examples* — 9
- *Exercises* — 15

2. Set Theory — **19–40**

- 2.0 Introduction — 19
- 2.1 Sets — 19

2.2	Types of Sets	20
2.3	Cardinality of a Set	21
2.4	Subset and Superset	22
2.5	Comparability of Sets	23
2.6	Power Set	23
2.7	Operations on Sets	23
2.8	Disjoint Sets	28
2.9	Application of Set Theory	28
2.10	Product of Sets	29
2.11	Fundamental Products	30
	Solved Examples	30
	Exercises	39
3.	**Binary Relation**	**41–68**
3.0	Introduction	41
3.1	Binary Relation	41
3.2	Inverse Relation	42
3.3	Graph of Relation	43
3.4	Kinds of Relation	44
3.5	Arrow Diagram	45
3.6	Void Relation	45
3.7	Identity Relation	45
3.8	Universal Relation	46
3.9	Relation Matrix (Matrix of the Relation)	46
3.10	Composition of Relations	46
3.11	Types of Relations	48
3.12	Types of Relations and Relation Matrix	50
3.13	Equivalence Relation	52
3.14	Partial Order Relation	53
3.15	Total Order Relation	54
3.16	Closures of Relations	55
3.17	Equivalence Classes	56
3.18	Partitions	57
	Solved Examples	58
	Exercises	66
4.	**Function**	**69–92**
4.0	Introduction	69
4.1	Function	69
4.2	Equality of Functions	71
4.3	Types of Function	71
4.4	Graph of Function	72

4.5 Composition of Functions 74
4.6 Inverse Function 76
4.7 Some Important Functions 77
4.8 Hash Function 79
Solved Examples 80
Exercises 89

5. Generating Function and Recurrence Relation 93–140

5.0 Introduction 93
5.1 Generating Functions 93
5.2 Partitions of Integers 95
5.3 Recurrence Relations 97
5.4 Models of Recurrence Relation 97
5.5 Linear Recurrence Relation with Constant Coefficients 100
5.6 Different Methods of Solution 101
5.7 Homogeneous Solutions 102
5.8 Particular Solution 104
5.9 Total Solution 105
5.10 Solution by Generating Function 105
5.11 Analysis of the Algorithms 106
Solved Examples 112
Exercises 136

6. Combinatorics 141–171

6.0 Introduction 141
6.1 Fundamental Principle of Counting 141
6.2 Factorial Notation 142
6.3 Permutation 142
6.4 Combination 144
6.5 The Binomial Theorem 147
6.6 Binomial Theorem for Rational Index 150
6.7 The Catalan Numbers 151
6.8 Ramsey Number 153
Solved Examples 155
Exercises 168

7. Group Theory 173–197

7.0 Introduction 173
7.1 Binary Operation on a Set 173
7.2 Algebraic Structure 173
7.3 Group 174
7.4 Subgroup 178

7.5 Cyclic Group 181
7.6 Cosets 183
7.7 Homomorphism 183
Solved Examples 184
Exercises 197

8. Codes and Group Codes 199–204

8.0 Introduction 199
8.1 Terminologies 199
8.2 Error Correction 200
8.3 Group Codes 200
8.4 Weight of Code Word 200
8.5 Distance Between the Code Words 200
8.6 Error Correction for Block Code 201
8.7 Cosets 202
Solved Examples 202
Exercises 204

9. Ring Theory 205–226

9.0 Introduction 205
9.1 Ring 205
9.2 Special Types of Ring 207
9.3 Ring without Zero Divisor 209
9.4 Integral Domain 209
9.5 Division Ring 209
9.6 Field 209
9.7 The Pigeonhole Principle 211
9.8 Characteristics of a Ring 211
9.9 Sub Ring 213
9.10 Homomorphism 213
9.11 Kernel of Homomorphism of Ring 215
9.12 Isomorphism 215
Solved Examples 216
Exercises 225

10. Boolean Algebra 227–253

10.0 Introduction 227
10.1 Gates 227
10.2 More Logic Gates 229
10.3 Combinatorial Circuit 231
10.4 Boolean Expression 232
10.5 Equivalent Combinatorial Circuits 233

10.6	Boolean Algebra	234
10.7	Dual of a Statement	237
10.8	Boolean Function	238
10.9	Various Normal Forms	239
	Solved Examples	240
	Exercises	249
11.	**Introduction to Lattices**	**255–267**
11.0	Introduction	255
11.1	Lattices	255
11.2	Hasse Diagram	256
11.3	Principle of Duality	257
11.4	Distributive Lattice	259
11.5	Bounded Lattice	261
11.6	Complemented Lattice	262
11.7	Some Special Lattices	263
	Solved Examples	263
	Exercises	266
12.	**Graph Theory**	**269–308**
12.0	Introduction	269
12.1	Graph	270
12.2	Kinds of Graph	271
12.3	Digraph	272
12.4	Weighted Graph	272
12.5	Degree of a Vertex	273
12.6	Path	273
12.7	Complete Graph	274
12.8	Regular Graph	274
12.9	Cycle	274
12.10	Pendant Vertex	275
12.11	Acyclic Graph	275
12.12	Matrix Representation of Graphs	276
12.13	Connected Graph	279
12.14	Graph Isomorphism	280
12.15	Bipartite Graph	280
12.16	Subgraph	281
12.17	Walks	282
12.18	Operations on Graphs	283
12.19	Fusion of Graphs	285
	Solved Examples	289
	Exercises	303

13. Tree **309–347**

13.0 Introduction 309
13.1 Tree 309
13.2 Fundamental Terminologies 310
13.3 Binary Tree 311
13.4 Bridge 312
13.5 Distance and Eccentricity 313
13.6 Central Point and Centre 314
13.7 Spanning Tree 315
13.8 Searching Algorithms 317
13.9 Shortest Path Algorithms 321
13.10 Cut Vertices 323
13.11 Euler Graph 324
13.12 Hamiltonian Path 326
13.13 Closure of a Graph 326
13.14 Travelling Salesman Problem 327
Solved Examples 328
Exercises 343

14. Fuzzy Set Theory **349–372**

14.0 Introduction 349
14.1 Fuzzy Versus Crisp 349
14.2 Fuzzy Sets 350
14.3 Basic Definitions 352
14.4 Basic Operations on Fuzzy Sets 355
14.5 Properties of Fuzzy Sets 357
14.6 Interval Valued Fuzzy Set 358
14.7 Operations on i-v Fuzzy Sets 359
14.8 Fuzzy Relations 360
14.9 Operations on Fuzzy Relations 362
14.10 Fuzzy Logic 364
Solved Examples 366
Exercises 371

References **373–380**

Index **381–386**

List of Symbols

Mathematical Logic

$(P \wedge Q)$	: P and Q
$(P \vee Q)$	: P or Q
$\neg P$	: negation of P
$P \rightarrow Q$	: if P, then Q
$P \leftrightarrow Q$	: P if and only if Q
$P \bar{\vee} Q$	: P exclusive or Q
$P \uparrow Q$	: P NAND Q
$P \downarrow Q$	: P NOR Q
$P \equiv Q$	: P and Q are logically equivalent

Set Theory

$\{a, b, c, d\}$	: set containing the elements a, b, c, d.
$x \in A$	: x is an element of A
$x \notin A$	: x is not an element of A
$\{x \mid P(x)\}$	: $P(x)$ is the property that describes the elements x of a set.
ϕ	: null set
$\lvert A \rvert$	: cardinality of A
$A \approx B$	: A similar to B (A and B containing equal number of elements)
$A \subseteq B$	: A is a subset of B
$A = B$	: set equality (A and B have same elements)
$A \subset B$	: A is a proper subset of B
$A \not\subset B$	: A is not a subset of B
$A \neq B$	: A and B are not equal
$P(A)$	: power set of A
$A \cup B$	: A union B
$A \cap B$	: A intersection B
$A - B$	: set difference
$A \oplus B$	: set symmetric difference
$A \Delta B$	: set symmetric difference
A^c	: complement of A
$\overline{A}$	: complement of A
A'	: complement of A
$n(A)$	: number of distinct elements of A.
$A \times B$	: Cartesian product of A and B
(x, y)	: Order pair
$(x_1, x_2, x_3, \cdots, x_n)$	: n-tuple of $x_1, x_2, x_3, \cdots, x_n$
$\bigcup_{i=1}^{n} A_i$	: union of $A_1, A_2, A_3, \cdots, A_n$
$\bigcap_{i=1}^{n} A_i$	: intersection of $A_1, A_2, A_3, \cdots, A_n$

Binary Relation

$x R y$	: $(x, y) \in R$ *i.e.*, [x is related to y]
$x \not{R} y$	: $(x, y) \notin R$ *i.e.*, [x is not related to y]
D(R)	: domain of the relation R
dom. R	: domain of the relation R
R(R)	: range of the relation R
rng. R	: range of the relation R
R^{-1}	: inverse of the relation R
I_A	: identity relation

M(R) : matrix of the relation R.
R_1R_2 : composition of relations R_1 and R_2
r(R) : reflexive closure of the relation R
s(R) : symmetric closure of the relation R
t(R) : transitive closure of the relation R
$[x]$: equivalence class containing x.

Function

$f : A \to B$: function from A to B
R(f) : range of a function f
$f(x)$: value assigned to x
$g \circ f$: composition of f and g
$g f$: composition of f and g
f^{-1} : inverse function
$|x|$: absolute value function
$[x]$: greatest integer function
$\lfloor x \rfloor$: floor function
$\lceil x \rceil$: ceiling function
χ_A : characteristic function of A
$R_y(x)$: remainder function
sgn (x) : signum function
H(n) : hash function

Generating Function and Recurrence Relation

G(x) : generating function for the sequence of real numbers $a_0, a_1, a_2, \cdots, a_n, \cdots$
E(x) : exponential generating function
P(n) : partition function [number of partitions of an integer n]
s_n^h : homogeneous solution to the linear difference equation
s_n^p : particular solution to the linear difference equation
$s_n^h + s_n^p$: total solution to the linear difference equation

$\theta(n^2)$: theta of n^2.
lg$n$: logarithm to the base 2 of n
lg2 : logarithm to the base 2 of 2

Combinatorics

$n!$: factorial of a natural number n
$P(n, r)$: permutation of r elements out of n elements
$C(n, r)$: combination of r elements out of n elements
$c(x)$: generating function for the Catalan numbers
T_{r+1} : general term in the binomial expansion.
C_n : n^{th} Catalan number
$R(m, n)$: Ramsey number

Group Theory

$(a \circ b)$: a binary operation b
(A, $\circ$) : algebraic structure
O(G) : order of the group G [number of elements of G]
O(a) : order of an element a
aH : left coset of H in G
Ha : right coset of H in G

Codes and Group Codes

(G, $\oplus$) : group code
ω(X) : weight of the code word X
d(X, Y) : distance between two code words X and Y
P(X_i | Y) : probability that X_i is transmitted when the received word is Y
(G $\oplus$ y) : coset with respect y where (G, $\oplus$) is a group code and y is a word

Ring Theory

Z_p : commutative ring
Ch(R) : characteristic of a ring
I(ϕ) : the kernel of homomorphism

Boolean Algebra

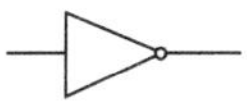 : NOT gate (inverter)

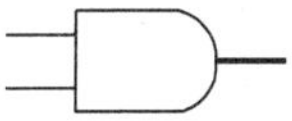 : AND gate

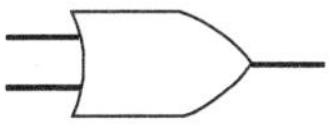 : OR gate

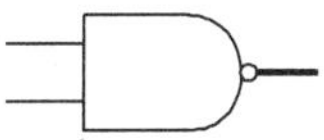 : NAND gate

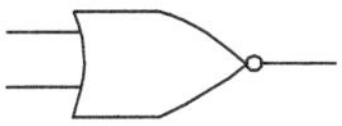 : NOR gate

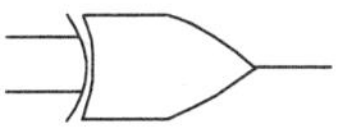 : XOR gate

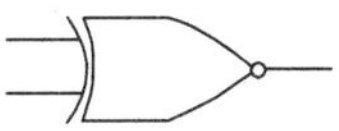 : XNOR gate

x' : not x

$\bar{x}$: not x

$(x_1 \wedge x_2)$: x_1 and x_2

$(x_1 \vee x_2)$: x_1 or x_2

$(x_1 \oplus x_2)$: exclusive-OR of x_1 and x_2

$\{S, \wedge, \vee, ', 0, 1\}$: Boolean algebra

$f : B^n \to B$: Boolean function if $f(x_1, x_2, ..., x_n) = X(x_1, x_2, ..., x_n)$

Introduction to Lattices

L.U.B : least upper bound

G.L.B : greatest lower bound

$(a \vee b)$: join of a and b

$(a \wedge b)$: meet of a and b

$D(n)$: set of all positive divisors of n

$a | b$: a divides b

Graph Theory

G(V, E) : graph with finite set of vertices V and a finite set of edges E

$|V|$: order of graph G

$|E|$: size of graph G

degree (v) : degree of the vertex v

$e = (u, v)$: edge of a graph

$w(e)$: weight of the edge e

K_n : complete graph with n vertices

$A[a_{ij}]$: adjacency matrix of the graph G

$I[a_{ij}]$: incidence matrix of the graph G

$P[p_{ij}]$: path matrix of the graph G

$(G_1 \cup G_2)$: union of two graphs and G_1 and G_2

$(G_1 \cap G_2)$: intersection of two graphs and G_1 and G_2

$\bar{G}$: complement of the graph G

$(G_1 \times G_2)$: product of graphs and G_1 and G_2

$G_1[G_2]$: composition of two graphs and G_1 and G_2

Tree

$d(u, v)$: distance between two vertices u and v

$e(v)$: eccentricity of the vertex v

rad (G) : radius of the graph G

diam (G) : diameter of the graph G

BFS : Breadth First Search

DFS : Depth First Search

C(G) : closure of graph G

Fuzzy Set Theory

$\tilde{A}$: fuzzy set of the universe of discourse X

$\mu_{\tilde{A}}(x) : X \to [0, 1]$: membership function of $\tilde{A}$

$\tilde{A} - \tilde{B}$: equality of fuzzy sets

$\tilde{A} \subseteq \tilde{B}$: containment

Support $(\tilde{A})$: Support of a fuzzy set

A_α : α-level cut

$k.\tilde{A}$: product of a fuzzy set by a crisp number

$\tilde{A}^m$: m power of a fuzzy set

CON : concentration

DIL : dilation

$\tilde{A} \cup \tilde{B}$: union of fuzzy sets

$\tilde{A} \cap \tilde{B}$: intersection of fuzzy sets

$\tilde{A}^c$: complement of $\tilde{A}$

$\tilde{A} \cdot \tilde{B}$: product of two fuzzy sets

$\tilde{A} - \tilde{B}$: difference of two fuzzy sets

$\tilde{A} \oplus \tilde{B}$: disjunctive sum

i–v fuzzy set : interval valued fuzzy set

$L_{\tilde{A}}$: lower bound of $\tilde{A}$

$U_{\tilde{A}}$: upper bound of $\tilde{A}$

D[0, 1] : family of all closed intervals contained in the interval [0, 1]

$\overline{\mu_{\tilde{A}}}(x)$: $X \to D[0, 1]$: membership function of $i - v$ fuzzy set

$\tilde{R} = \tilde{A} \times \tilde{B}$: fuzzy Cartesian product

$\tilde{R} \circ \tilde{S}$: composition of fuzzy relations $\tilde{R}$ and $\tilde{S}$

$\tilde{P}$: fuzzy proposition

$T(\tilde{P})$: truth value of the fuzzy proposition

Miscellaneous

$\forall$: for all

$|$: such that

$\Rightarrow$: implies

$\sum_{n=0}^{\infty} s_n$: $s_0 + s_1 + s_2 + \cdots + s_n + \cdots$

$\prod_{i=0}^{n} a_i$: $a_0 . a_1 . a_2 . a_3 \ldots a_n$

$\Leftrightarrow$: if and only if

1

Mathematical Logic

1.0 INTRODUCTION

Mathematics is considered to be a deductive science. We infer things from certain premises through logical reasoning. Consider an example.

Three caps problem. A certain father had three sons: Smith, Clark and Jones. The father brought three caps of different colours; say Red, Blue and Black. He showed them the caps. After which they are folded blind. The father put three caps on the heads of three sons. Then the sons were taken away from his father to another room. Few minutes after father called Jones and removed the blindfold of Jones and asked him to tell the colour of his cap. Jones said he could not infer about the colour of his own cap. Then he called Clark and removed the blindfold of Clark and asked him to tell the colour of his cap by looking at the colour of the cap of Jones. He too could not infer. Then he called Smith and asked him to tell the colour of his cap without removing the blindfold of Smith. Smith replied he could tell the colour of the cap on his own head.

How Smith come to that conclusion? Let us see. Smith asked two questions one to Jones and another to Clark. He asked to Jones about the colour of Clark's cap and asked to Clark about the colour of Jones's cap. By the way he got two colours of the cap. As a result Smith got the colour of his own cap.

In the above reasoning we have certain premises and we conclude from them by a pure deductive reasoning. In the following passages we shall formalize the process of deduction.

1.1 STATEMENT (PROPOSITION)

A statement is a declarative sentence which is either true or false but not both. The statement is also known as proposition. The truth value True and False are denoted by the symbols **T** and **F** respectively. Sometimes it is also denoted by **1** and **0**, where **1** stands for true and **0** stands for false. As it depends on only two possible truth values, we call it as two-valued logic or bi-valued logic.

Consider the following examples

(*a*) Man is mortal.

(*b*) Sun rises in the east.
(*c*) Two is less than five.
(*d*) May God bless you!
(*e*) x is a Dog.
(*f*) Lopez is a nice Cat.
(*g*) It is too cold today.
(*h*) 6 is a composite number.

From the above example, it is very clear that (*a*), (*b*), (*c*) and (*h*) are statements as they declare a definite truth value T or F. The other example (*d*), (*e*), (*f*), and (*g*) are not statements as they do not declare any truth value T or F.

Consider the sentence 111011 + 11 = 111110

The above sentence is a statement but its truth value depends on the context. If we consider the binary number system, the statement is True (T) but in decimal number system the statement is False (F).

1.2 LOGICAL CONNECTIVES

Another important aspect is that logical connectives. We use some logical connectives to connect several statements into a single statement. The most basic and fundamental connectives are Negation, Conjunction and Disjunction.

1.2.1 Negation

It is observed that the negation of a statement is also a statement. We use the connective **Not** for negation. Usually the statements are denoted by single letters P, Q, R etc. If P be a statement, then the negation of P is denoted as $\neg$ P.

Consider the example of a statement.

P: New York is the capital of France.

$\neg$ P: New York is not the capital of France.

As we all know that Paris is the capital of France, the truth value for the statements P is false (F) and $\neg$ P is true (T). From the above it is clear that P and $\neg$ P has opposite truth values. $\neg$ P can also be written as

$\neg$ P: It is not true that New York is the capital of France.

Rule: If P is true, then $\neg$ P is false and if P is false, then $\neg$ P is true.

Truth Table (Negation)

P	$\neg$ P
T	F
F	T

1.2.2 Conjunction

The conjunction of two statements P and Q is also a statement denoted by (P $\wedge$ Q). We use the connective **And** for conjunction.

Consider the example where P and Q are two statements.

P : 2 + 3 = 5

Q : 5 is a composite number.

So, (P ∧ Q) : 2 + 3 = 5 and 5 is a composite number.

As another example if P: Smith went to the college and Q: Mary went to the college, then (P ∧ Q): Smith and Mary went to the college.

It is clear that (P ∧ Q) stand for P and Q. In order to make (P ∧ Q) true, P and Q have to be simultaneously true.

Rule: (P ∧ Q) is true if both P and Q are true, otherwise false.

Truth Table (Conjunction)

P	Q	(P ∧ Q)
T	T	T
T	F	F
F	T	F
F	F	F

1.2.3 Disjunction

The disjunction of two statements P and Q is also a statement denoted by (P∨ Q). We use the connective **Or** for disjunction. Consider the example where P and Q are two statements

P : 2 + 3 is not equal to 5

Q : 5 is a prime number

So, (P ∨ Q) : 2 + 3 is not equal to 5 or 5 is a prime number.

It is observed that (P ∨ Q) is true when P may be true or Q may be true and this also includes the case when both are true, that is the truth value of one statement is not assumed in exclusion of the truth value of the other statement. We call it as also inclusive or.

Rule: (P ∨ Q) is true, if either P or Q is true and it is false when both P and Q are false.

Truth Table (Disjunction)

P	Q	(P ∨ Q)
T	T	T
T	F	T
F	T	T
F	F	F

1.3 CONDITIONAL

Let P and Q be any two statements. Then the statement P → Q is called a conditional statement. This can be put in any one of the following forms.

(*a*) If P, then Q
(*b*) P only if Q
(*c*) P implies Q
(*d*) Q if P

In an implication P → Q, P is called the antecedent (hypothesis) and Q is called the consequent (conclusion). To explain the conditional statement, consider the example

A boy promises a girl, "I will take you boating on Sunday, if it is not raining".

Now if it is raining, then the boy would not be deemed to have broken his promise. The boy would be deemed to have broken his promise only when it is not raining and the boy did not take the girl for boating on Sunday.

Let us break the above conditional statement to symbolic from.

P: It is not raining.

Q: I will take you boating on Sunday.

So, the above statement reduces to $P \to Q$.

From the above discussion it is clear that if P is false, then $P \to Q$ is true, whatever be the truth value of Q. The conditional $P \to Q$ is false, if P is true and Q is false.

Rule: An implication (conditional) $P \to Q$ is false only when the hypothesis (P) is true and conclusion (Q) is false, otherwise true.

Truth Table (Conditional)

P	Q	$(P \to Q)$
T	T	T
T	F	F
F	T	T
F	F	T

■ 1.4 BI-CONDITIONAL

Let P and Q be any two statements. Then the statement $P \leftrightarrow Q$ is called a bi-conditional statement. This $P \leftrightarrow Q$ can be put in any one of the following forms.

(*a*) P if and only if Q

(*b*) P is necessary and sufficient of Q

(*c*) P is necessary and sufficient for Q

(*d*) P is implies and implied by Q

The bi-conditional (double implication) $P \leftrightarrow Q$ is defined as

$$(P \leftrightarrow Q): (P \to Q) \wedge (Q \to P)$$

From the truth table discussed below, it is clear that $P \leftrightarrow Q$ has the truth value T whenever both P and Q have identical truth values.

Truth Table (Bi-Conditional)

P	Q	$P \to Q$	$Q \to P$	$(P \leftrightarrow Q)$
T	T	T	T	T
T	F	F	T	F
F	T	T	F	F
F	F	T	T	T

Rule: $(P \leftrightarrow Q)$ is true only when both P and Q have identical truth values, otherwise false.

1.5 CONVERSE

Let P and Q be any two statements. The converse statement of the conditional $P \rightarrow Q$ is given as $Q \rightarrow P$.

Consider the example "all concurrent triangles are similar". The above statement can also be written as "if triangles are concurrent, then they are similar".

Let P: Triangles are concurrent
Q: Triangles are similar

So, the statement becomes $P \rightarrow Q$. The converse statement is given as "if triangles are similar, then they are concurrent" or all similar triangles are concurrent.

1.6 INVERSE

Let P and Q be any two statements. The inverse statement of the conditional $(P \rightarrow Q)$ is given as $(\neg P \rightarrow \neg Q)$.

Consider the example "all concurrent triangles are similar". The above statement can also be written as "if triangles are concurrent, then they are similar".

Let P: Triangles are concurrent
Q: Triangles are similar

So, the statement becomes $P \rightarrow Q$. The inverse statement is given as "if triangles are not concurrent, then they are not similar".

1.7 CONTRA POSITIVE

Let P and Q be any two statements. The contra positive statement of the conditional $(P \rightarrow Q)$ is given as $(\neg Q \rightarrow \neg P)$. Consider the Example "all concurrent triangles are similar". The above statement can also be written as "if triangles are concurrent, then they are similar".

Let P: Triangles are concurrent and
Q: Triangles are similar

So, the statement becomes $P \rightarrow Q$. The contra positive statement is given as "if triangles are not similar, then they are not concurrent".

Truth Table (Contra Positive)

P	Q	$P \rightarrow Q$	$\neg Q$	$\neg P$	$(\neg Q \rightarrow \neg P)$
T	T	T	F	F	T
T	F	F	T	F	F
F	T	T	F	T	T
F	F	T	T	T	T

From the truth table it is observed that both conditional $(P \rightarrow Q)$ and contra positive $(\neg Q \rightarrow \neg P)$ have same truth values.

1.8 EXCLUSIVE OR

Let P and Q be any two statements. The exclusive OR of two statements P and Q is denoted by $(P \veebar Q)$. We use the connective XOR for exclusive OR. The exclusive OR $(P \veebar Q)$ is true if either P or Q is true but not both. The exclusive OR is also termed as exclusive disjunction.

Consider the example where P and Q be two statements such that $P \equiv 2 + 3 = 5$ and $Q \equiv 5 - 3 = 2$. Here both the statements are true. Therefore $(P \veebar Q)$ is false.

Rule: $(P \veebar Q)$ is true if either P or Q is true but not both, otherwise false.

Truth Table (Exclusive OR)

P	Q	$(P \veebar Q)$
T	T	F
T	F	T
F	T	T
F	F	F

1.9 NAND

The word NAND stands for NOT and AND. The connective NAND is denoted by the symbol ↑. If P and Q be two statements, then NAND of P and Q is given as $(P \uparrow Q)$ defined by

$$(P \uparrow Q) \equiv \neg (P \wedge Q).$$

Rule: $(P \uparrow Q)$ is true if either P or Q is false, otherwise false.

Truth Table (NAND)

P	Q	$(P \uparrow Q)$
T	T	F
T	F	T
F	T	T
F	F	T

1.10 NOR

The word NOR stands for NOT and OR. The connective NOR is denoted by the symbol ↓. If P and Q be two statements, then NOR of P and Q is given as $(P \downarrow Q)$ defined by

$$(P \downarrow Q) \equiv \neg (P \vee Q)$$

Rule: $(P \downarrow Q)$ is true only when both P and Q are false, otherwise false.

Truth Table (NOR)

P	Q	$(P \downarrow Q)$
T	T	F
T	F	F
F	T	F
F	F	T

1.11 TAUTOLOGY

If the truth values of a composite statement are always true irrespective of the truth values of the atomic (individual) statements, then it is called a tautology.

For example the composite statement $(P \wedge (P \to Q)) \to Q$ is a tautology. To verify this draw the truth table with composite statement as $(P \wedge (P \to Q)) \to Q$

Truth Table

P	Q	$(P \to Q)$	$P \wedge (P \to Q)$	$(P \wedge (P \to Q)) \to Q$
T	T	T	T	T
T	F	F	F	T
F	T	T	F	T
F	F	T	F	T

So, $(P \wedge (P \to Q)) \to Q$ is a tautology.

1.12 CONTRADICTION

If the truth values of a composite statement are always false irrespective of the truth values of the atomic statements, then it is called a contradiction or unsatisfiable.

For example the composite statement $\neg (P\to (Q \to (P \wedge Q)))$ is a contradiction.

To verify this draw the truth table of $\neg (P \to (Q \to (P \wedge Q)))$. Let $R \equiv P \to (Q \to (P \wedge Q))$

Truth Table

P	Q	$(P \wedge Q)$	$Q \to (P \wedge Q)$	$(P \to (Q \to (P \wedge Q))$	$\neg R$
T	T	T	T	T	F
T	F	F	T	T	F
F	T	F	F	T	F
F	F	F	T	T	F

So, $\neg R \equiv \neg (P \to (Q \to (P \wedge Q)))$ is a contradiction.

1.13 SATISFIABLE

If the truth values of a composite statement are some times true and some times false irrespective of the truth values of the atomic statements, then it is called a satisfiable.

Consider the composite statement $(P \to Q) \to (Q \to P)$. To verify this draw the truth table of $(P \to Q) \to (Q \to P)$.

Truth Table

P	Q	$P \to Q$	$Q \to P$	$(P \to Q) \to (Q \to P)$
T	T	T	T	T
T	F	F	T	T
F	T	T	F	F
F	F	T	T	T

So, the composite statement $(P \to Q) \to (Q \to P)$ is satisfiable.

1.14 DUALITY LAW

Two formulae P and P* are said to be duals of each other if either one can be obtained from the other by interchanging $\wedge$ by $\vee$ and $\vee$ by $\wedge$. The two connectives $\wedge$ and $\vee$ are called dual to each other.

Consider the formulae $P \equiv (P \vee Q) \wedge R$ and $P^* \equiv (P \wedge Q) \vee R$ which are dual to each other.

1.15 ALGEBRA OF PROPOSITIONS

If P, Q and R be three statements, then the following laws hold good.

(*a*) Commutative Laws: $P \wedge Q \equiv Q \wedge P$ and
$P \vee Q \equiv Q \vee P$

(*b*) Associative Laws: $P \wedge (Q \wedge R) \equiv (P \wedge Q) \wedge R$ and
$P \vee (Q \vee R) \equiv (P \vee Q) \vee R$

(*c*) Distributive Laws: $P \wedge (Q \vee R) \equiv (P \wedge Q) \vee (P \wedge R)$ and
$P \vee (Q \wedge R) \equiv (P \vee Q) \wedge (P \vee R)$

(*d*) Idempotent Laws: $P \wedge P \equiv P$ and
$P \vee P \equiv P$

(*e*) Absorption Laws: $P \vee (P \wedge Q) \equiv P$ and
$P \wedge (P \vee Q) \equiv P$

1.15.1 de Morgan's Laws

If P and Q be two statements, then

(*i*) $\neg (P \wedge Q) \Leftrightarrow (\neg P) \vee (\neg Q)$ and
(*ii*) $\neg (P \vee Q) \Leftrightarrow (\neg P) \wedge (\neg Q)$

1.16 MATHEMATICAL INDUCTION

Generally direct methods are adopted for proving theorems and propositions. Sometimes it is too difficult and tedious. As a result the other methods are developed for proving theorems and propositions. These are (*i*) method of contra positive, (*ii*) method of contradiction and (*iii*) method of induction. Here, we will discuss method of induction. The method of induction is otherwise known as mathematical induction.

Suppose that n be a natural number. Our aim is to show that some statement P(n) involving n is true for any n. The following steps are used in mathematical induction.

1. Suppose that P(n) be a statement.
2. Show that P(1) and P(2) are true. *i.e.*, P(n) is true for $n = 1$ and $n = 2$.
3. Assume that P(k) is true. *i.e.*, P(n) is true for $n = k$.
4. Show that P($k + 1$) follows from P(k).

Consider an example $1 + 2 + 3 + \ldots + n = \dfrac{n(n+1)}{2}$

Suppose that $P(n) \equiv 1 + 2 + 3 + \ldots + n = \dfrac{n(n+1)}{2}$

So, $$P(1) \equiv 1 = \frac{1(1+1)}{2}$$

and $$P(2) \equiv 1 + 2 = 3 = \frac{2(2+1)}{2}$$

So, P(1) and P(2) are true.

Assume that P(k) is true. So,

$$1 + 2 + 3 + \ldots + k = \frac{k(k+1)}{2}$$

So,
$$P(k+1) \equiv 1 + 2 + 3 + \ldots + k + (k+1)$$
$$= \frac{k(k+1)}{2} + (k+1) \qquad [\because \text{ P}(k) \text{ is true.}]$$
$$= \left(\frac{k+1}{2}\right)(k+2) = \frac{(k+1)(k+2)}{2}$$

which shows that P(k + 1) is also true. Hence, P(n) is true for all n.

SOLVED EXAMPLES

Example 1 *Find the negation of $P \rightarrow Q$.*

Solution: $P \rightarrow Q$ is equivalently written as $(\neg P \vee Q)$

So, negation of $$P \rightarrow Q \equiv \neg(\neg P \vee Q)$$
$$\equiv \neg(\neg P) \wedge (\neg Q), \text{ (By de Morgan's Law)}$$
$$\equiv P \wedge (\neg Q)$$

Hence the negation of $P \rightarrow Q$ is $P \wedge (\neg Q)$.

Example 2 *Construct the truth table for $(P \rightarrow Q) \leftrightarrow (\neg P \vee Q)$.*

Solution: The given compound statement is $(P \rightarrow Q) \leftrightarrow (\neg P \vee Q)$ where P and Q are two atomic statements.

P	**Q**	**¬ P**	**P→ Q**	**¬ P ∨ Q**	**(P → Q) ↔ (¬ P ∨ Q)**
T	T	F	T	T	T
T	F	F	F	F	T
F	T	T	T	T	T
F	F	T	T	T	T

Example 3 *Construct the truth table for $P \rightarrow (Q \leftrightarrow P \wedge Q)$.*

Solution: The given compound statement is $P \rightarrow (Q \leftrightarrow P \wedge Q)$, where P and Q are two atomic statements.

P	**Q**	**P ∧ Q**	**Q ↔ P ∧ Q**	**P → (Q ↔ P ∧ Q)**
T	T	T	T	T
T	F	F	T	T
F	T	F	F	T
F	F	F	T	T

Example 4 *Find the negation of the following statement. "If Cows are Crows then Crows are four legged".*

Solution: Let P: Cows are Crows

Q: Crows are four legged

Given statement : If Cows are Crows then Crows are four legged.

$$\equiv P \rightarrow Q$$

So, the negation is given as $P \wedge (\neg Q)$ *i.e.* Cows are Crows and Crows are not four legged.

Example 5 *Find the negation of the following statement.*

He is rich and unhappy.

Solution: Let $P \equiv$ He is rich

$Q \equiv$ He is unhappy

Given statement: He is rich and unhappy

$$\equiv P \wedge Q$$

By de Morgan's law $\neg(P \wedge Q) \equiv \neg P \vee \neg Q$

$\equiv$ He is neither rich nor unhappy.

Example 6 *Prove by constructing truth table*

$$P \rightarrow (Q \vee R) \equiv (P \rightarrow Q) \vee (P \rightarrow R)$$

Solution: Our aim to prove $P \rightarrow (Q \vee R) \equiv (P \rightarrow Q) \vee (P \rightarrow R)$

Let P, Q and R be three atomic statements.

P	Q	R	$Q \vee R$	$P \rightarrow (Q \vee R)$	$P \rightarrow Q$	$P \rightarrow R$	$(P \rightarrow Q) \vee (P \rightarrow R)$
T	T	T	T	T	T	T	T
T	F	F	F	F	F	F	F
F	T	F	T	T	T	T	T
F	F	T	T	T	T	T	T
F	T	T	T	T	T	T	T
T	F	T	T	T	F	T	T
T	T	F	T	T	T	F	T
F	F	F	F	T	T	T	T

From the truth table it is clear that $P \rightarrow (Q \vee R) \equiv (P \rightarrow Q) \vee (P \rightarrow R)$.

Example 7 *Find the negation of $P \leftrightarrow Q$.*

Solution: $P \leftrightarrow Q$ is equivalently written as $(P \rightarrow Q) \wedge (Q \rightarrow P)$

So, $\neg(P \leftrightarrow Q) \equiv \neg((P \rightarrow Q) \wedge (Q \rightarrow P))$

$\equiv \neg(P \rightarrow Q) \vee \neg(Q \rightarrow P)$; (de Morgan's Law)

$\equiv \neg(\neg P \vee Q) \vee \neg(\neg Q \vee P)$

$\equiv (P \wedge \neg Q) \vee (Q \wedge \neg P)$; (de Morgan's Law)

Hence, $\neg(P \leftrightarrow Q) \equiv (P \wedge \neg Q) \vee (Q \wedge \neg P)$.

Example 8 *With the help of truth table prove that $\neg(P \wedge Q) \equiv \neg P \vee \neg Q$.*

Solution: Our claim is $\neg(P \wedge Q) \equiv \neg P \vee \neg Q$.

Let P and Q be two atomic statements.

P	Q	$P \wedge Q$	$\neg(P \wedge Q)$	$\neg P$	$\neg Q$	$\neg P \vee \neg Q$
T	T	T	F	F	F	F
T	F	F	T	F	T	T
F	T	F	T	T	F	T
F	F	F	T	T	T	T

From the truth table it is clear that $\neg(P \wedge Q) \equiv \neg P \vee \neg Q$.

Example 9 *Show that $(P \to Q) \leftrightarrow (\neg P \vee Q)$ is a tautology.*

Solution: Let P and Q be two atomic statements. Our aim is to show $(P \to Q) \leftrightarrow (\neg P \vee Q)$ is a tautology.

P	Q	$P \to Q$	$\neg P$	$\neg P \vee Q$	$(P \to Q) \leftrightarrow (\neg P \vee Q)$
T	T	T	F	T	T
T	F	F	F	F	T
F	T	T	T	T	T
F	F	T	T	T	T

Hence $(P \to Q) \leftrightarrow (\neg P \vee Q)$ is a tautology.

Example 10 *Show that the following statements are equivalent.*

Statement 1: Good food is not cheap

Statement 2: Cheap food is not good.

Solution: Let $P \equiv$ Food is good and $Q \equiv$ Food is cheap

Statement 1: Good food is not cheap

i.e., $P \to \neg Q$

Statement 2: Cheap food is not good

i.e., $Q \to \neg P$

Truth Table

P	Q	$\neg P$	$\neg Q$	$P \to \neg Q$	$Q \to \neg P$
T	T	F	F	F	F
T	F	F	T	T	T
F	T	T	F	T	T
F	F	T	T	T	T

From truth table it is clear that both statements are equivalent.

Example 11 *Express $P \to Q$ using $\downarrow$ and $\uparrow$ only.*

Solution:

$$\begin{aligned} P \to Q &\equiv \neg P \vee Q \\ &= \neg P \vee \neg(\neg Q) \\ &\equiv \neg(P \wedge \neg Q) \equiv P \uparrow \neg Q \\ &\equiv P \uparrow (\neg Q \vee \neg Q) \\ &\equiv P \uparrow \neg(Q \wedge Q) \equiv P \uparrow (Q \uparrow Q) \end{aligned}$$

i.e., $P \to Q \equiv P \uparrow (Q \uparrow Q)$

Example 12 *Prove that $(P \wedge Q) \wedge \neg(P \vee Q)$ is a contradiction.*

Solution: Truth table for $(P \wedge Q) \wedge \neg (P \vee Q)$

P	Q	$(P \wedge Q)$	$(P \vee Q)$	$\neg (P \vee Q)$	$(P \wedge Q) \wedge \neg (P \vee Q)$
T	T	T	T	F	F
T	F	F	T	F	F
F	T	F	T	F	F
F	F	F	F	T	F

Hence, $(P \wedge Q) \wedge \neg (P \vee Q)$ is a contradiction.

Example 13 *Express $P \leftrightarrow Q$ using $\downarrow$ and $\uparrow$ only.*

Solution:

$$\begin{aligned}
P \leftrightarrow Q &\equiv (P \to Q) \wedge (Q \to P)\\
&\equiv (\neg P \vee Q) \wedge (P \vee \neg Q)\\
&\equiv ((\neg P \vee Q) \wedge P) \vee ((\neg P \vee Q) \wedge \neg Q)\\
&\equiv \neg (\neg ((\neg P \vee Q) \wedge P) \wedge \neg ((\neg P \vee Q) \wedge \neg Q))\\
&\equiv \neg ((\neg P \vee Q) \wedge P) \uparrow \neg ((\neg P \vee Q) \wedge \neg Q)\\
&\equiv ((\neg P \vee Q) \uparrow P) \uparrow ((\neg P \vee Q) \uparrow \neg Q)\\
&\equiv (\neg (P \wedge \neg Q) \uparrow P) \uparrow (\neg (P \wedge \neg Q) \uparrow \neg Q)\\
&\equiv ((P \uparrow \neg Q) \uparrow P) \uparrow ((P \uparrow \neg Q) \uparrow \neg Q)\\
&\equiv ((P \uparrow (\neg Q \vee \neg Q) \uparrow P) \uparrow ((P \uparrow (\neg Q \vee \neg Q)) \uparrow (\neg Q \vee \neg Q))\\
&\equiv ((P \uparrow \neg (Q \wedge Q) \uparrow P) \uparrow ((P \uparrow \neg (Q \wedge Q)) \uparrow \neg (Q \wedge Q))\\
&\equiv ((P \uparrow (Q \uparrow Q) \uparrow P) \uparrow ((P \uparrow (Q \uparrow Q)) \uparrow (Q \uparrow Q))
\end{aligned}$$

Note: These expressions are not unique.

Alternative Solution:

$$\begin{aligned}
P \leftrightarrow Q &\equiv (P \to Q) \wedge (Q \to P)\\
&\equiv (\neg P \vee Q) \wedge (P \vee \neg Q)\\
&\equiv ((\neg P \vee Q) \wedge P) \vee ((\neg P \vee Q) \wedge \neg Q)\\
&\equiv ((\neg P \wedge P) \vee (Q \wedge P)) \vee ((\neg P \wedge \neg Q) \vee (Q \wedge \neg Q))\\
&\equiv (Q \wedge P) \vee (\neg P \wedge \neg Q)\\
&\equiv \neg (\neg (Q \wedge P)) \vee \neg (P \vee Q)\\
&\equiv \neg (\neg (Q \wedge P) \wedge (P \vee Q))\\
&\equiv \neg (Q \wedge P) \uparrow (P \vee Q)\\
&\equiv (Q \uparrow P) \uparrow \neg (\neg (P \vee Q))\\
&\equiv (Q \uparrow P) \uparrow \neg (\neg P \wedge \neg Q))\\
&\equiv (Q \uparrow P) \uparrow (\neg P \uparrow \neg Q)\\
&\equiv (Q \uparrow P) \uparrow ((\neg P \vee \neg P) \uparrow (\neg Q \vee \neg Q))\\
&\equiv (Q \uparrow P) \uparrow (\neg (P \wedge P) \uparrow \neg (Q \wedge Q))\\
&\equiv (Q \uparrow P) \uparrow ((P \uparrow P) \uparrow (Q \uparrow Q))
\end{aligned}$$

Example 14 *Prove that $n (n + 1)$ is an even natural number.*

Solution: Suppose that $P(n) \equiv n (n + 1)$ is even.

So, $P(1) \equiv 1(1 + 1) = 2$, which is even and

$P(2) \equiv 2 (2 + 1) = 6$, which is also even.

Hence, P(1) and P(2) are true.

Assume that $P(k) \equiv k\,(k + 1)$ is even

i.e., $k\,(k + 1) = 2m;\ m \in N$

So, $P(k + 1) \equiv (k + 1)\,(k + 2) = k\,(k + 1) + 2\,(k + 1)$

$= 2m + 2\,(k + 1)$ [$\therefore$ P(k) is true.]

$= 2(m + k + 1)$, which is even.

Which shows that P(k + 1) is also true.

So, P(n) is true for all n.

Example 15 *Show by truth table the following statements are equivalent.*

Statement 1: Rich men are unhappy.

Statement 2: Men are unhappy or poor.

Solution: Let P ≡ Men are Rich and Q ≡ Men are unhappy.

Statement 1: Rich men are unhappy.

i.e., If men are rich then they are unhappy.

i.e., $P \rightarrow Q$.

Statement 2: Men are unhappy or poor.

i.e., $Q \vee \neg P$; (Here poor indicates not rich)

Truth Table

P	Q	P → Q	¬ P	Q ∨ ¬ P
T	T	T	F	T
T	F	F	F	F
F	T	T	T	T
F	F	T	T	T

So, it is clear that both statements are equivalent.

Example 16 *A boy promises a girl "I will take you park on Monday if it is not raining". When the boy would be deemed to have broken his promise. Explain with the help of truth table.*

Solution: Let P: I will take you park on Monday

Q: It is raining.

Given statement: I will take you park on Monday if it is not raining

i.e., P if $\neg$ Q

i.e., $\neg Q \rightarrow P$

Truth Table

P	Q	¬ Q	¬ Q → P
T	T	F	T
T	F	T	T
F	T	F	T
F	F	T	F

It indicates that if ¬ Q is true and P is false, then the boy is deemed to have broken his promise. *i.e.* when it is not raining and the boy does not take her park on Monday, then the boy is deemed to have broken his promise.

Example 17 *Prove by method of induction*

$$1^3 + 2^3 + 3^3 + \ldots + n^3 = \left(\frac{n(n+1)}{2}\right)^2$$

Solution: Suppose that $P(n) \equiv 1^3 + 2^3 + 3^3 + \ldots + n^3 = \left(\frac{n(n+1)}{2}\right)^2$

So, $$P(1) = 1^3 = 1 = \left(\frac{1(1+1)}{2}\right)^2$$

and $$P(2) = 1^3 + 2^3 = 9 = \left(\frac{2(2+1)}{2}\right)^2$$

Hence, P(1) and P(2) are true.
Assume that $P(k)$ is true, so

$$P(k) \equiv 1^3 + 2^3 + 3^3 + \ldots\ldots + k^3 = \left(\frac{k(k+1)}{2}\right)^2$$

$$P(k+1) \equiv 1^3 + 2^3 + 3^3 + \ldots\ldots + k^3 + (k+1)^3$$

$$= \left(\frac{k(k+1)}{2}\right)^2 + (k+1)^3 \qquad [\because\ P(k) \text{ is true.}]$$

$$= (k+1)^2 (k^2 + 4(k+1)) / 4$$

$$= \left(\frac{(k+1)(k+2)}{2}\right)^2$$

Which shows that $P(k+1)$ is also true.
So, $P(n)$ is true for all n .

Example 18 *Show by method of induction*

$$\frac{1}{1*2} + \frac{1}{2*3} + \frac{1}{3*4} + \ldots + \frac{1}{n*(n+1)} = \frac{n}{n+1}$$

Solution: Suppose that

$$P(n) \equiv \frac{1}{1*2} + \frac{1}{2*3} + \frac{1}{3*4} + \ldots + \frac{1}{n*(n+1)} = \frac{n}{n+1}$$

So, $$P(1) \equiv \frac{1}{1*2} = \frac{1}{2} = \frac{1}{1+1}$$

and $$P(2) \equiv \frac{1}{1*2} + \frac{1}{2*3} = \frac{1}{2} + \frac{1}{6} = \frac{2}{3} = \frac{2}{2+1}$$

Assume that P(k) is true. So,

$$P(k) \equiv \frac{1}{1*2}+\frac{1}{2*3}+\frac{1}{3*4}+\ldots+\frac{1}{k*(k+1)}=\frac{k}{k+1}$$

$\therefore$

$$P(k+1) \equiv \frac{1}{1*2}+\frac{1}{2*3}+\frac{1}{3*4}+\ldots+\frac{1}{k*(k+1)}+\frac{1}{(k+1)*(k+2)}$$

$$=\frac{k}{k+1}+\frac{1}{(k+1)*(k+2)}; \qquad [\because \ P(k) \text{ is true}]$$

$$=\frac{1}{(k+1)}\left(k+\frac{1}{(k+2)}\right)$$

$$=\frac{1}{(k+1)}\left(\frac{k^2+2*k+1}{(k+2)}\right)$$

$$=\frac{(k+1)^2}{(k+1)(k+2)}=\frac{k+1}{k+2}$$

Which shows that P(k + 1) is also true.
So, P(n) is true for all n.

EXERCISES

1. Find the negation of the following statements.
   (*a*) Today is Sunday or Monday.
   (*b*) If I am tired and busy, then I cannot study.
   (*c*) Either it is raining or some one left the shower on.
   (*d*) The moon rises in the west.
   (*e*) The triangles are equilateral is necessary and sufficient for three equal sides.
   (*f*) $2 + 3 \neq 18$.
2. Prove the following by using truth table.
   (*a*) $P \vee (Q \wedge R) \equiv (P \vee Q) \wedge (P \vee R)$
   (*b*) $P \wedge (Q \vee R) \equiv (P \wedge Q) \vee (P \wedge R)$
   (*c*) $\neg (P \vee Q) \equiv \neg P \wedge \neg Q$
   (*d*) $P \to (Q \wedge R) \equiv (P \to Q) \wedge (P \to R)$
   (*e*) $(P \wedge Q) \wedge R \equiv P \wedge (Q \wedge R)$
   (*f*) $P \vee Q = \neg(\neg P \wedge \neg Q)$
   (*g*) $P \overline{\vee} Q \equiv (P \vee Q) \wedge \neg (P \wedge Q)$
   (*h*) $(P \downarrow Q) \downarrow (P \downarrow Q) \equiv P \vee Q$
   (*i*) $P \wedge Q \equiv (P \downarrow P) \downarrow (Q \downarrow Q)$
   (*j*) $\neg (P \vee Q) \vee (\neg P \wedge Q) \equiv \neg P$
3. For each of the following formulas tell whether it is (*i*) tautology, (*ii*) satisfiable, or (*iii*) contradiction.
   (*a*) $(P \to (Q \to R)) \to ((P \to Q) \to (P \to R))$
   (*b*) $(P \to (Q \to R)) \leftrightarrow ((P \wedge Q) \to R)$
   (*c*) $P \wedge \neg Q$
   (*d*) $(P \vee Q) \to P$
   (*e*) $\neg (P \to Q) \to (P \wedge \neg Q)$
   (*f*) $(P \to Q) \to (Q \to P)$
   (*g*) $((P \to Q) \leftrightarrow Q) \to P$
   (*h*) $\neg P \wedge (P \vee Q) \to P$
   (*i*) $P \to (P \wedge Q)$
   (*j*) $P \to (Q \to (P \wedge Q))$
   (*k*) $(P \vee Q) \leftrightarrow (Q \wedge P)$
   (*l*) $(P \to (Q \to (P \wedge Q))) \leftrightarrow P$

4. Prove by using different laws.

(*a*) $\neg(P \vee Q) \vee (\neg P \wedge Q) \equiv \neg P$ (*b*) $P \vee (P \wedge Q) \equiv P$

(*c*) $(P \vee Q) \wedge \neg P \equiv \neg P \wedge Q$

5. Write each of the following in symbolic form by indicating statements.

(*a*) Brown is rich and unhappy.

(*b*) Jackson speaks English or French.

(*c*) I am hungry and I can study.

(*d*) I am tired if and only if I work hard.

(*e*) If New York is a city, then it is the capital of US.

(*f*) 5 + 2 = 7 if 7 – 2 = 5.

6. Write the truth value of each of the following statements.

(*a*) Sun rises in the south.

(*b*) Man is mortal.

(*c*) London is the capital of UK.

(*d*) If three sides of a triangle are equal, then it is an equilateral triangle.

(*e*) $(11101)_2 + (1)_2 = (11110)_2$

(*f*) $(11101)_{10} + (1)_{10} = (11110)_{10}$

(*g*) $(11111)_2 + (1)_2 = (100000)_2$ and $(111)_2 = (7)_{10}$

(*h*) $(270)_8 + (5)_8 = (184)_{10}$ or $(11101)_2 + (111)_2 = (100101)_2$

(*i*) $2^2 = 9$ if and only if $2 \neq 3$

(*j*) $(111)_2 + (010)_2 = (1001)_2$ if and only if $(1001)_2 - (010)_2 = (111)_2$.

7. Write the converse, inverse and contra positive of the following statement by indicating the conditional statement.

(*a*) In binary number system 1 + 1 = 10.

(*b*) Good food are not cheap.

(*c*) If $9x + 36 = 9$, then $x \neq 17$.

(*d*) If $\cos(x) = 1$, then $x = 0$.

(*e*) Two sets are similar, if they contains equal number of elements.

8. Prove by using method of induction.

(*a*) $1^2 + 2^2 + 3^2 + \ldots + n^2 = \dfrac{n(n+1)(2n+1)}{6}$

(*b*) $1 + r + r^2 + \ldots + r^{n-1} = \dfrac{1 - r^n}{1 - r}; r \neq 1$

(*c*) $1 + r + r^2 + \ldots + r^n = \dfrac{1 - r^{n+1}}{1 - r}; r \neq 1$

(*d*) $a + ar + ar^2 + \ldots + ar^n = \dfrac{a\left(1 - r^{n+1}\right)}{1 - r}; r \neq 1$

(*e*) $a + (a + d) + (a + 2d) + \ldots + (a + (n-1)d) = \dfrac{n(2a + (n-1)d)}{2}$

(*f*) $3 + 7 + 11 + \ldots + (4n - 1) = n\,(2n + 1)$

(*g*) $2 + 4 + 6 + \ldots + 2n = n\,(n + 1)$

(*h*) $1^2 + 4^2 + 7^2 + \ldots + (3n - 2)^2 = \dfrac{n\left(6n^2 - 3n - 1\right)}{2}$

(*i*) $3 * 6 + 6 * 9 + \ldots + 3n\,(3n + 3) = 3n\,(n + 1)\,(n + 2)$

(*j*) $1 * 2 + 2 * 3 + 3 * 4 + \ldots + n(n + 1) = \dfrac{n\,(n+1)\,(n+2)}{3}$

(*k*) $1 * 2 * 3 + 2 * 3 * 4 + \ldots + n(n + 1)\,(n + 2) = \dfrac{n\,(n+1)\,(n+2)\,(n+3)}{4}$

(*l*) $1 + 2 * 3 + 3 * 5 + \ldots + n(2n - 1) = \dfrac{n\,(n+1)\,(4n-1)}{6}$

(*m*) $1 * 3 * 5 + 3 * 5 * 7 + \ldots + (2n - 1)\,(2n + 1)(2n + 3) = n\,(n + 2)(2n^2 + 4n - 1).$

(*n*) $1^2 + (1^2 + 2^2) + (1^2 + 2^2 + 3^2) + \ldots + (1^2 + 2^2 + \ldots + n^2) = \dfrac{n\,(n+1)^2\,(n+2)}{12}$

(*o*) $1 * 2^2 + 2 * 3^2 + \ldots + n\,(n + 1)^2 = \dfrac{n\,(n+1)\,(n+2)\,(3n+5)}{12}$

(*p*) $3 * 8 + 6 * 11 + \ldots + 3n\,(3n + 5) = 3n\,(n + 1)\,(n + 3)$

(*q*) $1 + (1 + 4) + (1 + 4 + 7) + \ldots + (1 + 4 + 7 + \ldots + (3n - 2)) = \dfrac{n^2\,(n+1)}{2}$

(*r*) $2 + 6 + 12 + 20 + \ldots + \dfrac{n\,(2n+2)}{2} = \dfrac{n\,(n+1)\,(n+2)}{3}$

(*s*) $1 + \dfrac{1}{2} + \dfrac{1}{4} + \ldots + \dfrac{1}{2^{n-1}} = \dfrac{2^n - 1}{2^{n-1}}$

(*t*) $1 * 4 + 2 * 7 + 3 * 10 + \ldots + n\,(3n + 1) = n\,(n + 1)^2.$

2
Set Theory

■ 2.0 INTRODUCTION

An ordinary understanding of a set is a collection of objects. In our day-to-day life we use phrases like a set of utensils, a bunch of flowers, a set of books, a herd of cattle, a set of birds and etc., which are all examples of sets.

In the 19th century the German Mathematician George Cantor developed the theory of sets to define numbers and to base mathematics on a solid logical foundation. In late 19th century, Frege developed these ideas further, but his work did not attract much attention. In 20th century Bertrand Russell rediscovered his analysis independently. His works in 1903 led to the monumental work with North Whitehead the principia Mathematica a landmark in the foundations of mathematics. It was observed in 1940s that all mathematics could develop from the idea of sets and mathematics was systematized.

In this chapter we try to impart fundamental concepts and approach to the problem, that is how to proceed for the expected solution as for as set theory is concerned. By the way, we will study and learn about the basic concepts of sets, some of the operations on sets, Venn diagrams, Cartesian product of sets and its applications.

■ 2.1 SETS

Collection of well defined objects is called a set. Well defined means distinct and distinguishable. The objects are called as elements of the set. The ordering of elements in a set does not change the set, *i.e.*, the ordering of elements can not play a vital role in the set theory. For example

$$A = \{a, b, c, d\} \text{ and } B = \{b, a, d, c\} \text{ are equal sets.}$$

The symbol $\in$ stands for 'belongs to'. $x \in A$ means x is an element of the set A. It is observed that if A be a set and x is any object, then either $x \in A$ or $x \notin A$ but not both. Generally sets are denoted by capital letters A, B, C and etc.

Consider the examples of set:

$$A = \{2, 4, 6, 8, 10, 12, 14, 16, 18\}$$

$$B = \{x, y, z, u, v, w\}$$
$$N = \{1, 2, 3, ...\}$$
$$I = \{..., -2, -1, 0, 1, 2, 3, ...\}$$

In general the set can be expressed in two ways, Tabular method (Roster method) and Set-builder method (Specification method).

2.1.1 Tabular Method

Expressing the elements of a set within a parenthesis where the elements are separated by commas is known as tabular method, roster method or method of extension.

Consider the example

$$A = \{1, 3, 5, 7, 9, 11, 13, 15\}$$

2.1.2 Set Builder Method

Expressing the elements of a set by a rule or formula is known as set-builder method, specification method or method of intension. Mathematically

$$S = \{x \mid P(x)\}$$

where P(x) is the property that describes the elements of the set. The symbol | stands for 'such that'. It is not possible to write every set in tabular form. Consider an example

$$S = \{x \mid x \text{ is an Italian}\}$$

The above set S can not be expressed in tabular form as it is impossible to list all Italians. Consider the examples

$$A = \{x \mid x = 2n + 1; 0 \le n \le 7; n \in I\}$$
$$= \{1, 3, 5, 7, 9, 11, 13, 15\}$$

and

$$B = \{x \mid x = 1, x = a, x = \text{Book}, x = \text{Pen}\}$$
$$= \{1, a, \text{Book}, \text{Pen}\}$$

From the second example given above it is clear that the elements of a set do not have any common property also.

■ 2.2 TYPES OF SETS

On considering real life problems, it is observed that the sets are of different types. Keeping in view to these problems, we discuss different types of sets in this section.

2.2.1 Finite Set

A set which contains finite number of elements is known as finite set. Consider the example of finite set as —

$$A = \{a, b, c, d, e\}$$

2.2.2 Infinite Set

A set which contains infinite number of elements is known as infinite set. Consider the example of infinite set as

$$N = \{1, 2, 3, 4, ...\}$$
$$I = \{..., -3, -2, -1, 0, 1, 2, 3, ...\}$$

2.2.3 Singleton Set

A set which contains only one element is known as a singleton set. Consider the example

$$S = \{9\}$$

2.2.4 Pair Set

A set which contains only two elements is known as a pair set. Consider the examples

$$S = \{e, f\}$$
$$S = \{\{a\}, \{1, 3, 5\}\}$$

2.2.5 Empty Set

A set which contains no element is known as empty set. The empty set is also known as void set or null set. Generally denoted by ϕ. Consider the examples

(i) $\phi = \{x : x \neq x\}$
(ii) $\phi = \{x : x$ is a month of the year containing 368 days$\}$

2.2.6 Set of Sets

A set which contains sets is known as set of sets. Consider the example

$$A = \{\{a, b\}, \{1\}, \{1, 2, 3, 4\}, \{u, v\}, \{\text{Book, Pen}\}\}$$

2.2.7 Universal Set

A set which is superset of all the sets under consideration or particular discussion is known as universal set. Generally denoted by U or E or Ω. Generally, the universal set can be chosen arbitrarily for discussion, but once chosen it is fixed for discussion. Consider the example:

Let $A = \{a, b, c\}$
$B = \{a, e, i, o, u\}$
$C = \{p, q, r, s\}$

So, we can take the universal set U as $\{a, b, c, ..., z\}$

i.e., $U = \{a, b, c, d, e, ..., z\}$

2.3 CARDINALITY OF A SET

If S be a set, then the number of elements present in the set S is known as cardinality of S and is denoted by $|S|$. Mathematically if $S = \{s_1, s_2, s_3, ..., s_k\}$, then $|S| = k; k \in N$.

Consider the example

Let $A = \{2, 4, 8, 16, 32, 64, 128, 256\}$

So, $|A| = 8$

2.3.1 Equivalent Sets

Two sets A and B are said to be equivalent if they contains equal number of elements. In other words A and B are said to be equivalent if they have same cardinality, *i.e.* $|A| = |B|$. The equivalent sets are also known as similar sets and is denoted as $A \approx B$.

Consider the example of two sets.

$$A = \{a, e, i, o, u\}$$
$$B = \{7, 9, 11, 13, 15\}$$

Here, $|A| = 5 = |B|$. Thus A and B are similar.

2.4 SUBSET AND SUPERSET

Set A is said to be a subset of B or set B is said to be the superset of A if each element of A is also an element of the set B. We write $A \subseteq B$.

i.e., $A \subseteq B \leftrightarrow \{x \in A \rightarrow x \in B; \forall x \in A\}$

Consider the examples

(*i*) Let $A = \{1, 2, 3, 4, 5, 6\}$
$B = \{1, 2, 3, 4, 5, 6, 7, 8\}$
So $A \subseteq B$.

(*ii*) Let $A = \{a, b, c\}$
$B = \{b, c, a\}$
so, $A \subseteq B$ and $B \subseteq A$.

(*iii*) Let $A = \{\ \}$ and $B = \{1, 2, 3\}$
So, $A \subseteq B$.

2.4.1 Equal Sets

Two sets A and B are said to be equal if and only if every element of A is in B and every element of B is in A, *i.e.* $A \subseteq B$ and $B \subseteq A$. Mathematically

$$A = B \leftrightarrow \{A \subseteq B \text{ and } B \subseteq A\}$$

i.e.,
$$A = B \leftrightarrow \{x \in A \leftrightarrow x \in B\}$$

Consider the example: Let $A = \{x, y, z, p, q, r\}$

$$B = \{p, q, r, x, y, z\}$$

So, $B \subseteq A$ and $A \subseteq B$. Thus $A = B$.

2.4.2 Proper Subset

Set A is said to be a proper subset of B if each element of A is also an element of B and set B has at least one element which is not an element of set A. We write $A \subset B$.

Mathematically

$$A \subset B \leftrightarrow \{x \in A \rightarrow x \in B \text{ and for at least one } y \in B \rightarrow y \notin A\}.$$

Consider an example

Let
$$A = \{a, b, c, d\}$$
$$B = \{a, b, c, d, e, f, g\}$$

Here for $x \in A$ we have $x \in B$ and $y = e \in B$ such that $y = e \notin A$. Thus $A \subset B$.

Note: 1. Every set is a subset of itself, *i.e.* $A \subseteq A$.

2. Empty set is a subset of every set, *i.e.* $\phi \subseteq A$.

■ 2.5 COMPARABILITY OF SETS

Two sets A and B are said to be comparable if any one of the following relation holds.

i.e., (*i*) $A \subset B$ or (*ii*) $B \subset A$ or (*iii*) $A = B$.

Consider the following sets

$$A = \{a, b, c, d, e\};\ B = \{2, 3, 5\} \text{ and } C = \{c, d, e\}.$$

It is clear that $A \not\subset B$, $B \not\subset A$ and $A \neq B$. So, A and B are not comparable. Similarly $B \not\subset C$, $C \not\subset B$ or $C \neq B$. So, B and C are also not comparable. At the same time it is clear that, $C \subset A$, thus A and C are comparable.

■ 2.6 POWER SET

Power set is of great importance while studying finite state systems such as non-deterministic finite automation. Here, we present the concept of power set that will be useful while studying finite state systems.

If A be a set, then the set of all subsets of A is known as power set of A and is denoted as P(A).

Mathematically, $P(A) = \{X : X \subseteq A\}$

Consider the example:

Let $A = \{a\}$

$\Rightarrow$ $P(A) = \{\Phi, \{a\}\}$

Let $A = \{a, b\}$

$\Rightarrow$ $P(A) = \{\{a\}, \{b\}, \{a, b\}, \Phi\}$

Let $A = \{a, b, c\}$

$\Rightarrow$ $P(A) = \{\{a\}, \{b\}, \{c\}, \{a, b\}, \{b, c\}, \{c, a\}, \{a, b, c\}, \Phi\}$

From the above examples it is clear that if a set A contains n elements, then the power set of A, *i.e.* P(A) contains 2^n elements.

i.e., $|A| = n \Rightarrow |P(A)| = 2^n$.

■ 2.7 OPERATIONS ON SETS

It is observed that set theory is a tool to solve many real life problems. In order to solve these problems, it is essential to study different set operations. Here we discuss certain operations such as union, intersection and difference in order to develop an algebra of sets.

2.7.1 Union

If A and B be two sets, then the union $(A \cup B)$ is defined as a set of all those elements which are either in A or in B or in both.

Symbolically,

$$A \cup B = \{x : x \in A \text{ or } x \in B\}$$

Venn diagram

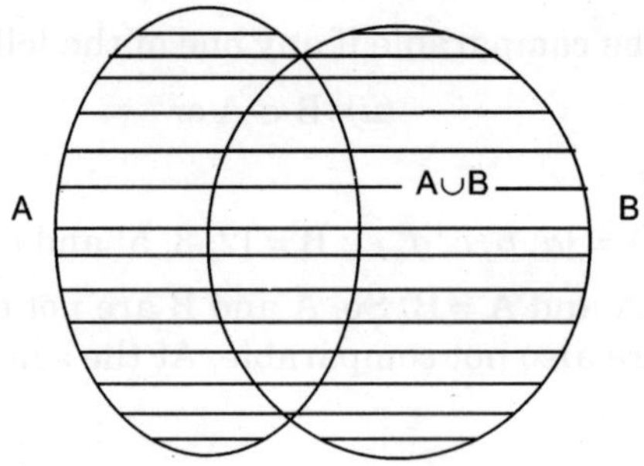

Consider the example:

Let $A = \{a, b, c, d, e\}$

$B = \{a, e, i, o, u\}$

Therefore, $(A \cup B) = \{a, b, c, d, e, i, o, u\}$

2.7.2 Intersection

If A and B be two sets, then the intersection $(A \cap B)$ is defined as a set of all those elements which are common to both the sets. Symbolically

$$(A \cap B) = \{x : x \in A \text{ and } x \in B\}$$

Venn diagram

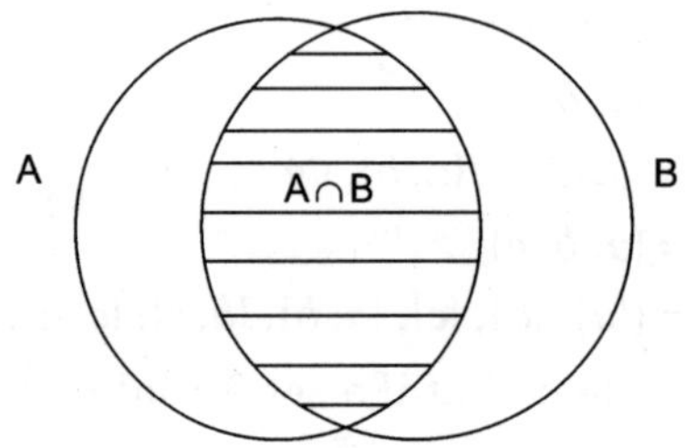

Consider the example:

Let $A = \{a, b, c, d, e\}$

$B = \{a, e, i, o, u\}$

Therefore, $(A \cap B) = \{a, e\}$

2.7.3 Difference

If A and B be two sets, then the difference $(A - B)$ is defined as a set of all those elements of A which are not in B. Symbolically, $(A - B) = \{x \mid x \in A \text{ and } x \notin B\}$

Venn diagram

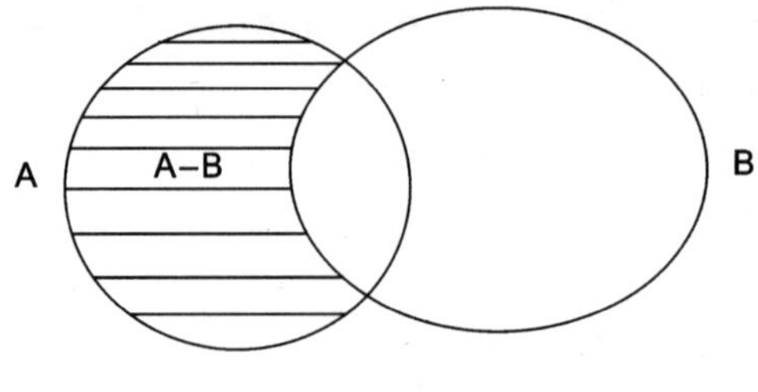

Consider the example:

Let $A = \{a, b, c, d, e, f\}$

$B = \{a, c, i, o, u, k\}$

Therefore, $(A - B) = \{b, d, e, f\}$

2.7.4 Symmetric Difference

If A and B be two sets, then the symmetric difference $(A \Delta B)$ or $(A \oplus B)$ is defined as a set of all those elements which are either in A or in B but not in both.

Symbolically,

$$(A \oplus B) = (A - B) \cup (B - A)$$

Venn diagram

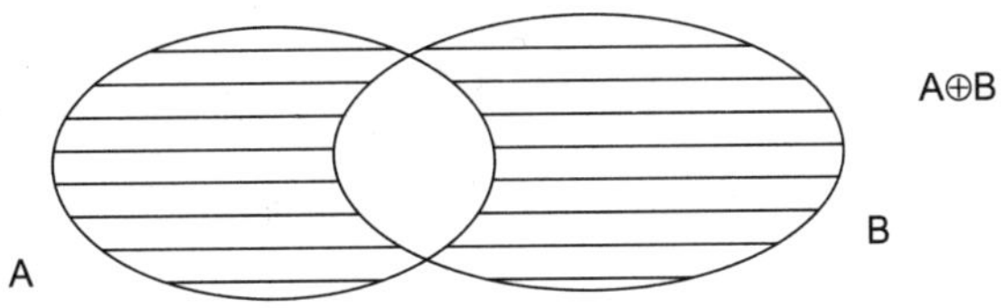

Consider the example:

Let $A = \{a, b, c, k, p, q, r, s\}$

$B = \{b, k, q, m, n, o, t\}$

So, $(A - B) = \{a, c, p, r, s\}$

and $(B - A) = \{m, n, o, t\}$

Therefore, $(A \oplus B) = (A - B) \cup (B - A)$

$= \{a, c, p, r, s, m, n, o, t\}$

2.7.5 Complement of a Set

If A be a set, then the complement of A is given as A^c, A' or $\overline{A}$ and is defined as a set of all those elements of the universal set U which are not in A. Symbolically,

$$A^c = \{x \mid x \in U \text{ and } x \notin A\}$$

Venn diagram

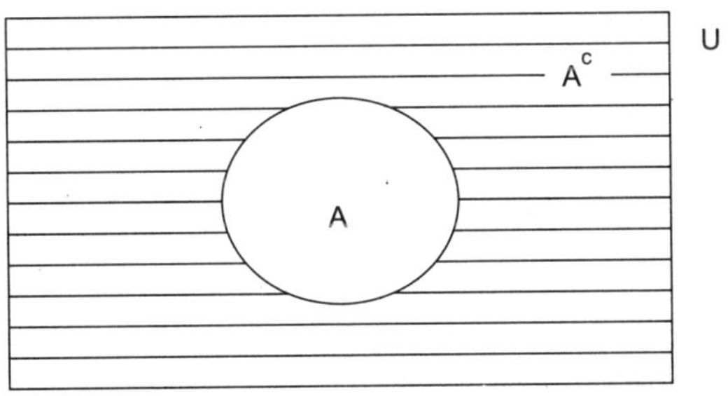

Consider the example:

Let $A = \{b, c, k, d, i, p, q, r, s, t\}$

So, we can take the universal set $U = \{a, b, c, \ldots, x, y, z\}$.

Therefore, $A^c = U - A$

$= \{a, e, f, g, h, j, l, m, n, o, u, v, w, x, y, z\}$

2.7.6 Theorem

Let A, B and C be subsets of the universal set U. Then the following important laws hold.

(*a*) Commutative laws:
$(A \cup B) = (B \cup A)$; $(A \cap B) = (B \cap A)$

(*b*) Associative laws:
$A \cup (B \cup C) = (A \cup B) \cup C$; $A \cap (B \cap C) = (A \cap B) \cap C$

(*c*) Idempotent laws:
$(A \cup A) = A$; $(A \cap A) = A$

(*d*) Identity laws:
$(A \cup \phi) = A$; $(A \cap U) = A$

(*e*) Bound laws:
$(A \cup U) = U$; $(A \cap \phi) = \phi$

(*f*) Absorption laws:
$A \cup (A \cap B) = A$; $A \cap (A \cup B) = A$

(*g*) Complement laws:
$(A \cup A^c) = U$; $(A \cap A^c) = \phi$

(*h*) Involution law:
$(A^c)^c = A$

(*i*) Distributive laws :
(*i*) $A \cup (B \cap C) = (A \cup B) \cap (A \cup C)$
(*ii*) $A \cap (B \cup C) = (A \cap B) \cup (A \cap C)$

Proof: Proofs of (*a*), (*b*), (*c*), (*d*), (*e*), (*f*), (*g*) and (*h*) are immediate consequences of the definitions. We prove only the distributive laws.

(*i*) $x \in A \cup (B \cap C)$
$\Leftrightarrow$ $x \in A$ or $x \in (B \cap C)$
$\Leftrightarrow$ $x \in A$ or ($x \in B$ and $x \in C$)
$\Leftrightarrow$ ($x \in A$ or $x \in B$) and ($x \in A$ or $x \in C$)
$\Leftrightarrow$ $x \in (A \cup B)$ and $x \in (A \cup C)$
$\Leftrightarrow$ $x \in (A \cup B) \cap (A \cup C)$
So, $A \cup (B \cap C) = (A \cup B) \cap (A \cup C)$

(*ii*) $x \in A \cap (B \cup C)$
$\Leftrightarrow$ $x \in A$ and $x \in (B \cup C)$
$\Leftrightarrow$ $x \in A$ and ($x \in B$ or $x \in C$)
$\Leftrightarrow$ ($x \in A$ and $x \in B$) or ($x \in A$ and $x \in C$)
$\Leftrightarrow$ $x \in (A \cap B)$ or $x \in (A \cap C)$
$\Leftrightarrow$ $x \in (A \cap B) \cup (A \cap C)$
So, $A \cap (B \cup C) = (A \cap B) \cup (A \cap C)$

2.7.7 Theorem

Let A, B and C be subsets of the universal set U. Then the following properties hold.

(*a*) $(A \Delta A) = \phi$
(*b*) $(A \Delta B) = (B \Delta A)$
(*c*) $A \cap (B \Delta C) = (A \cap B) \Delta (A \cap C)$
(*d*) $(A \Delta B) = (A \cup B) - (A \cap B)$

Proof: Proofs of (*a*) and (*b*) are immediate consequences of definitions. Here, we prove (*c*) and (*d*).

(*c*) $x \in A \cap (B \Delta C)$

$\Leftrightarrow$ $x \in A$ and $x \in (B \Delta C)$

$\Leftrightarrow$ $x \in A$ and $x \in ((B - C) \cup (C - B))$

$\Leftrightarrow$ $x \in A$ and $(x \in (B - C)$ or $x \in (C - B))$

$\Leftrightarrow$ $(x \in A$ and $x \in (B - C))$ or $(x \in A$ and $x \in (C - B))$

$\Leftrightarrow$ $(x \in A$ and $(x \in B$ and $x \notin C))$

or $(x \in A$ and $(x \in C$ and $x \notin B))$

$\Leftrightarrow$ $((x \in A$ and $x \in B)$ and $(x \in A$ and $x \notin C))$

or $((x \in A$ and $x \in C)$ and $(x \in A$ and $x \notin B))$

$\Leftrightarrow$ $(x \in (A \cap B)$ and $x \notin (A \cap C))$ or

$(x \in (A \cap C)$ and $x \notin (A \cap B))$

$\Leftrightarrow$ $x \in ((A \cap B) - (A \cap C))$ or $x \in ((A \cap C) - (A \cap B))$

$\Leftrightarrow$ $x \in ((A \cap B) - (A \cap C)) \cup ((A \cap C) - (A \cap B))$

$\Leftrightarrow$ $x \in (A \cap B) \Delta (A \cap C)$

So, $A \cap (B \Delta C) = (A \cap B) \Delta (A \cap C)$.

(*d*) $x \in (A \cup B) - (A \cap B)$

$\Leftrightarrow$ $x \in (A \cup B)$ and $x \notin (A \cap B)$

$\Leftrightarrow$ $x \in (A \cup B)$ and $(x \notin A$ or $x \notin B)$

$\Leftrightarrow$ $(x \in (A \cup B)$ and $x \notin A)$ or $(x \in (A \cup B)$ and $x \notin B)$

$\Leftrightarrow$ $((x \in A$ or $x \in B)$ and $x \notin A)$

or $((x \in A$ or $x \in B)$ and $x \notin B)$

$\Leftrightarrow$ $((x \in A$ and $x \notin A)$ or $(x \in B$ and $x \notin A))$

or $((x \in A$ and $x \notin B)$ or $(x \in B$ and $x \notin B))$

$\Leftrightarrow$ $(x \in \phi$ or $x \in (B - A))$ or $(x \in (A - B)$ or $x \in \phi)$

$\Leftrightarrow$ $x \in (\phi \cup (B - A))$ or $x \in ((A - B) \cup \phi)$

$\Leftrightarrow$ $x \in (B - A) \cup (A - B)$ [By identity law]

$\Leftrightarrow$ $x \in (B \Delta A)$

$\Leftrightarrow$ $x \in (A \Delta B)$ [By commutative law]

So, $(A \Delta B) = (A \cup B) - (A \cap B)$

2.7.8 de Morgan's Law

Let A and B be subsets of the universal set U. Then

(*a*) $(A \cup B)^c = (A^c \cap B^c)$

(*b*) $(A \cap B)^c = (A^c \cup B^c)$

Proof: (*a*) $x \in (A \cup B)^c$

$\Leftrightarrow$ $x \notin (A \cup B)$

$\Leftrightarrow$ $x \notin A$ and $x \notin B$

$\Leftrightarrow$ $x \in A^c$ and $x \in B^c$

$\Leftrightarrow$ $x \in A^c \cap B^c$

So, $(A \cup B)^c = (A^c \cap B^c)$

(*b*) $x \in (A \cap B)^c$

$\Leftrightarrow$ $x \notin (A \cap B)$

$\Leftrightarrow$ $x \notin A$ or $x \notin B$

$\Leftrightarrow$ $x \in A^c$ or $x \in B^c$

$\Leftrightarrow$ $x \in A^c \cup B^c$

So, $(A \cap B)^c = (A^c \cup B^c)$

■ 2.8 DISJOINT SETS

Two sets A and B are called disjoint or non-overlapping if both sets have no common element. Mathematically, $(A \cap B) = \phi$.
Venn diagram

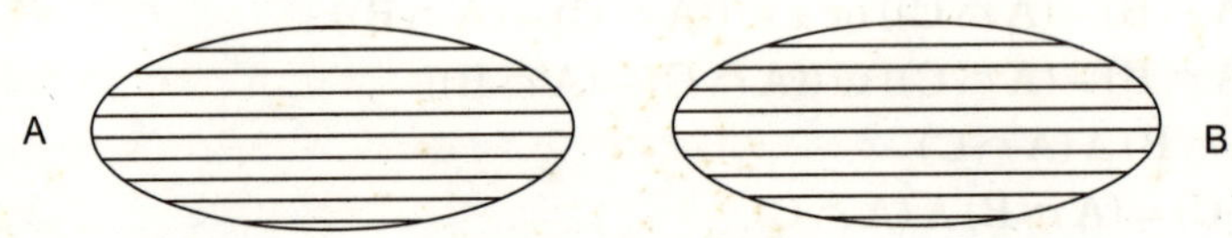

■ 2.9 APPLICATION OF SET THEORY

Let A and B be finite sets. Let $n(A)$ be the number of distinct elements of the set A. Then

$$n(A \cup B) = n(A) + n(B) - n(A \cap B).$$

Further if A and B are disjoint, then

$$n(A \cup B) = n(A) + n(B)$$

Proof: A and B be finite sets and $n(A)$ represent the number of distinct elements of the set A.

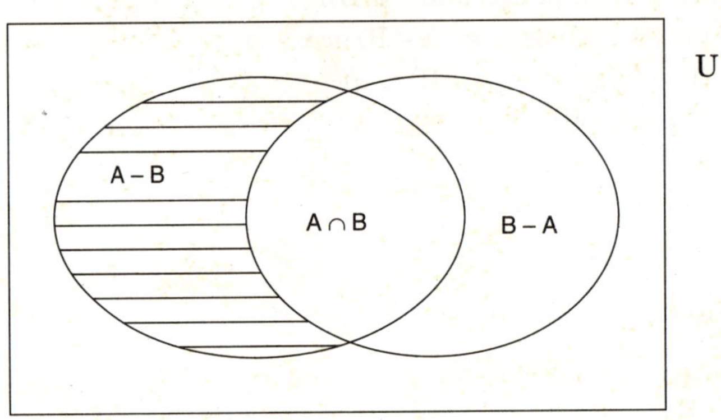

From the above Venn diagram it is clear that

$$n(A) = n(A - B) + n(A \cap B)$$

and $$n(B) = n(B - A) + n(A \cap B)$$

and $$n(A \cup B) = n(A - B) + n(A \cap B) + n(B - A)$$

$$= n(A) - n(A \cap B) + n(A \cap B) + n(B) - n(A \cap B)$$

$$= n(A) + n(B) - n(A \cap B)$$

i.e., $$n(A \cup B) = n(A) + n(B) - n(A \cap B)$$

If A and B are disjoint, then $(A \cap B) = \phi$ *i.e.*, $n(A \cap B) = 0$

Therefore, $n(A \cup B) = n(A) + n(B)$.

■ 2.10 PRODUCT OF SETS

The product of sets is defined with the help of an order pair. An order pair is usually denoted by (x, y) such that $(x, y) \neq (y, x)$ whenever $x \neq y$. The product of two sets A and B is the set of all those order pairs whose first coordinate is an element of A and the second coordinate is an element of B. The set is denoted by $(A \times B)$. Mathematically,

$$(A \times B) = \{(x, y) \mid x \in A \text{ and } x \in B\}$$

Consider the example:

Let $A = \{1, 2, 3, 5, 7\}$

$B = \{4, 9, 25\}$

So, $(A \times B) = \{(1,4), (1, 9), (1, 25), (2, 4), (2, 9), (2, 25), (3, 4), (3, 9), (3, 25), (5, 4), (5, 9), (5, 25), (7, 4), (7, 9), (7, 25)\}$

Note: The product of sets can be extendable for n sets $A_1, A_2, A_3, \ldots\ldots., A_n$.

Thus, $A_1 \times A_2 \times A_3 \times \ldots \times A_n$ can be defined as

$A_1 \times A_2 \times A_3 \times \ldots \times A_n = \{(x_1, x_2, x_3, \ldots, x_n) \mid x_1 \in A_1 \text{ and } x_2 \in A_2 \text{ and } x_3 \in A_3 \text{ and } \ldots \text{ and } x_n \in A_n\}$where $(x_1, x_2, x_3, \ldots, x_n)$ is called as n-tuple of $x_1, x_2, x_3, \ldots, x_n$. To explain this consider the example in which $A = \{a, b, c\}$; $B = \{1, 2\}$ and $C = \{\alpha, \beta\}$.

Therefore, $A \times B \times C = \{(a, 1, \alpha), (a, 1, \beta), (a, 2, \alpha), (a, 2, \beta), (b, 1, \alpha), (b, 1, \beta), (b, 2, \alpha), (b, 2, \beta), (c, 1, \alpha), (c, 1, \beta), (c, 2, \alpha), (c, 2, \beta)\}$.

From the above example, it is very clear that $|A \times B \times C| = |A| \times |B| \times |C|$.

In general, $|A_1 \times A_2 \times A_3 \times \ldots \times A_n| = |A_1| \times |A_2| \times |A_3| \times \ldots \times |A_n|$.

2.10.1 Theorem

Let A, B and C be three subsets of the universal set U. Then

(*a*) $A \times (B \cup C) = (A \times B) \cup (A \times C)$

(*b*) $A \times (B \cap C) = (A \times B) \cap (A \times C)$

Proof: (*a*) $(x, y) \in A \times (B \cup C)$

$\Leftrightarrow \quad x \in A \text{ and } y \in (B \cup C)$

$\Leftrightarrow \quad x \in A \text{ and } (y \in B \text{ or } y \in C)$

$\Leftrightarrow \quad (x \in A \text{ and } y \in B) \text{ or } (x \in A \text{ and } y \in C)$

$\Leftrightarrow (x, y) \in (A \times B) \text{ or } (x, y) \in (A \times C)$

$\Leftrightarrow (x, y) \in (A \times B) \cup (A \times C)$

Therefore, $A \times (B \cup C) = (A \times B) \cup (A \times C)$.

(*b*) $(x, y) \in A \times (B \cap C)$

$\Leftrightarrow \quad x \in A \text{ and } y \in (B \cap C)$

$\Leftrightarrow \quad x \in A \text{ and } (y \in B \text{ and } y \in C)$

$\Leftrightarrow \quad (x \in A \text{ and } y \in B) \text{ and } (x \in A \text{ and } y \in C)$

$\Leftrightarrow (x, y) \in (A \times B) \text{ and } (x, y) \in (A \times C)$

$\Leftrightarrow (x, y) \in (A \times B) \cap (A \times C)$

Therefore, $A \times (B \cap C) = (A \times B) \cap (A \times C)$.

2.11 FUNDAMENTAL PRODUCTS

Let $A_1, A_2, A_3, \ldots, A_n$ be n sets. A fundamental product of these n sets is an expression of the form $(B_1 \cap B_2 \cap B_3 \cap \ldots \cap B_n)$ where B_i is either A_i or A_i^c.

Consider an example with three sets A, B and C. The fundamental products of these three sets are as follows, which are 2^3 in number.

i.e., $A \cap B \cap C$; $A^c \cap B \cap C$; $A \cap B^c \cap C$; $A \cap B \cap C^c$;
$A^c \cap B^c \cap C$; $A \cap B^c \cap C^c$; $A^c \cap B \cap C^c$; $A^c \cap B^c \cap C^c$.

SOLVED EXAMPLES

Example 1 *Let A, B and C be any three subsets of the universal set U. Then prove that*

(a) $A - (B \cup C) = (A - B) \cap (A - C)$
(b) $A - (B \cup C) = (A - B) - C$
(c) $(A \cap B) - C = A \cap (B - C)$

Solution: (a) $x \in A - (B \cup C)$

$\Leftrightarrow \quad x \in A$ and $x \notin (B \cup C)$
$\Leftrightarrow \quad x \in A$ and $(x \notin B$ and $x \notin C)$
$\Leftrightarrow \quad (x \in A$ and $x \notin B)$ and $(x \in A$ and $x \notin C)$
$\Leftrightarrow \quad x \in (A - B)$ and $x \in (A - C)$
$\Leftrightarrow \quad x \in (A - B) \cap (A - C)$

Therefore, $A - (B \cup C) = (A - B) \cap (A - C)$

(b) $x \in A - (B \cup C)$

$\Leftrightarrow \quad x \in A$ and $x \notin (B \cup C)$
$\Leftrightarrow \quad x \in A$ and $(x \notin B$ and $x \notin C)$
$\Leftrightarrow \quad (x \in A$ and $x \notin B)$ and $x \notin C$
$\Leftrightarrow \quad x \in (A - B)$ and $x \notin C$
$\Leftrightarrow \quad x \in (A - B) - C$

Therefore, $A - (B \cup C) = (A - B) - C$

(c) $x \in (A \cap B) - C$

$\Leftrightarrow \quad (x \in A$ and $x \in B)$ and $x \notin C$
$\Leftrightarrow \quad x \in A$ and $(x \in B$ and $x \notin C)$
$\Leftrightarrow \quad x \in A$ and $x \in (B - C)$
$\Leftrightarrow \quad x \in A \cap (B - C)$

Therefore, $(A \cap B) - C = A \cap (B - C)$

Example 2 *Show that* $A - \bigcup_{i=1}^{n} B_i = \bigcap_{i=1}^{n} (A - B_i)$

Solution: $x \in A - \bigcup_{i=1}^{n} B_i$

$\Leftrightarrow \quad x \in A$ and $x \notin \bigcup_{i=1}^{n} B_i$
$\Leftrightarrow \quad x \in A$ and $x \notin (B_1 \cup B_2 \cup B_3 \cup \ldots \cup B_n)$
$\Leftrightarrow \quad x \in A$ and $(x \notin B_1$ and $x \notin B_2$ and $x \notin B_3$ and $\ldots$ and $x \notin B_n)$

$\Leftrightarrow \quad (x \in A \text{ and } x \notin B_1) \text{ and } (x \in A \text{ and } x \notin B_2) \text{ and } \ldots. \text{ and } (x \in A \text{ and } x \notin B_n)$

$\Leftrightarrow \quad x \in (A - B_1) \text{ and } x \in (A - B_2) \text{ and } \ldots \text{ and } x \in (A - B_n)$

$\Leftrightarrow \quad x \in (A - B_1) \cap (A - B_2) \cap \ldots \cap x \in (A - B_n)$

$$\Leftrightarrow \quad x \in \bigcap_{i=1}^{n} (A - B_i)$$

Therefore, $A - \bigcup_{i=1}^{n} B_i = \bigcap_{i=1}^{n} (A - B_i)$

Example 3 *If A and B subsets of the universal set U, then show that*

(a) $(A^c)^c = A$

(b) $A - B = A \cap B^c$

(c) $(A - B) \cap B = \phi$

Solution: (*a*) $x \in (A^c)^c$

$\Leftrightarrow \quad x \notin A^c$

$\Leftrightarrow \quad x \in A$

So, $(A^c)^c = A$

(*b*) $x \in (A - B)$

$\Leftrightarrow \quad x \in A \text{ and } x \notin B$

$\Leftrightarrow \quad x \in A \text{ and } x \in B^c$

$\Leftrightarrow \quad x \in (A \cap B^c)$

So, $(A - B) = (A \cap B^c)$

(*c*) $x \in (A - B) \cap B$

$\Leftrightarrow \quad x \in (A - B) \text{ and } x \in B$

$\Leftrightarrow \quad (x \in A \text{ and } x \notin B) \text{ and } x \in B$

$\Leftrightarrow \quad x \in A \text{ and } (x \notin B \text{ and } x \in B)$

$\Leftrightarrow \quad x \in A \text{ and } x \in \phi$

$\Leftrightarrow \quad x \in (A \cap \phi)$

$\Leftrightarrow \quad x \in \phi$

So, $(A - B) \cap B = \phi$

Example 4 *Let A, B be the subsets of the universal set U, then prove that*

(a) $A - (A \cap B) = A \cap B^c$

(b) $(A \cap B^c)^c = A^c \cup B$

Solution: (*a*) $x \in A - (A \cap B)$

$\Leftrightarrow \quad x \in A \text{ and } x \notin (A \cap B)$

$\Leftrightarrow \quad x \in A \text{ and } (x \notin A \text{ or } x \notin B)$

$\Leftrightarrow \quad (x \in A \text{ and } x \notin A) \text{ or } (x \in A \text{ and } x \notin B)$

$\Leftrightarrow \quad x \in \phi \text{ or } (x \in A \text{ and } x \in B^c)$

$\Leftrightarrow \quad x \in \phi \text{ or } x \in (A \cap B^c)$

$\Leftrightarrow \quad x \in \phi \cup (A \cap B^c)$

$\Leftrightarrow \quad x \in (A \cap B^c)$

So, $A - (A \cap B) = A \cap B^c$

(*b*) $x \in (A \cap B^c)^c$

$\Leftrightarrow$ $x \notin (A \cap B^c)$

$\Leftrightarrow$ $x \notin A$ or $x \notin B^c$

$\Leftrightarrow$ $x \in A^c$ or $x \in B$

$\Leftrightarrow$ $x \in (A^c \cup B)$

So, $(A \cap B^c)^c = (A^c \cup B)$

Example 5 *Let A, B and C be three subsets of the universal set U. Then show that*

$$n(A \cup B \cup C) = n(A) + n(B) + n(C) - n(A \cap B) - n(B \cap C) - n(C \cap A) + n(A \cap B \cap C)$$

Solution: Let $(B \cup C) = D$. So, we have

$$\begin{aligned} n(A \cup B \cup C) &= n(A \cup D) \\ &= n(A) + n(D) - n(A \cap D) \\ &= n(A) + n(B \cup C) - n(A \cap (B \cup C)) \\ &= n(A) + n(B) + n(C) - n(B \cap C) - n((A \cap B) \cup (A \cap C)) \\ &= n(A) + n(B) + n(C) - n(B \cap C) - n(A \cap B) - n(A \cap C) + n(A \cap B \cap C) \end{aligned}$$

Therefore, $n(A \cup B \cup C) = n(A) + n(B) + n(C) - n(A \cap B) - n(B \cap C) - n(C \cap A) + n(A \cap B \cap C)$.

Example 6 *In the IEEE conference held at New York, 500 delegates attended. 200 of them could take tea, 350 could take coffee and 10 did not take either coffee or tea. Then answer the following questions.*

(*a*) *How many can take both tea and coffee.*
(*b*) *How many can take tea only and*
(*c*) *How many can take coffee only.*

Solution: Let T: Set of persons who take tea.
C: Set of persons who take coffee.
U: Total number of delegates.

Hence we have $n(U) = 500$; $n(T) = 200$; $n(C) = 350$

Number of delegates did not take either coffee or tea = 10

Therefore number of delegates who take either coffee or tea = 500 –10 = 490

i.e., $n(T \cup C) = 490$

i.e., $n(T) + n(C) - n(T \cap C) = 490$

i.e., $n(T \cap C) = n(T) + n(C) - 490 = 200 + 350 - 490 = 60$

So, the number of persons who take both coffee and tea = $n(T \cap C) = 60$

Number of persons take tea only = $n(T) - n(T \cap C) = 140$

Number of persons take coffee only = $n(C) - n(T \cap C) = 290$.

Example 7 *If 65% of students like apples where 75% like grapes, then what percentage of students likes both apples and grapes?*

Solution: Let $n(S)$: Total number of students = 100
$n(A)$: Total number of students who like apples = 65
$n(B)$: Total number of students who like grapes = 75

Therefore, $n(S) = n(A \cup B) = n(A) + n(B) - n(A \cap B)$

i.e., $100 = 65 + 75 - n(A \cap B)$

i.e., $n(A \cap B) = 40$

So, 40% of students like both apples and grapes.

Example 8 *If A = {2, 3, 4, 5, 6}, B = {3, 4, 5, 6, 7} and C = {4, 5, 6, 7, 8} then find the followings.*

(i) $(A \cup B) \cap (A \cup C)$ (ii) $(A \cap B) \cup (A \cap C)$

(iii) $A - (B - C)$ (iv) $(A \Delta B)$.

Solution: Given A = {2, 3, 4, 5, 6}, B = {3, 4, 5, 6, 7} and C = {4, 5, 6, 7, 8}

(i) $(A \cup B) = \{2, 3, 4, 5, 6, 7\}$

$(A \cup C) = \{2, 3, 4, 5, 6, 7, 8\}$

Therefore, $(A \cup B) \cap (A \cup C) = \{2, 3, 4, 5, 6, 7\}$

(ii) $(A \cap B) = \{3, 4, 5, 6\}$

$(A \cap C) = \{4, 5, 6\}$

Therefore, $(A \cap B) \cup (A \cap C) = \{3, 4, 5, 6\}$

(iii) $(B - C) = \{3\}$

Therefore, $A - (B - C) = \{2, 4, 5, 6\}$

(iv) $(A \cup B) = \{2, 3, 4, 5, 6, 7\}$

$(A \cap B) = \{3, 4, 5, 6\}$

Therefore, $(A \Delta B) = (A \cup B) - (A \cap B) = \{2, 7\}$

Example 9 *Find the power sets of the following sets.*

(i) {0}

(ii) {1, {1, 2}} and

(iii) {4, 1, 8}.

Solution: (i) Let A = {0}

Therefore, $P(A) = \{\{0\}, \phi\}$

(ii) Let $A = \{1, \{1, 2\}\}$

So, $P(A) = \{\{1\}, \{\{1, 2\}\}, A, \phi\}$

(iii) Let $A = \{4, 1, 8\}$

So, $P(A) = \{\{4\}, \{1\}, \{8\}, \{4, 1\}, \{4, 8\}, \{1, 8\}, A, \phi\}$.

Example 10 *If A = {4, 5}, B = {7, 8} and C = {9, 10}, then find the followings.*

(a) $(A \times B) \cup (B \times C)$ and (b) $A \times (B \cup C)$.

Solution: Given A = {4, 5}, B = {7, 8} and C = {9, 10}

(a) $(A \times B) = \{(4, 7), (4, 8), (5, 7), (5,8)\}$

$(B \times C) = \{(7, 9), (7, 10), (8, 9), (8, 10)\}$

So, $(A \times B) \cup (B \times C) = \{(4, 7), (4, 8), (5, 7), (5,8), (7, 9), (7, 10), (8, 9), (8, 10)\}$

(b) $(B \cup C) = \{7, 8, 9, 10\}$

So, $A \times (B \cup C) = \{(4, 7), (4, 8), (4, 9), (4,10), (5,7), (5, 8), (5, 9), (5, 10)\}$.

Example 11 *If A = {1, 2, 3}, B = {2, 3, 4} and C = {3, 4, 5}, then verify the product laws.*

Solution: Given A = {1, 2, 3}, B = {2, 3, 4} and C = {3, 4, 5}

Therefore, $(B \cup C) = \{2, 3, 4, 5\}$ and

$A \times (B \cup C) = \{(1, 2), (1, 3), (1, 4), (1, 5), (2, 2), (2, 3), (2, 4), (2, 5), (3, 2), (3, 3), (3, 4), (3, 5)\}$

$(A \times B) = \{(1, 2), (1, 3), (1, 4), (2, 2), (2, 3), (2, 4), (3, 2), (3, 3), (3, 4)\}$

$(A \times C) = \{(1, 3), (1, 4), (1, 5), (2, 3), (2, 4), (2, 5), (3, 3), (3, 4), (3, 5)\}$

Thus, $(A \times B) \cup (A \times C) = \{(1, 2), (1, 3), (1, 4), (1, 5), (2, 2), (2, 3), (2, 4), (2, 5), (3, 2), (3, 3), (3, 4), (3, 5)\}$

$= A \times (B \cup C)$

Similarly, the second product law $A \times (B \cap C) = (A \times B) \cap (A \times C)$ can be verified.

Example 12 *If P = {a, c, e}, Q ={100, 101, 102} and R = {m, c, e, 101}. Compute ((Q ∪ P) – (P ∩ Q)) × R, where ∩, ∪, – and × are well known set theoretic binary operations.*

Solution: Given P = {a, c, e}; Q = {100, 101, 102} and R = {m, c, e, 101}.

So, (Q ∪ P) = {100, 101, 102, a, c, e} and (P ∩ Q) = ϕ

Therefore, ((Q ∪ P) – (P ∩ Q)) = {100, 101, 102, a, c, e}

Thus, ((Q ∪ P) – (P ∩ Q)) × R = {(100, m), (100, c), (100, e), (100, 101), (101, m), (101, c), (101, e), (101, 101), (102, m), (102, c), (102, e), (102, 101), (a, m), (a, c), (a, e), (a, 101), (c, m), (c, c), (c, e), (c, 101), (e, m), (e, c), (e, e), (e, 101)}.

Example 13 *Show the following sets by Venn diagram.*

(a) $(A - B)^c$ *(b)* $A^c \cap B$ *(c)* $A \cap B \cap C$

Solution:

(*a*) $(A - B)^c$

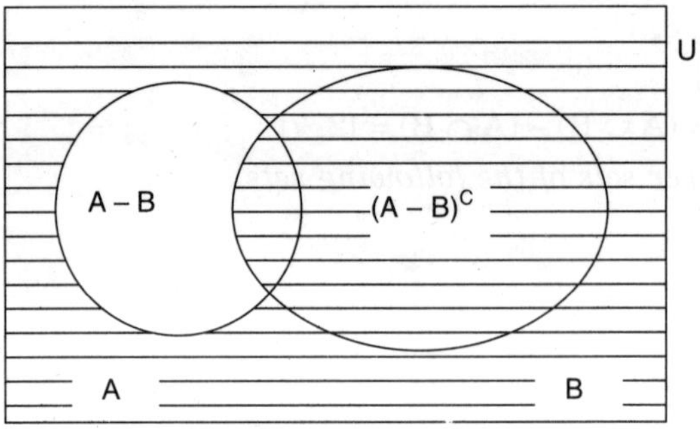

(*b*) $(A^c \cap B)$

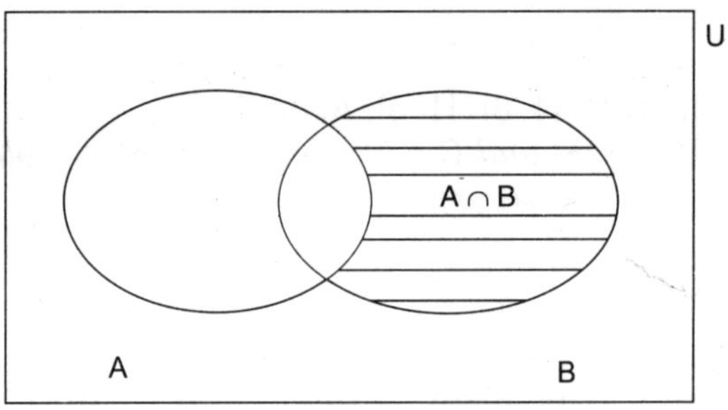

(*c*) $A \cap B \cap C$

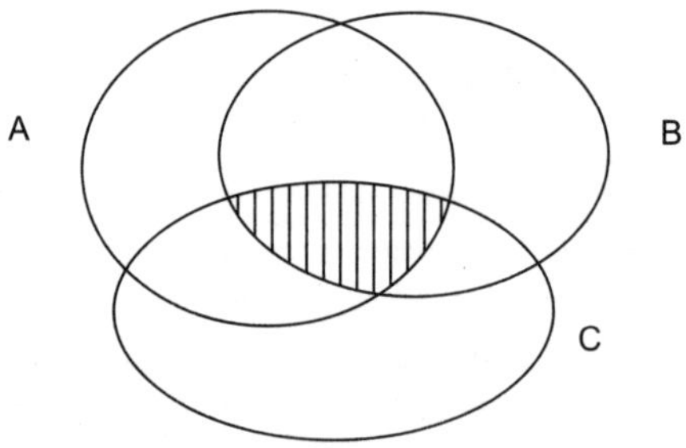

Example 14 *In a group of 64 students 26 can speak French only, 14 can speak English only. How many can speak both French and English?*

Solution: Let F: Set of students who can speak French.

E: Set of students who can speak English.

Let n(S): Total number of students = 64

i.e., $n(S) = n(F \cup E) = 64$

Given: $n(F - E)$: Number of students speak French only = 26

and $n(E - F)$: Number of students speak English only = 14

Therefore, $n(F \cup E) = n(F - E) + n(E - F) + n(F \cap E)$

i.e., $n(F \cap E) = 64 - 26 - 14 = 24$

So, 24 students can speak both French and English.

Example 15 *Draw a Venn diagram to represent the following facts for the sets P, Q, R and S. $(P \cap Q) \neq \phi$, $S \subseteq Q \subseteq R$ and $(P \cap S) = \phi$.*

Solution: Given conditions are $(P \cap Q) \neq \phi$, $S \subseteq Q \subseteq R$ and $(P \cap S) = \phi$. The Venn diagram for the above facts is given below.

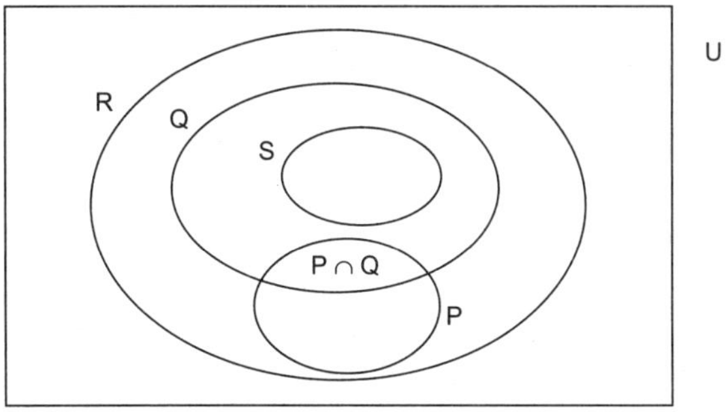

Example 16 *If in a city 60% of the residents can speak German and 50% can speak French. What percentage of residents can speak both the languages, if 20% residents can not speak any of these two languages?*

Solution: Let n(S) : Total number of residents = 100

n(G) : Total number of residents who speak German = 60

n(F) : Total number of residents who speak French = 50

$n(G \cup F)^c$: Total number of residents who cannot speak any of these two languages = 20

So, $n(G \cup F) = n(S) - n(G \cup F)^c = 100 - 20 = 80$

i.e., $n(G) + n(F) - n(G \cap F) = 80$

i.e., $n(G \cap F) = 60 + 50 - 80 = 30$

Therefore, 30% of the residents can speak both the languages German and French.

Example 17 *In a survey about liking for colours, it was found that everyone who was surveyed had a liking for at least one of the three colours namely Red, Green and Blue. Further 30% liked Red; 40% liked Green and 50% liked Blue. Further 10% people liked both Red and Green, 5% liked both Green and Blue and 10% liked both Red and Blue. Find the percentage of the surveyed people who like all the colours.*

Solution: Let R: Set of people who like Red colour

G: Set of people who like Green colour

B: Set of people who like Blue colour

and S: Set of all people who was surveyed.

Therefore, $n(S) = 100$; $n(R) = 30$; $n(G) = 40$; $n(B) = 50$; $n(R \cap G) = 10$; $n(G \cap B) = 5$; $n(R \cap B) = 10$.

Thus $n(S) = n(R \cup G \cup B) = 100$

i.e., $n(R) + n(G) + n(B) - n(R \cap G) - n(G \cap B) - n(R \cap B) + n(R \cap G \cap B) = 100$

i.e., $n(R \cap G \cap B) = 100 - 30 - 40 - 50 + 10 + 5 + 10 = 5$

So, 5% of the surveyed people like all the colours, i.e. Red, Green and Blue.

Example 18 *If $A \subset B$ and $B \subset C$, then show that $A \subset C$.*

Solution: Given $B \subset C$, *i.e.* $x \in B \Rightarrow x \in C$

Again $A \subset B$, *i.e.* $x \in A \Rightarrow x \in B$ $\quad \forall x \in A$

i.e., $x \in A \Rightarrow x \in B \Rightarrow x \in C$

i.e., $x \in A \Rightarrow x \in C$

Therefore, $A \subset C$.

Example 19 *For all sets A and B prove that $\overline{A \times B} = \overline{A} \times \overline{B}$.*

Solution: $(x, y) \in \overline{A \times B}$

$\Leftrightarrow$ $(x, y) \notin A \times B$

$\Leftrightarrow$ $x \notin A$ and $y \notin B$

$\Leftrightarrow$ $x \in \overline{A}$ and $y \in \overline{B}$

$\Leftrightarrow$ $(x, y) \in \overline{A} \times \overline{B}$

So, $(x, y) \in \overline{A \times B} \Leftrightarrow (x, y) \in \overline{A} \times \overline{B}$

Therefore, $\overline{A \times B} = \overline{A} \times \overline{B}$

Example 20 *For all sets A, B and C prove that $A \times (B - C) = (A \times B) - (A \times C)$.*

Solution: $(x, y) \in A \times (B - C)$

$\Leftrightarrow$ $x \in A$ and $y \in (B - C)$

$\Leftrightarrow$ $x \in A$ and ($y \in B$ and $y \notin C$)

$\Leftrightarrow$ ($x \in A$ and $y \in B$) and ($x \in A$ and $y \notin C$)

$\Leftrightarrow$ $(x, y) \in (A \times B)$ and $(x, y) \notin (A \times C)$

$\Leftrightarrow$ $(x, y) \in (A \times B) - (A \times C)$

Therefore, $A \times (B - C) = (A \times B) - (A \times C)$.

Example 21 *In a group of 191 students, 10 are taking English, Computer Science and Music; 36 are taking English and Computer Science; 20 are taking English and Music; 18 are taking Computer Science and Music; 65 are taking English; 76 are taking Computer Science and 63 are taking Music. Then answer the followings*

(a) How many are taking English and Music but not Computer Science.
(b) How many are taking Computer Science and Music but not English.
(c) How many are taking Computer Science and neither English nor Music.
(d) How many are taking none of the three subjects.

Solution: Let S: Set of students
E: Set of students taking English
C: Set of students taking Computer Science
M: Set of students taking Music.

Given that $n(S) = 191$; $n(E) = 65$; $n(C) = 76$; $n(M) = 63$; $n(E \cap C \cap M) = 10$; $n(E \cap C) = 36$; $n(E \cap M) = 20$; $n(C \cap M) = 18$

(*a*) Number of students taking English and Music but not Computer Science

$$= n(E \cap M) - n(E \cap C \cap M) = 20 - 10 = 10$$

(*b*) Number of students taking Computer Science and Music but not English

$$= n(C \cap M) - n(E \cap C \cap M) = 18 - 10 = 8$$

(*c*) Number of students taking Computer Science and neither English nor Music

$$= n(C) - n(E \cap C) - n(C \cap M) + n(E \cap C \cap M) = 76 - 36 - 18 + 10 = 32$$

(*d*) Number of students taking none of the three subjects

$$= n(E \cup C \cup M)^c$$
$$= n(S) - n(E \cup C \cup M)$$
$$= n(S) - \{n(E) + n(C) + n(M) - n(E \cap C) - n(C \cap M) - n(E \cap M) + n(E \cap C \cap M)\}$$
$$= 191 - (65 + 63 + 76 - 20 - 36 - 18 + 10)$$
$$= 51.$$

Example 22 *Examine whether the following sets are equivalent or not.*

(*a*) $A = \{x \mid x^2 - 7x + 12 = 0; x \in N\}$ (*b*) $B = \{x \mid x = a \text{ and } x = b\}$

(*c*) $C = \{a, b, c, d, e\}$ (*d*) $D = \{x \mid x^2 - 4 = 0; x \in I\}$

Solution: Given that $A = \{x \mid x^2 - 7x + 12 = 0; x \in N\}$

Therefore $A = \{3, 4\}$

i.e., $|A| = 2$

Similarly $B = \{x \mid x = a \text{ and } x = b\}$

$= \{a, b\}$

i.e., $|B| = 2$

Also $C = \{a, b, c, d, e\}$

i.e., $|C| = 5$

Again $D = \{x \mid x^2 - 4 = 0; x \in I\} = \{2, -2\}$

i.e., $|D| = 2$

Therefore, $|A| = |B| = |D| = 2 \neq |C| = 5$; So A, B and D are equivalent.

Example 23 *For all Sets A and B prove that $(A \cap B) \cup (B - A) = B$.*

Solution: $(A \cap B) \cup (B - A) = (A \cap B) \cup (B \cap A^c)$

$= ((A \cap B) \cup B) \cap ((A \cap B) \cup A^c)$ [Distributive law]

$= B \cap ((A \cap B) \cup A^c)$ [Absorption law]

$= B \cap ((A \cup A^c) \cap (B \cup A^c))$ [Distributive law]

$= B \cap (U \cap (B \cup A^c))$ [Complement law]

$= B \cap (B \cup A^c)$

$= B$ [Absorption law]

Example 24 *By applying properties of sets prove that $(A - B) \cap (B - A) = \phi$ for all sets A and B.*

Solution: $(A - B) \cap (B - A) = (A \cap B^c) \cap (B \cap A^c)$

$= A \cap (B^c \cap (B \cap A^c))$ [Associative law]

$= A \cap ((B^c \cap B) \cap A^c)$ [Associative law]

$= A \cap (\phi \cap A^c)$ [Complement law]

$= (A \cap \phi)$ [Bound law]

$= \phi$ [Bound law]

Example 25 *For all sets X, Y and Z prove that* $X \cap (Y - Z) = (X \cap Y) - (X \cap Z)$.

Solution: $x \in X \cap (Y - Z)$

$\Leftrightarrow$ $x \in X$ and $x \in (Y - Z)$

$\Leftrightarrow$ $x \in X$ and ($x \in Y$ and $x \notin Z$)

$\Leftrightarrow$ ($x \in X$ and $x \in Y$) and ($x \in X$ and $x \notin Z$)

$\Leftrightarrow$ $x \in (X \cap Y)$ and $x \notin (X \cap Z)$

$\Leftrightarrow$ $x \in (X \cap Y) - (X \cap Z)$

Therefore, $X \cap (Y - Z) = (X \cap Y) - (X \cap Z)$

Example 26 *For all sets X, Y and Z prove that* $X - (Y \cup Z) = (X - Y) \cap Z^c$.

Solution: $x \in X - (Y \cup Z)$

$\Leftrightarrow$ $x \in X$ and $x \notin (Y \cup Z)$

$\Leftrightarrow$ $x \in X$ and ($x \notin Y$ and $x \notin Z$)

$\Leftrightarrow$ ($x \in X$ and $x \notin Y$) and $x \notin Z$

$\Leftrightarrow$ $x \in (X - Y)$ and $x \in Z^c$

$\Leftrightarrow$ $x \in (X - Y) \cap Z^c$

Therefore, $X - (Y \cup Z) = (X - Y) \cap Z^c$.

Example 27 *Determine the equality for the following pair of sets.*

A = {1, 2, 3} and $B = \{x \mid x \in N;\ x^3 - 6x^2 + 11x - 6 = 0\}$

Solution: Given A = {1, 2, 3} and

$$B = \{x \mid x \in N;\ x^3 - 6x^2 + 11x - 6 = 0\}$$
$$= \{x \mid x \in N;\ (x - 1)(x - 2)(x - 3) = 0\}$$
$$= \{1, 2, 3\}$$

Therefore, sets A and B are equal as $A \subseteq B$ and $B \subseteq A$.

Example 28 *Express* $A \cup (B - C)$ *as the union of fundamental products.*

Solution: The figure given below represents the Venn diagram for $A \cup (B - C)$. From this it is clear that $A \cup (B - C)$ consists of the five areas of the Venn diagram corresponding to the fundamental products $(A \cap B \cap C)$, $(A \cap B \cap C^c)$, $(A \cap B^c \cap C)$, $(A^c \cap B \cap C^c)$ and $(A \cap B^c \cap C^c)$.

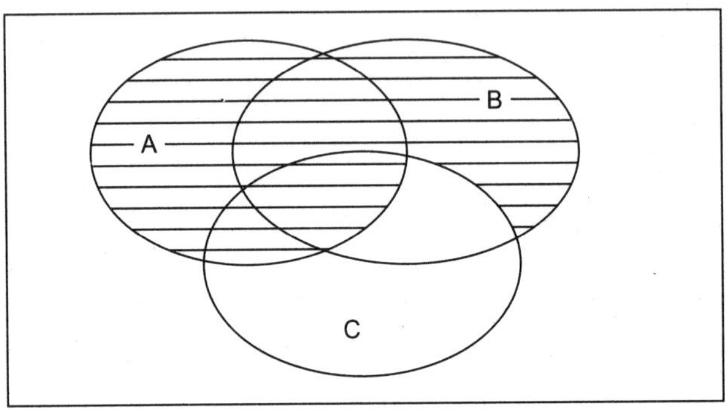

$A \cup (B - C)$ is shaded

Thus, $A \cup (B - C) = (A \cap B \cap C) \cup (A \cap B \cap C^c) \cup (A \cap B^c \cap C) \cup (A^c \cap B \cap C^c) \cup (A \cap B^c \cap C^c)$.

EXERCISES

1. Express the following sets in tabular form.

(*a*) A = {x | x is a letter in the word MATHEMATICS}
(*b*) B = {x | $x = 2n + 1$; $1 \le n < 5$; $n \in$ N}
(*c*) C = {x | x = Book and $x = 1$ and $x = a$ and x = Pen}
(*d*) D = {x | x is an even integer and $1 \le x \le 15$}
(*e*) E = {x | $x \in$ I and $x^2 + x - 20 = 0$}

2. Express the following sets in set builder form.

(*a*) A = {1, 8, 27, 64, 125}
(*b*) B = {a, e, i, o, u}
(*c*) C = {2, 9, 28, 65, 126}
(*d*) D = {a, b, 2, 4, 6, Book}
(*e*) E = {1, 2, 3, 4, 5, 6, 7, ...}
(*f*) F = {1, 3}

3. Find the power sets of the following sets.

(*a*) {ϕ}
(*b*) {k, l, m, n}
(*c*) {x | $x \in$ N and $x^2 - 4x + 3 = 0$}
(*d*) {1, {1, 2}, {1, 2, 3}}
(*e*) {x | x is a letter of the word wolf}

4. Let the universal set U = {1, 2, 3, 4, 5, 6, 7, 8, 9, 10}, Let A = {1, 2, 3, 4, 5}, B = {2, 4, 6, 8} and C = {1, 4, 7, 10}, then find the followings.

(*i*) $(A \cup B)$
(*ii*) $(A \cup B) \cap C$
(*iii*) $(A \cup B) \cap (A \cup C)$
(*iv*) $(A \cap B) \cup (A \cap C)$
(*v*) $A - (B \cup C)$
(*vi*) $(A \cap B) - C$
(*vii*) $B^c - (C - A)$
(*viii*) $A^c \cap B^c$
(*ix*) AB
(*x*) A^c
(*xi*) $C - B$
(*xii*) $(A \cup B) - (C - B)$

5. Draw the Venn diagram and indicate the region for the given sets.

(*a*) $A \cup (B \cap C)$
(*b*) $A \cap (B \cup C)$
(*c*) $A^c - B$
(*d*) $(A \cup B) - B$
(*e*) $(A^c \cup B) \cap (C^c - A)$
(*f*) $B \cap (C \cup A)^c$
(*g*) $(B \cup C) - A$
(*h*) $(A \cup B \cup C)^c$

6. In a group of 1000 people, there are 800 people who can speak English and 500 people who can speak German. Except 100 people in the group, each person speaks at least one of English and German. Find how many people can speak both English and German.

7. If P = {a, c, e}, Q = {100, 101, 102} and R = {m, c, e, 101}, then compute $((P \cup R) - (P \cap R)) \times Q$.

8. If G = {p, q, r}, H = {20, 70, 90} and K = {r, 70, s}, then compute $(G - K) \times (K - H)$.

9. Let X = {a, b, c} and Y = {1, 2}, then compute the followings.

(*a*) $X \times Y$
(*b*) $Y \times X$
(*c*) $Y \times Y$
(*d*) $X \times X$
(*e*) $(X \Delta Y) \times Y$

10. If $B_1, B_2, \ldots, B_n$ and A are sets, then prove that $A - \bigcap_{i=1}^{n} B_i = \bigcup_{i=1}^{n} (A - B_i$

11. If $B_1, B_2, \ldots, B_n$ are sets, then prove the following de Morgan's laws.

(*a*) $\left(\bigcup_{i=1}^{n}\right)' = \bigcap_{i=1}^{n} B_i'$
(*b*) $\left(\bigcap_{i=1}^{n} B_i\right)' = \bigcup_{i=1}^{n} B_i'$

12. Let X, Y and Z be three sets. Show that $X - (Y \cap Z) = (X - Y) \cup (X - Z)$.

13. In a class of 120 students, 80 students study Mathematics, 45 study History and 20 students neither study History nor study Mathematics. What is the number of students who study both Mathematics and History?

14. Let $A = \{1, 2\}$, $B = \{\alpha\}$ and $C = \{\alpha, \beta\}$, then compute the followings.
   (*a*) $A \times B \times C$ (*b*) $A \times B \times B$
   (*c*) $B \times A \times C$ (*d*) $A \times A \times A$
   (*e*) $(A - B) \times C$.

15. Examine the comparability with the following sets.
   (*i*) $A = \{a, b, c\}$ (*ii*) $B = \{a, e, i, o, u\}$
   (*iii*) $C = \{b, c, o, u\}$ (*iv*) $D = \{b, c, i, o, u, k\}$.

16. In a class containing 100 students, 30 play tennis; 40 play cricket; 40 do athletics; 6 play tennis and cricket; 12 play cricket and do athletics; and 10 play tennis and do athletics; while 14 play no game or do athletics at all. How many play cricket, tennis and do athletics?

17. If in a city 70% of the residents can speak French and 50% can speak English, what percentage of residents can speak both the languages, if 10% residents cannot speak any of these two languages?

18. Let X, Y, Z and T be four sets. Then prove that $(X \cap Z) \times (Y \cap T) = (X \times Y) \cap (Z \times T)$.

19. Write the following sets as the union of fundamental products.
   (*a*) $A \cap (B \cup C)$ (*b*) $A^c \cap (B \cup C)$
   (*c*) $A \cup (B \cap C)$ (*d*) $A \cup (B - C)$.

20. Identify the smallest set X containing the sets.
   {Book, Pen}; {Pen, Pencil, Box}; {Book, Box, Ball}.

21. One hundred students were asked whether they had taken courses in any of the three subjects, Mathematics, Computer Science and Information Technology. The results were given below. 45 had taken Mathematics; 18 had taken Mathematics and Computer Science; 38 had taken Computer Science; 21 had taken Information Technology; 9 had taken Mathematics and Information Technology; 4 had taken Computer Science and Information Technology and 23 had taken no courses in any of the subjects. Draw a Venn diagram that will show the results of the survey.

3

Binary Relation

■ 3.0 INTRODUCTION

After the development of set theory we shall try to develop another concept based on it. In this chapter we will introduce an important modeling in mathematics known as relation. This has tremendous application in Computer Science. The relations which are used in Mathematics and Computer Science are "less than", "is a subset of", "is perpendicular to", "is equal to", and so on.

Table

Student Names	*Subjects Taken*
Mary	Computer Science
Smith	Mathematics
Loreena	Computer Science
Finzi	Human Resource
Adams	Marketing
Brown	Mathematics
Mary	Mathematics

A relation can be thought of as a table. Consider the Table given above in which the first column represent the student names and the second column represent the subject taken by the students. From the table it is clear that Mary is taking Computer Science and Mathematics, Loreena is taking Computer Science whereas Smith is taking Mathematics. This is nothing but a set of ordered pairs. We define a relation to be a set of ordered pairs.

Mostly the relations we come across are defined with two entities. We call such relation as binary relation or simply relation.

■ 3.1 BINARY RELATION

Let A and B be two sets. Then any subset R of the Cartesian product (A × B) is a relation (binary relation) from the set A to the set B. Symbolically R (A × B).

i.e., $$R = \{(x, y) \mid x \in A \text{ and } y \in B\}$$

If $(x, y) \in R$, then we write $x \, R \, y$ and say that x is related to y. If $(x, y) \notin R$, then we write $x \not R \, y$ and say that x is not related to y. If A = B, then R is a relation (binary relation) on A.

Consider the example A = {1, 2, 3, 4, 5} and B = {5, 6, 7, 8, 9} and let the relation R from the set A to the set B as

$$R = \{(x, y) \mid x \in A \text{ and } y = 2x + 3 \in B\}$$

i.e., $$R = \{(1, 5), (2, 7), (3, 9), (4, 11), (5, 13)\}$$

i.e., $$R \subseteq A \times B$$

3.1.1 Domain of a Relation

Let R be a relation from the set A to the set B. Then the set of all first constituents of the ordered pairs present in the relation R is known as domain of R . Denoted by dom. R or D(R).

Mathematically,

$$D(R) = \{x \mid (x, y) \in R, \text{ for } x \in A\}$$

i.e., $$D(R) \subseteq A$$

3.1.2 Range of a Relation

Let R be a relation from the set A to the set B. Then the set of all second constituents of the ordered pairs present in the relation R is known as range of R. Denoted by rng.R or R(R).

Mathematically,

$$R(R) = \{y \mid (x, y) \in R, \text{ for } y \in B\}$$

i.e., $$R(R) \subseteq B$$

Consider the example: Let $A = \{a, b, c, d\}$ and B = {5, 6, 7}. Let us define a relation R from the set A to the set B as below.

$$R = \{(a, 5), (a, 6), (c, 6), (d, 6)\}$$

So, $$D(R) = \{a, c, d\} \text{ and } R(R) = \{5, 6\}$$

3.2 INVERSE RELATION

Let R be a relation from the set A to the set B. Then the inverse of the relation R is a relation from the set B to the set A. It is denoted by $R^{-1}$and is defined as

$$R^{-1} = \{(y, x) \mid (x, y) \in R\}$$

Consider the example: Let A = {1, 2, 3, 4, 5}

and B = {4, 9, 16, 17, 25}

Let us consider the relation R from the set A to the set B as R = {(2, 4), (3, 9), (4, 16), (3, 17)}

Therefore, R^{-1}= {(4, 2), (9, 3), (16, 4), (17, 3)}.

3.2.1 Theorem

If R be a relation from the set A to the set B, then (*i*) $D(R) = R(R^{-1})$ and (*ii*) $R(R) = D(R^{-1})$.

Proof: Given that R be a relation from the set A to the set B. *i.e.,* $R \subseteq (A \times B)$. Thus

$$R = \{(x, y) \mid x \in A \text{ and } y \in B\}$$

Let $x \in D(R)$. Then there exists $x \in A$ and $y \in B$ such that

$(x, y) \in R$

This implies $(y, x) \in R^{-1}$

i.e., $x \in R(R^{-1})$

So, $x \in D(R) \Rightarrow x \in R(R^{-1})$

Thus, $D(R) \subseteq R(R^{-1})$ ...(1)

Again let $x \in R(R^{-1})$. Then there exists $x \in A$ and $y \in B$ such that $(y, x) \in R^{-1}$.

This implies $(x, y) \in R$

i.e., $x \in D(R)$

So, $x \in R(R^{-1}) \Rightarrow x \in D(R)$

Thus, $R(R^{-1}) \subseteq D(R)$ ...(2)

Therefore from equations (1) and (2) it is clear that $D(R) = R(R^{-1})$

Similarly, let $y \in R(R)$, Then there exists $x \in A$ and $y \in B$ such that $(x, y) \in R$

This implies $(y, x) \in R^{-1}$

i.e., $y \in D(R^{-1})$

So, $y \in R(R) \Rightarrow y \in D(R^{-1})$

Thus, $R(R) \subseteq D(R^{-1})$ ...(3)

Again let $y \in D(R^{-1})$, Then there exists $x \in A$ and $y \in B$ such that $(y, x) \in R^{-1}$

This implies $(x, y) \in R$

i.e., $y \in R(R)$

So, $y \in D(R^{-1}) \Rightarrow y \in R(R)$

Thus, $D(R^{-1}) \subseteq R(R)$ ...(4)

Therefore from equations (3) and (4) it is clear that $R(R) = D(R^{-1})$

Note: Let R be a relation from the set A to the set B . Then $(R^{-1})^{-1} = R$.

Proof: Given that R be a relation from the set A to the set B. *i.e.* $R \subseteq (A \times B)$

Let $(x, y) \in (R^{-1})^{-1}$

$\Leftrightarrow$ $(y, x) \in R^{-1}$

$\Leftrightarrow$ $(x, y) \in R$

So, $(x, y) \in (R^{-1})^{-1} \Leftrightarrow (x, y) \in R$

Therefore, $(R^{-1})^{-1} = R$

■ 3.3 GRAPH OF RELATION

Let R be a relation from the set A to the set B; that is R is a subset of $(A \times B)$. Since $(A \times B)$ can be represented by the set of points on the coordinate diagram of $(A \times B)$, we can picture R by emphasizing those points in the plane which belong to R. The pictorial representation of the relation R on the coordinate diagram of $(A \times B)$ is known as graph of the relation.

Consider the example: Let $A = \{-3, -2, -1, 1, 2, 3\}$ and $B = \{1, 2, 3, 4, 5, 6, 7, 8, 9\}$ and $x \, R \, y$ such that $y = x^2$. Thus we have

$$R = \{(-1, 1), (1, 1), (-2, 4), (2, 4), (-3, 9), (3,9)\}$$

So, the graph of R is represented on the coordinate diagram of $(A \times B)$ as shown in the following Fig. 1.

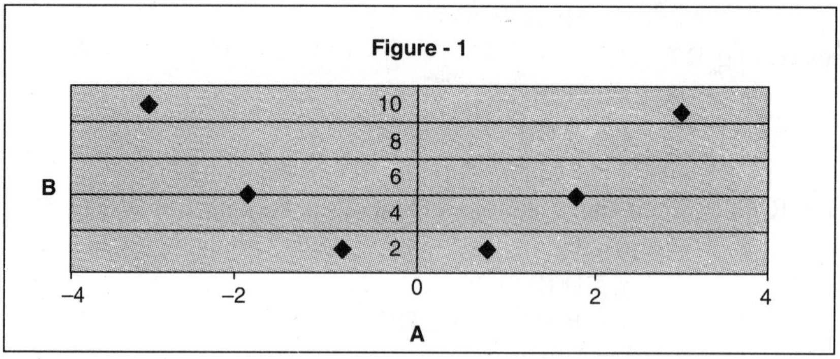

Consider another example: Let A = $\{x \mid x$ is a real number$\}$ and x R y such that $2x + 3y \le 6$. Thus, we have R = $\{(x, y) \mid 2x + 3y \le 6$ and $x, y \in$ A$\}$.

So, the graph of R is represented on the coordinate diagram of (A × A) as shown in the following Fig. 2.

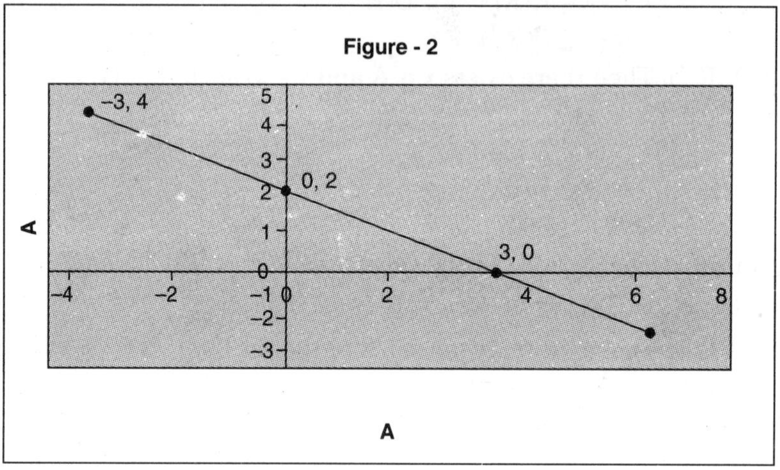

3.4 KINDS OF RELATION

In the study of database systems, kinds of relation play a vital role. In order to get a clear idea on database systems, here we discuss different kinds of relation. A relation R from a set A to a set B may be of four kinds.

(*a*) One-One (*b*) One-Many

(*c*) Many-One (*d*) Many-Many

The relation R from the set A to the set B is said to be One-One relation if $(x_1, y_1) \in$ R, $(x_2, y_2) \in$ R, then $y_1 = y_2 \Rightarrow x_1 = x_2$

The relation R from the set A to the set B is said to be One-Many relation if $(x_1, y_1) \in$ R, $(x_1, y_2) \in$ R for some $x_1 \in$ A and y_1 , $y_2 \in$ B with $y_1 \ne y_2$

The relation R from the set A to the set B is said to be Many-One relation if $(x_1, y_1) \in$ R, $(x_2, y_1) \in$ R for some $y_1 \in$ B and x_1 , $x_2 \in$ A with $x_1 \ne x_2$

The relation R from the set A to the set B is said to be Many-Many relation if $(x_1, y_1) \in$ R, $(x_1, y_2) \in$ R, $(x_2, y_1) \in$ R, and $(x_2, y_2) \in$ R for some $x_1, x_2 \in$ A and $y_1, y_2 \in$ B with $x_1 \ne x_2$ and $y_1 \ne y_2$

■ 3.5 ARROW DIAGRAM

We use arrow diagrams to represent relations. Write down the elements of the set A and the elements of the set B in two disjoint sets, and then draw an arrow from $x \in A$ to $y \in B$ whenever xRy.

Consider the example: Let A = {1, 2, 3, 4, 5} and B = {2, 4, 6, 8}. Let us define the relations from the set A to the set B as

$$R_1 = \{(1, 2), (3, 6), (4, 8)\}$$

$$R_2 = \{(2, 4), (2, 6), (2, 8), (1, 2)\}$$

$$R_3 = \{(1, 4), (2, 4), (3,4), (5, 8)\}$$

and

$$R_4 = \{(1, 4), (2, 4), (1, 8), (2, 8), (5, 2)\}$$

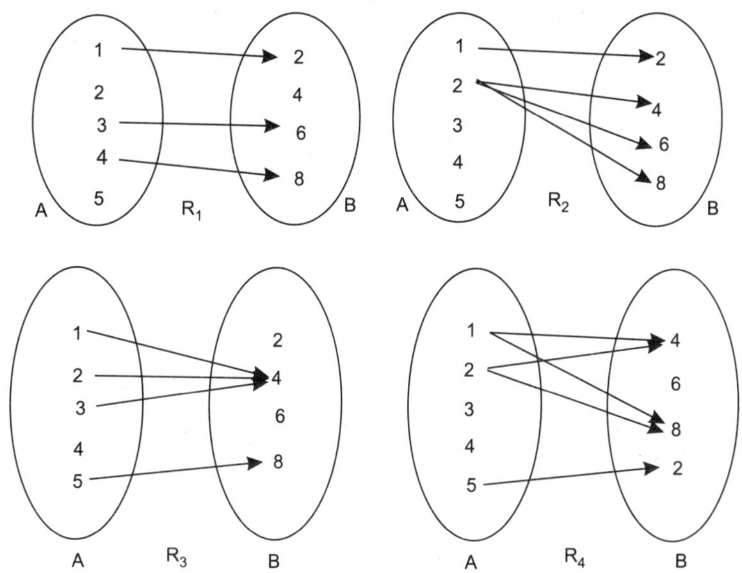

The arrow diagrams for the above relations are given above. From the above diagrams it is clear that R_1, R_2, R_3 and R_4 are One-One, One-Many, Many-One and Many-Many relations respectively.

■ 3.6 VOID RELATION

A relation R from a set A to a set B is said to be a void relation or empty relation if $R = \phi$.

Consider the example: Let A = {3, 5, 7}; B = {2, 4, 8}; $R \subseteq A \times B$ and $x\,R\,y \mid x$ divides y; $x \in A$, $y \in B$. Hence, we observe that $R = \phi \subseteq A \times B$ is a void relation from the set A to the set B.

■ 3.7 IDENTITY RELATION

Let R be a relation on a set A; that is R is a subset of $(A \times A)$. Then the relation R is said to be an identity relation if $(x, x) \in R$. Generally denoted by I_A. Mathematically,

$$I_A = \{(x, x) \mid x \in A\}$$

Consider the example: Let A = $\{a, b, c\}$ and I_A be a relation on A such that $I_A = \{(a, a), (b, b), (c, c)\}$. This is an identity relation on A.

■ 3.8 UNIVERSAL RELATION

A relation R from a set A to a set B is said to be an universal relation if R is equal to (A × B). That is R = (A × B).

Let A = {1, 2, 3} and B = $\{a, b\}$. Therefore the universal relation R from the set A to the set B is given as

$$R = \{(1, a), (1, b), (2, a), (2, b), (3, a), (3, b)\}.$$

■ 3.9 RELATION MATRIX (MATRIX OF THE RELATION)

A matrix is a convenient way to represent a relation R. Such a representation can be used by a computer to analyze the relation.

Let $$A = \{a_1, a_2, a_3, \ldots, a_i, \ldots, a_k\}$$
and $$B = \{b_1, b_2, b_3, \ldots, b_j, \ldots, b_l\}$$
be two finite sets and R be a relation from the set A to the set B. Then the matrix of the relation R, *i.e.*, M(R) is defined as

$$M(R) = [m_{ij}] \text{ of order } (k \times l)$$

where $$m_{ij} = \begin{cases} 1; & \text{if } a_i \, R \, b_j \\ 0; & \text{if } a_i \not{R} \, b_j \end{cases}$$

In other words label the rows of rectangular array by the elements of A and the columns by the elements of B. Each position of the array is to be filled with *a* 1 (one) or 0 (zero) according as $a \in A$ is related or not related to $b \in B$. Consider the example:

Let A = {1, 2, 3}; B = $\{a, b, c, d, e\}$ and R ⊆ (A × B) such that R = $\{(1, a), (1, d), (2, b), (3, c), (3, d)\}$.

So the matrix of the above relation R is given as

$$M(R) = \begin{array}{c} \\ 1 \\ 2 \\ 3 \end{array} \begin{array}{c} \begin{array}{ccccc} a & b & c & d & e \end{array} \\ \begin{bmatrix} 1 & 0 & 0 & 1 & 0 \\ 0 & 1 & 0 & 0 & 0 \\ 0 & 0 & 1 & 1 & 0 \end{bmatrix} \end{array}$$

■ 3.10 COMPOSITION OF RELATIONS

Let R_1 be a relation from the set A to the set B and R_2 be a relation from the set B to the set C. That is R_1 is a subset of (A × B) and R_2 is a subset of (B × C). Then the composition of R_1 and R_2 is given by R_1R_2 and is defined by

$$R_1R_2 = \{(x, z) \in (A \times C) \mid \text{for some } y \in B, (x, y) \in R_1 \text{ and } (y, z) \in R_2\}$$

Consider the example: Let A = {1, 2, 4, 5, 7};

$$B = \{a, b, c, d, e\}$$
and $$C = \{1, 4, 16, 25\}$$

Consider the relations R_1: A → B and R_2: B → C as

$R_1 = \{(1, a), (1, c), (2, d), (2, e), (5, d)\}$ and $R_2 = \{(c, 1), (d, 4), (e, 25)\}$. The arrow diagram is given as

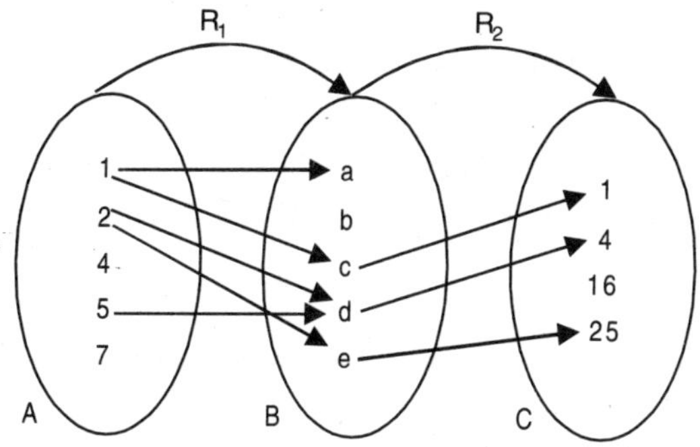

So, $$R_1R_2 = \{(1, 1), (2, 4), (2, 25), (5, 4)\}$$

3.10.1 Composition of Relations and Relation Matrix

Let R_1 be a relation from the set A to the set B and R_2 be a relation from the set B to the set C. That is R_1 is a subset of $(A \times B)$ and R_2 is a subset of $(B \times C)$. Then the composition of R_1 and R_2 is given by R_1R_2 and the matrix of the composition R_1R_2 is defined as

$$M(R_1R_2) = M(R_1)\,M(R_2)$$

and replace all nonzero entries by 1 in $M(R_1R_2)$, where $M(R_1)$ is the matrix of the relation R_1 and $M(R_2)$ is the matrix of the relation R_2.

Consider the same example stated above; we have

$$M(R_1) = \begin{bmatrix} 1 & 0 & 1 & 0 & 0 \\ 0 & 0 & 0 & 1 & 1 \\ 0 & 0 & 0 & 0 & 0 \\ 0 & 0 & 0 & 1 & 0 \\ 0 & 0 & 0 & 0 & 0 \end{bmatrix} \text{ and } M(R_2) = \begin{bmatrix} 0 & 0 & 0 & 0 \\ 0 & 0 & 0 & 0 \\ 1 & 0 & 0 & 0 \\ 0 & 1 & 0 & 0 \\ 0 & 0 & 0 & 1 \end{bmatrix}$$

So, $$M(R_1R_2) = M(R_1)\,M(R_2)$$

$$= \begin{bmatrix} 1 & 0 & 1 & 0 & 0 \\ 0 & 0 & 0 & 1 & 1 \\ 0 & 0 & 0 & 0 & 0 \\ 0 & 0 & 0 & 1 & 0 \\ 0 & 0 & 0 & 0 & 0 \end{bmatrix} \begin{bmatrix} 0 & 0 & 0 & 0 \\ 0 & 0 & 0 & 0 \\ 1 & 0 & 0 & 0 \\ 0 & 1 & 0 & 0 \\ 0 & 0 & 0 & 1 \end{bmatrix} = \begin{bmatrix} 1 & 0 & 0 & 0 \\ 0 & 1 & 0 & 1 \\ 0 & 0 & 0 & 0 \\ 0 & 1 & 0 & 0 \\ 0 & 0 & 0 & 0 \end{bmatrix}$$

Therefore, $R_1R_2 = \{(1, 1), (2, 4), (2, 25), (5, 4)\}$.

3.10.2 Theorem

Let R_1 and R_2 are relations from the set A to the set B. Let R_3 and R_4 are relations from the set B to the set C. If $R_1 \subseteq R_2$ and $R_3 \subseteq R_4$ then $R_1R_3 \subseteq R_2R_4$.

Proof: Given R_1 and R_2 are relations from the set A to the set B. R_3 and R_4 are relations from the set B to the set C.

Suppose that $R_1 \subseteq R_2$ and $R_3 \subseteq R_4$.

Let $(x, z) \in R_1 R_3$

Then for some $y \in B$, we have $(x, y) \in R_1$ and $(y, z) \in R_3$. Therefore we have $(x, y) \in R_1 \subseteq R_2$ and $(y, z) \in R_3 \subseteq R_4$.

i.e., $(x, y) \in R_2$ and $(y, z) \in R_4$.

This implies $(x, z) \in R_2 R_4$.

Hence $(x, z) \in R_1R_3 \Rightarrow (x, z) \in R_2 R_4$

i.e., $R_1R_3 \subseteq R_2R_4$.

3.10.3 Theorem

Let R_1 be relation from the set A to the set B and R_2 be a relation from the set B to the set C. Then,

$$(R_1R_2)^{-1} = R_2^{-1}R_1^{-1}.$$

Proof: Let R_1 be a relation from the set A to the set B and R_2 be a relation from the set B to the set C.

Our claim: $(R_1R_2)^{-1} = R_2^{-1}R_1^{-1}$.

i.e., $(R_1R_2)^{-1} \subseteq R_2^{-1}R_1^{-1}$ and $R_2^{-1}R_1^{-1} \subseteq (R_1R_2)^{-1}$

Let $(x, z) \in (R_1R_2)^{-1}$

This implies $(z, x) \in R_1R_2$. Then for some $y \in$ B we have

$(z, y) \in R_1$ and $(y, x) \in R_2$

$\Rightarrow$ $(y, z) \in R_1^{-1}$ and $(x, y) \in R_2^{-1}$

i.e., $(x, y) \in R_2^{-1}$ and $(y, z) \in R_1^{-1}$

This implies $(x, z) \in R_2^{-1}R_1^{-1}$

Therefore, $(x, z) \in (R_1R_2)^{-1} \Rightarrow (x, z) \in R_2^{-1}R_1^{-1}$

i.e., $(R_1R_2)^{-1} \subseteq R_2^{-1}R_1^{-1}$ ...(*i*)

Again let $(x, z) \in R_2^{-1}R_1^{-1}$. Then for some $y \in$ B we have

$(x, y) \in R_2^{-1}$ and $(y, z) \in R_1^{-1}$

$\Rightarrow$ $(y, x) \in R_2$ and $(z, y) \in R_1$

i.e., $(z, y) \in R_1$ and $(y, x) \in R_2$

This implies $(z, x) \in R_1R_2$

i.e., $(x, z) \in (R_1R_2)^{-1}$

Therefore, $(x, z) \in R_2^{-1}R_1^{-1} \Rightarrow (x, z) \in (R_1R_2)^{-1}$

i.e., $R_2^{-1}R_1^{-1} \subseteq (R_1R_2)^{-1}$ ...(*ii*)

Thus from equations (*i*) and (*ii*) we get $(R_1R_2)^{-1} = R_2^{-1}R_1^{-1}$.

3.11 TYPES OF RELATIONS

This section discusses a number of different important types of relations on a set A that are important for the study of finite state systems.

3.11.1 Reflexive Relations

A relation R defined on a set A is said to be reflexive if $(x, x) \in R$ for every element $x \in A$.

i.e., $x \, R \, x \quad \forall x \in A$

Consider the following relations on the set A = {1, 3, 5, 7}

$$R_1 = \{(1, 1), (1, 3), (1, 5), (5, 5), (5, 7)\}$$

$$R_2 = \{(1, 3), (1, 5), (5, 7), (3, 7)\}$$
$$R_3 = \{(1, 1), (1, 3), (3, 3), (5, 5), (5, 7), (1, 7), (7, 7)\}$$

From the above relations it is clear that R_3 is a reflexive relation. R_1 is not a reflexive relation as $(3, 3) \notin R_1$ and $(7, 7) \notin R_1$. Similarly, R_2 is also not reflexive.

3.11.2 Symmetric Relations

A relation R defined on a set A is said to be symmetric if $(x, y) \in R$ then $(y, x) \in R$.

i.e., $x \; R \; y \Rightarrow y \; R \; x$.

Consider the following relations on the set A = {1, 3, 5, 7}

$$R_1 = \{(1, 1), (1, 3), (3, 5), (3, 1), (5, 3), (5, 5)\}$$
$$R_2 = \{(1, 1), (1, 3), (3, 1), (3, 5), (5, 3), (5, 7), (7, 7)\}$$

From the above relations it is clear that R_1 is a symmetric relation, but R_2 is not a symmetric relation as $(5, 7) \in R_2 \Rightarrow (7, 5) \notin R_2$.

3.11.3 Transitive Relations

A relation R defined on a set A is said to be transitive if $(x, y) \in R$ and $(y, z) \in R$ then $(x, z) \in R$.

i.e., $x \; R \; y$ and $y \; R \; z \Rightarrow x \; R \; z$

Consider the following relations on the set A = {1, 3, 5, 7}.

$$R_1 = \{(1, 1), (1, 3), (1, 5), (1, 7), (3, 3), (3, 5), (3, 7), (5, 3), (5, 5), (5, 7)\}$$
$$R_2 = \{(1, 1), (1, 3), (3, 5), (5, 5), (7, 7)\}$$

From the above relations it is clear that R_1 is a transitive relation. The relation R_2 is not transitive as $(1, 3) \in R_2, (3, 5) \in R_2 \Rightarrow (1, 5) \notin R_2$.

3.11.4 Anti-Reflexive Relations

A relation R defined on a set A is said to be anti-reflexive or irreflexive if $(x, x) \notin R$ for every element $x \in A$.

i.e., $x \not{R} \; x \quad \forall \; x \in A$

Consider the following relations on the set A = {1, 3, 5, 7}

$$R_1 = \{(1, 1), (1, 3), (1, 7), (3, 3), (5, 5), (5, 7), (7, 7)\}$$
$$R_2 = \{(1, 3), (1, 5), (5, 7), (3, 7)\}$$
$$R_3 = \{(1, 1), (1, 3), (1, 5), (7, 7)\}$$

From the above relations it is clear that R_2 is an anti-reflexive relation. R_3 is not an anti-reflexive relation as $(1, 1) \in R_3$ and $(7, 7) \in R_3$. Similarly, R_1 is not an anti-reflexive relation.

3.11.5 Asymmetric Relations

A relation R defined on a set A is said to be asymmetric if $(x, y) \in R$ then $(y, x) \notin R$.

i.e., $x \; R \; y \Rightarrow y \not{R} \; x$

Consider the following relations on the set A = {1, 3, 5, 7}

$$R_1 = \{(1, 3), (3, 5), (3, 7), (5, 7)\}$$
$$R_2 = \{(1, 3), (3, 5), (3, 7), (5, 3), (5, 7)\}$$

From the above relations it is clear that R_1 is an asymmetric relation. R_2 is not an asymmetric relation as $(3, 5) \in R_2 \Rightarrow (5, 3) \in R_2$.

3.11.6 Anti-Symmetric Relations

A relation R defined on a set A is said to be anti-symmetric relation if $(x, y) \in R$ and $(y, x) \in R$, then $x = y$.

i.e., $x \, R \, y$ and $y \, R \, x \Rightarrow x = y$.

Consider the following relations on the set A = {1, 3, 5, 7}

$$R_1 = \{(1,1), (1, 3), (3, 5), (5, 5), (5, 7)\}$$
$$R_2 = \{(1, 1), (3, 3), (7, 7)\}$$
$$R_3 = \{(3, 3), (3, 5), (5, 3), (5, 7), (7, 5), (7, 7)\}$$

From the above relations it is clear that R_1 and R_2 are anti-symmetric. R_3 is not an anti-symmetric relation as $(3, 5) \in R$ and $(5, 3) \in R$, but $3 \neq 5$. Similarly $(5, 7) \in R$ and $(7, 5) \in R$, but $5 \neq 7$.

■ 3.12 TYPES OF RELATIONS AND RELATION MATRIX

Let $A = \{a_1, a_2, \dots , a_i, \dots , a_j, \dots\dots , a_n\}$ be a non-empty set and R be a relation defined on the set A. Hence the matrix of the relation R relative to the ordering $a_1, a_2, \dots , a_i, \dots , a_j, \dots\dots , a_n$ is defined as

$$M(R) = [\, m_{ij} \,]_{n \times n}$$

where
$$m_{ij} = \begin{cases} 1 & \text{if } a_i \, R \, a_j \\ 0 & \text{if } a_i \not{R} \, a_j \end{cases}$$

3.12.1 Reflexive Relations

The relation R is said to be reflexive if $m_{ii} = 1 \; \forall \; 1 \leq i \leq n$

i.e., all elements of the main diagonal in the relation matrix M(R) are 1.

3.12.2 Symmetric Relations

The relation R is said to be symmetric if $m_{ij} = m_{ji} \; \forall \; 1 \leq i \leq n$ and $1 \leq j \leq n$.

In other words the relation R is said to be symmetric if $M(R) = [\,M(R)]^T$, where $[M(R)]^T$ represents the transpose of the relation matrix M(R).

3.12.3 Transitive Relation

The relation R is said to be transitive if $m_{ij} = 1$ and $m_{jk} = 1$, then $m_{ik} = 1$ for $1 \leq i \leq n$; $1 \leq j \leq n$ and $1 \leq k \leq n$.

In other words the relation R is said to be transitive if and only if $R^2 \subseteq R$. *i.e.* Whenever entry i, j in $[M(R)]^2$ is non-zero, entry i, j in M(R) is also non-zero.

Let R be a relation on the set A and R is transitive.

Let $(x, z) \in R^2 = R \cdot R$.

So, there exists $y \in A$ such that $(x, y) \in R$ and $(y, z) \in R$

Thus $(x, z) \in R$ [$\because$ R is transitive]

i.e., $(x, z) \in R^2 \Rightarrow (x, z) \in R$

Therefore, $R^2 \subseteq R$.

Conversely, suppose that $R^2 \subseteq R$.

Let $(x, y) \in R$ and $(y, z) \in R$

This implies $\qquad (x, z) \in R \cdot R = R^2$

i.e., $\qquad (x, z) \in R^2 \subseteq R$

i.e., $\qquad (x, z) \in R$

Therefore, R is transitive.

3.12.4 Anti-Reflexive Relations

The relation R is said to be anti-reflexive if $m_{ii} = 0 \ \forall \ 1 \leq i \leq n$
i.e., all elements of the main diagonal in relation matrix M(R) are 0 (zero).

3.12.5 Asymmetric Relations

The relation R is said to be asymmetric if $m_{ij} = 1$, then $m_{ji} = 0$ and $m_{ii} = 0$.

3.12.6 Anti-Symmetric Relations

The relation R is said to be anti-symmetric if $a_i \neq a_j$ then either $m_{ij} = 0$ or $m_{ji} = 0$ and $m_{ij} = 1 = m_{ji}$ implies $a_i = a_j$.

Consider the following relations on the set A = {1, 3, 5, 7}

$$R_1 = \{(1, 1), (1, 3), (1, 7), (3, 3), (3, 7), (5, 5), (5, 7), (7, 7)\}$$
$$R_2 = \{(1, 1), (1, 5), (1, 7), (3, 5), (3, 7), (5, 1), (5, 3), (7, 1), (7, 3)\}$$
$$R_3 = \{(1, 1), (1, 3), (1, 5), (1, 7), (3, 1), (3, 3), (3, 5), (3, 7), (5, 7)\}$$
$$R_4 = \{(1, 3), (1, 7), (3, 7), (5, 7), (7, 1)\}$$
$$R_5 = \{(1, 3), (3, 5), (5, 7), (7, 1), (7, 3)\}$$
$$R_6 = \{(1,1), (1, 7), (7, 5), (7, 3), (5, 3)\}$$

Relative to the ordering 1, 3, 5, 7, we get

$$M(R_1) = \begin{bmatrix} 1 & 1 & 0 & 1 \\ 0 & 1 & 0 & 1 \\ 0 & 0 & 1 & 1 \\ 0 & 0 & 0 & 1 \end{bmatrix}; \quad M(R_2) = \begin{bmatrix} 1 & 0 & 1 & 1 \\ 0 & 0 & 1 & 1 \\ 1 & 1 & 0 & 0 \\ 1 & 1 & 0 & 0 \end{bmatrix};$$

$$M(R_3) = \begin{bmatrix} 1 & 1 & 1 & 1 \\ 1 & 1 & 1 & 1 \\ 0 & 0 & 0 & 1 \\ 0 & 0 & 0 & 0 \end{bmatrix}; \quad M(R_4) = \begin{bmatrix} 0 & 1 & 0 & 1 \\ 0 & 0 & 0 & 1 \\ 0 & 0 & 0 & 1 \\ 1 & 0 & 0 & 0 \end{bmatrix};$$

$$M(R_5) = \begin{bmatrix} 0 & 1 & 0 & 0 \\ 0 & 0 & 1 & 0 \\ 0 & 0 & 0 & 1 \\ 1 & 1 & 0 & 0 \end{bmatrix}; \quad M(R_6) = \begin{bmatrix} 1 & 0 & 0 & 1 \\ 0 & 0 & 0 & 0 \\ 0 & 1 & 0 & 0 \\ 0 & 1 & 1 & 0 \end{bmatrix};$$

From the above matrices it is clear that $m_{ii} = 1$ in $M(R_1)$ and $m_{ii} = 0$ in $M(R_4)$ and $M(R_5)$. Thus the relation R_1 is reflexive whereas the relations R_4 and R_5 are anti-reflexive. Again

$$[M(R_2)]^T = \begin{bmatrix} 1 & 0 & 1 & 1 \\ 0 & 0 & 1 & 1 \\ 1 & 1 & 0 & 0 \\ 1 & 1 & 0 & 0 \end{bmatrix} = M(R_2)$$

So, the relation R_2 is symmetric. Also $[M(R_1)]^T \neq M(R_1)$, and hence the relation R_1 is not symmetric. Similarly it can be shown that the relations R_3, R_4, R_5 and R_6 are not symmetric.

Now in $M(R_1)$, $M(R_2)$, $M(R_3)$ and $M(R_6)$, we see that $m_{ii} \neq 0$, so the relations R_1, R_2, R_3 and R_6 are not asymmetric. In $M(R_4)$ we see that $m_{ii} = 0$, but $m_{14} = 1 = m_{41}$. This violate the conditions of asymmetric relation hence not asymmetric. It is also observed that in $M(R_5)$, $m_{ii} = 0$; $m_{12} = 1, m_{21} = 0; m_{23} = 1, m_{32} = 0; m_{34} = 1, m_{43} = 0; m_{41} = 1, m_{14} = 0$ and $m_{42} = 1, m_{24} = 0$. Thus the relation R_5 is asymmetric. Again

$$[M(R_3)]^2 = \begin{bmatrix} 1 & 1 & 1 & 1 \\ 1 & 1 & 1 & 1 \\ 0 & 0 & 0 & 1 \\ 0 & 0 & 0 & 0 \end{bmatrix} \begin{bmatrix} 1 & 1 & 1 & 1 \\ 1 & 1 & 1 & 1 \\ 0 & 0 & 0 & 1 \\ 0 & 0 & 0 & 0 \end{bmatrix} = \begin{bmatrix} 2 & 2 & 2 & 3 \\ 2 & 2 & 2 & 3 \\ 0 & 0 & 0 & 0 \\ 0 & 0 & 0 & 0 \end{bmatrix}$$

We see that whenever i, j in $[M(R_3)]^2$ is non-zero, entry i, j in $M(R_3)$ is also non-zero. So the relation R_3 is transitive. It is also cleared that $[M(R_i)]^2 \not\subset M(R_i)$ for $i = 1, 2, 4, 5, 6$. Thus the relations R_1, R_2, R_4, R_5 and R_6 are not transitive. Also it can be shown that the relation R_6 is anti-symmetric.

3.13 EQUIVALENCE RELATION

The concept of equivalence relation is of great importance in knowledge representation and artificial intelligence. Here, in this section we discuss the concept of equivalence relation. A relation R defined on a set A is said to be an equivalence relation in A if and only if R is reflexive, symmetric and transitive.

Consider the relation R in the real numbers defined by $x = y$ *i.e.*, x R $y : x = y$

Reflexive: For all $x \in R$

$x = x$

i.e., x R x

i.e., R is reflexive.

Symmetric: Suppose x R y

i.e., $x = y$

i.e., $y = x$

i.e., y R x

i.e., R is symmetric.

Transitive: Suppose x R y and y R z

i.e., $x = y$ and $y = z$

This implies $x = z$

i.e., x R z

i.e., R is transitive.

So, the relation R in the real numbers defined by $x = y$ is an equivalence relation.

3.13.1 Theorem

If R be an equivalence relation defined in a set A, then R^{-1} is also an equivalence relation in the set A.

Proof: Let R be an equivalence relation defined in a set A. Thus R is reflexive, symmetric and transitive.

Our claim: R^{-1} is an equivalence relation in the set A.

Reflexive: For all $x \in A$

$(x, x) \in R$ [$\because$ R is reflexive]

$\Rightarrow$ $(x, x) \in R^{-1}$

So, $(x, x) \in R^{-1} \forall x \in A$

Symmetric: Suppose $(x, y) \in R^{-1}$

$\Rightarrow$ $(y, x) \in R$

$\Rightarrow$ $(x, y) \in R$ [$\because$ R is symmetric]

$\Rightarrow$ $(y, x) \in R^{-1}$

i.e., R^{-1} is symmetric.

Transitive: Suppose $(x, y) \in R^{-1}$ and $(y, z) \in R^{-1}$

$\Rightarrow$ $(y, x) \in R$ and $(z, y) \in R$

i.e., $(z, y) \in R$ and $(y, x) \in R$

$\Rightarrow$ $(z, x) \in R$ [$\because$ R is transitive]

$\Rightarrow$ $(x, z) \in R^{-1}$

i.e., R^{-1} is transitive.

Therefore, R^{-1} is an equivalence relation in the set A.

3.14 PARTIAL ORDER RELATION

Let R be a relation defined on a set A. Then the relation R is said to be a partial order relation in A if R is reflexive, transitive and anti-symmetric. It is of great use while studying lattices.

Consider the relation R in the real numbers defined by $x \le y$. *i.e.,* $x \, R \, y : x \le y$.

Reflexive: For all $x \in R$, $x \le x$

i.e., $x \, R \, x$

i.e., R is reflexive.

Transitive: Suppose that $x \, R \, y$ and $y \, R \, z$

i.e., $x \le y$ and $y \le z$

This implies $x \le z$

i.e., $x \, R \, z$

i.e., R is transitive.

Anti-symmetric: Suppose that $x \, R \, y$ and $y \, R \, x$

i.e., $x \le y$ and $y \le x$

This implies $x = y$

i.e., R is anti-symmetric.

So, the relation R in the real numbers defined by $x \le y$ is a partial order relation.

3.14.1 Theorem

Let A be a set and R be a partial order relation on A. Then R^{-1} is also a partial order relation on A.

Proof : Let R be a partial order relation defined in a set A. Therefore R is reflexive, transitive and anti-symmetric.

Our claim: R^{-1} is a partial order relation.

Reflexive: For all $x \in A$

$(x, x) \in R$ [$\because$ R is reflexive]

This implies $(x, x) \in R^{-1}$

i.e., R^{-1} is reflexive.

Transitive: Suppose that $(x, y) \in R^{-1}$ and $(y, z) \in R^{-1}$

This implies$(y, x) \in R$ and $(z, y) \in R$

i.e., $(z, y) \in R$ and $(y, x) \in R$

This implies $(z, x) \in R$ [$\because$ R is transitive]

i.e., $(x, z) \in R^{-1}$

i.e., R^{-1} is transitive.

Anti-symmetric: Suppose that $(x, y) \in R^{-1}$ and $(y, x) \in R^{-1}$

This implies$(y, x) \in R$ and $(x, y) \in R$

This implies $x = y$ [$\because$ R is anti-symmetric]

i.e., R^{-1} is anti-symmetric.

Therefore, R^{-1} is a partial order relation in the set A.

■ 3.15 TOTAL ORDER RELATION

Let R be a relation defined on a set A. Then the relation R is said to be a total order relation in A if R is a partial order relation and for any two elements x, y in A either $x < y$, $x = y$ or $x > y$ holds.

Consider the relation R in D(6) defined by $x \leq y$, where D(6) is the set of all positive divisors of 6.

Therefore, $D(6) = \{1, 2, 3, 6\}$ and $x \, R \, y : x \leq y$

i.e., $R = \{(1, 1), (1, 2), (1, 3), (1, 6), (2, 2), (2, 3), (2, 6), (3, 3), (3, 6), (6, 6)\}$

So, R is reflexive, transitive and anti-symmetric. *i.e.* R is a partial order relation in D(6).

Besides this for any two elements x, y belongs D(6), one of the relations $x \leq y$ or $y \leq x$ holds. Thus the relation R in D(6) defined by $x \leq y$ is a total order relation.

Consider another relation R in A = {1, 2, 3, ... , 10} defined by x is a multiple of y.

i.e., $x \, R \, y$: x is a multiple of y

Reflexive: For all $x \in A$

x is a multiple of x

i.e., $x \, R \, x$

i.e., R is reflexive.

Transitive:Suppose $x \, R \, y$ and $y \, R \, z$

i.e., x is a multiple of y and y is a multiple of z

$\Rightarrow$ $x = K_1 y$ and $y = K_2 z$ for $K_1, K_2 \in I$; $K_1, K_2 \neq 0$

$\Rightarrow$ $x = K_1 K_2 z$; $K_1, K_2 \in I$; $K_1 K_2 \neq 0$

i.e., x is a multiple of z

i.e., $x \text{ R } z$

i.e., R is transitive.

Anti-symmetric: Suppose $x \text{ R } y$ and $y \text{ R } x$

i.e., x is a multiple of y and y is a multiple of x

$\Rightarrow \quad x = K_1 y$ and $y = K_2 x$ for $K_1, K_2 \in I; K_1, K_2 \neq 0$

$\Rightarrow \quad x = K_1K_2 x$

$\Rightarrow \quad K_1K_2 = 1$

$\Rightarrow \quad K_1 = K_2 = 1 \qquad [\because K_1, K_2 \neq 0 \text{ and } K_1, K_2 \in I]$

So, $x = y$, *i.e.*, R is anti-symmetric. Therefore, the relation in A defined by x is a multiple of y is a partial order relation.

Now R = {(1, 1), (1, 2), (1, 3), (1, 4), (1, 5), (1, 6), (1, 7), (1, 8), (1, 9), (1, 10), (2, 2), (2, 4), (2, 6), (2, 8), (2, 10), (3, 3), (3, 6), (3, 9), (4, 8), (5, 10)}

Again for 2 and 5 belongs to A either of the relations $2 \leq 5$ or $2 \geq 5$ do not hold because 2 is not a multiple of 5. Therefore, R is not a total order relation.

■ 3.16 CLOSURES OF RELATIONS

If R be a relation defined on A, then the closure of the relation R is the smallest relation R′ that includes all the pairs of R and possesses the required properties of the closure.

3.16.1 Reflexive Closure

Let R be a relation defined on the set A. Then, the reflexive closure $r(\text{R})$ is defined by

(*i*) If $(x, y) \in \text{R}$ then $(x, y) \in r(\text{R})$

(*ii*) If $x \in \text{A}$, then $(x, x) \in r(\text{R})$

(*iii*) Nothing is in $r(\text{R})$ unless it is so follows from (*i*) and (*ii*).

Consider the following relation on the set A = {2, 4, 6, 8}

$$\text{R} = \{(2, 2), (2, 4), (6, 8), (6, 6), (6, 4)\}$$

Therefore, $r(\text{R}) = \{(2, 2), (2, 4), (6, 8), (6, 6), (6, 4), (4, 4), (8, 8)\}$

3.16.2 Symmetric Closure

Let R be a relation defined on the set A. Then, the symmetric closure $s(\text{R})$ is defined by

(*i*) If $(x, y) \in \text{R}$ then $(x, y) \in s(\text{R})$

(*ii*) If $(x, y) \in \text{R}$, then $(y, x) \in s(\text{R})$

(*iii*) Nothing is in s(R) unless it is so follows from (*i*) and (*ii*).

Consider the following relation on the set A = {2, 4, 6, 8}

$$\text{R} = \{(2, 2), (2, 4), (2, 6), (4, 2), (4, 6), (6, 4), (6, 8), (8, 2)\}$$

Therefore, $s(\text{R}) = \{(2, 2), (2, 4), (2, 6), (4, 2), (4, 6), (6, 4), (6, 8), (8, 2), (6, 2), (8, 6), (2, 8)\}$

3.16.3 Transitive Closure

Let R be a relation defined on the set A. Then the transitive closure $t(\text{R})$ is defined by

(*i*) If $(x, y) \in \text{R}$, then $(x, y) \in t(\text{R})$

(*ii*) If $(x, y) \in R$, $(y, z) \in R$ then $(x, z) \in t(R)$
(*iii*) Nothing is in $t(R)$ unless it is so follows from (*i*) and (*ii*).

Consider the following relation on the set A = {2, 4, 6, 8}

$$R = \{(2, 2), (2, 4), (4, 6), (4, 8), (2, 8)\}$$

Therefore, $$t(R) = \{(2, 2), (2, 4), (4, 6), (4, 8), (2, 8), (2, 6)\}$$

3.17 EQUIVALENCE CLASSES

The study of equivalence classes are used in knowledge representation and finite state systems. Apart from these two, it has many applications in computer science. Here, we discuss the basic idea of an equivalence class.

Let A be a non-empty set. R be an equivalence relation in A. For each $x \in A$, the sets $[x]$ are called equivalence classes of A given by the relation R defined as

$$[x] = \{y \in A \mid y \, R \, x\}$$

Consider the equivalence relation R defined on the set A = {1, 3, 5, 7, 9} as

R = {(1, 1), (1, 3), (1, 5), (3, 1), (3, 3), (3, 5), (5, 1), (5, 3), (5, 5), (7, 7), (7, 9), (9, 7), (9, 9)}

So, the equivalence classes are given as

$$[1] = [3] = [5] = \{1, 3, 5\}$$
$$[7] = [9] = \{7, 9\}$$

3.17.1 Theorem

Let R be an equivalence relation defined on a non-empty set A and x, y be arbitrary elements in A. Then

(*i*) $x \in [x]$ and (*ii*) If $y \in [x]$, then $[x] = [y]$

Proof: Let R be an equivalence relation defined on a non-empty set A.

(*i*) Let $x \in A$. Therefore, $[x] = \{y \in A \mid y \, R \, x\}$

As R is reflexive in A, we have $x \, R \, x$. *i.e.*, $x \in [x]$

(*ii*) Suppose that $y \in [x]$

$\Rightarrow$ $y \, R \, x$ [By definition]

$\Rightarrow$ $x \, R \, y$ [$\because$ R is symmetric]

Let $a \in [x]$; this implies $a \, R \, x$

So, $a \, R \, x$ and $x \, R \, y$

This implies $a \, R \, y$ [$\because$ R is transitive]

i.e., $a \in [y]$

Therefore, $a \in [x] \Rightarrow a \in [y]$ *i.e.*, $[x] \subseteq [y]$ ...(*i*)

Similarly, Let $b \in [y]$

This implies $b \, R \, y$ [By definition]

So, $b \, R \, y$ and $y \, R \, x$.

$\Rightarrow$ $b \, R \, x$ [$\because$ R is transitive]

i.e., $b \in [x]$.

Therefore, $b \in [y] \Rightarrow b \in [x]$ *i.e.*, $[y] \subseteq [x]$ ...(*ii*)

Therefore from equations (*i*) and (*ii*), we have $[x] = [y]$.

3.17.2 Theorem

Let A be a non-empty set and R be an equivalence relation defined in A. Let x, y be two arbitrary elements of A. Then $[x] = [y]$ if and only if $x \mathrel{R} y$.

Proof: Let R be an equivalence relation defined in A, and let $x, y \in A$. Assume that $[x] = [y]$. Our claim is $x \mathrel{R} y$.

As R is reflexive, we have $x \mathrel{R} x$

i.e., $x \in [x]$

$\Rightarrow$ $x \in [x] = [y]$

$\Rightarrow$ $x \in [y]$

i.e., $x \mathrel{R} y$

Conversely, suppose that $x \mathrel{R} y$,

i.e., $y \mathrel{R} x$ [$\because$ R is symmetric]

Our claim is $[x] = [y]$

Let $a \in [x]$ this implies $a \mathrel{R} x$

i.e., $a \mathrel{R} x$ and $x \mathrel{R} y$

This implies $a \mathrel{R} y$ [$\because$ R is transitive]

i.e., $a \in [y]$

Therefore, $a \in [x]$ implies $a \in [y]$, *i.e.,* $[x] \subseteq [y]$ ...(*i*)

Again $a \in [y]$ this implies $a \mathrel{R} y$

i.e., $a \mathrel{R} y$ and $y \mathrel{R} x$

This implies $a \mathrel{R} x$ [$\because$ R is transitive]

i.e., $a \in [x]$

Therefore, $a \in [y]$ implies $a \in [x]$, *i.e.,* $[y] \subseteq [x]$ ...(*ii*)

Thus, from equations (*i*) and (*ii*) we get $[x] = [y]$.

3.17.3 Theorem

Let A be a non-empty set and R be an equivalence relation in A. Let $x, y \in A$. Then the equivalence classes $[x]$ and $[y]$ are either equal or disjoint.

Proof: Let A be a non-empty set and R be an equivalence relation defined in A. Let $x, y \in A$.

Assume that the equivalence classes $[x]$ and $[y]$ are not disjoint, *i.e.,* $[x] \cap [y] \neq \phi$

Thus, there exists at least one element a in $[x] \cap [y]$.

i.e., $a \in [x] \cap [y]$

i.e., $a \mathrel{R} x$ and $a \mathrel{R} y$

i.e., $x \mathrel{R} a$ and $a \mathrel{R} y$ [$\because$ R is symmetric]

This implies $x \mathrel{R} y$ [$\because$ R is transitive]

Hence by previous theorem 3.17.2, it is clear that $[x] = [y]$. Therefore, it is clear that if two equivalence classes $[x]$ and $[y]$ are either disjoint or equal.

■ 3.18 PARTITIONS

In real life, changing data into knowledge is not a straight forward task. A set of data is generally disorganized whereas knowledge is just the opposite, but expressed by means of a proper language. In order to get better knowledge, one should have better classifying power. Therefore, partition plays a vital role in knowledge representation.

Let A be a non-empty set. A partition P of A is a collection $\{A_i\}$ of non-empty subsets of A with the following two properties.

(i) $\bigcup_i A_i = A$ and

(ii) $A_i \cap A_j = \phi$ for $A_i \neq A_j$

In other words a partition of A is a collection of non-empty disjoint subset of A whose union is A.

Consider the relation $x \equiv y \pmod 3$ defined on the set of integers I. The above relation is an equivalence relation in I. The set of three equivalence classes are [0], [1] and [2]. Where

$$[0] = \{\ldots, -6, -3, 0, 3, 6, 9, \ldots\ldots\}$$

$$[1] = \{\ldots\ldots, -5, -2, 1, 4, 7, 10, \ldots\}$$

$$[2] = \{\ldots\ldots, -4, -1, 2, 5, 8, 11, \ldots\ldots\}$$

It is clear that [0], [1] and [2] are non-empty subsets of I with $[0] \cup [1] \cup [2] = I$, and [0], [1] and [2] are pair-wise disjoint. Thus {[0], [1], [2]} is a partition of I.

SOLVED EXAMPLES

Example 1 *Show that the relation $x \equiv y \pmod 5$ defined on the set of integers I is an equivalence relation.*

Solution: Given that the relation is $x \equiv y \pmod 5$

i.e., $(x - y)$ is divisible by 5

i.e., $(x - y) = 5k; k \in I$

i.e., $x\,R\,y : (x - y) = 5k; k \in I$

Reflexive: For all $x \in I$ we have $(x - x) = 0$

i.e., $(x - x) = 5k; k = 0 \in I$

i.e., $x\,R\,x$

i.e., R is reflexive.

Symmetric: Suppose that $x\,R\,y$

i.e., $(x - y) = 5k$

$\Rightarrow$ $(y - x) = -5k$

i.e., $(y - x) = 5(-k)$

i.e., $y\,R\,x$

So, $x\,R\,y$ implies $y\,R\,x$.

i.e., R is symmetric.

Transitive: Suppose that $x\,R\,y$ and $y\,R\,z$

i.e., $(x - y) = 5k_1$ and $(y - z) = 5k_2$; $k_1, k_2 \in I$

$\Rightarrow$ $(x - y) + (y - z) = 5(k_1 + k_2)$; $(k_1 + k_2) \in I$

$\Rightarrow$ $(x - z) = 5(k_1 + k_2)$

i.e., $x\,R\,z$

i.e., R is transitive.

So, the relation R on I defined by $x \equiv y \pmod 5$ is an equivalence relation.

Example 2 *Is every relation which is symmetric and transitive on a set A, always reflexive? Why or why not?*

Solution: Let R be a symmetric and transitive relation on A.

Let $x, y \in$ R and x R y

As R is symmetric, $x \text{ R } y \Rightarrow y \text{ R } x$

Again x R y and y R $x \Rightarrow x$ R x [$\because$ R is transitive]

Therefore R is reflexive, but the argument is not true.

Consider an example: A = {1, 2, 3, 4, 5}

Let R be a relation defined on A such that

$$R = \{(2, 3), (3, 4), (2, 4), (3, 2), (4, 3), (4, 2), (2, 2), (3, 3), (4, 4)\}$$

Which is symmetric and transitive but not reflexive. Therefore every relation which is symmetric and transitive on a set A is not always reflexive.

Example 3 *Let R be the relation in A = {1, 2, 3, 4, 5, 6} defined by 'x and y are relative prime'. Find the relation R and draw R on a coordinate diagram of (A ×A).*

Solution: Given A = {1, 2, 3, 4, 5, 6} and R $\subseteq$ (A × A) defined by x R y : x and y are relative prime.

i.e., R = {(1, 1), (1, 2), (1, 3), (1, 4), (1, 5), (1, 6), (2, 1), (2, 3), (2, 5), (3, 1), (3, 2), (3, 4), (3, 5), (4, 1), (4, 3), (4, 5), (5, 1), (5, 2), (5, 3), (5, 4), (5, 6), (6, 1), (6, 5)}

The coordinate diagram of *R* is given below.

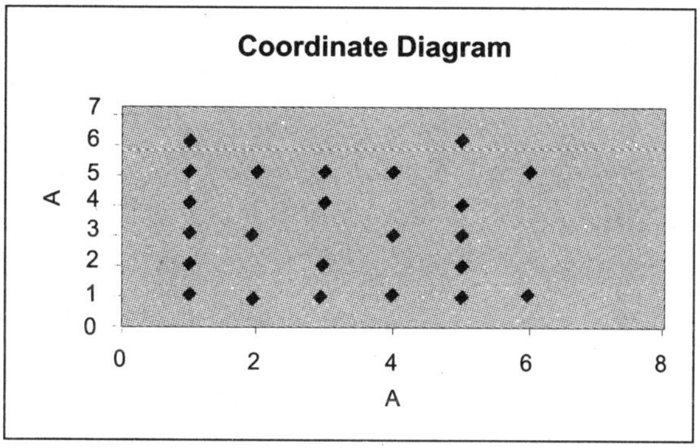

Example 4 *Prove that a relation R on a set A is symmetric if and only if $R^{-1} = R$.*

Solution: Suppose that a relation R on a set A is symmetric. Our claim is $R^{-1} = R$.

Let $(x, y) \in R^{-1}$

$\Rightarrow$ $(y, x) \in R$

$\Rightarrow$ $(x, y) \in R$ [$\because$ R is symmetric]

i.e., $(x, y) \in R^{-1} \Rightarrow (x, y) \in R$

i.e., $R^{-1} \subseteq R$ ... (*i*)

Again, let $(x, y) \in R$

$\Rightarrow$ $(y, x) \in R$ [$\because$ R is symmetric]

$\Rightarrow$ $(x, y) \in R^{-1}$

i.e., $(x, y) \in R \Rightarrow (x, y) \in R^{-1}$

i.e., $R \subseteq R^{-1}$ ...(*ii*)

Hence, we have $R = R^{-1}$

Conversely, suppose that $R = R^{-1}$.

Our claim is R on A is symmetric.

Let $(x, y) \in R \Rightarrow (y, x) \in R^{-1} = R$

i.e., R is symmetric.

Example 5 *Let N be the set of all natural numbers. R be a relation in N defined by x R y if and only if x + 3y = 12. Examine the relation for (i) reflexive, (ii) symmetric and (iii) transitive.*

Solution: Let R be a relation in N defined by

$$x \, R \, y: x + 3y = 12$$

Reflexive: Assume that $x + 3y = 12$ for $y = x$.

This implies $4x = 12$

i.e., $x = 3$

i.e., $x \, R \, x$ for $x = 3$ only.

i.e., $x \, \mathbf{R} \, x \; \forall \; x \in N$

Hence R is not reflexive.

Symmetric: Assume that $x \, R \, y$

i.e., $x + 3y = 12$

i.e., $y + 3x$ may or may not equal to 12.

i.e., $y \, \mathbf{R} \, x$.

Hence R is not symmetric.

Transitive: Assume that $x \, R \, y$ and $y \, R \, z$.

i.e., $x + 3y = 12$ and $y + 3z = 12$

This holds only when $x = y = z = 3 \in N$

i.e., $x + 3z = 12$

i.e., $x \, R \, z$

So, R is transitive.

Example 6 *For a relation R on a set A = {1, 2, 3, 4, 5} given by R = {(1, 3), (1, 2), (2, 2), (3, 4)}, find reflexive closure, symmetric closure and transitive closure of R on the given set A.*

Solution: Given A = {1, 2, 3, 4, 5} and the relation

$$R = \{(1, 3), (1, 2), (2, 2), (3, 4)\}.$$

Therefore, $r(R) = \{(1, 3),(1, 2), (1, 1), (2, 2), (3, 3),(3, 4), (4, 4),(5, 5)\}$

$s(R) = \{(1, 3),(1, 2), (2, 2),(3, 4), (3, 1), (2, 1),(4, 3)\}$

$t(R) = \{(1, 3),(1, 2), (2, 2), (3, 4), (1, 4)\}$

Example 7 *Let N be the set of all natural numbers. R be a relation in N defined by x R y if and only if x + y = 18. Show that R is symmetric but neither reflexive nor transitive.*

Solution: Let R be a relation in N defined by

$$x \, R \, y: x + y = 18$$

Reflexive: Assume that $x + y = 18$ for $y = x$

$\Rightarrow$ $2x = 18$

$\Rightarrow$ $x = 9$

i.e., $x \, R \, x$ for $x = 9$ only. So, R is not reflexive.

Symmetric: Suppose that x R y

i.e., $$x + y = 18$$

$\Rightarrow$ $$y + x = 18$$

i.e., y R x, *i.e.* R, is symmetric.

Transitive: Assume that x R y and y R z

i.e., $$x + y = 18 \text{ and } y + z = 18$$

i.e., $(x + z)$ may or may not equal to 18.

For example let $x = 4$, $y = 14$ and $z = 4$. Hence, we have $(x + y) = 18$ and $(y + z) = 18$, but $(x + z) = 8 \neq 18$

Therefore, $x \not\mathrel{R} z$, *i.e.* R is not transitive.

Example 8 *A relation R defined on the set of natural numbers N by x R y if and only if $(x \cdot y) > 0$ for $x, y \in N$ is an equivalence relation.*

Solution: Given R be a relation in N defined by

$$x \text{ R } y : (x \cdot y) > 0 \text{ for } x, y \in \text{N}$$

Reflexive: For all $x \in$ N we have

$$(x \cdot x) = x^2 > 0$$

i.e., x R x. Thus R is reflexive.

Symmetric: Suppose that x R y

$\Rightarrow$ $$(x \cdot y) > 0$$

$\Rightarrow$ $$(y \cdot x) > 0$$

i.e., $$y \text{ R } x$$

Thus, R is symmetric.

Transitive: Suppose that x R y and y R z

i.e., $$(x \cdot y) > 0 \text{ and } (y \cdot z) > 0$$

This implies $$(x \cdot y)(y \cdot z) > 0$$

i.e., $$(x \cdot z) y^2 > 0$$

As $y^2 > 0$ for all $y \in$ N we have $(x \cdot z) > 0$.

i.e., $$x \text{ R } z$$

Hence, x R y and y R $z \Rightarrow x$ R z

Thus, R is transitive.

Therefore, the relation R in N defined by $(x \cdot y) > 0$ is an equivalence relation.

Example 9 *Let A = {2, 4, 6, 8}; B = {1, 5, 7, 9} and Let R be a relation from A to B defined as x R y if and only if $x \leq y$. Find the domain, range and inverse of the relation R.*

Solution: Given that A = {2, 4, 6, 8}; B = {1, 5, 7, 9} and R be a relation from A to B defined as x R y if and only if $x \leq y$.

Therefore, R = {(2, 5), (2, 7), (2, 9), (4, 5), (4, 7), (4, 9), (6, 7), (6, 9), (8, 9)}

Thus, D(R) = {2, 4, 6, 8}; R(R) = {5, 7, 9} and R^{-1} = {(5, 2), (7, 2), (9, 2), (5, 4), (7, 4), (9, 4), (7, 6), (9, 6), (9, 8)}

Example 10 *Let I be the set of all integers and R be a relation defined on I such that x R y if and only if $x \geq y$. Show that R is reflexive, transitive but not symmetric.*

Solution: Given R be a relation in I defined by

$$x \mathrm{R} y : x \geq y \text{ for } x, y \in \mathrm{I}$$

Reflexive: For all $x \in \mathrm{I}$ we have

$$x \geq x$$

i.e., $x \mathrm{R} x.$

Thus R is reflexive.

Transitive:Suppose that $x \mathrm{R} y$ and $y \mathrm{R} z$

i.e., $x \geq y$ and $y \geq z$

This implies $x \geq z$

i.e., $x \mathrm{R} z$

Thus R is transitive.

Symmetric: Suppose that $x \mathrm{R} y$

i.e., $x \geq y$

This implies $y \ngeq x$

i.e., $y \ \mathbb{R}\ x$

Thus R is not symmetric.

Therefore the relation $x \geq y$ defined in I is reflexive, transitive but not symmetric.

Example 11 *Show that the relation $x \leq y$ defined on the set of integers is a partial order relation.*

Solution: Let R be a relation in I defined by

$$x \mathrm{R} y : x \leq y \text{ for } x, y \in \mathrm{I}$$

Reflexive: For all $x \in \mathrm{I}$ we have

$$x \leq x$$

i.e., $x \mathrm{R} x.$

Thus R is reflexive.

Transitive: Suppose that $x \mathrm{R} y$ and $y \mathrm{R} z$

i.e. $x \leq y$ and $y \leq z$

This implies $x \leq z$

i.e., $x \mathrm{R} z$

Thus, R is transitive.

Anti-symmetric: Suppose that $x \mathrm{R} y$ and $y \mathrm{R} x$

i.e., $x \leq y$ and $y \leq x$

This implies $x = y$

Thus, R is anti-symmetric.

Therefore, the relation $x \leq y$ defined in I is reflexive, transitive and anti-symmetric. So, $x \leq y$ is a partial order relation.

Example 12 *Let R be the relation on the set {1, 2, 3, 4, 5} defined by the rule $(x, y) \in R$ if $x + y \leq 6$. Find the followings.*

(*a*) List the elements of R
(*b*) List the elements of R^{-1}
(*c*) Domain of R
(*d*) Range of R
(*e*) Range of R^{-1}
(*f*) Domain of R^{-1}

Check that domain of R is equal to range of R^{-1} and range of R is equal to domain of R^{-1}.

Solution: Let A = {1, 2, 3, 4, 5} and R = $\{(x, y) \in R \mid x + y \leq 6 ; x, y \in A\}$

(*a*) R = {(1, 1), (1, 2), (1, 3), (1, 4), (1, 5), (2, 1), (2, 2), (2, 3), (2, 4), (3, 1), (3, 2), (3, 3), (4, 1), (4, 2), (5, 1)}

(*b*) R^{-1} = {(1, 1), (2, 1), (3, 1), (4, 1), (5, 1), (1, 2), (2, 2), (3, 2), (4, 2), (1, 3), (2, 3), (3, 3), (1, 4), (2, 4), (1, 5)}

(*c*) Domain of R *i.e.*, D(R) = {1, 2, 3, 4, 5}

(*d*) Range of R *i.e.*, R(R) = {1, 2, 3, 4, 5}

(*e*) Range of R^{-1} *i.e.*, $R(R^{-1})$ = {1, 2, 3, 4, 5}

(*f*) Domain of R^{-1} *i.e.*, $D(R^{-1})$ = {1, 2, 3, 4, 5}

From this it is clear that $D(R) = R(R^{-1})$ and $R(R) = D(R^{-1})$.

Example 13 *Consider a relation R on {1, 2, 3, 4} as R = {(1, 3), (1, 4), (2, 2), (3, 3), (4, 1)}. Examine the relation for reflexive, symmetric and transitive with the help of relation matrix.*

Solution: Given that the relation R = {(1, 3), (1, 4), (2, 2), (3, 3), (4, 1)}. Relative to the ordering 1, 2, 3, 4 we get

$$M(R) = \begin{bmatrix} 0 & 0 & 1 & 1 \\ 0 & 1 & 0 & 0 \\ 0 & 0 & 1 & 0 \\ 1 & 0 & 0 & 0 \end{bmatrix}$$

From the above matrix it is clear that $m_{11} \neq 1$ and $m_{44} \neq 1$. So, the relation R is not reflexive. Again R is not symmetric because

$$[M(R)]^T = \begin{bmatrix} 0 & 0 & 0 & 1 \\ 0 & 1 & 0 & 0 \\ 1 & 0 & 1 & 0 \\ 1 & 0 & 0 & 0 \end{bmatrix} \neq M(R)$$

Also we have

$$[M(R)]^2 = \begin{bmatrix} 0 & 0 & 1 & 1 \\ 0 & 1 & 0 & 0 \\ 0 & 0 & 1 & 0 \\ 1 & 0 & 0 & 0 \end{bmatrix} \begin{bmatrix} 0 & 0 & 1 & 1 \\ 0 & 1 & 0 & 0 \\ 0 & 0 & 1 & 0 \\ 1 & 0 & 0 & 0 \end{bmatrix} = \begin{bmatrix} 1 & 0 & 1 & 0 \\ 0 & 1 & 0 & 0 \\ 0 & 0 & 1 & 0 \\ 0 & 0 & 1 & 1 \end{bmatrix}$$

From this it is clear that the 1st row and 1st column entry in $[M(R)]^2$ is non-zero whereas the 1st row and 1st column entry in M(R) is zero. Therefore the relation R is not transitive. Hence the relation R is neither reflexive nor symmetric and transitive.

Example 14 *Let A = {1, 2, 3, 4, 5}; B = {a, b, c, d} and C = {1, 4, 9, 16, 25}. Consider the relations R_1 from A to B and R_2 from B to C as R_1 = {(1, a), (1, b), (2, c), (2, d), (3, b), (5, d)} and R_2 = {(a, 1), (d, 4), (b, 9), (d, 25)}. Find the composition R_1R_2 with the help of relation matrix.*

Solution: Let A = {1, 2, 3, 4, 5}; B = $\{a, b, c, d\}$ and C = {1, 4, 9, 16, 25}. Given $R_1 \subseteq (A \times B)$ and $R_2 \subseteq (B \times C)$ with

$$R_1 = \{(1, a), (1, b), (2, c), (2, d), (3, b), (5, d)\}$$

and

$$R_2 = \{(a, 1), (d, 4), (b, 9), (d, 25)\}.$$

Therefore, we get

$$M(R_1) = \begin{bmatrix} 1 & 1 & 0 & 0 \\ 0 & 0 & 1 & 1 \\ 0 & 1 & 0 & 0 \\ 0 & 0 & 0 & 0 \\ 0 & 0 & 0 & 1 \end{bmatrix} \text{ and } M(R_2) = \begin{bmatrix} 1 & 0 & 0 & 0 & 0 \\ 0 & 0 & 1 & 0 & 0 \\ 0 & 0 & 0 & 0 & 0 \\ 0 & 1 & 0 & 0 & 1 \end{bmatrix}$$

So,

$$M(R_1R_2) = M(R_1)M(R_2)$$

$$= \begin{bmatrix} 1 & 1 & 0 & 0 \\ 0 & 0 & 1 & 1 \\ 0 & 1 & 0 & 0 \\ 0 & 0 & 0 & 0 \\ 0 & 0 & 0 & 1 \end{bmatrix} \begin{bmatrix} 1 & 0 & 0 & 0 & 0 \\ 0 & 0 & 1 & 0 & 0 \\ 0 & 0 & 0 & 0 & 0 \\ 0 & 1 & 0 & 0 & 1 \end{bmatrix} = \begin{bmatrix} 1 & 0 & 1 & 0 & 0 \\ 0 & 1 & 0 & 0 & 1 \\ 0 & 0 & 1 & 0 & 0 \\ 0 & 0 & 0 & 0 & 0 \\ 0 & 1 & 0 & 0 & 1 \end{bmatrix}$$

Thus, $R_1R_2 = \{(1, 1), (1, 9), (2, 4), (2, 25), (3, 9), (5, 4), (5, 25)\}$.

Example 15 *Let R_1 and R_2 be the relations on $\{1, 2, 3, 4\}$ given by $R_1=\{(1, 1), (1, 2), (3, 4), (4, 2)\}$and $R_2 = \{(1, 1), (2, 1), (3, 1), (4, 4), (2, 2)\}$ List the elements of R_1R_2 and R_2R_1. Show that $R_1R_2 \neq R_2R_1$.*

Solution: Given R_1 and R_2 be relations on $\{1, 2, 3, 4\}$ as

$R_1 = \{(1, 1), (1, 2), (3, 4), (4, 2)\}$ and $R_2 =\{(1, 1), (2, 1), (3, 1), (4, 4), (2, 2)\}$ Relative to the ordering $\{1, 2, 3, 4\}$ we have

$$M(R_1) = \begin{bmatrix} 1 & 1 & 0 & 0 \\ 0 & 0 & 0 & 0 \\ 0 & 0 & 0 & 1 \\ 0 & 1 & 0 & 0 \end{bmatrix} \text{and } M(R_2) = \begin{bmatrix} 1 & 0 & 0 & 0 \\ 1 & 1 & 0 & 0 \\ 1 & 0 & 0 & 0 \\ 0 & 0 & 0 & 1 \end{bmatrix}$$

Therefore, $M(R_1R_2) = M(R_1)\, M(R_2)$

$$= \begin{bmatrix} 1 & 1 & 0 & 0 \\ 0 & 0 & 0 & 0 \\ 0 & 0 & 0 & 1 \\ 0 & 1 & 0 & 0 \end{bmatrix} \begin{bmatrix} 1 & 0 & 0 & 0 \\ 1 & 1 & 0 & 0 \\ 1 & 0 & 0 & 0 \\ 0 & 0 & 0 & 1 \end{bmatrix} = \begin{bmatrix} 2 & 1 & 0 & 0 \\ 0 & 0 & 0 & 0 \\ 0 & 0 & 0 & 1 \\ 1 & 1 & 0 & 0 \end{bmatrix}$$

Replacing all non-zero entries by 1 in $M(R_1R_2)$ we have $M(R_1R_2) = \begin{bmatrix} 1 & 1 & 0 & 0 \\ 0 & 0 & 0 & 0 \\ 0 & 0 & 0 & 1 \\ 1 & 1 & 0 & 0 \end{bmatrix}$

i.e., $R_1R_2 = \{(1, 1), (1, 2), (3, 4), (4, 1), (4, 2)\}$

Similarly, $M(R_2R_1) = M(R_2)\, M(R_1)$

$$= \begin{bmatrix} 1 & 0 & 0 & 0 \\ 1 & 1 & 0 & 0 \\ 1 & 0 & 0 & 0 \\ 0 & 0 & 0 & 1 \end{bmatrix} \begin{bmatrix} 1 & 1 & 0 & 0 \\ 0 & 0 & 0 & 0 \\ 0 & 0 & 0 & 1 \\ 0 & 1 & 0 & 0 \end{bmatrix} = \begin{bmatrix} 1 & 1 & 0 & 0 \\ 1 & 1 & 0 & 0 \\ 1 & 1 & 0 & 0 \\ 0 & 1 & 0 & 0 \end{bmatrix}$$

i.e., $R_2R_1 = \{(1, 1), (1, 2), (2, 1), (2, 2), (3, 1), (3, 2), (4, 2)\}$

Therefore, $R_1R_2 \neq R_2R_1$

Example 16 *Let R be the relation on the set $\{1, 2, 3, 4, 5\}$ defined by the rule $(x, y) \in R$ if $x = y - 1$. Find R in terms of relation matrix. Check the relation R for symmetric and irreflexive.*

Solution: Let $A = \{1, 2, 3, 4, 5\}$

and $R = \{(x, y): x = y - 1\ x, y \in A\}$

i.e., $R = \{(1, 2), (2, 3), (3, 4), (4, 5)\}$

Relative to the ordering $\{1, 2, 3, 4, 5\}$, we have

$$M(R) = \begin{bmatrix} 0 & 1 & 0 & 0 & 0 \\ 0 & 0 & 1 & 0 & 0 \\ 0 & 0 & 0 & 1 & 0 \\ 0 & 0 & 0 & 0 & 1 \\ 0 & 0 & 0 & 0 & 0 \end{bmatrix}$$

Now, $$[M(R)]^T = \begin{bmatrix} 0 & 0 & 0 & 0 & 0 \\ 1 & 0 & 0 & 0 & 0 \\ 0 & 1 & 0 & 0 & 0 \\ 0 & 0 & 1 & 0 & 0 \\ 0 & 0 & 0 & 1 & 0 \end{bmatrix} \neq M(R)$$

So, R is not symmetric. It is clear that R is irreflexive as $m_{ii} = 0$ for all $1 \le i = 5$. Therefore, R is irreflexive but not symmetric.

Example 17 *Let R and S be the following relations on A = {2, 4, 5, 6}. R = {(2, 4), (2, 5), (2, 6), (4, 2), (4, 4)} and S = {(5, 4), (5, 5), (5, 6), (6, 2), (6, 4)}. Find $(R \cup S)^c$ and R^2.*

Solution: Given R = {(2, 4), (2, 5), (2, 6), (4, 2), (4, 4)} and S = {(5, 4), (5, 5), (5, 6), (6, 2), (6, 4)}

(*i*) $(R \cup S)$ = {(2, 4), (2, 5), (2, 6), (4, 2), (4, 4), (5, 4), (5, 5), (5, 6), (6, 2), (6, 4)}
This implies $(R \cup S)^c$ = {(2, 2), (4, 5), (4, 6), (5, 2), (6, 5), (6, 6)}

(*ii*) Relative to the ordering 2, 4, 5, 6, we have

$$M(R) = \begin{bmatrix} 0 & 1 & 1 & 1 \\ 1 & 1 & 0 & 0 \\ 0 & 0 & 0 & 0 \\ 0 & 0 & 0 & 0 \end{bmatrix}$$

So, $$M(R^2) = M(R)\,M(R)$$

$$= \begin{bmatrix} 0 & 1 & 1 & 1 \\ 1 & 1 & 0 & 0 \\ 0 & 0 & 0 & 0 \\ 0 & 0 & 0 & 0 \end{bmatrix} \begin{bmatrix} 0 & 1 & 1 & 1 \\ 1 & 1 & 0 & 0 \\ 0 & 0 & 0 & 0 \\ 0 & 0 & 0 & 0 \end{bmatrix} = \begin{bmatrix} 1 & 1 & 0 & 0 \\ 1 & 2 & 1 & 1 \\ 0 & 0 & 0 & 0 \\ 0 & 0 & 0 & 0 \end{bmatrix}$$

Replacing all non-zero entries by 1 in $M(R^2)$, we get

$$M(R^2) = \begin{bmatrix} 1 & 1 & 0 & 0 \\ 1 & 1 & 1 & 1 \\ 0 & 0 & 0 & 0 \\ 0 & 0 & 0 & 0 \end{bmatrix}$$

So, R^2 = {(2, 2), (2, 4), (4, 2), (4, 4), (4, 5), (4, 6)}

Example 18 *Let R be the relation on the set {2, 3, 4, 5, 6} defined by the rule $(x, y) \in R$ if $x + 2y \le 12$. Find the relation R. Also find the reflexive, symmetric and transitive closure of R.*

Solution: Let A = {2, 3, 4, 5, 6} and R = $\{(x, y) \in R \text{ if } x + 2y \le 12; x, y \in A\}$

i.e., R = {(2, 2), (2, 3), (2, 4), (2, 5), (3, 2), (3, 3), (3, 4), (4, 2), (4, 3), (4, 4), (5, 2), (5, 3), (6, 2), (6, 3)}

The reflexive closure of R *i.e.*, r(R) = {(2, 2), (2, 3), (2, 4), (2, 5), (3, 2), (3, 3), (3, 4), (4, 2), (4, 3), (4, 4), (5, 2), (5, 3), (5, 5), (6, 2), (6, 3), (6, 6)}

The symmetric closure of R *i.e.*, s(R) = {(2, 2), (2, 3), (2, 4), (2, 5), (2, 6), (3, 2), (3, 3), (3, 4), (3, 5), (3, 6), (4, 2), (4, 3), (4, 4), (5, 2), (5, 3), (6, 2), (6, 3)}

The transitive closure of R *i.e.*, t(R) = {(2, 2), (2, 3), (2, 4), (2, 5), (3, 2), (3, 3), (3, 4), (4, 2), (4, 3), (4, 4), (5, 2), (5, 3), (6, 2), (6, 3), (3, 5), (4, 5), (5, 4), (5, 5), (6, 4), (6, 5)}

Example 19 *Let R be the relation in the integers I defined by $(x - y)$ is an even integer. Prove that R is an equivalence relation and find the disjoint equivalence classes.*

Solution: Let R be the relation in the integers I defined by $(x - y)$ is an even integer. *i.e.*, $(x - y)$ is divisible by 2.

Reflexive: For all $x \in I$ we have $(x - x) = 0$

i.e., $$(x - x) = 2k;\ k = 0 \in I$$

i.e., $$x\,R\,x\ \forall\ x \in I$$

i.e., R is reflexive.

Symmetric: Suppose that x R y

i.e., $(x - y)$ is divisible by 2

i.e., $(x - y) = 2k;\ k \in$ I

$\Rightarrow$ $(y - x) = 2(-k);\ -k \in$ I

i.e., y R x

Thus, R is symmetric.

Transitive: Suppose that x R y and y R z

i.e., $(x - y)$ and $(y - z)$ are even integer.

i.e., $(x - y) = 2k_1$ and $(y - z) = 2k_2\ ;\ k_1, k_2 \in$ I.

$\Rightarrow$ $(x - y) + (y - z) = 2(k_1 + k_2);\ (k_1 + k_2) \in$ I

i.e., $(x - z)$ is an even integer.

i.e., x R z

Thus, R is transitive.

Therefore, the relation R on I defined by $(x - y)$ is even integer is an equivalence relation. The disjoint equivalence classes are

$$[0] = \{\ldots\ldots, -4, -2, 0, 2, 4, 6, \ldots\}$$
$$[1] = \{\ldots\ldots, -3, -1, 1, 3, 5, 7, \ldots\ldots\}$$

Example 20 *Let R be a relation defined in A = {1, 2, 3, 5, 7, 9} as R = {(1, 1), (1, 3), (1, 5), (1, 7), (3, 1), (3, 3), (3, 5), (3, 7), (5, 1), (5, 3), (5, 5), (5, 7), (7, 1), (7, 3), (7, 5), (7, 7), (9, 9), (2, 2)}. Find the partitions of A based on the equivalence relation R.*

Solution: Given A = {1, 2, 3, 5, 7, 9} and

R = {(1, 1), (1, 3), (1, 5), (1, 7), (3, 1), (3, 3), (3, 5), (3, 7), (5, 1), (5, 3), (5, 5), (5, 7), (7, 1), (7, 3), (7, 5), (7, 7), (9, 9), (2, 2)}

The disjoint equivalence classes are

$$[1] = \{1, 3, 5, 7\};\ [9] = \{9\} \text{ and } [2] = \{2\}.$$

Obviously

(*i*) The sets [1], [2] and [9] are non-empty

(*ii*) $[1] \cap [2] = \phi$; $[1] \cap [9] = \phi$ and $[2] \cap [9] = \phi$

(*iii*) $[1] \cup [2] \cup [9] = A$

Hence, { [1], [2], [9]} is a partition of A.

Example 21 *Find the number of relations from the set A to the set B if $|A| = m$ and $|B| = n$.*

Solution: Given $|A| = m$ and $|B| = n$

Therefore, $|A \times B| = mn$.

A relation R from the set A to the set B is a subset of (A × B). So, the number of subsets of (A × B) is equal to 2^{mn}. Therefore total number of relations from the set A to the set B is 2^{mn}.

EXERCISES

1. Let A = {p, q, r, s} and R be an universal relation on A. Write down the relation R. Find out the smallest and largest subset of the universal relation which is an equivalence relation.

2. Let A = {1, 2, 3, 4, 5} and B = {1, 2, 4, 6, 7}. Find the relation from A to B defined by

(*a*) Greater than (*b*) Less than

(*c*) Greater than equal to (*d*) Less than equal to

(*e*) Equal to

3. For the above No. 2, determine the domains and ranges of each of the cases.

4. For the above No. 2, determine the inverse relation in each of the above cases.

5. Prove that the relation $x \equiv y$ mod(3) on the set of integers Z is an equivalence relation.

6. Give an example of a relation which is

(*a*) Reflexive, Symmetric but not Transitive.

(*b*) Reflexive, Transitive but not Symmetric.

(*c*) Symmetric, Transitive but not Reflexive.

(*d*) Reflexive but not Symmetric and Transitive.

(*e*) Symmetric but not Reflexive and Transitive.

(*f*) Transitive but not Reflexive and Symmetric.

(*g*) Neither Reflexive nor Symmetric and Transitive.

(*h*) Reflexive, Symmetric and Transitive.

(*i*) Symmetric and Anti-symmetric.

(*j*) Anti-symmetric but not Reflexive.

7. Prove that the relation on the set of natural numbers N determined by x R y if and only if x divides y is reflexive, transitive but not symmetric.

8. Consider the relations R_1 and R_2 on $\{a, b, c, d, e\}$ as $R_1 = \{(a, b), (a, c), (b, b), (c, d), (c, c), (c, e)\}$ and $R_2 = \{(a, a), (a, d), (d, b), (d, e), (d, d), (e, c)\}$. Find the reflexive, symmetric and transitive closures of R_1 and R_2.

9. Write the following relations as a table

(*a*) $R_1 = \{(1, 1), (2, 4), (3, 9), (4, 16), (6, 36), (7, 49)\}$

(*b*) $R_2 = \{(2, 5), (5, 8), (8, 11), (11, 14)\}$

(*c*) $R_3 = \{(8, i), (1, l), (4, o), (1, v), (4, e), (0, u)\}$

(*d*) R_4 = {(Bapa, Comp. Sc.), (Megha, Math), (Suni, Math)}

10. Let R be the relation in the natural numbers N defined by $(a - b)$ is divisible by 8. Show that R is an equivalence relation.

11. Let L be the set of lines in the Euclidean plane and let R be the relation in L defined by l_1 R l_2 if and only if l_1 is parallel to l_2. Show that R is an equivalence relation.

12. Write the following relations as a set of order pairs.

(*a*)

Cloth Material	*Price in Rupees*
Cotton	55
Teri cot	60
Woolen	50
Fancy	45

(*b*)

Names	*Course*
Mary	Comp. Sc.
Smith	Math
Smith	Comp. Sc.
Finzi	Chemistry
Loreena	Economics

(*c*)

Number	*Square*
5	25
4	16
3	9
2	4
1	1

(*d*)

Alphabet	*Number*
a	1
c	3
e	5
z	26
m	13

13. Let A = {5, 6, 7, 8, 9}; B = {x, y, z, p, q, r}; C = {5, 7, 25, 36, 81} and let R_1 = {(5, p), (5, r), (6, z), (7, y), (9, x), (9, z)} and R_2 = { (p, 25), (x, 81), (z, 36), (y, 7), (r, 5)}. Find the composition R_1R_2 with the help of relation matrix.

14. For the relation R on the set {5, 6, 7, ,8, 9} defined by the rule (x, y) if $x + 2y \leq 20$. Find the followings.

(*a*) Elements of R (*b*) Elements of R^{-1}
(*c*) Domain of R (*d*) Range of R

15. Let S = {1, 2, 3, 4, 5} and let R be a relation defined by a rule (x, y) if $(x - y)$ is an even natural number. Find the followings.

(*a*) Elements of R (*b*) Inverse relation of R
(*c*) Domain of R (*d*) Range of R^{-1}.

16. Let R be a relation defined on the set S = {1, 2, 3, 4, 5} by a rule (x, y) if $x^2 + y^2 \leq 16$. Find the reflexive, symmetric and transitive closures of R.

17. Let R be a relation on {a, b, c, d} defined as R = {(a, a), (a, b), (a, c), (b, a), (b, b), (b, c), (c, a), (c, b), (c, c), (d, d)}. Show that the relation R is an equivalence relation using relation matrix.

18. Let N be the set of all natural numbers. Define a relation R in N by x R y if and only if $(x - y) = 34$. Show that R is anti-symmetric.

19. Let R be a relation defined by a rule A R B if and only if A $\subseteq$ B. Show that R is a partial order relation.

20. Test the following relations on N for being reflexive, symmetric and transitive. Let $x, y \in$ N.

(*a*) $x + y$ is even (*b*) $\frac{x}{y}$ is a power of 2. (*c*) $x + y \leq 20$

21. Examine the following relations on the set of integers I for partial order relations. Let $(x, y) \in$ R if and only if

(*a*) $x = y$ (*b*) $x \geq y$ (*c*) $x = y^2$ (*d*) $x < y$.

22. Let R_1 and R_2 be relations on the set S. Show that $(R_1 \cup R_2)$ is reflexive if both R_1 and R_2 are reflexive.

23. Let R_1 be an anti-symmetric relation on the set S. Prove that R^{-1} is also an anti-symmetric relation on the set S.

24. Show that if R_1 and R_2 be transitive relations on a set A, then $(R_1 \cup R_2)$ is not necessarily transitive on A.

25. Find the equivalence classes determined by the equivalence relation R on Z defined by a R b if and only if $a \equiv b$ mod (5) for $a, b \in$ Z.

26. Sketch each of the following relations on R.

(*a*) $x^2 + y^2 < 25$ (*b*) $x^2 + 4y^2 = 16$ (*c*) $x^2 - 4y^2 = 16$ (*d*) $3x + 2y \geq 6$

27. Let R be a relation in the natural number N defined by a R b if and only if 'a is a multiple of b' for $a, b \in$ N. Examine the above relation for reflexive, symmetric, anti-symmetric, transitive and anti-reflexive.

28. Let R be a relation in the natural number N defined by x R y if and only if '$x^2 = y^2$' for $x, y \in$ N. Examine the above relation for reflexive, symmetric, anti-symmetric, transitive and partial order.

29. Let A be the set of non zero integers and R be a relation in A defined by (a, b) R (c, d) if and only if $a + d = b + c$. Prove that R is an equivalence relation.

30. Let A be the set of non-zero integers and R be a relation in A defined by (a, b) R (c, d) if and only if $ad = bc$. Show that R is an equivalence relation.

4

Function

4.0 INTRODUCTION

One of the most important concepts in mathematics is that of a function. It is being used in our day-to-day life. At every moment by knowingly or unknowingly. German Mathematician Leibnitz was first to use the term function (1646–1716). The terms mapping, map and transformation mean the same thing. Computer Science has many applications of function. Hashing function is one of that.

Consider a computing device that accepts any real number, multiplies it by 5 and adds 3 with the product, and gives the output.

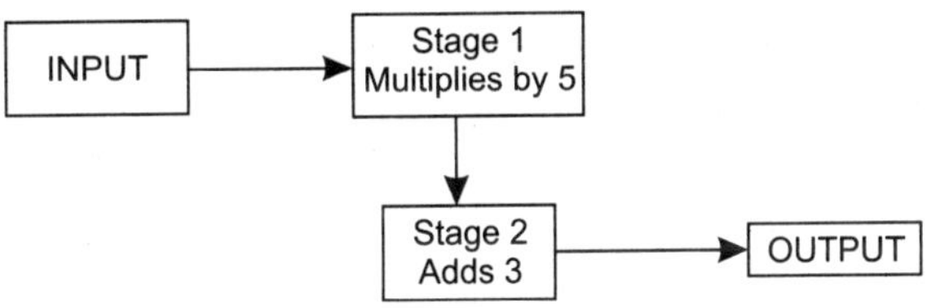

(Computing Device)

If the input is 1, then the output is 8. If the input is $\frac{1}{5}$, then the output is 4. If the input is 10, then the output is $(10 \times 5 + 3) = 53$. This clear indicates that if the input is $x, x \in$ R, then the output is $(5x + 3)$. As a result the computing device pairs off the elment $x \in$ R as $(x, 5x + 3)$ in a definite way or principle. This is nothing but a function.

4.1 FUNCTION

Let A and B be two non-empty sets. A relation f from the set A to the set B is said to be a function if it satisfies the following two conditions.

(*i*) $D(f) = A$ and

(*ii*) if $(x_1, y_1) \in f$ and $(x_2, y_2) \in f$, then $y_1 = y_2$.

In other words a relation f from the set A to the set B is said to be a function if for each element x in A there exists unique element y in B. A function from A to B is sometimes denoted as $f : A \to B$.

Consider the following relations from the set A = {1, 2, 3, 4} to the set B = {1, 4, 6, 9, 16, 18}.

$$f_1 = \{(1, 1), (2, 6), (4, 9), (4, 18)\}$$
$$f_2 = \{(1, 1), (2, 6), (3, 9), (4, 9), (4, 16)\}$$
$$f_3 = \{(1, 1), (2, 4), (3, 9), (4, 16)\}$$

and
$$f_4 = \{(1, 1), (2, 4), (3, 9), (4, 9)\}$$

Now, $D(f_1) = \{1, 2, 4\} \neq A$. Therefore f_1 is not a function from the set A to the set B. Further $D(f_2) = \{1, 2, 3, 4\} = A$; but $(4, 9) \in f_2$ and $(4, 16) \in f_2$ with $9 \neq 16$. This implies f_2 can not be a function from the set A to the set B.

Again $D(f_3) = \{1, 2, 3, 4\} = A$ and for every element $x \in A$ there exists unique $y \in B$. Therefore f_3 is a function from the set A to the set B. Similarly f_4 is also a function. The arrow diagrams are given below.

Note: From the above discussions it is clear that One-Many and Many-Many relations are not functions.

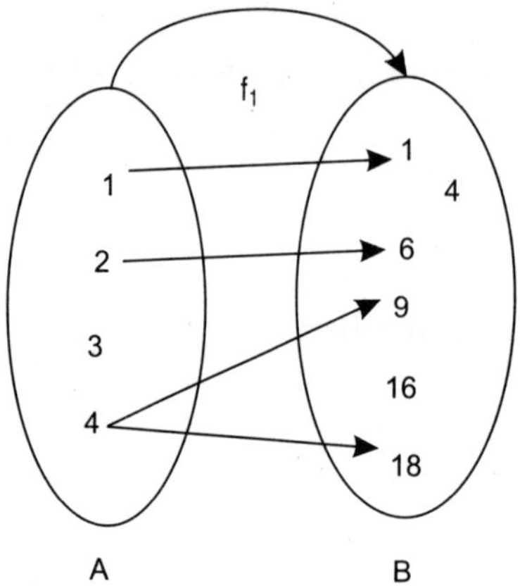

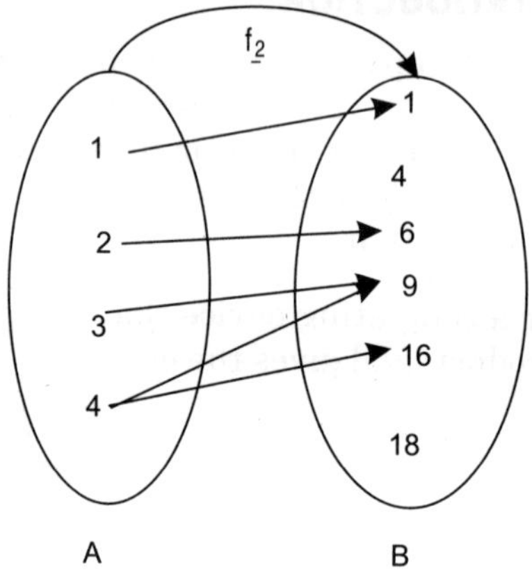

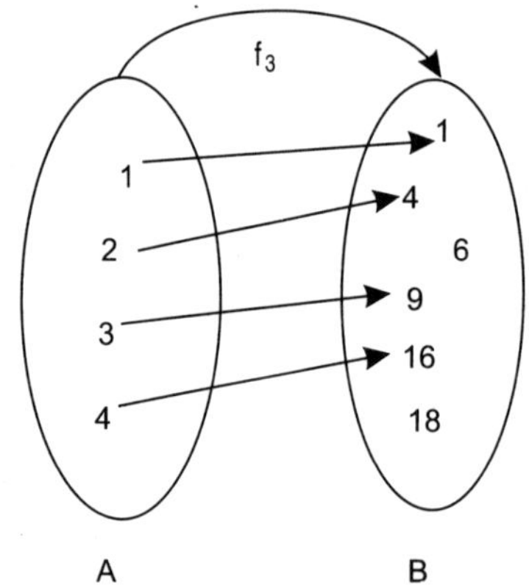

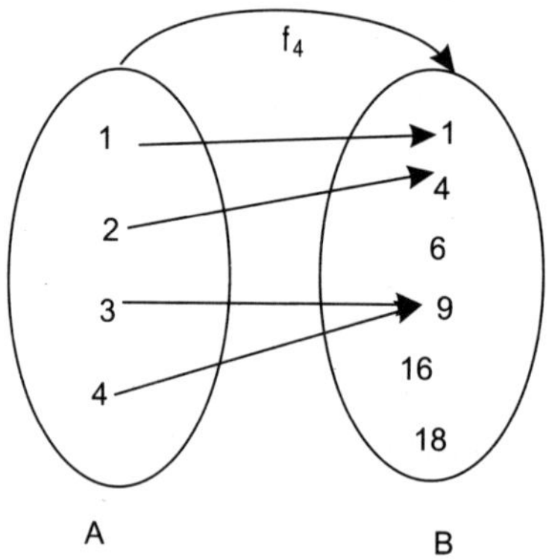

4.1.1 Domain and Co-domain of a Function

Suppose that f be a function from the set A to the set B. The set A is called the domain of the function f where as the set B is called the co-domain of the function f.

Consider the function f from the set A = $\{a, b, c, d\}$ to the set B = {1, 2, 3, 4} as

$$f = \{(a, 1), (b, 2), (c, 2), (d, 4)\}$$

Therefore, domain of $f = \{a, b, c, d\}$ and co-domain of $f = \{1, 2, 3, 4\}$. *i.e.*, $D(f) = \{a, b, c, d\}$ and Co-domain $f = \{1, 2, 3, 4\}$.

4.1.2 Range of a Function

Let f be a function from the set A to the set B. The element $y \in B$ which the function f associates to an element $x \in A$ is called the image of x or the value of the function f for x. From the definition of function it is clear that each element of A has an unique image on B. Therefore the range of a function $f : A \to B$ is defined as the image of its domain A. Mathematically,

$$R(f) \text{ or rng } (f) = \{y = f(x) : x \in A\}$$

It is clear that $R(f) \subseteq B$.

Consider the function f from A = $\{a, b, c\}$ to B = {1, 3, 5, 7, 9} as $f = \{(a, 3), (b, 5), (c, 5)\}$. Therefore, $R(f) = \{3, 5\}$.

■ 4.2 EQUALITY OF FUNCTIONS

If f and g are functions from A to B, then they are said to be equal *i.e.*, $f = g$ if the following conditions hold.

(*a*) $D(f) = D(g)$ (*b*) $R(f) = R(g)$

(*c*) $f(x) \neq g(x)$ $\forall x \in A$.

Consider $f(x) = 3x^2 + 6 : R \to R$ and $g(x) = 3x^2 + 6 : C \to C$, where R and C are the set of real numbers and complex numbers respectively. Now it is clear that $D(f) \neq D(g)$. Therefore $f(x) \neq g(x)$.

Let us consider A = {1, 2, 3, 4}; B = {1, 2, 7, 8, 17, 18, 31, 32} and the function $f : A \to B$ defined by $f = \{(1, 2), (2, 8), (3, 18), (4, 32)\}$. Consider another function $g : A \to N$ defined by $g(x) = 2x^2$. Now it is clear that $D(f) = \{1, 2, 3, 4\}$ with $f(1) = 2, f(2) = 8, f(3) = 18, f(4) = 32$.

Similarly $D(g) = A = \{1, 2, 3, 4\}$ with $g(1) = 2, g(2) = 8, g(3) = 18, g(4) = 32$. Therefore, we get

(*a*) $D(f) = \{1, 2, 3, 4\} = D(g)$ (*b*) $R(f) = \{2, 8, 18, 32\} = R(g)$ and

(*c*) $f(x) = g(x)$ $\forall x \in \{1, 2, 3, 4\}$.

This implies f and g are equal, *i.e.*, $f = g$.

■ 4.3 TYPES OF FUNCTION

In this section we will discuss different types of function that are highly useful in computer science and its applications. These are basically one-one, onto, one-one onto and into functions.

4.3.1 One-One Function

A function $f : A \to B$ is said to be an one-one function or injective if $f(x_1) = f(x_2)$, then $x_1 = x_2$ for $x_1, x_2 \in A$. *i.e.*, $x_1 \neq x_2 \Rightarrow f(x_1) \neq f(x_2)$

Consider a function $f: Q \to Q$ defined by $f(x) = 4x + 3$; $x \in Q$.

Suppose that $f(x_1) = f(x_2)$ for $x_1, x_2 \in$ Q.

$$\Rightarrow \quad 4x_1 + 3 = 4x_2 + 3$$
$$\Rightarrow \quad 4x_1 = 4x_2$$
$$\Rightarrow \quad x_1 = x_2$$

i.e., $f(x_1) = f(x_2) \Rightarrow x_1 = x_2$. So, $f(x) = (4x + 3)$: Q → Q is One-One.

Consider another function f : R → R defined by $f(x) = x^2$; $x \in$ R. Suppose that $f(x_1) = f(x_2)$

$$\Rightarrow \quad x_1^2 = x_2^2$$
$$\Rightarrow \quad x_1 = \pm x_2$$
$$\Rightarrow \quad x_1 \neq x_2$$

i.e., $f(x_1) = f(x_2) \Rightarrow x_1 \neq x_2$. It is also clear that $f(1) = 1 = f(-1)$; but $1 \neq -1$. Therefore $f(x) = x^2$: R → R; $x \in$ R is not One-One.

4.3.2 Onto Function

A function f: A → B is said to be an onto function or surjective if for every $y \in$ B there exists at least one element $x \in$ A such that $f(x) = y$.

In other words a function f: A → B is said to be an Onto function if f(A) = B. *i.e.*, range of f is equal to co-domain of f.

Consider a function f: Q → Q defined by $f(x) = 4x + 3$, $x \in$ Q. Then for every $y \in$ co-domain set Q there exists $x = \dfrac{y-3}{4}$ belongs to domain set Q. Therefore, $f(x) = 4x + 3$ is an Onto function.

4.3.3 One-One Onto Function

A function f: A → B is said to be an one-one onto function or bijective if f is both one-one and onto function.

Consider a function f: Q → Q defined by $f(x) = 4x + 3$, $x \in$ Q. From the above discussions it is clear that $f(x) = 4x + 3$, $x \in$ Q is an one-one onto function.

4.3.4 Into Function

A function f: A → B is said to be an into function if for at least one $y \in$ B there exists no element $x \in$ A such that $f(x) = y$. In other words

A function f: A → B is said to be an into function if $f(A) \subset B$, *i.e.*, range of f is a proper subset of co-domain of f.

Consider a function f: Q → R defined by $f(x) = x + 4$, $x \in$ Q. Hence, it is clear that for $y = \sqrt{3} \in$ R there exists no element $x = \sqrt{3} - 4 \in$ Q. Therefore, $f(x) = x + 4$: Q → R is an into function.

■ 4.4 GRAPH OF FUNCTION

Let f be a function from A to B *i.e.*, for every $x \in$ A there exists unique $y \in$ B such that $y = f(x)$. Further note that using the functional notation, f can be expressed as

$$f = \{(x, f(x)) : x \in A\}.$$

This representation is known as the graph of the function f.

Consider the functions f_1: R → R defined by $f_1(x) = x + 1$; and

f_2: R → {– 2, 2} defined by $f_2(x) = \begin{cases} -2 & \text{if } x > 0 \\ 2 & \text{if } x < 0 \end{cases}$

The graphs of above functions are given below.

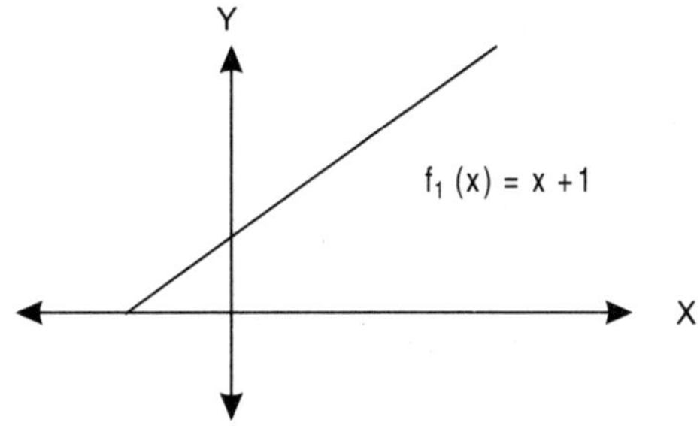

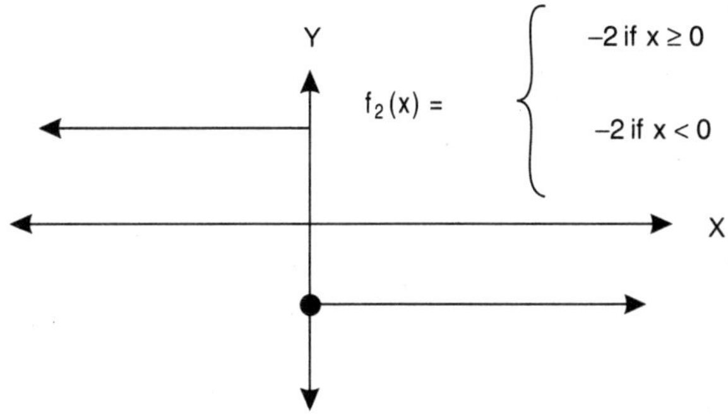

Now consider the relations f_1: [– 4, 4] → [– 4, 4] defined by $[f_1(x)]^2 = 16 - x^2$; $x \in [-4, 4]$ and f_2 : R → R defined by $[f_2(x)]^2 = 16x$; $x \in$ R. The graphs of above relations are figure-1 and figure-2 respectively. These are nothing but a circle and parabola respectively, where in figure –1, $y = f_1(x)$ and in figure – 2, $y = f_2(x)$. From the graph it is clear that for one value of x in the domain set leads to two values in the range set. Hence, these relations are not functions.

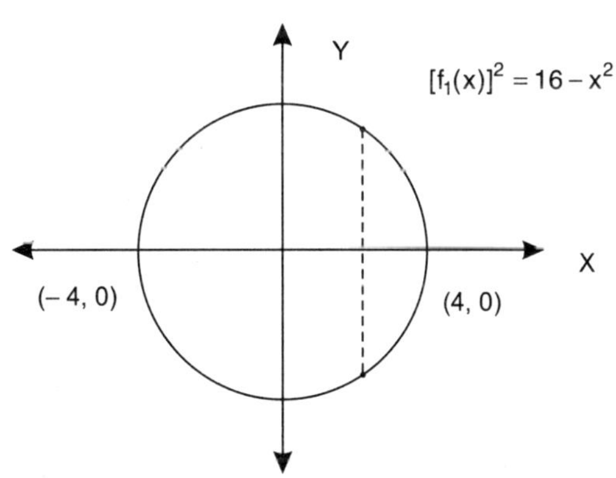

Figure – 1

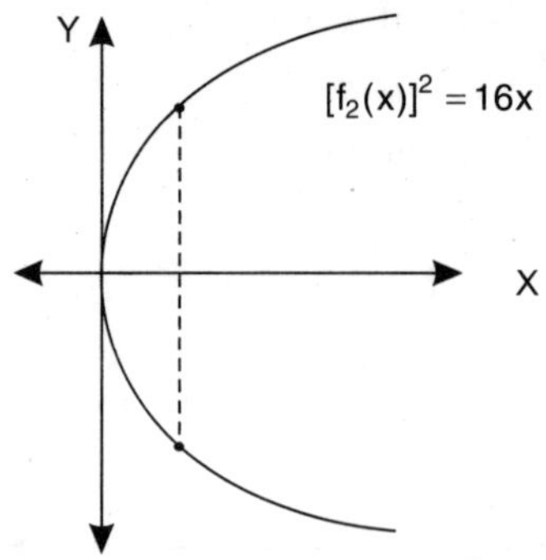

Figure – 2

■ 4.5 COMPOSITION OF FUNCTIONS

Let f be a function from the set A to the set B and g be a function from the set B to the set C. Then the composition of the functions f and g is given as $(g_o f)$ or gf. This is a function from the set A to the set C. It may also be noted that domain of g is equal to co-domain of f.

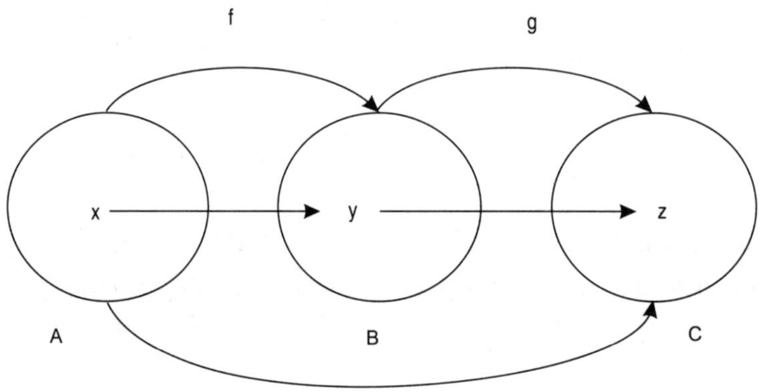

$(g_o f)$

As f is a function from the set A to the set B, then for every $x \in$ A there exists unique $y \in$ B such that $y = f(x)$. Similarly g is a function from the set B to the set C, then for every $y \in$ B there exists unique $z \in$ C such that $z = g(y)$. Again $(g_o f)$ is a function from the set A to the set C, so we get $(g_o f)(x) = z$ for all $x \in$ A.

i.e., $(g_o f)(x) = g(y)$

i.e., $(g_o f)(x) = g(f(x))$

Consider two functions $f(x) = 2x + 5$ and $g(x) = 3x$.

Therefore $(g_o f)(x) = g(f(x))$

$$= g(2x + 5)$$
$$= 3(2x + 5)$$

i.e., $(g_o f)(x) = 6x + 15$

Similarly, $(f_o g)(x) = f(g(x))$

$$= f(3x)$$
$$= 2(3x) + 5$$

i.e., $(f_o g)(x) = 6x + 5$

4.5.1 Theorem

Let $f: A \to B$ and $g: B \to C$ be two functions. Then $(g_{\,0} f)$ is one-one if both f and g are one-one and $(g_{\,0} f)$ is onto if both f and g are onto.

Proof: Let $f: A \to B$ and $g: B \to C$ be two functions. Since f is a function from the set A to the set B, then for every $x \in A$ there exists unique $y \in B$ such that $y = f(x)$. Similarly g is a function from the set B to the set C, then for every $y \in B$ there exists unique $z \in C$ such that $z = g(y)$.

Suppose that f and g are both one-one. Our claim is $(g_{\,0} f)$ is one-one.

Since $f: A \to B$ and $g: B \to C$ we have $(g_{\,0} f): A \to C$.

Let $x_1, x_2 \in A$ and $(g_{\,0} f)(x_1) = (g_{\,0} f)(x_2)$

This implies $\quad g(f(x_1)) = g(f(x_2))$

i.e., $\quad f(x_1) = f(x_2) \quad$ [$\because$ g is one-one]

i.e., $\quad x_1 = x_2 \quad$ [$\because$ f is one-one]

Therefore, $g(f(x_1)) = g(f(x_2))$ implies $x_1 = x_2$. So $(g_{\,0} f)$ is one-one.

Suppose that both f and g are onto. Since g is onto, for every $z \in C$ there is at least one $y \in B$ such that $g(y) = z$. Again as f is onto, for every $y \in B$ there exists at least one $x \in A$ such that $f(x) = y$.

As a result for every $z \in C$ there is at least one $x \in A$ such that $(g_{\,0} f)(x) = z$. Therefore $(g_{\,0} f)$ is onto.

4.5.2 Theorem

If $f: A \to B$; $g: B \to C$ and $h: C \to D$, then $h_{\,0} (g_{\,0} f) = (h_{\,0} g)_{\,0} f$, *i.e.*, composition of functions holds the associative law.

Proof : Let $f: A \to B$, $g: B \to C$ and $h: C \to D$ be three functions. So, $(g_{\,0} f): A \to C$. Therefore $h_{\,0} (g_{\,0} f): A \to D$.

Further $(h_{\,0} g): B \to D$. So, $(h_{\,0} g)_{\,0} f: A \to D$. Therefore both $h_{\,0} (g_{\,0} f)$ and $(h_{\,0} g)_{\,0} f$ are functions from $A \to D$.

Since $f: A \to B$, then for every $x \in A$ there exists unique $y \in B$ such that $f(x) = y$. Further $g: B \to C$, then for every $y \in B$ there exists unique $z \in C$ such that $g(y) = z$. Again $h: C \to D$, then for every $z \in C$ there exists unique $t \in D$ such that $h(z) = t$.

Then
$$\begin{aligned} h_{\,0} (g_{\,0} f)(x) &= h(g_{\,0} f(x)) \\ &= h(g(f(x))) \\ &= h(g(y)) = h(z) = t. \end{aligned}$$

Further,
$$\begin{aligned} (h_{\,0} g)_{\,0} f(x) &= (h_{\,0} g)(f(x)) \\ &= (h_{\,0} g)(y) \\ &= h(g(y)) = h(z) = t. \end{aligned}$$

Therefore for $x \in A$ we have $h_{\,0} (g_{\,0} f)(x) = (h_{\,0} g)_{\,0} f(x)$ for all $x \in A$. *i.e.*, $h_{\,0} (g_{\,0} f) = (h_{\,0} g)_{\,0} f$.

4.6 INVERSE FUNCTION

Let $f : A \to B$ be a bijective function. Then the inverse of f, *i.e.* f^{-1} be a function from B to A. Since f is a function from A to B, for every $x \in A$, there exists unique $y \in B$ such that $f(x) = y$.

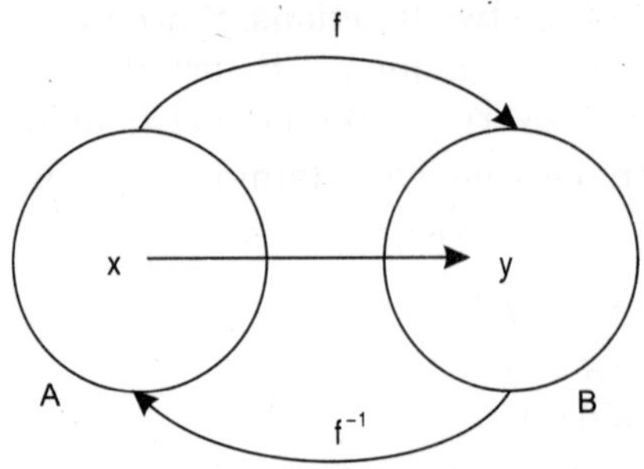

Since f^{-1}: B → A for every $y \in B$ there exists unique $x \in A$ such that $f^{-1}(y) = x$, *i.e.*, $f^{-1}(f(x)) = x$.

4.6.1 Theorem

If $f : A \to B$ is bijective, then the function f posses inverse mapping.

Proof: Suppose that $f : A \to B$ is not bijective and posses an inverse mapping, *i.e.*, (*i*) f is onto but not one-one. (*ii*) f is one-one but not onto or (*iii*) f is neither one-one nor onto.

Case (*i*) Suppose that f is onto but not one-one.

As f is onto, so for every $y_1 \in B$ there exists at least one $x_1 \in A$ such that $f(x_1) = y_1$ and $R(f) = B$. Again as f is not one-one we have $x_1 \neq x_2$, $x_1, x_2 \in A$ implies $y_1 = f(x_1) = f(x_2) = y_2$.

Since $f^{-1}: B \to A$, so $D(f^{-1}) = R(f) = B$, *i.e.*, $D(f^{-1}) = B$. Also $(x_1, y_1), (x_2, y_2) \in f$ implies $(y_1, x_1), (y_2, x_2) \in f^{-1}$ with $x_1 \neq x_2$ as $y_1 = y_2$. Hence, f^{-1} can not be a function.

Case (*ii*) Suppose that f is one-one but not onto.

As f is not onto, so for at least one $y_1 \in B$ there exists no $x_1 \in A$ such that $f(x_1) = y_1$ and $R(f) \neq B$. Since, $f^{-1}: B \to A$, so

$D(f^{-1}) = R(f) \neq B$, *i.e.*, $D(f^{-1}) \neq B$. Hence f^{-1} can not be a function.

Case (*iii*) Similarly it can be proved that f^{-1} can not be a function if f is neither onto nor one-one.

Therefore, it is a contradiction. So our supposition is wrong. Hence, f: B → A must be bijective to posses a inverse mapping.

4.6.2 Theorem

Let $f : A \to B$ and $g : B \to C$ be two functions. If both f and g are invertible, then

$$(g \circ f)^{-1} = f^{-1} \circ g^{-1}.$$

Proof: Suppose that both f and g are invertible. This indicates that both f and g are bijective functions. So by theorem 4.5.1, $(g \circ f)$ is also bijective and hence invertible.

As $f : A \to B$ and $g : B \to C$ we have $(g \circ f) : A \to C$ *i.e.*, $(g \circ f)^{-1} : C \to A$. Also $f^{-1} : B \to A$ and $g^{-1} : C \to B$ we have $f^{-1} \circ g^{-1} : C \to A$.

Hence first of all it is evident that both $(g \circ f)^{-1}$ are $f^{-1} \circ g^{-1}$ are functions from the set C to the set A and $(g \circ f)^{-1}(z) = x$ for $z \in C$ and $x \in A$.

Again g^{-1}: C → B, so for every $z \in$ C there exists unique $y \in$ B such that $g^{-1}(z) = y$. Similarly, f^{-1}: B → A, so for every $y \in$ B there exists $x \in$ A such that $f^{-1}(y) = x$. Further

$$f^{-1} \circ g^{-1}(z) = f^{-1}(g^{-1}(z)) = f^{-1}(y) = x.$$

i.e., $f^{-1} \circ g^{-1}(z) = x = (g \circ f)^{-1}(z)$. Therefore $(g \circ f)^{-1} = f^{-1} \circ g^{-1}$.

4.6.3 Theorem

If f: A → B is a bijective function, then f^{-1}: B → A is also a bijective function.

Proof: Let f : A → B is a bijective function, *i.e.*, f is one-one and onto function. Since f is one-one and onto, for every $y \in$ B there exists unique $x \in$ A such that $f(x) = y$. Again f^{-1}: B → A such that $f^{-1}(y) = x$.

Let $y_1, y_2 \in$ B with $f^{-1}(y_1) = f^{-1}(y_2)$

This implies $x_1 = x_2$

i.e., $f(x_1) = f(x_2)$ [∵ f is one-one]

i.e., $y_1 = y_2$

i.e., $f^{-1}(y_1) = f^{-1}(y_2) \Rightarrow y_1 = y_2$. Thus, f^{-1} is one-one. Besides this $R(f^{-1}) = D(f) = A$. *i.e.*, $R(f^{-1})$ = A. This indicates that f^{-1} is onto. Therefore f^{-1} is both one-one and onto. *i.e.*, f^{-1} is bijective.

4.7 SOME IMPORTANT FUNCTIONS

In this section we will discuss some important functions that are useful in the context of computer science and applications.

4.7.1 Identity Function

Let A be a set. The function f: A → A is said to be an identity function if for every $x \in$ A, $f(x) = x$. Mathematically $f(x) = x \ \forall \ x \in$ A.

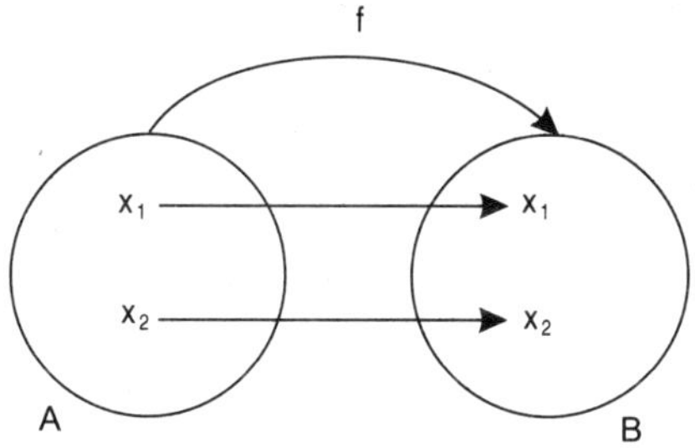

4.7.2 Constant Function

The function f: A → B is said to be a constant function if for every $x \in$ A there exists unique $y \in$ B such that $f(x) = y$. Mathematically,

$$f(x) = y \quad \forall \ x \in A$$

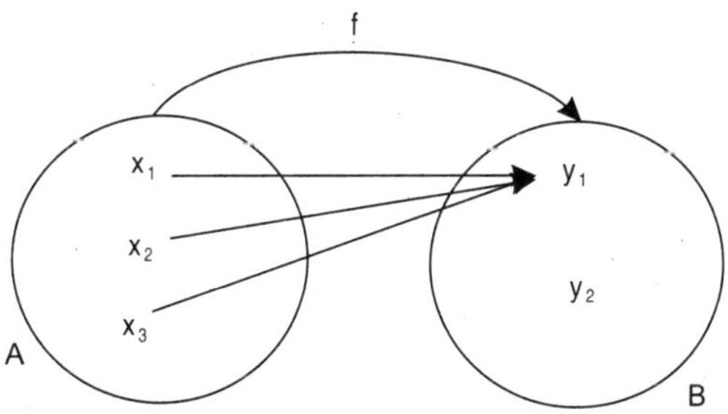

Consider a function f: R $\rightarrow$ I defined by $f(x) = 2$ for $x \in$ R. Which is a constant function.

4.7.3 Absolute Function

The absolute function or absolute value function $f(x) = |x|$ is defined as

$$|x| = \begin{cases} x; & \text{if } x \geq 0 \\ -x; & \text{if } x < 0 \end{cases}$$

The graph of $f = \{(x, |x|): x \in \text{R}\}$ is shown in the following figure.

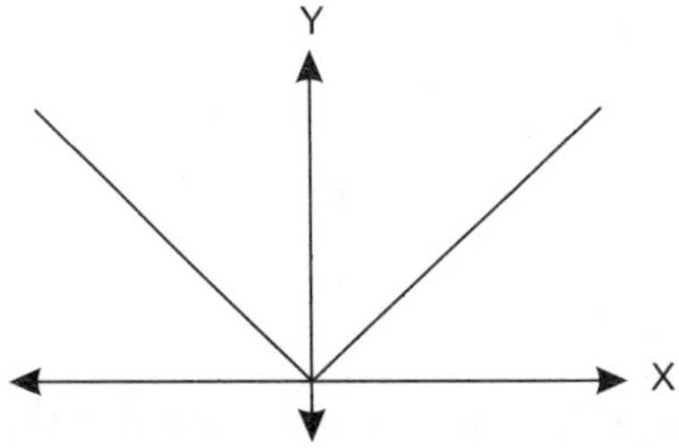

4.7.4 Greatest Integer Function

The greatest integer function $f(x) = [x]$ is defined as the greatest integer less than or equal to x. The value of $f(x) = [x]$ is equal to n if $n \leq x < (n + 1)$; $n \in$ Z.

Consider the examples [5] = 5; [5.7] = 5; [– 3.9] = – 4; [– 2.2] = – 3 and [6.1] = 6.

4.7.5 Floor and Ceiling Function

The floor function $f(x) = \lfloor x \rfloor$ is defined as the greatest integer less than or equal to x. The ceiling function $f(x) = \lceil x \rceil$ is defined as the least integer greater than or equal to x.

Let x be any real number, then x lies between two integers called floor of x and ceiling of x.

Consider the following examples. $\lfloor 3.5 \rfloor = 3$; $\lfloor 5 \rfloor = 5$; $\lfloor -7.2 \rfloor = -8$; $\lceil 3.5 \rceil = 4$; $\lceil 5 \rceil = 5$; $\lceil -7.2 \rceil = -7$.

Note: From the above discussion it is clear that $\lceil x \rceil = \lfloor x \rfloor + 1$ if x is not an integer otherwise $\lceil x \rceil = \lfloor x \rfloor$.

4.7.6 Even and Odd Functions

A real function $y = f(x)$ is said to be even if $f(-x) = f(x)$ and odd if $f(-x) = -f(x)$.

Consider the function $f(x) = 5x^6 + 2x^4 - x^2$.

Therefore, $f(-x) = 5(-x)^6 + 2(-x)^4 - (-x)^2 = 5x^6 + 2x^4 - x^2 = f(x)$.

Hence, $f(x) = 5x^6 + 2x^4 - x^2$ is an even function.

Similarly consider another function $f(x) = \sin x - 5x^3$.

Therefore, $f(-x) = \sin(-x) - 5(-x)^3 = -\sin x + 5x^3 = -(\sin x - 5x^3) = -f(x)$.

Hence, $f(x) = \sin x - 5x^3$ is an odd function.

Note: It is to be noted that a function can neither be even nor odd. Consider the example

$$f(x) = x^4 + x^3 + x^2 - x.$$

Therefore, $f(-x) = (-x)^4 + (-x)^3 + (-x)^2 - (-x) = x^4 - x^3 + x^2 + x$. This implies neither $f(-x) = f(x)$ nor $f(-x) = -f(x)$.

Therefore, $f(x) = x^4 + x^3 + x^2 - x$ is neither even nor odd function.

4.7.7 Characteristic Function

Suppose A be any subset of the universal set U. The characteristic function of A *i.e.*, χ_A is a real valued function χ_A: U $\to$ {0, 1} defined by

$$\chi_A(x) = \begin{cases} 1; & \text{if } x \in A \\ 0; & \text{if } x \notin A \end{cases}$$

Consider the example where A = {2, 5, 7} and U = {1, 2, 3, 4, 5,7}. Then we have $\chi_A(1) = 0$, $\chi_A(2) = 1$, $\chi_A(3) = 0$, $\chi_A(4) = 0$, $\chi_A(5) = 1$, $\chi_A(7) = 1$. The arrow diagram is given below.

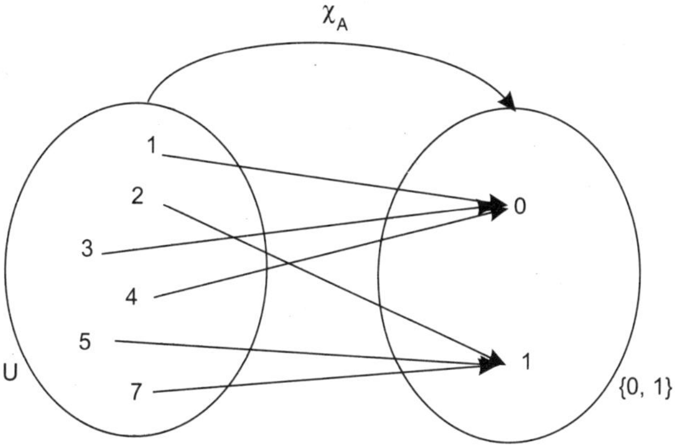

4.7.8 Remainder Function

Let x be a non-negative integer and y be a positive integer. We define x mod y or $R_y(x)$ to be the remainder when x is divided by y. Thus R_y is a function on Z.

Consider the following examples

8 mod 2 = 0,	15 mod 4 = 3,	251 mod 2 = 1,	177 mod 3 = 0
i.e., $R_2(8) = 0$,	$R_4(15) = 3$,	$R_2(251) = 1$,	$R_3(177) = 0$.

4.7.9 Signum Function

The signum function sgn(x) on R is defined as

$$\text{sgn}(x) = \begin{cases} 0\ , & \text{if } x = 0 \\ \dfrac{x}{|x|}, & \text{if } x \neq 0 \end{cases}$$

The range of this function is {–1, 0, 1}.

■ 4.8 HASH FUNCTION

Suppose that we have cells in a computer memory indexed from 0 to 16. This is given in the following figure.

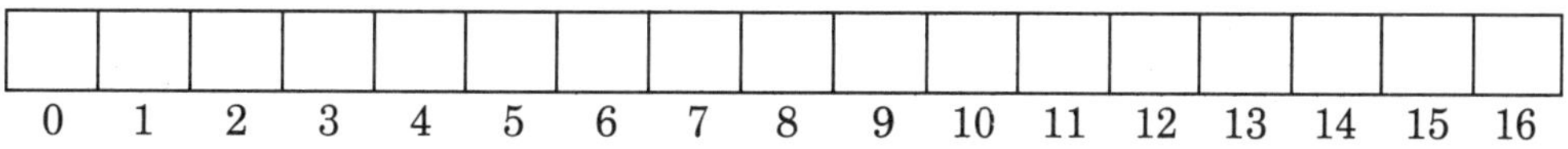

We wish to store and retrieve arbitrary positive integers in these empty cells. One way is to use a hash function. A hash function takes a data item to be stored or retrieved and computes the first choice for a location for the data item by the relation

$$H(n) = n \bmod k.$$

Where n is the data item (number) to be stored or retrieved. k is the size of the computer memory (preferably prime). If the first choice for a location is already occupied, then we say that a collision has occurred. To handle collisions, a collision resolution policy is required. One simple policy is to find the next highest unoccupied cell.

If we want to locate a stored value n, compute $m = H(n)$ and begin looking at location m. If n is not at this location, move forward in the next highest location. In this context we used one collision resolution policy. Besides this there are several other methods to handle collision, which is beyond the scope of this Book.

Consider an example in which the data item 15, 286, 77, 18, 5, 572, 102, 257 and 55 are to be stored in order in a computer memory indexed from 0 to 16. Here $k = 17$. It is clear that

H(15) = 15 mod 17 = 15, H(286) = 286 mod 17 = 14. Similarly H(77) = 9, H(18) = 1, H(5) = 5, H(572) = 11, H(102) = 0, H(257) = 2, H(55) = 4. Thus the allocation in the computer memory is given in the following figure.

102	18	257		55	5	89			77		572			286	15	
0	1	2	3	4	5	6	7	8	9	10	11	12	13	14	15	16

Now suppose we want to store 89. Since H(89) = 89 mod 17 = 4, 89 should be stored at location 4; but this position is already occupied. If we use the collision resolution policy discussed earlier, we would store 89 at location 6, which is as shown in the above figure.

SOLVED EXAMPLES

Example 1 *Let A = {a, b, c, d} and B = {7, 8, 9}. Find whether the following subsets of (A × B) are functions from A to B.*

(i) $f_1 = \{(a, 7), (b, 8), (c, 8)\}$

(ii) $f_2 = \{(a, 7), (a, 8), (b, 9), (c, 9), (d, 9)\}$

(iii) $f_3 = \{(a, 7), (b, 8), (c, 9), (d, 9)\}$

(iv) $f_4 = \{(a, 7), (b, 7), (c, 9), (d, 8)\}$

Solution: Given that A = $\{a, b, c, d\}$ and B = {7, 8, 9}.

(i) Given $f_1 = \{(a, 7), (b, 8), (c, 8)\}$

This implies $D(f_1) = \{a, b, c\} \neq A$. Hence f_1 can not be a function.

(ii) Given $f_2 = \{(a, 7), (a, 8), (b, 9), (c, 9), (d, 9)\}$

This implies $D(f_2) = \{a, b, c, d\} = A$ and $(a, 7) \in f_2$, $(a, 8) \in f_2$ with $7 \neq 8$. Thus, f_2 can not be a function.

(iii) Given $f_3 = \{(a, 7), (b, 8), (c, 9), (d, 9)\}$

This implies $D(f_3) = \{a, b, c, d\} = A$ and there is no such order pair $(x, y) \in f_3$, $(x, z) \in f_3$ such that $y = z$. So f_3 is a function.

(iv) Given $f_4 = \{(a, 7), (b, 7), (c, 9), (d, 8)\}$

This implies $D(f_4) = \{a, b, c, d\} = A$ and there is no such order pair $(x, y) \in f_4$, $(x, z) \in f_4$ such that $y = z$. So f_4 is a function.

Example 2 *Give an example of a function which is*

(*a*) *Injective but not surjective.*
(*b*) *Surjective but not injective.*
(*c*) *Bijective*
(*d*) *Neither injective nor surjective.*
(*e*) *Constant.*

Explain with the help of arrow diagrams.

Solution: (*a*) Let A = {a, b, c, d, e}and B = {1, 2, 3, 4, 5, 6}. Consider a function f_1 from A to B as

$$f_1 = \{(a, 1), (b, 2), (c, 3), (d, 4), (e, 5)\}$$

Now R (f_1) = {1, 2, 3, 4, 5} ≠ B. Hence f_1 is not a surjective function but injective. The arrow diagram is given below.

(*b*) Let A = {a, b, c, d, e} and B = {1, 2, 3}
Consider a function f_2 from A to B as $f_2 = \{(a, 1), (b, 2), (c, 3), (d, 3), (e, 3)\}$
Here R(f_2) = {1, 2, 3} = B and $(c, 3) \in f_2$, $(d, 3) \in f_2$, $(e, 3) \in f_2$ such that $c \neq d \neq e$.
Thus f_2 is surjective but not injective. The arrow diagram is given below.

(*c*) Let A = {a, b, c} and B = {1, 2, 3}
Consider a function f_3 from A to B as $f_3 = \{(a, 2), (b, 3), (c, 1)\}$. Here R ($f_3$) = {1, 2, 3} = B.
Therefore f_3 is bijective. The arrow diagram is given below.

(*d*) Let A = {a, b, c, d} and B = {1, 2, 3, 4, 5}.
Consider a function f_4 from A to B as $f_4 = \{(a, 2), (b, 3), (c, 4), (d, 4)\}$.
Here R(f_4) = {2, 3, 4} ≠ B and $(c, 4) \in f_4$, $(d, 4) \in f_4$ such that $c \neq d$.
Therefore, f_4 is neither injective nor surjective. The arrow diagram is given below.

(*e*) Let A = {a, b, c, d} and B = {1, 2, 3}
Consider a function f_5 from A to B as $f_5 = \{(a, 2), (b, 2), (c, 2), (d, 2)\}$. This implies $f_5(x) = 2$ for every $x \in$ A. Therefore f_5 is constant. The arrow diagram is given below.

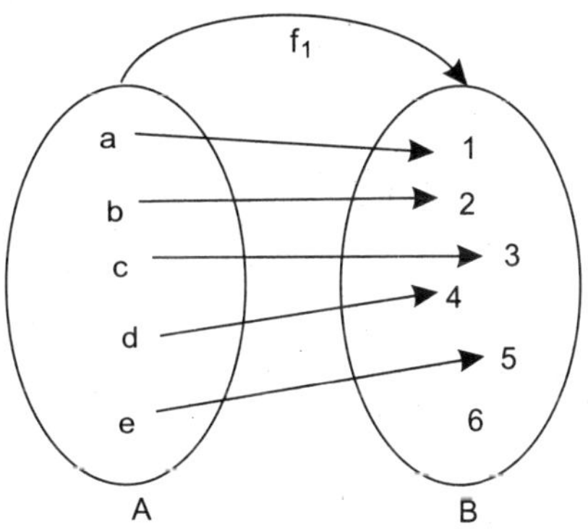

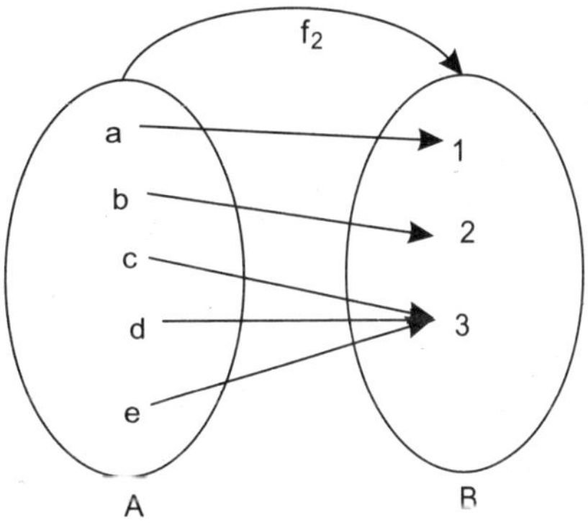

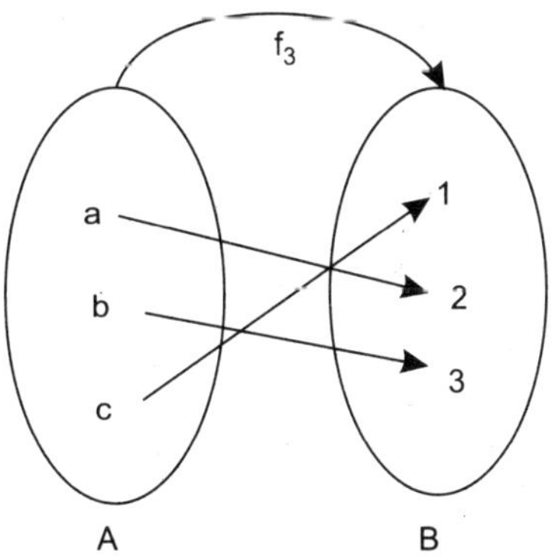

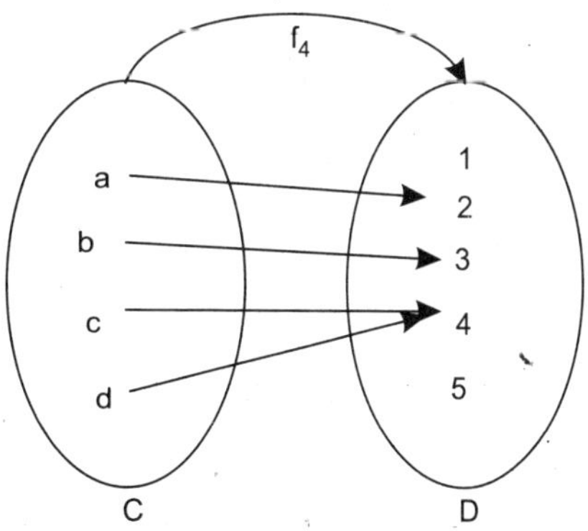

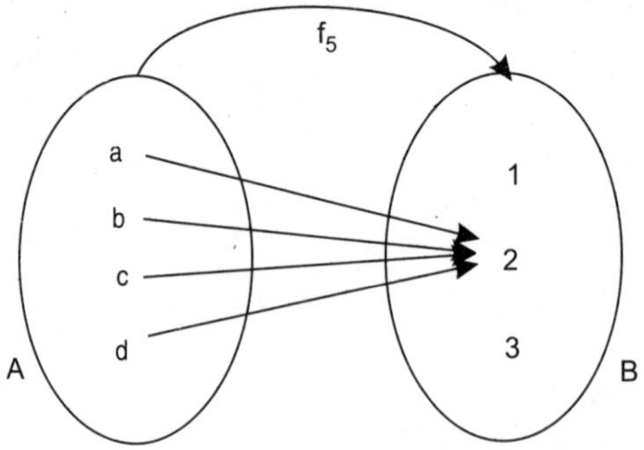

Example 3 *If $f: x \to 2x$; $g: x \to x^2$ and $h: x \to (x + 1)$, then find $(f_0 g)_0 h$ and $f_0 (g_0 h)$. Show that $(f_0 g)_0 h = f_0 (g_0 h)$.*

Solution: Let $f: x \to 2x$; $g: x \to x^2$ and $h: x \to (x + 1)$.

i.e., $f(x) = 2x$, $g(x) = x^2$ and $h(x) = x + 1$.

So, $(f_0 g)(x) = f(g(x))$
$= f(x^2) = 2x^2$

Therefore, $(f_0 g)_0 h(x) = (f_0 g)(h(x)) = (f_0 g)(x + 1) = 2(x + 1)^2$.

Again $(g_0 h)(x) = g(h(x))$
$= g(x + 1) = (x + 1)^2$

Therefore $f_0 (g_0 h)(x) = f(x + 1)^2 = 2(x + 1)^2$.

Hence $(f_0 g)_0 h = f_0 (g_0 h)$.

Example 4 *Let $f(x)$ be any real function. Show that $g_1(x) = \dfrac{f(x) + f(-x)}{2}$ is always an even function whereas $g_2(x) = \dfrac{f(x) - f(-x)}{2}$ is always an odd function.*

Solution: Let $$g_1(x) = \frac{f(x) + f(-x)}{2}$$

Therefore $$g_1(-x) = \frac{f(x) + f(-x)}{2} = g_1(x), \text{ i.e., } g_1(-x) = g_1(x).$$

Also let $$g_2(x) = \frac{f(x) - f(-x)}{2}.$$

Therefore $g_2(-x) = \dfrac{f(-x) - f(x)}{2} = -\dfrac{f(x) - f(-x)}{2} = -g_2(x)$. This implies $g_1(x)$ is an even function whereas $g_2(x)$ is an odd function.

Example 5 *Find the composition $(f_0 g)$ and $(g_0 f)$ in the following cases.*

(i) $f(x) = \sin^2 x$ *and* $g(x) = x^2 + 1$
(ii) $f(x) = e^x$ *and* $g(x) = x^3$
(iii) $f(x) = 2x^2 + x$ *and* $g(x) = x^2 + 1$

Hence show that $(f_0 g) \neq (g_0 f)$.

Solution: (*i*) Let $f(x) = \sin^2 x$ and $g(x) = x^2 + 1$

Therefore $(f_0 g)(x) = f(g(x))$
$= f(x^2 + 1) = \sin^2(x^2 + 1)$

Similarly $(g_o f)(x) = g(f(x)) = g(\sin^2 x) = \sin^4 x + 1$

So, $(f_o g) \neq (g_o f)$.

(*ii*) Let $f(x) = e^x$ and $g(x) = x^3$

Therefore $(f_o g)(x) = f(g(x)) = f(x^3) = e^{x^3}$

Similarly $(g_o f)(x) = g(f(x)) = g(e^x) = (e^x)^3 = e^{3x}$.

So, $(f_o g) \neq (g_o f)$.

(*iii*) Let $f(x) = 2x^2 + x$ and $g(x) = x^2 + 1$

Therefore $(f_o g)(x) = f(g(x)) = f(x^2 + 1)$

$= 2(x^2 + 1)^2 + (x^2 + 1) = 2x^4 + 5x^2 + 3$

$(g_o f)(x) = g(f(x)) = g(2x^2 + x) = (2x^2 + x)^2 + 1$

$= 4x^4 + 4x^3 + x^2 + 1$

So, $(f_o g)(x) \neq (g_o f)(x)$.

Example 6 *Determine whether the given functions are one-one, onto or bijective.*

(*a*) *$f: R^+ \to R^+$ defined by $f(x) = |x|$*

(*b*) *$f: I \to R^+$ defined by $f(x) = 2x + 7$*

(*c*) *$f: R \to R$ defined by $f(x) = |x|$*

Solution: (*a*) Given $f: R^+ \to R^+$ defined by $f(x) = |x|$.

Suppose that $f(x_1) = f(x_2)$

$\Rightarrow |x_1| = |x_2|$; *i.e.*, $x_1 = x_2$

So, $f: R^+ \to R^+$ defined by $f(x) = |x|$ is one-one. Again $f(x) = |x|$; $x \in R^+$

This implies $y = |x| = x$ [$\because x \in R^+$]. This indicates that for every $y \in R^+$ there exists $x \in R^+$ such that $y = f(x) = |x|$. Hence, $f: R^+ \to R^+$ defined by $f(x) = |x|$ is onto. Therefore, bijective.

(*b*) $f: I \to R^+$ defined by $f(x) = 2x + 7$

Assume that $f(x_1) = f(x_2)$

This implies $2x_1 + 7 = 2x_2 + 7$; *i.e.*, $x_1 = x_2$.

So, $f(x) = 2x + 7$; $x \in I$, is One-One.

Again $f(x) = 2x + 7$; $x \in I$

$\Rightarrow$ $y = 2x + 7$ $[\because y = f(x)]$

$\Rightarrow$ $x = \dfrac{y - 7}{2}$

It is clear that for $y = 5$ we get $x = -1 \notin I$ (Set of positive integers). Hence $f(x) = 2x + 7$ is not onto. Thus $f(x) = 2x + 7$ is One-One only.

(*c*) Given $f: R \to R$ defined by $f(x) = |x|$

Suppose that $f(x_1) = f(x_2)$

$\Rightarrow$ $|x_1| = |x_2|$

$\Rightarrow$ $\pm x_1 = x_2$

$\Rightarrow$ $x_1 \neq x_2$.

So, $f: R \to R$ defined by $f(x) = |x|$ is not One-One. Again for $f(x) = y = -5$ in the co-domain R there exists no element x *in* the domain R. Hence $f: R \to R$ defined by $f(x) = |x|$ is neither One-One nor onto.

Example 7 *Let A = {1, 2, 3, 4}, B = {x, y, z, t} and C = {2, 4, 9}. Let f = {(1, x), (2, z), (3, y), (4, t)} and g = {(x, 2), (y, 2), (z, 4), (t, 9)} be two functions from A → B and B → C respectively. Find the composition $(g_o f)$.*

Solution: Let A = {1, 2, 3, 4}, B = {x, y, z, t} and C = {2, 4, 9}. Let f = {(1, x), (2, z), (3, y), (4, t)} and

$$g = \{(x, 2), (y, 2), (z, 4), (t, 9)\}$$

Therefore, $g\ _o f$ = {(1, 2), (3, 2), (2, 4), (4, 9)}. Which is a function from A → C . The arrow diagram is given below.

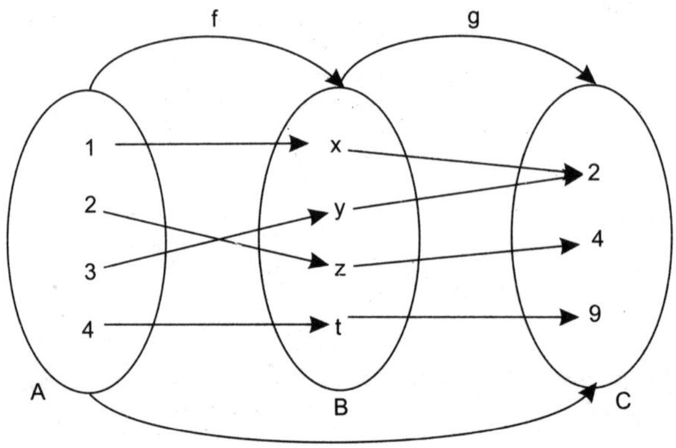

(g$_0$f)

Example 8 *Let Q be the set of rational numbers. Show that the function $f: Q \to Q$ defined by $f(x) = 2x + 7$, $x \in Q$ is a bijective function. Find $f^{-1}(0)$, $f^{-1}(1)$ and $f^{-1}(2)$.*

Solution: f: Q → Q defined by $f(x) = 2x + 7$; $x \in$ Q

Let $x_1, x_2 \in$ Q such that $f(x_1) = f(x_2)$

$$\Rightarrow \quad 2x_1 + 7 = 2x_2 + 7$$

$$\Rightarrow \quad x_1 = x_2.$$

i.e., $f(x_1) = f(x_2) \Rightarrow x_1 = x_2$. So, $f(x) = 2x + 7$; $x \in$ Q is One-One. Again for every $y \in$ Q (co-domain set) there exists $x = \dfrac{y-7}{2}$ in the domain set Q such that $y = f(x) = 2x + 7$. Hence $f(x) = 2x + 7$ is onto. Therefore $f(x) = 2x + 7$, $x \in$ Q is a bijective function.

To compute the inverse we have $f^{-1}(y) = x$

i.e.,
$$f^{-1}(y) = \frac{y-7}{2}$$

In general $f^{-1}(x) = \dfrac{x-7}{2}$; $x \in$ Q. Hence we have $f^{-1}(0) = \dfrac{-7}{2}$, $f^{-1}(1) = -3$ and $f^{-1}(2) = \dfrac{-5}{2}$.

Example 9 *Let A = R – {3} and B = R – {1}, where R is the set of real numbers. Let f: A → B defined by $f(x) = \dfrac{x-2}{x-3}$, $x \in$ A. Show that f is One-One and onto. Find the inverse function of f.*

Solution: Let A = R – {3} and B = R – {1}, where R is the set of real numbers. Let f: A → B defined by $f(x) = \dfrac{x-2}{x-3}$, $x \in$ A. Let $x_1, x_2 \in$ A such that $f(x_1) = f(x_2)$

$$\Rightarrow \quad \frac{x_1 - 2}{x_1 - 3} = \frac{x_2 - 2}{x_2 - 3}$$

$$\Rightarrow \quad x_1x_2 - 2x_2 - 3x_1 + 6 = x_1x_2 - 3x_2 - 2x_1 + 6$$

$$\Rightarrow \quad -2x_2 - 3x_1 = -3x_2 - 2x_1$$

$$\Rightarrow \quad x_1 = x_2$$

i.e.,
$$f(x_1) = f(x_2)$$

$$\Rightarrow \quad x_1 = x_2$$

So, $f(x)$ is One-One.

Again for every $y \in$ B, there exists $x = \dfrac{3y-2}{y-1}$ in A such that $y = f(x) = \dfrac{x-2}{x-3}$.

Hence, $\qquad f(x) = \dfrac{x-2}{x-3}$ is onto.

Therefore, $f(x) = \dfrac{x-2}{x-3}$, $x \in$ A is a bijective function. To compute inverse we have $f^{-1}(y) = x$.

i.e., $\qquad f^{-1}(y) = \dfrac{3y-2}{y-1}$. In general $f^{-1}(x) = \dfrac{3x-2}{x-1}$; $x \in$ B

Example 10 *Let A = {1, 2, 4, 6}; B = {3, 5, 7, 9}, C = {1, 2, 4, 6} and f : A → B defined by f = {(1, 3), (2, 5), (4, 7), (6, 9)}; g : B → C defined by g = {(5, 6), (3, 2), (7, 1), (9, 4)} be two functions. Find the compositions* $(f_0 g)$ *and* $(g_0 f)$. *Show that* $(f_0 g) \neq (g_0 f)$.

Solution: Let A = {1, 2, 4, 6}; B = {3, 5, 7, 9}and C = {1, 2, 4, 6}. Also given f = {(1, 3), (2, 5), (4, 7), (6, 9)} with g = {(5, 6), (3, 2), (7, 1), (9, 4)}. Consider the arrow diagram to compute $(g_0 f)$.

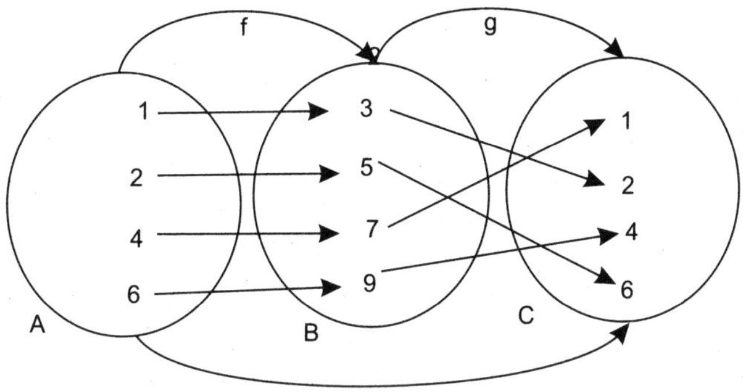

$(g_0 f)$

Therefore, $(g_0 f)$ = {(1, 2), (2, 6), (4, 1), (6, 4)}.

Similarly, to compute $(f_0 g)$ the arrow diagram becomes

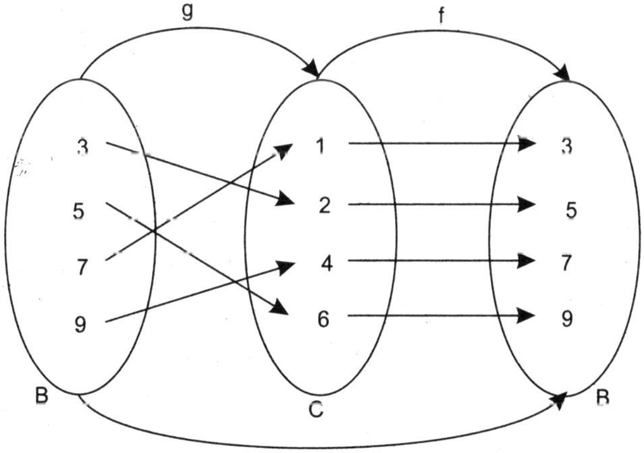

$(f_0 g)$

Therefore, $(f_0 g)$ = {(5, 9), (3, 5), (7, 3), (9, 7)}. Hence, it is clear that $(f_0 g) \neq (g_0 f)$.

Example 11 *Let f: R → (– 1, 1) defined by* $f(x) = \dfrac{x}{1+x^2}$, $x \in R$. *Find the inverse of above function if exists, where R is the set of real numbers.*

Solution: Let $f : R \to (-1, 1)$ defined by $f(x) = \dfrac{x}{1+x^2}$, $x \in R$. Let $x_1, x_2 \in R$ such that $f(x_1) = f(x_2)$

$$\Rightarrow \quad \frac{x_1}{1+x_1^2} = \frac{x_2}{1+x_2^2}$$

$$\Rightarrow \quad x_1 + x_1x_2^2 - x_2 - x_2x_1^2 = 0$$

$$\Rightarrow \quad (x_1 - x_2)(1 - x_1x_2) = 0$$

$$\Rightarrow \quad x_1 = x_2 \text{ or } (1/x_2)$$

So, $f(x) = \dfrac{x}{1+x^2}$, $x \in R$ is not One-One, hence not bijective. Therefore, inverse does not exists.

Example 12 *Find the characteristic function for the set A. Where the universal set U = {1, 2, 3, 4, 5, 6, 7, 8} and A = {1, 4, 7, 8}.*

Solution: Given universal set U = {1, 2, 3, 4, 5, 6, 7, 8} and A = {1, 4, 7, 8}. The characteristic function for the set A is given as $\chi_A(1) = 1$, $\chi_A(2) = 0$, $\chi_A(3) = 0$, $\chi_A(4) = 1$, $\chi_A(5) = 0$, $\chi_A(6) = 0$, $\chi_A(7) = 1$, $\chi_A(8) = 1$. The arrow diagram is given below.

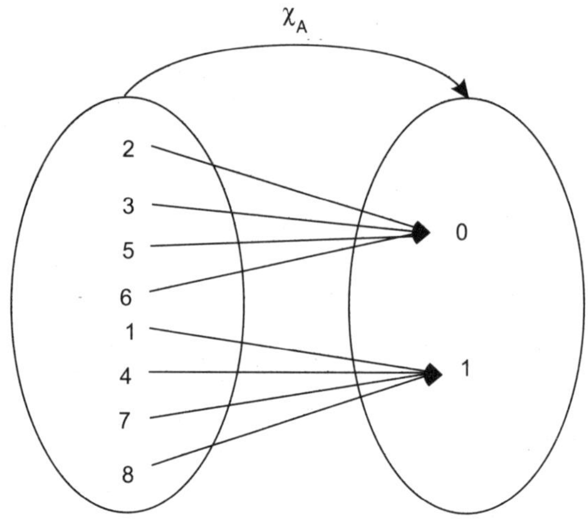

Example 13 *If f(x) and g(x) are both even or both odd, then prove that f(x) g(x) is even.*

Solution: Suppose that $f(x)$ and $g(x)$ are both even.

i.e., $f(-x) = f(x)$ and $g(-x) = g(x)$.

Let $h(x) = f(x)\,g(x)$

Therefore, $h(-x) = f(-x)\,g(-x) = f(x)\,g(x) = h(x)$.

This indicates that $f(x)\,g(x)$ is even. Similarly it can be proved that if both $f(x)$ and $g(x)$ are odd, then $f(x)\,g(x)$ is even.

Example 14 *Sketch the graph of*

$$f(x) = \begin{cases} \dfrac{1}{x}; & \text{if } x \neq 0 \\ 0; & \text{if } x = 0 \end{cases}$$

Solution: Given that $f(x) = \begin{cases} \dfrac{1}{x}; & \text{if } x \neq 0 \\ 0; & \text{if } x = 0 \end{cases}$;

It is clear that if x is very large, then $f(x)$ is nearly equals to 0 and $f(x)$ is very large when x is nearly equals to zero. The graph is given below.

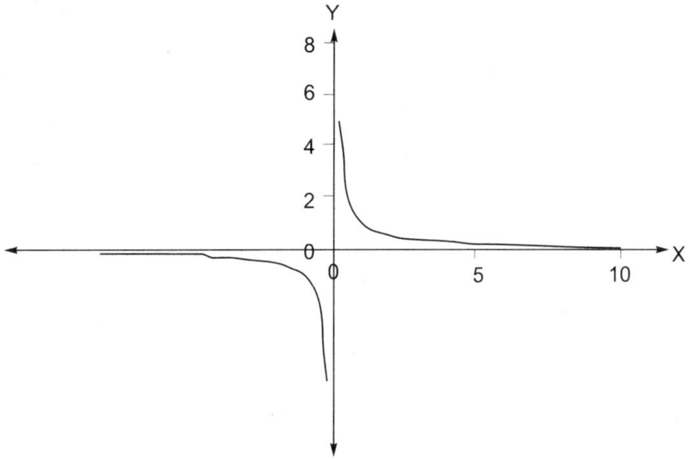

Example 15 *Let $f(x)$ and $g(x)$ are both even functions. Prove that $(f_0 g)$ is also an even function.*

Solution: Suppose that $f(x)$ and $g(x)$ are both even.

i.e., $f(-x) = f(x)$ and $g(-x) = g(x)$.

Now $(f_0 g)(-x) = f(g(-x)) = f(g(x)) = (f_0 g)(x)$

i.e., $(f_0 g)(-x) = (f_0 g)(x)$. Therefore $(f_0 g)(x)$ is an even function.

Example 16 *Prove that $(f_0 g)(x)$ is an odd function if both $f(x)$ and $g(x)$ are odd functions.*

Solution: Suppose that $f(x)$ and $g(x)$ are odd functions.

i.e., $f(-x) = -f(x)$ and $g(-x) = -g(x)$

Now

$$\begin{aligned} (f_0 g)(-x) &= f(g(-x)) \\ &= f(-g(x)) && [\because\ g(x) \text{ is an odd function}] \\ &= -f(g(x)) && [\because f(x) \text{ is an odd function}] \\ &= -(f_0 g)(x) \end{aligned}$$

i.e., $(f_0 g)(-x) = -(f_0 g)(x)$. Therefore $(f_0 g)(x)$ is an odd function.

Example 17 *If $g(x) = e^x$ and $(f_0 g)$ is an identity function, prove that $f(x) = \ln x$.*

Solution: Let $g(x) = e^x$ and $(f_0 g)$ is an identity function. *i.e.,* $(f_0 g)(x) = x$

i.e., $f(g(x)) = x$

i.e., $f(e^x) = x = \ln(e^x)$

Therefore, in general $f(x) = \ln x$.

Example 18 *Let $f(x) = 2x + 1$ and $g(x) = x^2 + 2$. Find the values of $(g_o f)(4)$ and $(f_o g)(4)$. Show that they are not equal.*

Solution: Given that $f(x) = 2x + 1$ and $g(x) = x^2 + 2$

Therefore, $(g_o f)(4) = g(f(4)) = g(9)$ [∵ $f(4) = 2(4) + 1 = 9$]

$= 83$ [∵ $g(9) = 9^2 + 2 = 83$]

Again $(f_o g)(4) = f(g(4)) = f(18)$ [∵ $g(4) = 4^2 + 2 = 18$]

$= 37$ [∵ $f(18) = 2(18) + 1 = 37$]

Therefore, $(f_o g)(4) \neq (g_o f)(4)$

Example 19 *Let $f: R \to R$ be defined as $f(x) = x^2 - 3x$, if $x < 2$ and $x + 2$, if $x \geq 2$. Find $f(5)$, $f(2)$, $f(0)$ and $f(-2)$.*

Solution:

Given that
$$f(x) = \begin{cases} x^2 - 3x; & \text{if } x < 2 \\ x + 2; & \text{if } x \geq 2 \end{cases}$$

So,
$$f(5) = 5 + 2 = 7$$
$$f(0) = 0 - 0 = 0$$
$$f(-2) = (-2)^2 - 3(-2) = 10$$
$$f(2) = 2 + 2 = 4.$$

Example 20 *Find the domain $D(f)$ of each of the following functions. (i) $f(x) = \sqrt{16 - x^2}$; (ii) $f(x) = \dfrac{1}{x-4}$; (iii) $f(x) = x^2 - 5x + 6$*

Solution: (*i*) Given $f(x) = \sqrt{16 - x^2}$; it is clear that $f(x)$ is not defined for $16 - x^2 \leq 0$. *i.e.*, $-4 \leq x \leq 4$. So, $f(x)$ is not defined for $x \in [-4, 4]$. Therefore $D(f) = R - [-4, 4]$.

(*ii*) Given $f(x) = \dfrac{1}{(x-4)}$; it is clear that $f(x)$ is not defined for $(x - 4) = 0$, *i.e.*, $f(x)$ is not defined at $x = 4$. Therefore $D(f) = R - \{4\}$.

(*iii*) Given $f(x) = x^2 - 5x + 6$. It is clear that $f(x)$ is defined for every real number R. Therefore, $D(f) = R$.

Example 21 *Find the graph of the following functions.*

(*a*) $f(x) = \begin{cases} 3x - 1 & \text{if } x > 3 \\ x^2 - 2 & \text{if } -2 \leq x \leq 3 \\ 2x + 3 & \text{if } x < -2 \end{cases}$ (*b*) $f(x) = \begin{cases} x + 6 & \text{if } x \leq -1 \\ 5 - x & \text{if } x > -1 \end{cases}$

Solution: (*a*) Given that

$$f(x) = \begin{cases} 3x - 1 & \text{if } x > 3 \\ x^2 - 2 & \text{if } -2 \leq x \leq 3 \\ 2x + 3 & \text{if } x < -2 \end{cases}$$

The graph of above function is given below.

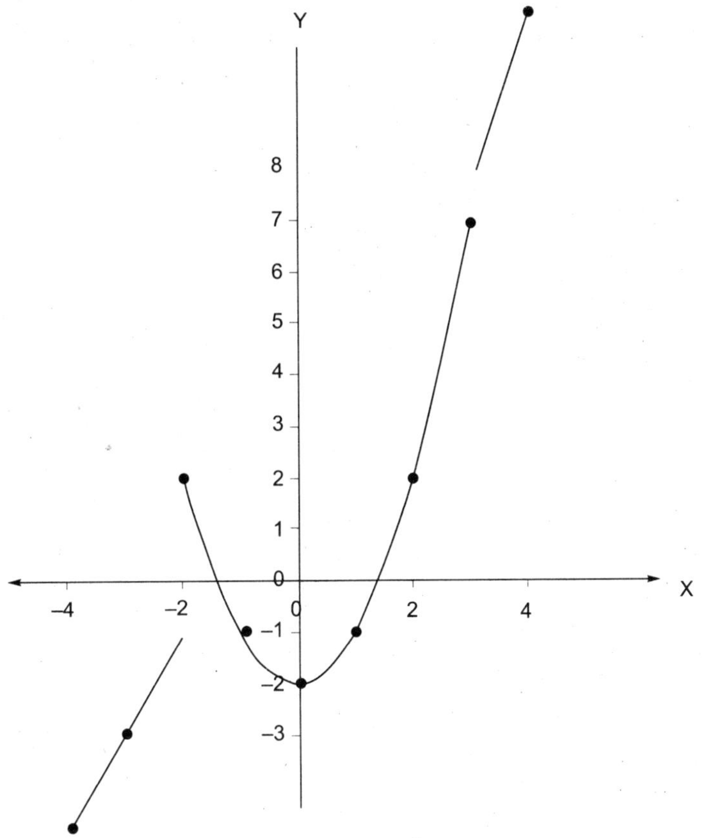

(*b*) Given that $f(x) = \begin{cases} x + 6, & \text{if } x \leq -1 \\ 5 - x, & \text{if } x > -1 \end{cases}$

The graph is given below.

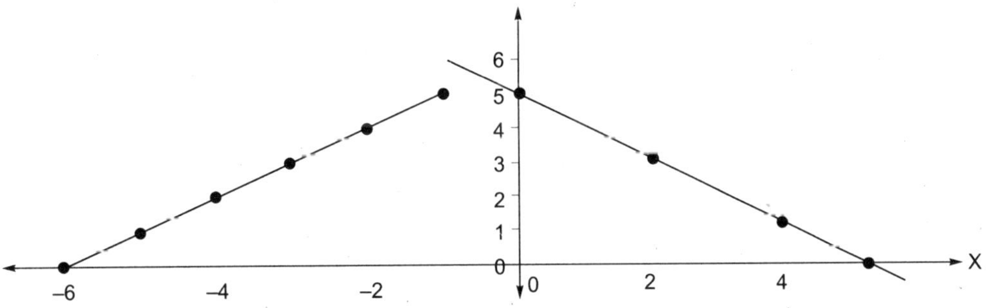

EXERCISES

1. Let A = {1, 2, 3, 4, 5, 6} and B = {a, b, c, d, e, f}. Determine whether each relation given below is a function from A to B. If it is a function, find domain, range. Draw the arrow diagram of each relation.

(*a*) {(1, d), (1, e), (2, a), (3, b), (4, e), (5, e), (6, f)}

(*b*) {(3, d), (4, e), (5, e)}

(c) {(1, b), (2, b), (3, c), (4, c), (5, f), (6, f)}
(d) {(1, c), (2, c), (3, c), (4, c), (5, c), (6, c)}
(e) {(1, a), (2, b), (3, c), (4, d), (5, e), (6, f)}
(f) {(1, a), (2, b), (3, c), (3, d), (3, f)}
(g) {(1, e), (4, e), (5, e), (6, e)}

2. Let f be a function from the set R to the set R, where R is a set of real numbers. Determine whether the following are One-One, Onto or both.

(a) $f(x) = \cos x$ (b) $f(x) = 7x + 3$
(c) $f(x) = x^3 + 27$ (d) $f(x) = 3x^2 - 3x + 1$
(e) $f(x) = 3^x + 2$ (f) $f(x) = e^x - 4$
(g) $f(x) = \dfrac{2x+3}{2x-4}$

3. Let f and g be functions from I to I, where I is the set of positive integers. Find the compositions $(f_0 g)$ and $(g_0 f)$.

(a) $f(x) = 2x + 7$; $g(x) = \cos x$ (b) $f(x) = x^2 + 2$; $g(x) = 3^x + 5$
(c) $f(x) = \log x$; $g(x) = 5x + 2$ d) $f(x) = x + 4$; $g(x) = |x|$
(e) $f(x) = 2^x + 2$; $g(x) = x^2$

4. Let A = {a, b, c, d}, B = {1, 2, 3}, C = {4, 5, 6} and f : A → B defined by f = {(a, 1), (b, 1), (c, 2), (d, 2)}; g: B → C defined by g = {(1, 4), (2, 5), (3, 6)} be two functions. Find $(g_0 f)$. Is $(f_0 g)$ defined?

5. Let A = {1, 8, 27, 64}; B = {a, b, c, d, e}; C = {1, 8, 27, 64} and f : A → B defined by f = {(1, a), (8, d), (27, b), (64, e)}; g : B → C defined by g = {(a, 1), (b, 64), (c, 8), (d, 27), (e, 8)} be two functions. Find both $(f_0 g)$ and $(g_0 f)$. Show that $(f_0 g) \neq (g_0 f)$.

6. Let U = {a, e, I, o, u} be the universal set. Find the characteristic function for the set A = {e, o, u}.

7. Sketch the functions given below on R → R.

(a) $f(x) = [x]$; $-2 \leq x \leq 3$ (b) $f(x) = 3^x$
(c) $f(x) = 3x + 2$ (d) $f(x) = \begin{cases} x^2 \text{ if } x \geq 0 \\ 6 \text{ if } x < 0 \end{cases}$
(e) $f(x) = \begin{cases} 2x+1 & \text{if } 0 < x < 2 \\ -2 & \text{if } x \leq 0 \\ x+4 & \text{if } x \geq 2 \end{cases}$

8. Find the domain of each of the following functions.

(a) $f(x) = \dfrac{1}{(x-2)(x-3)}$ (b) $f(x) = \dfrac{1}{x^2 - 7x + 12}$
(c) $f(x) = x^2 - 7x + 12$ (d) $f(x) = x^2$; $0 \leq x \leq 2$
(e) $f(x) = \sqrt{36 - x^2}$

9. Let f : R → R defined by $f(x) = x^2 + x - 6$. Find $f^{-1}(14)$ and $f^{-1}(-8)$.

10. Draw the graph of following functions.

(a) $f(x) = x^3 - 3x + 2$ (b) $f(x) = x^4 - 10x^2 + 9$
(c) $f(x) = \dfrac{x}{2} + 1$

11. If $f(x) = 2x - 3$ and $g(x) = x^2 + 3x + 5$, find $(f_0 g)(5)$ and $(g_0 f)(5)$. Show that they are not equal.

12. If $f(x) = 2x - 3$ and $g(x) = x^2 + 2x + 1$. Find the compositions $(f_0 g)$ and $(g_0 f)$. Find $(f_0 g)(2)$ and $(g_0 f)(2)$.

13. If $f(x) = 5x + 1$. Find a formula for the composition function f^3. [**Hint :** $f^3 = (f_0 f_0 f)$]

14. Let A = $\{x, y, z\}$ and B = $\{2, 4, 6, 8\}$. State whether or not each diagram given below defines a function from A into B.

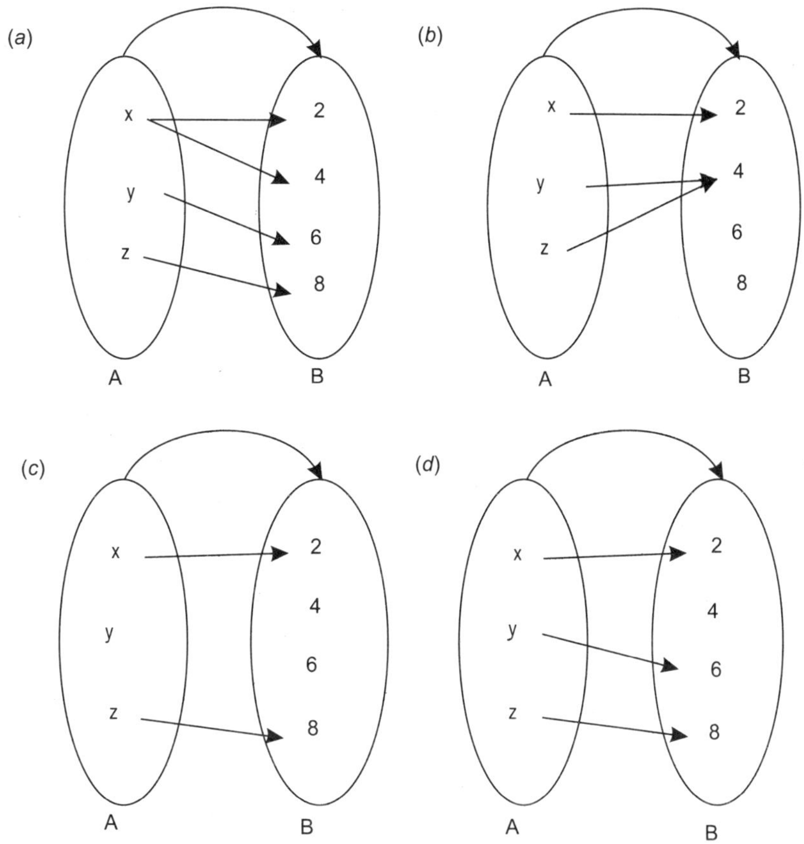

15. Let $f : \mathrm{R} \to \mathrm{R}$ defined by $f(x) = 2x + 5$. Show that $f(x)$ is invertible. Find the values of $f^{-1}(2)$, $f^{-1}(4)$ and $f^{-1}(5)$.

16. Let f and g be functions from the positive integers to the positive integers defined by $f(x) = 3x + 1$, and $g(x) = 2x + 1$. Find the compositions $(f_0 f)$, $(f_0 g)$, $(g_0 f)$ and $(g_0 g)$.

17. Let A = R – {2} and B = R – $\left\{\frac{3}{5}\right\}$, where R is the set of real numbers. Let $f : \mathrm{A} \to \mathrm{B}$ defined by $f(x) = \dfrac{3x - 9}{5x - 10}$; $x \in \mathrm{R}$. Show that f is bijective and hence find the inverse of f.

18. For each Hash function, show how the data would be inserted in the order given in initially empty cells. Use collision resolution policy if required.

(*a*) $h(n) = n \bmod 11$; cells indexed 0 to 10; data 55, 15, 285, 743, 375, 22, 10, 800.

(*b*) $h(n) = n \bmod 13$; cells indexed 0 to 12; data 714, 635, 26, 775, 42, 30, 10, 136, 509.

19. Show that the inverse of $f(x) = x^2 - 1$ does not exists in general, but $f: [0, \infty) \to [-1, \infty)$ has an inverse given by $f^{-1}(x) = \sqrt{x+1}$ and $f^{-1}: [-1, \infty) \to [0, \infty)$.

20. Let f be a function from $X \to X$; $X = \{1, 2, 3, 4, 5, 6\}$ defined by $f(x) = 3x \bmod 5$. Write the function and draw the arrow diagram.

21. Let $f: X \to X$; $X = \{0, 1, 2, 3, 4, 5, 6\}$ defined by $f(x) = 4x \bmod 5$. Write the function f as a set of order pairs. With the help of arrow diagram check whether or not f is one-to-one or onto.

22. Let $f: A \to B$ be a function. Show that f is injective if and only if $f^{-1}(f(X)) = X$ for all $X \subseteq A$.

23. Let $f: R - \{0\} \to R - \{0\}$ defined by $f(x) = \dfrac{1}{x}$. Show that f is bijective and its inverse is given by $f^{-1}(x) = \dfrac{1}{x}$.

24. Show that the function $f(x) = \dfrac{x}{x^2+1}: R \to R$ is neither one-one nor onto.

5

Generating Function and Recurrence Relation

5.0 INTRODUCTION

Function that we have discussed in Chapter 4 is a binary relation that assigns a unique value to each element in the domain. The class of functions whose domain is the set of natural numbers and range is the set of real numbers is known as numeric functions or discrete numeric functions. These numeric functions are highly useful in digital computation. In this chapter we discuss generating functions, recurrence relations, solutions to recurrence relations and its application to computer science.

5.1 GENERATING FUNCTIONS

An important idea in discrete mathematics is to establish a relation between two fields so as to apply knowledge in one field to the other field. This leads to the development of generating function. This generating function is an alternative way of representing the numeric functions. These are important tools in discrete mathematics to solve various types of counting problems and it is not limited to solve the recurrence relations. It was introduced by famous mathematician de Moivre.

The generating function for the sequence of real numbers $a_0, a_1, a_2, \ldots, a_n, \ldots$ is the expression G(x) which usually contains infinitely many non-zero terms such that

$$\begin{aligned} G(x) &= a_0 + a_1x + a_2x^2 + \ldots + a_nx^n + \ldots \\ &= \sum_{i=0}^{\infty} a_i x^i \end{aligned}$$

We call the above expression G(x), where x is an indeterminate, as generating function because in some sense it generates its coefficients. It deviates from a polynomial in the fact that the coefficients a_i are all zero after a certain point.

For example the generating function of the arithmetic sequence 4, 9, 14, 19, ... is given as

$$G(x) = 4 + 9x + 14x^2 + 19x^3 + \ldots$$

5.1.1 Properties of Generating Function

In this subsection we will discuss the important properties known as shifting properties of generating function.

1. Let $G(x) = a_0 + a_1x + a_2x^2 + ... + a_nx^n + ...$ be a generating function. Thus, $xG(x)=a_0x+a_1x^2+a_2x^3+a_3x^4...+a_nx^{n+1}+...$ It indicates that, if $G(x)$ generates the sequence , $a_0,a_1,a_2...,a_n,...$, then $xG(x)$ generates the sequence $0, a_0, a_1, a_2,..., a_n,...$. Similarly, $x^2G(x)$ generates the sequence $0, 0, a_0, a_1, a_2,..., a_n,...$. In general, $x^kG(x)$ generates the sequence, $0,0,...,0, a_0, a_1, a_2,..., a_n,...$, where there are k zeros before a_0.
2. Let $G(x) = a_0 + a_1x + a_2x^2 + ... + a_nx^n + ...$ be a generating function. Therefore, $G(x) - a_0 = a_1x + a_2x^2 + ... + a_nx^n + ...$. It indicates that, if $G(x)$ generates the sequence $a_0, a_1, a_2,..., a_n,...$, then $G(x) - a_0$ generates the sequence $0,a_1,a_2,...,a_n,...$. Similarly, we have $G(x) - a_0 - a_1x$ generates the sequence $0,0,a_2,...,a_n,...$. Proceeding in this manner we will get $G(x) - a_0 - a_1x - ... - a_{k-1}x^{k-1}$ generates the sequence $0,0,...,0,a_k,a_{k+1},a_{k+2},...$, where there are k zeros before a_k.
3. Let $G(x) = a_0 + a_1x + a_2x^2 + ... + a_nx^n + ...$ be a generating function. Thus, $\frac{G(x) - a_0}{x} = a_1 + a_2x + a_3x^2 + ... + a_nx^{n-1} + ...$. Therefore, dividing by powers of x shifts the sequence to the left. It implies that if $G(x)$ generates the sequence $a_0, a_1, a_2,..., a_n,...$, then $\frac{G(x) - a_0}{x}$ generates the sequence $a_1, a_2,..., a_n,...$. Similarly, we have $\frac{G(x) - a_0 - a_1x}{x^2}$ generates the sequence $a_2, a_3,..., a_n,...$. In general $\frac{G(x) - a_0 - a_1x - ... - a_{k-1}x^{k-1}}{x^k}$, generates the sequence $a_k, a_{k+1}, a_{k+2},...$ for $k \geq 1$.

5.1.2 Addition of Two Generating Functions

Two generating functions can be added term by term as in the case of polynomials. This operation addition provides us to create a new generating function from old ones. Let $G(x)$ and $F(x)$ be two generating functions such that

$$G(x) = a_0 + a_1x + a_2x^2 + ... + a_nx^n + ...$$
$$F(x) = b_0 + b_1x + b_2x^2 + ... + b_nx^n + ...$$

The addition of two generating functions $G(x)$ and $F(x)$ is defined as

$$G(x) + F(x) = (a_0 + b_0) + (a_1 + b_1)x + (a_2 + b_2)x^2 + ... + (a_n + b_n)x^n + ...$$
$$= \sum_{i=0}^{\infty} (a_i + b_i)x^i$$

For example, if $F(x) = 2 + 3x + 5x^2$ and $G(x) = 4 + 6x + 8x^2$, then

$$F(x) + G(x) = (2 + 4) + (3 + 6)x + (5 + 8)x^2 = 6 + 9x + 13x^2.$$

5.1.3 Multiplication of Two Generating Functions

Two generating functions can be multiplied term by term as in the case of polynomials. This operation multiplication provides us to create a new generating function from old ones. Let G(x) and F(x) be two generating functions such that

$$G(x) = a_0 + a_1x + a_2x^2 + \ldots + a_nx^n + \ldots$$
$$F(x) = b_0 + b_1x + b_2x^2 + \ldots + b_nx^n + \ldots$$

The multiplication of two generating functions G(x) and F(x) is defined as

$$\begin{aligned} G(x)\cdot F(x) &= a_0b_0 + (a_1b_0 + a_0b_1)x + (a_0b_2 + a_1b_1 + a_2b_0)x^2 \\ &\quad + (a_0b_3 + a_1b_2 + a_2b_1 + a_3b_0)x^3 + \cdots + \left(\sum_{i=0}^{k} a_ib_{k-i}\right)x^k + \ldots \\ &= \sum_{k=0}^{\infty}\left(\sum_{i=0}^{k} a_ib_{k-i}\right)x^k \end{aligned}$$

For example, if $F(x) = 2 + 3x + 5x^2$ and $G(x) = 4 + 6x + 8x^2$, then

$$F(x)G(x) = 8 + 24x + 54x^2 + 54x^3 + 40x^4.$$

5.1.4 The Exponential Generating Function

The basic idea of exponential generating function is developed from binomial expansion for positive integral index. The exponential generating function E(x) for the sequence of real numbers $a_0, a_1, a_2, \ldots, a_n, \ldots$ is given as

$$\begin{aligned} E(x) &= a_0 + a_1x + a_2\frac{x^2}{2!} + a_3\frac{x^3}{3!} + \cdots + a_n\frac{x^n}{n!} + \cdots \\ &= \sum_{k=0}^{\infty} a_k\frac{x^k}{k!} \end{aligned}$$

5.2 PARTITIONS OF INTEGERS

Partitioning of integers is very rich and quite deep. Covering every aspect of partitions is beyond the scope of this book. In this section, we provide an overview of partitions, and discuss a powerful tool generating function dealing with partitions.

A partition of a positive integer, n, in number theory is a way of representing n as a sum of positive integers where the order of the addends is immaterial. For an integer n, we denote the partition function P(n) as the number of partitions of n. For example P(4) = 5, since

$$\begin{aligned} 4 &= 4 \\ &= 3 + 1 \\ &= 2 + 2 \\ &= 2 + 1 + 1 \\ &= 1 + 1 + 1 + 1 \end{aligned}$$

Similarly, it can be shown that P(1) = 1, P(2) = 2, P(3) = 3, P(5) = 7 and P(6) = 11 etc. A recursive procedure to find the partition of a given integer n is given below.

5.2.1 Generating Function for Partitions

In number theory many methods are available to find the partitions of an integer. In this sub-section we will discuss how generating function is used to compute the value of P(n) for an integer n. On introducing generating function to compute P(n) can reduce the difficulty of complex problems. This is because generating function can be manipulated more easily than combinatorial quantities. The concept is based on power series and was first applied to partitions by Euler. This is because the coefficient a_n of x^n, in the power series $G(x) = \sum_{i=0}^{\infty} a_i x^i = a_0 + a_1 x + a_2 x^2 + \ldots + a_n x^n + \ldots$ represents the number of ways that the event n can happen.

In order to find a generating function for the number of partitions of a number n, our aim is to find out the number of ones, twos, threes and so on are there in the partition. In each partition, one can occur 0, 1, 2, ... times; thus contributing a factor of $(1 + x + x^2 + x^3 + \ldots)$ to the generating function G(x). Again, two can occur 0, 1, 2, ... times; thus contributing a factor of $(1 + x^2 + x^4 + x^6 + \ldots)$ to G(x). Similarly, three can occur 0, 1, 2, 3, ... times; thus contributing a factor of $(1 + x^3 + x^6 + x^9 + \ldots)$ to G(x). Proceeding in this manner, we find that the generating function for the number of partitions of an integer is:

$$
\begin{aligned}
G(x) &= (1 + x + x^2 + x^3 + \ldots)(1 + x^2 + x^4 + x^6 + \ldots)(1 + x^3 + x^6 + x^9 + \ldots)\ldots \\
&= \frac{1}{(1-x)}\,\frac{1}{(1-x^2)}\,\frac{1}{(1-x^3)}\,\frac{1}{(1-x^4)}\ldots;\ |x| < 1 \\
&= \prod_{k=1}^{\infty} \frac{1}{(1-x^k)};\ |x| < 1
\end{aligned}
$$

The above function G(x) generates the sequence P(0), P(1), P(2), P(3), ... , where we define P(0) = 1. It implies that, if we consider the generating function $G(x) = \prod_{k=1}^{m} \frac{1}{(1-x^k)}$ for some fixed m, then the coefficient of x^n is the number of partitions of n into summands that do not exceed m. Some values of the partition function are given below in the form of a table for reference.

n	1	2	3	4	5	6	7	8	9	10
P(n)	1	2	3	5	7	11	15	22	30	42

For example, to get the value of P(3), we have to compute the coefficient of x^3 in the generating function $G(x) = \prod_{k=1}^{3} \frac{1}{(1-x^k)}$. Thus we have

$$G(x) = (1+x+x^2+x^3+...)(1+x^2+x^4+...)(1+x^3+x^6+...)$$
$$= 1+x+2x^2+3x^3+4x^4+...$$

Therefore, P(3) = 3.

5.3 RECURRENCE RELATIONS

Recurrence relation begins with some very elementary relations that are intuitively effective. It provides a few methods for constructing more complicated relations based on simpler relations. For example, if we ask a person about the age of his oldest son Blake, then he could tell us, 17 years directly or he could tell us he is 5 years older than his second son Smith whereas Smith is 4 years older than his only sister Loreena. When he tells us that his only daughter is 8 years old then we have no difficulty to calculate the age of his oldest son Blake, who is 17 years old. Therefore, a recurrence relation expresses the nth element of a sequence in terms of its predecessors. In this section we discuss an introduction to recurrence relations that are highly useful in design and analysis of algorithms.

A recurrence relation for the sequence $s_0, s_1, s_2, \cdots, s_n$ is an equation that expresses s_n in terms of its predecessors $s_0, s_1, s_2, \cdots, s_{n-1}$ for all integer $n \geq m$, where m is a positive integer. Thus, it is clear that a recurrence relation is defined with certain conditions known as initial conditions or boundary conditions of recurrence relation. A recurrence relation is otherwise known as difference equation.

For example, the recurrence relation of the sequence <7, 12, 17, 22, ...> is

$$s_n = s_{n-1} + 5,\ n \geq 2;\ s_1 = 2$$

The condition $s_1 = 2$ is known as initial condition.

5.4 MODELS OF RECURRENCE RELATION

Recurrence relation can be used to model variety of real life problems both inside and outside computer science. In this section we will discuss few important real world problems that use recurrence relation.

5.4.1 Towers of Hanoi

The puzzle Towers of Hanoi is fashioned after the ancient Tower Brahma ritual as shown below. According to legend, at the time the world was created, there was a diamond tower (Tower A) with 64 golden disks. The disks were of decreasing in size from bottom to top. Apart from this, there were two other diamond towers (Tower B and C). Since the time of creating, Brahman priests have been trying to move the disks from tower A to tower C using tower B for intermediate storage. The disks are so heavy that they can be moved only one at a time and a larger disk cannot be placed over a smaller disk. According to legend, the world will come to an end when the priests have completed their task.

In this model our aim is to transfer the disks from one peg A to another peg C through intermediate peg B by moving one disk at a time. The only restriction is that a larger diameter peg can not be placed over a smaller diameter peg. The idea of the above model can be cleared from the following figure.

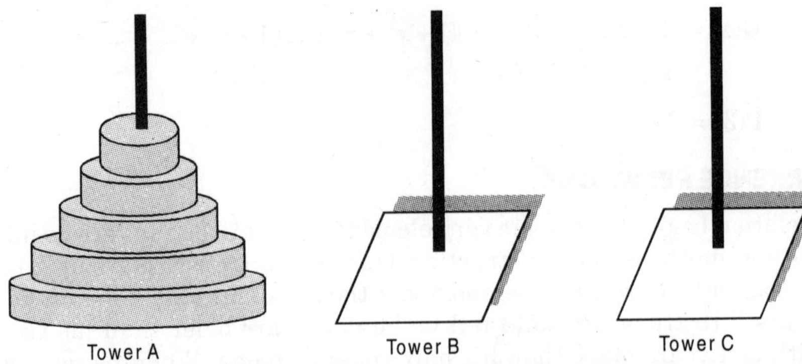

A very elegant solution results from the use of recursion. Let us consider the number of disks at tower A is n. Our aim is to get the larger disk at the bottom of tower C. In order to get this, we move the remaining $(n - 1)$ disks to tower B and then move the largest disk to tower C as shown in the figure given below. Now we have to move the disks from tower B to tower C. In order to get this we have tower A and C available. Therefore, our solution for an n disk problem leads to two $(n - 1)$ disk problems.

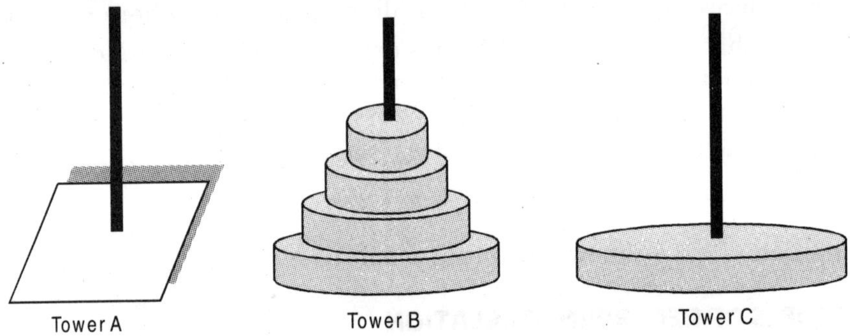

Let s_n denotes the number of moves required to solve the n disk puzzle. It is also clear that, if there is only one disk then we simply move it to the tower C. Therefore, we have the recurrence relation

$$s_n = 2s_{n-1} + 1,\ n > 1 \text{ with } s_1 = 1$$

We can use an iterative approach to solve this recurrence relation. Therefore, we get

$$\begin{aligned}
s_n &= 2s_{n-1} + 1 \\
&= 2\,(2s_{n-2} + 1) + 1 = 2^2 s_{n-2} + 2 + 1 \\
&= 2^2(2s_{n-3} + 1) + 2 + 1 = 2^3 s_{n-3} + 2^2 + 2 + 1 \\
&= 2^3(2s_{n-4} + 1) + 2^2 + 2 + 1 = 2^4 s_{n-4} + 2^3 + 2^2 + 2 + 1 \\
&= \cdots\ \cdots\ \cdots\ \cdots \\
&= \cdots\ \cdots\ \cdots\ \cdots \\
&= 2^{n-1} s_{n-(n-1)} + 2^{n-2} + \cdots + 2^3 + 2^2 + 2 + 1 \\
&= 2^{n-1} s_1 + 2^{n-2} + \cdots + 2^3 + 2^2 + 2 + 1 \\
&= 2^{n-1} + 2^{n-2} + \cdots + 2^3 + 2^2 + 2 + 1;\ [\because s_1 = 1] \\
&= 2^n - 1
\end{aligned}$$

From the above formula, the number of moves required to move 64 disks from tower A to tower C is equal to

$$s_{64} = 2^{64} - 1 = 18,\ 446,\ 744,\ 073,\ 709,\ 551,\ 615.$$

5.4.2 Rabbits and Fibonacci Numbers

A young pair of rabbits, one of each sex, is placed on an island. A pair of rabbits does not bread until they are two months old. After they are two months old, each pair produces another pair each month. Our aim is to find out number of pairs of rabbits on the island after n months, assuming that no rabbits ever die. This famous problem was posed by *Leonardo di pisa*, also known as Fibonacci.

A very elegant solution results from the use of recursion. First we consider tabulation and then we fit a recurrence relation. The tabulation of reproduction is given below.

Month (n)	*Reproduction Pairs*	*Young Pairs*	*Total Pairs* (s_n)
1	0	1	1
2	0	1	1
3	1	1	2 = 1 + 1
4	1	2	3 = 2 + 1
5	2	3	5 = 3 + 2
6	3	5	8 = 5 + 3

Let s_n denotes the number of pairs of rabbits after n months. From the above table it is clear that, at the end of the first month, the number of pairs of rabbits on the island is one *i.e.*, s_1=1. This is because this pair does not bread during the first two months. Therefore, s_2=1. In the third month the first pair produces one more pair of rabbits. So, the number of pairs is 2 *i.e.*, $s_3 = 2 = s_2 + s_1$. Therefore, the number of pairs of rabbits after n months is equal to the sum of members of rabbits in the previous month (s_{n-1}) and the number of new born pairs (s_{n-2}). Therefore, we get the recurrence relation as

$$s_n = s_{n-1} + s_{n-2} \quad \text{with}$$
$$s_1 = 1,\ s_2 = 1$$

The above recurrence relation is otherwise known as Fibonacci sequence.

5.4.3 Compound Interest

Another important real application of recurrence relation is to compute compound interest when principal, rate of interest and number of years are given. Suppose that Mr. Smith has deposited amount s_0 in a savings account at a bank. Our aim is to compute the total amount after n years if the rate of interest per year is r percent.

A very elegant solution results from the use of recurrence relation. Let s_n denote the amount in the account after n years. The amount in the account after n years is equal to the sum of amount in the account after $(n-1)$ years and the interest paid in the nth year. Therefore, we get

$$s_n = s_{n-1} + \left(\frac{r}{100}\right)s_{n-1} = \left(1+\frac{r}{100}\right)s_{n-1}$$

We use an iterative approach to solve this recurrence relation. Therefore,

$$\begin{aligned} s_n &= \left(1+\frac{r}{100}\right)s_{n-1} \\ &= \left(1+\frac{r}{100}\right)\left(1+\frac{r}{100}\right)s_{n-2} = \left(1+\frac{r}{100}\right)^2 s_{n-2} \\ &= \left(1+\frac{r}{100}\right)^2\left(1+\frac{r}{100}\right)s_{n-3} = \left(1+\frac{r}{100}\right)^3 s_{n-3} \\ &= \dots\dots\dots \\ &= \left(1+\frac{r}{100}\right)^n s_0 \end{aligned}$$

Hence, the amount in the account can be computed by using the above recurrence relation once the values s_0, r and n are known.

■ 5.5 LINEAR RECURRENCE RELATION WITH CONSTANT COEFFICIENTS

A linear recurrence relation with constant coefficients is of the form

$$c_0 s_n + c_1 s_{n-1} + c_2 s_{n-2} + c_3 s_{n-3} + \dots + c_k s_{n-k} = f(n)$$

where, c_i, $i = 0, 1, 2, \dots, k$ are constants. If both c_0 and c_k are non-zero, then the above equation is known as an kth order recurrence relation.

If $f(n) = 0$, then the above recurrence relation is termed as k^{th} order linear homogeneous recurrence relation with constant coefficients.

5.5.1 First Order Recurrence Relations

A first order linear homogeneous recurrence relation with constant coefficients is of the form

$$c_0 s_n = c_1 s_{n-1} \quad (n > 0) \text{ with } s_0 = \text{A}$$

where, c_0, c_1 are constants with the initial condition $s_0 = \text{A}$. The above equation can also be written as

$$s_n = r\ s_{n-1},\ (n > 0) \text{ with } s_0 = \text{A}$$

On applying recurrence repeatedly we get

$$s_n = r\ s_{n-1} = r^2 s_{n-2} = r^3 s_{n-3} = \dots\ \dots = r^n s_0 = \text{A}r^n$$

Therefore, the general solution to the recurrence relation is a geometric sequence with ratio r, *i.e.*, $s_n = Ar^n$

Similarly, a first order linear non-homogeneous recurrence relation with constant coefficients is given as

$$c_0 s_n = c_1 s_{n-1} + c_2 \quad (n > 0)$$

with $s_0 = \text{A}$

where, c_0, c_1, c_2 are constants with the initial condition $s_0 = \mathrm{A}$. The above equation can also be written as

$$s_n = r\ s_{n-1} + c_n \quad (n > 0) \text{ with } s_0 = \mathrm{A}$$

On applying recurrence repeatedly, we get

$$\begin{aligned} s_n &= r\ s_{n-1} + c_n = r^2 s_{n-2} + rc_{n-1} + c_n \\ &= \dots\ \dots \\ &= r^n s_0 + \sum_{i=1}^{n} r^{n-i}\ c_i = \mathrm{A}r^n + \sum_{i=1}^{n} r^{n-i}\ c_i \end{aligned}$$

If c_n is constant, (say k) then the solution is given as

$$\begin{aligned} s_n &= \mathrm{A}r^n + k\frac{r^n - 1}{r - 1} && \text{if } (r \neq 1) \\ &= \mathrm{A} + kn && \text{if } (r = 1) \end{aligned}$$

5.6 DIFFERENT METHODS OF SOLUTION

In this section we discuss the basic fundamental methods such as backtracking, forward chaining and summation method to solve first order linear recurrence relation with constant coefficients.

5.6.1 Backtracking Method

The fundamental technique to solve a recurrence relation is backtracking. In this method, to solve a recurrence relation we start with s_n and move backward towards s_1 to get a pattern by substituting s_{n-1}, s_{n-2}, ... and so on. For example, consider the recurrence relation

$$s_n = s_{n-1} + 2 \text{ with } s_1 = 3.$$

Thus, we have

$$\begin{aligned} s_n &= s_{n-1} + 2 \\ &= (s_{n-2} + 2) + 2 = s_{n-2} + 2 \times 2 \\ &= (s_{n-3} + 2) + 2 \times 2 = s_{n-3} + 3 \times 2 \\ &= (s_{n-4} + 2) + 3 \times 2 = s_{n-4} + 4 \times 2 \\ &= \dots\ \dots\ \dots \\ &= s_{n-(n-1)} + (n-1)\ 2 \\ &= s_1 + 2(n-1) \\ &= 3 + 2(n-1) \end{aligned}$$

Therefore, $s_n = 3 + 2(n-1)$

5.6.2 Forward Chaining Method

Another basic technique to solve recurrence relation is forward chaining. In this method, we start from initial condition and move forward towards s_n until we get a clear pattern. For example, consider the recurrence relation

$$s_n = s_{n-1} + 2 \text{ with } s_1 = 3.$$

Thus, we have

$$s_2 = s_1 + 2$$
$$s_3 = s_2 + 2 = (s_1 + 2) + 2 = s_1 + 2 \times 2$$
$$s_4 = s_3 + 2 = (s_1 + 2 \times 2) + 2 = s_1 + 2 \times 3$$
$$s_5 = s_4 + 2 = (s_1 + 2 \times 3) + 2 = s_1 + 2 \times 4$$
$$s_6 = s_5 + 2 = (s_1 + 2 \times 4) + 2 = s_1 + 2 \times 5$$
$$\ldots \ldots \ldots \ldots \ldots \ldots \ldots \ldots \ldots \ldots$$
$$s_n = s_{n-1} + 2 = (s_1 + 2(n-2)) + 2 = s_1 + 2(n-1)$$

Therefore, $s_n = 3 + 2(n-1)$

5.6.3 Summation Method

Another way of solving first order linear recurrence relation with constant coefficient is summation method. In this method, we rearrange the given recurrence relation in the form $s_n - ks_{n-1} = f(n)$ and then backtrack till the initial condition. As a result we get a number of equations and then add these equations in such a manner that all intermediate terms get cancelled. Finally, we get the required solution to the recurrence relation. For example, consider the first order recurrence relation

$$s_n = s_{n-1} + 2 \text{ with } s_1 = 3.$$

Thus, we get the following set of equations.

$$s_n - s_{n-1} = 2$$
$$s_{n-1} - s_{n-2} = 2$$
$$s_{n-2} - s_{n-3} = 2$$
$$s_{n-3} - s_{n-4} = 2$$
$$\ldots \ldots \ldots \ldots$$
$$s_2 - s_1 = 2$$

On adding these $(n - 1)$ equations we get $s_n - s_1 = 2(n-1)$.

Therefore, the solution to the recurrence relation is $s_n = 3 + 2(n-1)$.

■ 5.7 HOMOGENEOUS SOLUTIONS

The total solution of a linear recurrence relation with constant coefficients is the sum of homogeneous solution and particular solution. In this section, we discuss how to get the homogeneous solution of a linear difference equation with constant coefficients. The homogeneous solution to the linear difference equation with constant coefficients is obtained by making $f(n) = 0$. Consider the homogeneous linear difference equation

$$c_0 s_n + c_1 s_{n-1} + c_2 s_{n-2} + c_3 s_{n-3} + \ldots + c_k s_{n-k} = 0$$

A homogeneous solution of a linear difference equation with constant coefficients is of the form Ar_1^n, where r_1 is called a characteristic root and A is a constant obtained by the boundary conditions. On substituting $s_n = Ar^n$ in the above difference equation, we get

$$c_0Ar^n + c_1Ar^{n-1} + c_2Ar^{n-2} + c_3Ar^{n-3} + \ldots + c_kAr^{n-k} = 0$$

i.e., $$Ar^n[c_0 + c_1r^{-1} + c_2r^{-2} + c_3r^{-3} + \ldots + c_kr^{-k}] = 0$$

i.e., $$[c_0 + c_1r^{-1} + c_2r^{-2} + c_3r^{-3} + \ldots + c_kr^{-k}] = 0; \qquad [\because Ar^n \neq 0]$$

i.e., $$[c_0r^k + c_1r^{k-1} + c_2r^{k-2} + c_3r^{k-3} + \ldots + c_{k-1}r + c_k] = 0$$

The above equation is known as characteristic equation of the difference equation. Therefore, if r_1 is a root of the characteristic equation, Ar_1^n is a homogeneous solution to the difference equation. The above characteristic equation is of kth degree and thus it has k characteristic roots. So, there arise two cases, distinct roots and multiple roots.

Case – 1 If the roots of the characteristic equation are distinct, then the homogeneous solution to the difference equation is given as,

$$s_n^h = A_1r_1^n + A_2r_2^n + A_3r_3^n + \ldots + A_kr_k^n$$

where, $r_1, r_2, \cdots, r_k$ are distinct characteristic roots and the constants $A_1, A_2, \cdots, A_k$ are to be determined by the given boundary or initial conditions.

Case – 2 If some of the roots of the characteristic equation are multiple roots, then the homogeneous solution to the difference equation is given as,

$$s_n^h = (A_1 + A_2n + A_3n^2 + \ldots + A_mn^{m-1})r_1^n + B_1r_2^n + B_2r_3^n + \ldots + B_{k-m}r_{k-m+1}^n$$

where, r_1 be a root of multiplicity m and rest of the roots $r_2, r_3, \ldots, r_{k-m+1}$ are distinct. The constants $A_1, A_2, \ldots, A_m, B_1, B_2, \ldots, B_{k-m}$ are to be determined by the boundary conditions.

Consider the recurrence relation for the Fibonacci sequence

$$s_n = s_{n-1} + s_{n-2}$$

i.e., $$s_n - s_{n-1} - s_{n-2} = 0, \; n \geq 2; \; s_0 = 1, \; s_1 = 1$$

The characteristic equation is given as

$$r^2 - r - 1 = 0$$

On solving the above equation, we get the characteristic roots r_1 and r_2 as

$$r_1 = \frac{1+\sqrt{5}}{2} \quad \text{and} \quad r_2 = \frac{1-\sqrt{5}}{2}$$

They are distinct real roots and the ratio $\frac{1+\sqrt{5}}{2}$ is known as the Golden ratio. Thus, the general solution to the recurrence relation is given as

$$s_n = A_1\left(\frac{1+\sqrt{5}}{2}\right)^n + A_2\left(\frac{1-\sqrt{5}}{2}\right)^n$$

Using the boundary conditions we get the following equations.

$$A_1 + A_2 = 1 \text{ and } A_1 - A_2 = \frac{1}{\sqrt{5}}$$

On solving the above equations we get

$$A_1 = \frac{1+\sqrt{5}}{2\sqrt{5}} \text{ and } A_2 = -\frac{1-\sqrt{5}}{2\sqrt{5}}$$

Therefore, the general solution to the recurrence relation is given as

$$s_n = \left(\frac{1+\sqrt{5}}{2\sqrt{5}}\right)\left(\frac{1+\sqrt{5}}{2}\right)^n - \left(\frac{1-\sqrt{5}}{2\sqrt{5}}\right)\left(\frac{1-\sqrt{5}}{2}\right)^n$$

5.8 PARTICULAR SOLUTION

The solution that satisfies the difference equation with $f(n)$ on the right hand side of the difference equation is called as particular solution. There is no general procedure for determining the particular solution of a difference equation for every function $f(n)$. However, in simple cases this solution can be determined by the method of inspection called as trial sequence method. We consider a trial function depending on certain forms of $f(n)$. The unknown constants associated with the trial function are to be determined by substituting the trial function in the recurrence relation. We present a table for trial function based on $f(n)$ to get the particular solution of the difference equation.

$f(n)$	*Trial Function*
k, where k is a constant	P (Constant)
a^n, where a is a constant but not a characteristic root	Pa^n
a^n, where a is a constant but a characteristic root of multiplicity m.	$Pn^m a^n$
$a_0 n^m + a_1 n^{m-1} + a_2 n^{m-2} + \cdots + a_m$ $Ex: a_0 n^3 + a_1 n + a_2$	$P_0 n^m + P_1 n^{m-1} + \cdots + P_m$ $Ex: P_0 n^3 + P_1 n^2 + P_2 n + P_3$
$a^n(a_0 n^m + a_1 n^{m-1} + a_2 n^{m-2} + \cdots + a_m)$ $Ex: bna^n$	$a^n(P_0 n^m + P_1 n^{m-1} + \cdots + P_m)$ $Ex: a^n(P_0 n + P_1)$
$c^n(a_0 n^m + a_1 n^{m-1} + a_2 n^{m-2} + \cdots + a_m)$ where c is a characteristic root of multiplicity m.	$c^n n^m(P_0 n^m + P_1 n^{m-1} + \cdots + P_m)$
$a^n + b^n + c$, where a, b and c are constants.	$P_1 a^n + P_2 b^n + P_3$

5.9 TOTAL SOLUTION

The discrete numeric function that is the sum of homogeneous and particular solution, is known as total solution. Let s_n^h be the homogeneous solution to the difference equation whereas s_n^p is the particular solution. Therefore, we get

$$c_0 s_n^h + c_1 s_{n-1}^h + c_2 s_{n-2}^h + c_3 s_{n-3}^h + \cdots + c_k s_{n-k}^h = 0 \text{ and}$$

$$c_0 s_n^p + c_1 s_{n-1}^p + c_2 s_{n-2}^p + c_3 s_{n-3}^p + \cdots + c_k s_{n-k}^p = f(n)$$

From the above equations on adding we get

$$c_0 (s_n^h + s_n^p) + c_1 (s_{n-1}^h + s_{n-1}^p) + \cdots + c_k (s_{n-k}^h + s_{n-k}^p) = f(n)$$

Thus the total solution to the difference equation is $s_n = s_n^h + s_n^p$. It is observed that, the particular solution alone will not, in general, satisfy the boundary conditions, however, we can adjust the homogeneous solution so that the total solution satisfies the difference equation and boundary conditions.

Consider the difference equation, $s_n - 5s_{n-1} + 6s_{n-2} = 4$, $n \geq 2$ ...(*i*)

with $s_0 = 1$, $s_1 = 3$. The characteristic equation is given as

$$r^2 - 5r + 6 = 0 \quad i.e., \; r = 2, 3$$

The homogeneous solution is given as

$$s_n^h = A_1 (2)^n + A_2 (3)^n$$

Let the particular solution be $s_n^p = P$. Therefore, from equation (*i*), we get

$$P - 5P + 6P = 4; \qquad i.e., \qquad P = 2$$

Thus, the total solution to the difference equation is given as

$$s_n = s_n^h + s_n^p = A_1 (2)^n + A_2 (3)^n + 2$$

Using initial conditions we get

$$\begin{cases} A_1 + A_2 = -1 & for \;\; n = 0 \\ 2A_1 + 3A_2 = 1 & for \;\; n = 1 \end{cases}$$

On solving these equations we get $A_1 = -4$ and $A_2 = 3$. Therefore, the total solution to the difference equation is given as

$$s_n = -4(2)^n + 3(3)^n + 2 - 3^{n+1} - 2^{n+2} + 2$$

5.10 SOLUTION BY GENERATING FUNCTION

We can find a closed form for a linear recurrence relation by using generating function. Once the generating function is strong-minded, an expression for the value of the numeric function can be obtained. We illustrate the technique by the following example.

Consider the recurrence relation

$$s_n + 3s_{n-1} - 4s_{n-2} = 0,\ n \geq 2$$

$$s_0 = 3,\ s_1 = -2 \qquad \ldots(i)$$

Multiplying both sides of equation (*i*) by x^n, we get

$$s_n x^n + 3s_{n-1}x^n - 4s_{n-2}x^n = 0 \qquad \ldots(ii)$$

Summing equation (*ii*) for all n, $n \geq 2$, we get

$$\sum_{n=2}^{\infty} s_n x^n + \sum_{n=2}^{\infty} 3s_{n-1}x^n - \sum_{n=2}^{\infty} 4s_{n-2}x^n = 0 \qquad \ldots(iii)$$

Note that $G(x) = \sum_{n=0}^{\infty} s_n x^n$. Therefore, equation (*iii*) reduces to

$$(G(x) - s_0 - s_1 x) + 3x\sum_{n=2}^{\infty} s_{n-1}x^{n-1} - 4x^2 \sum_{n=2}^{\infty} s_{n-2}x^{n-2} = 0$$

i.e., $$(G(x) - s_0 - s_1 x) + 3x(G(x) - s_0) - 4x^2 G(x) = 0$$

i.e., $$G(x)(1 + 3x - 4x^2) = s_0 + s_1 x + 3s_0 x = 3 + 7x$$

$$G(x) = \frac{3+7x}{1+3x-4x^2} = \frac{3+7x}{(1-x)(1+4x)} = \frac{2}{1-x} + \frac{1}{1+4x}$$

$$= 2(1 + x + x^2 + \ldots) + (1 - 4x + (4x)^2 - \ldots)$$

$$= 2\sum_{n=0}^{\infty} x^n + \sum_{n=0}^{\infty} (-4x)^n$$

On equating the coefficient of x^n both sides, we get

$$s_n = 2 + (-4)^n,\ n \geq 0$$

■ 5.11 ANALYSIS OF THE ALGORITHMS

Recurrence relation is quite useful in computer science. We use recurrence relation to analyze the time complexity of several algorithms. The technique is to develop a recurrence relation and initial conditions that define a sequence $s_1, s_2, s_3, s_4, \ldots$, where s_n is the time required for an algorithm to execute an input of size n. We can determine the time required for an algorithm by solving the recurrence relation. In this section, we discuss selection sort, binary search, insertion sort and Strassen's matrix multiplication. Once, the idea is clear more complicated algorithms can be solved.

5.11.1 Selection Sort

This algorithm sorts the given sequence in increasing order. This technique finds the largest item and places it at the last position. After placing the largest item at the last position, the algorithm sorts the remaining elements recursively. Let us consider the input sequence of length n as $a_1, a_2, a_3, a_4, \ldots, a_n$. Our aim is to get the sequence in increasing order. Now, we present the algorithm.

Algorithm

```
1. selection_sort (a, n)
2. if (n = 1) then
3.        return
4. else
5.        max_index = 1
6.        for i = 1 to n do
7.                if (a_i > a_max_index) then
8.                        max_index = i
9.        swap(a_n, a_max_index)
10. selection_sort (a, n – 1)
```

In order to compute the time complexity, we count the number of comparisons s_n at line 7 required to sort n items. Here, immediately we obtain the initial condition $s_1 = 0$. In order to obtain the recurrence relation, we count the number of comparisons at each line and then add these numbers to get the total number of comparisons s_n. From the algorithm it is clear that the number of comparisons at lines 1 – 6 is zero. Again, we have $(n – 1)$ comparisons at line 7. Similarly, there is no comparison at lines 8 – 9. Line 10 again repeats the same algorithm with input size $(n – 1)$. But this algorithm requires s_{n-1} comparisons for input size $(n – 1)$. Therefore, the total number of comparisons is given as

$$s_n = s_{n-1} + (n-1) \text{ with } s_1 = 0.$$

This is the required recurrence relation. On solving this by iteration, we get

$$\begin{aligned} s_n &= s_{n-1} + (n-1) \\ &= s_{n-2} + (n-2) + (n-1) \\ &= s_{n-3} + (n-3) + (n-2) + (n-1) \\ &= \cdots \quad \cdots \quad \cdots \\ &= s_1 + 1 + \cdots + (n-3) + (n-2) + (n-1) \\ &= 1 + \cdots + (n-3) + (n-2) + (n-1) \\ &= \frac{(n-1)n}{2} = \theta(n^2) \end{aligned}$$

It is to be noted that the best-case, worst-case and average-case times are all the same for this algorithm.

5.11.2 Insertion Sort

This algorithm sorts a given sequence $a_1, a_2, a_3, a_4, \ldots, a_n$ in increasing order by recursively sorting the first $(n – 1)$ elements and then inserting n^{th} element in the correct position. So, the algorithm takes different amount of time to sort two input sequence of the same size. It is noted that, the algorithm works very firstly on less elements. Here, we present an algorithm for input sequence of length n.

Algorithm

```
1. Insertion_sort (a, n)
2. if (n = 1) then
3.         return
4. Insertion_sort (a, n − 1)
5.         i = n − 1
6.         temp = a_n
7.         while (i ≥ 1) and (a_i > temp) do
8.                 begin
9.                         a_{i+1} = a_i
10.                        i = i − 1
11.                end
12.        a_{i+1} = temp
13. end Insertion_sort
```

In order to compute the time complexity, we count the number of comparisons s_n at line 7 required to sort n elements. Here, directly we obtain the initial condition $s_1 = 0$. In order to obtain the recurrence relation, we count the number of comparisons at each line and then add these numbers to get the total number of comparisons s_n. From the algorithm it is clear that the number of comparisons at lines 1 – 3 is zero. Again, line 4 repeats the same algorithm with input size $(n – 1)$. But this algorithm requires s_{n-1} comparisons for input size $(n – 1)$. Similarly, there is no comparison at lines 5 – 6. The total number of comparisons at line 7 is either zero or maximum of n that depends on the input elements. Again there is no comparison at lines 8 –13. The worst-case occurs only when the elements in the sequence are given in reverse order, *i.e.*, in decreasing order. So, the total number of comparisons in worst-case is given as

$$s_n = s_{n-1} + n \text{ with } s_1 = 0$$

This is the required recurrence relation. On solving this by iteration we get

$$\begin{aligned} s_n &= s_{n-1} + n \\ &= s_{n-2} + (n-1) + n \\ &= \cdots \;\cdots \;\cdots \;\cdots \;\cdots \;\cdots \\ &= s_1 + 2 + \;\cdots\; + (n-2) + (n-1) + n \\ &= 2 + \;\cdots\; + (n-2) + (n-1) + n \\ &= \frac{n(n+1)}{2} - 1 = \theta(n^2) \end{aligned}$$

Similarly, the best-case occurs if the elements are already sorted. In this case the body of while loop at line 7 is never entered. This is because at each time when i has its initial value of $(n – 1)$ we find that $a_i \leq temp$ at line 7. Thus, $s_n = 1$ for $n = 2, 3, \ldots, n$ at line 4. Therefore, the best-case running time is given as $\theta(n)$.

5.11.3 Strassen's Matrix Multiplication

In this section, we discuss a new technique of matrix multiplication called Strassen's matrix multiplication. Let A and B be two $(r \times r)$ matrices. Our aim is to obtain the product matrix

AB. Let the product matrix be P. Therefore, P is also a matrix of order $(r \times r)$. Here we adopt a new strategy known as divide-and-conquer to get the product matrix P. For simplicity, we assume that r is a power of 2, *i.e.*, $r = 2^n$. If r is not a power of 2, then find a least suitable integer m greater than r and is a power of 2. Now, change both the input matrices A and B to be of order $(m \times m)$ by introducing $(m - r)$ number of rows and columns with entities as zero.

Algorithm

1. Partition both the given matrices A and B into $\left(\frac{r}{2} \times \frac{r}{2}\right)$ submatrices. Therefore, we get

$$A = \begin{bmatrix} A_{11} & \vdots & A_{12} \\ \cdots\cdots & \vdots & \cdots\cdots \\ A_{21} & \vdots & A_{22} \end{bmatrix} \text{ and } B = \begin{bmatrix} B_{11} & \vdots & B_{12} \\ \cdots\cdots & \vdots & \cdots\cdots \\ B_{21} & \vdots & B_{22} \end{bmatrix}$$

where $A_{11}, A_{12}, A_{21}, A_{22}, B_{11}, B_{12}, B_{21}, B_{22}$ are all $\left(\frac{r}{2} \times \frac{r}{2}\right)$ submatrices.

2. Compute fourteen matrices $A_1, A_2, \ldots, A_7, B_1, B_2, \ldots, B_7$ of order $\left(\frac{r}{2} \times \frac{r}{2}\right) r$ as below.

$$\begin{aligned} A_1 &= (A_{11} + A_{22}); & B_1 &= (B_{11} + B_{22}); \\ A_2 &= (A_{12} - A_{22}); & B_2 &= (B_{21} + B_{22}); \\ A_3 &= (A_{11} - A_{21}); & B_3 &= (B_{11} + B_{12}); \\ A_4 &= (A_{11} + A_{12}); & B_4 &= B_{22}; \\ A_5 &= (A_{21} + A_{22}); & B_5 &= B_{11}; \\ A_6 &= A_{11}; & B_6 &= (B_{12} - B_{22}); \\ A_7 &= A_{22}; & B_7 &= (B_{21} - B_{11}); \end{aligned}$$

3. Compute seven matrices $M_1, M_2, \ldots, M_7$ of order $\left(\frac{r}{2} \times \frac{r}{2}\right)$ such that

$$M_i = A_i B_i,\ i = 1, 2, \ldots, 7$$

4. The product AB can then be expressed as

$$P = AB = \begin{bmatrix} P_{11} & \vdots & P_{12} \\ \cdots\cdots & \vdots & \cdots\cdots \\ P_{21} & \vdots & P_{22} \end{bmatrix}$$

where $P_{11}, P_{12}, P_{21}, P_{22}$ are all $\left(\frac{r}{2} \times \frac{r}{2}\right)$ submatrices to be computed such that

$$\begin{aligned} P_{11} &= M_1 + M_2 - M_4 + M_7 & P_{21} &= M_5 + M_7 \\ P_{12} &= M_4 + M_6 & P_{22} &= M_1 - M_3 - M_5 + M_6 \end{aligned}$$

A quick count indicates that such an algorithm requires 7 multiplication operations and 18 addition operations of order $\left(\frac{r}{2} \times \frac{r}{2}\right)$. Let s_n denote the total number of arithmetic operations required to multiply two $(r \times r) = (2^n \times 2^n)$ matrices. Therefore, we have

$$s_n = 7s_{n-1} + 18 \times 2^{2n-2} \text{ with the initial condition } s_0 = 1$$

This is the required recurrence relation. The above recurrence relation can be written as

$$s_n - 7s_{n-1} = 18 \times 4^{n-1}, \quad s_0 = 1 \qquad \text{...}(i)$$

The characteristic equation is given as

$$r - 7 = 0 \qquad \textit{i.e.,}\ r = 7$$

The homogeneous solution is given as

$$s_n^h = A_1(7)^n$$

Let the particular solution be $s_n^p = P4^n$. Thus, from equation (i) we get

$$P4^n - 7P4^{n-1} = 18 \cdot 4^{n-1} \quad \textit{i.e.,}\ 4^n\left(P - \frac{7P}{4}\right) = 4^n\left(\frac{18}{4}\right)$$

On equating the coefficient of 4^n both sides we get P = 6. Therefore, the total solution to the difference equation is given as

$$s_n = s_n^h + s_n^p = A_1(7)^n + 6 \cdot 4^n$$

On using the initial condition $s_0 = 1$, we get $A_1 = 7$. Therefore, the total solution to the recurrence relation is given as

$$s_n = 7 \cdot 7^n + 6 \cdot 4^n = 7 \cdot 7^{\lg r} + 6 \cdot 4^{\lg r}; \qquad [\because n = \lg r]$$

$$= 7 \cdot r^{\lg 7} + 6 \cdot r^{\lg 4} = 7 \cdot r^{2 \cdot 81} + 6 \cdot r^2$$

Thus, the computing time of the algorithm presented above is $\theta(r^{2 \cdot 81})$. So, it takes less computing time than the ordinary matrix multiplication.

5.11.4 Binary Search

Let $a_i, a_{i+1}, a_{i+2}, \ldots, a_j$ be a list of elements that are sorted in increasing order. Consider a problem of determining whether a given element x is present in the list. If the element x is present in the list, then our aim is to determine the index 'j' such that $a_j = x$. If the element is not present in the list, then 'j' is to be set to zero. Let $P = (n, a_i, \ldots \ldots, a_j, x)$ denotes an arbitrary instance of this search problem, where n is the number of elements in the list $a_1, a_{i+1}, a_{i+2}, \ldots, a_j$, and x is the element to be searched. Here we present a recursive algorithm of this binary search.

Algorithm

```
Binary_search (a, i, j, x)
if (i > j) then
        return (0)
else
        k = ⌊(i + j)/2⌋
        if (x = a_k) then
```

```
7.      return (k)
8.           if (x < a_k)
9.                j = k − 1
10.          else
11.          j = k + 1
12.     Binary_search(a, i, j, x)
13. end Binary_search
```

Now, we have to compute the time complexity in the worst case. Let the total number of elements in the list be n. Denote s_n as the worst-case time. If $n = 1$, then the sequence contains only one element a_i and thus $i = j$. In worst-case, the element x will not be found at line 6, so the algorithm will be invoked a second time at line 12. However, in the second call of binary search we will have $i > j$ and the algorithm will terminate at line 3 by returning value 0. Therefore, it is clear that the algorithm invoked twice when $n = 1$. Therefore, the initial condition is given as

$$s_1 = 2.$$

If $n > 1$, then definitely $i < j$. Thus, the condition at line 2 is false. In the worst-case, the element will not be found at line 6 and thus the algorithm will be invoked at line 12. So, the time required by the invocation at line 12 is s_m, where m is the size of the sequence that is input at line 12. Thus, the sizes of the left and right sides of the original sequence are $\left\lfloor \frac{n-1}{2} \right\rfloor$ and $\left\lfloor \frac{n}{2} \right\rfloor$ respectively. As the worst-case occurs with larger sequence, so we get $s_m = s_{\lfloor \frac{n}{2} \rfloor}$.

Therefore, the recurrence relation to the binary search is given as

$$s_n = s_{\lfloor \frac{n}{2} \rfloor} + 1 \text{ with } s_1 = 2 \qquad \ldots(i)$$

Such type of difference equations are difficult to solve, however we solve this when n is a power of 2. Let us take $n = 2^k$. Therefore, above difference equation (i) reduces to

$$s_{2^k} = s_{2^{k-1}} + 1 \qquad \ldots(ii)$$

with $\qquad s_1 = 2$

Let us take $b_k = s_{2^k}$. Therefore, $s_1 = s_{2^0} = b_0$. So, the above equation (ii) with initial condition reduces to

$$b_k = b_{k-1} + 1 \text{ with } b_0 = 2$$

On solving by iterative method, we have

$$\begin{aligned} b_k &= b_{k-1} + 1 \\ &= b_{k-2} + 1 + 1 = b_{k-2} + 2 \times 1 \\ &= b_{k-3} + 1 + 2 \times 1 = b_{k-3} + 3 \times 1 \\ &= \cdots \; \cdots \; \cdots \; \cdots \\ &= b_{k-k} + k \times 1 = b_0 + k = 2 + k \end{aligned}$$

Therefore, we get $b_k = 2 + k$ ...(*iii*)

On substituting $n = 2^k$, we get $\lg n = k \lg 2 = k$. Therefore, the above equation (*iii*) reduces to

$$b_k = 2 + \lg\ n$$

i.e.,
$$s_n = 2 + \lg\ n$$

If n is not a power of 2, then n lies between two powers of 2. So, we get

$$2^{k-1} \le n \le 2^k$$

Since the given sequence s is increasing, we get

$$s_{2^{k-1}} \le s_n \le s_{2^k}$$

i.e.,
$$2 + (k-1) \le 2 + \lg n \le 2 + k \qquad [\because s_{2^k} = b_k = 2 + k]$$

Therefore, on combining these in-equations we get

$$\lg n < k + 1 = s_{2^{k-1}} \le s_n \le s_{2^k} = 2 + k < 3 + \lg n$$

Thus, $s_n = \theta(\lg n)$. This is the worst-case running time of binary search for input of size n.

SOLVED EXAMPLES

Example 1 *Find the generating function in closed form for the sequence of real numbers 0, 0, 0, 1, 1, 1, 1, ...*

Solution: We know that $1 + x + x^2 + x^3 + \cdots + x^n + \cdots = \dfrac{1}{1-x}$.

Therefore, the generating function of 1, 1, 1, ... is $\dfrac{1}{1-x}$. It implies that

$$\frac{x^3}{1-x} = x^3 + x^4 + x^5 + \cdots + x^{n+3} + \ldots.$$

Hence, the generating function for the sequence of real numbers 0, 0, 0, 1, 1, 1, ...is $\dfrac{x^3}{1-x}$.

Example 2 *Find the generating function in closed form for the sequence of real numbers 2, 2, 0, 2, 2, ...*

Solution: We know that $1 + x + x^2 + x^3 + \ldots + x^n + \ldots = \dfrac{1}{1-x}$.

Therefore,
$$2 + 2x + 2x^2 + 2x^3 + \ldots + 2x^n + \ldots = \frac{2}{1-x}.$$

It implies that $\dfrac{2}{1-x} - 2x^2 = 2 + 2x + 2x^3 + \ldots + 2x^n + \ldots = \displaystyle\sum_{\substack{k=0 \\ k \ne 2}}^{\infty} 2x^k$. So, the generating function of

2, 2, 0, 2, 2, ... is $\dfrac{2}{1-x} - 2x^2$.

Example 3 *Solve the recurrence relation* $s_n - 7s_{n-1} + 10s_{n-2} = 0,\ n \ge 2$. *It is given that the initial conditions are* $s_0 = 0$ *and* $s_1 = 3$.

Solution: Given that the recurrence relation $s_n - 7s_{n-1} + 10s_{n-2} = 0$

The characteristic equation is given as

$$r^2 - 7r + 10 = 0$$

On solving the above equation, we get the characteristic roots

$$r = 2 \text{ and } r = 5$$

Therefore, the solution to the recurrence relation is given as

$$s_n = A_1(2)^n + A_2(5)^n \qquad ...(i)$$

On imposing the initial conditions, we get

$$\begin{cases} A_1 + A_2 = 0 & [\because s_0 = 0] \\ 2A_1 + 5A_2 = 3 & [\because s_1 = 3] \end{cases}$$

On solving the above equations we get $A_1 = -1$ and $A_2 = 1$. Therefore, from equation (*i*), we get

$$s_n = -1(2)^n + 5^n.$$

Example 4 *Solve the recurrence relation* $s_n - 7s_{n-1} + 10s_{n-2} = 0$, $s_0 = 0$ *and* $s_1 = 3$ *by using generating function where,* $n \geq 2$.

Solution: Consider the recurrence relation

$$\begin{aligned} &s_n - 7s_{n-1} + 10s_{n-2} = 0, \ n \geq 2 \\ &s_0 = 0, \ s_1 = 3 \end{aligned} \qquad ...(i)$$

Multiplying both sides of equation (*i*) by x^n, we get

$$s_n x^n - 7s_{n-1}x^n + 10s_{n-2}x^n = 0 \qquad ...(ii)$$

Summing equation (*ii*) for all n, $n \geq 2$, we get

$$\sum_{n=2}^{\infty} s_n x^n - \sum_{n=2}^{\infty} 7s_{n-1}x^n + \sum_{n=2}^{\infty} 10s_{n-2}x^n = 0 \qquad ...(iii)$$

Note that $G(x) = \sum_{n \geq 0} s_n x^n$. Therefore, equation (*iii*) reduces to

$$(G(x) - s_0 - s_1 x) - 7x\sum_{n=2}^{\infty} s_{n-1}x^{n-1} + 10x^2 \sum_{n=2}^{\infty} s_{n-2}x^{n-2} = 0$$

i.e., $\quad (G(x) - s_0 - s_1 x) - 7x(G(x) - s_0) + 10x^2 G(x) = 0$

i.e., $\quad G(x)(1 - 7x + 10x^2) = s_0 + s_1 x - 7s_0 x = 3x$

i.e., $\quad G(x) = \dfrac{3x}{1 - 7x + 10x^2} = \dfrac{3x}{(1-2x)(1-5x)} = \dfrac{-1}{1-2x} + \dfrac{1}{1-5x}$

i.e., $\quad G(x) = -1(1-2x)^{-1} + (1-5x)^{-1} = -1\sum_{n \geq 0} (2x)^n + \sum_{n \geq 0} (5x)^n$

On equating the coefficient of x^n both sides, we get

$$s_n = (-1)2^n + 5^n, \qquad n \geq 0$$

Example 5 *Solve the recurrence relation* $s_n - 4s_{n-1} + 4s_{n-2} = 0$, $s_0 = 1$ *and* $s_1 = 6$ *by using generating function where,* $n \geq 2$.

Solution: Consider the recurrence relation

$$s_n - 4s_{n-1} + 4s_{n-2} = 0, \qquad n \geq 2 \qquad \text{...}(i)$$

$$s_0 = 1,\ s_1 = 6$$

Multiplying both sides of equation (i) by x^n, we get

$$s_n x^n - 4s_{n-1}x^n + 4s_{n-2}x^n = 0 \qquad \text{...}(ii)$$

Summing equation (ii) for all n, $n \geq 2$, we get

$$\sum_{n=2}^{\infty} s_n x^n - \sum_{n=2}^{\infty} 4s_{n-1}x^n + \sum_{n=2}^{\infty} 4s_{n-2}x^n = 0 \qquad \text{...}(iii)$$

Note that $G(x) = \sum_{n \geq 0} s_n x^n$. Therefore, equation (iii) reduces to

$$(G(x) - s_0 - s_1 x) - 4x\sum_{n=2}^{\infty} s_{n-1}x^{n-1} + 4x^2 \sum_{n=2}^{\infty} s_{n-2}x^{n-2} = 0$$

i.e., $$(G(x) - s_0 - s_1 x) - 4x(G(x) - s_0) + 4x^2 G(x) = 0$$

i.e., $$G(x)(1 - 4x + 4x^2) = s_0 + s_1 x - 4s_0 x = 1 + 2x$$

i.e., $$G(x) = \frac{1+2x}{1-4x+4x^2} = \frac{1+2x}{(1-2x)^2} = \frac{-1}{1-2x} + \frac{2}{(1-2x)^2}$$

i.e., $$G(x) = -1(1-2x)^{-1} + 2(1-2x)^{-2}$$

i.e., $$G(x) = -1(1 + 2x + (2x)^2 + \cdots) + 2(1 + 2(2x) + 3(2x)^2 + \cdots)$$

i.e., $$G(x) = -1\sum_{n \geq 0}(2x)^n + 2\sum_{n \geq 0}(n+1)(2x)^n$$

On equating the coefficient of x^n both sides, we get

$$s_n = (-1)2^n + (n+1)2^{n+1} = (1+2n)2^n, \quad n \geq 0$$

Example 6 *Solve the recurrence relation* $s_n - 4s_{n-1} + 4s_{n-2} = 0$, $s_0 = 1$ *and* $s_1 = 6$ *without using generating function where,* $n \geq 2$.

Solution: Consider the recurrence relation

$$s_n - 4s_{n-1} + 4s_{n-2} = 0, \qquad n \geq 2 \qquad \text{...}(i)$$

$$s_0 = 1,\ s_1 = 6$$

The characteristic equation to the equation (i) is given as

$$r^2 - 4r + 4 = 0,\ i.e.,\ (r-2)^2 = 0$$

The characteristic roots are given as $r = 2, 2$. Therefore, the solution to the recurrence relation is given as

$$s_n = (A_1 + A_2 n)2^n \qquad \text{...}(ii)$$

On imposing the initial conditions, we get

$$\begin{cases} A_1 = 1 & [\because s_0 = 1] \\ 2A_1 + 2A_2 = 6 & [\because s_1 = 6] \end{cases}$$

On solving these equations we get $A_1 = 1$ and $A_2 = 2$. Therefore, the solution to the difference equation is given as

$$s_n = (1 + 2n)2^n, \; n \geq 0$$

Example 7 *Solve the difference equation* $s_n - 7s_{n-1} + 10s_{n-2} = 3^n$, *given that* $s_0 = 0$ *and* $s_1 = 1$.

Solution: Consider the difference equation

$$s_n - 7s_{n-1} + 10s_{n-2} = 3^n \qquad \text{...}(i)$$

$$s_0 = 0, \; s_1 = 1$$

The characteristic equation to the difference equation (*i*) is given as

$$r^2 - 7r + 10 = 0$$

i.e.,

$$(r - 2)(r - 5) = 0$$

Therefore, the characteristic roots are $r = 2, 5$. So, the homogeneous solution to the equation (*i*) is given as

$$s_n^h = A_1(2)^n + A_2(5)^n \qquad \text{...}(ii)$$

Let the particular solution to the difference equation (*i*) be $s_n^p = P3^n$. Thus, from equation (*i*) we get

$$P3^n - 7P3^{n-1} + 10P3^{n-2} = 3^n$$

i.e.,

$$3^n\left(P - \frac{7P}{3} + \frac{10P}{9}\right) = 3^n$$

i.e.,

$$3^n\left(-\frac{2P}{9}\right) = 3^n$$

On equating the coefficient of 3^n both sides, we get $P = -\frac{9}{2}$. Therefore, the particular solution is given as $s_n^p = \left(-\frac{9}{2}\right) 3^n$. Hence, the solution to the difference equation (*i*) is given as

$$s_n = s_n^h + s_n^p = A_1(2)^n + A_2(5)^n + \left(-\frac{9}{2}\right)3^n$$

On imposing the initial conditions, we get

$$\begin{cases} A_1 + A_2 = \dfrac{9}{2} & [\because s_0 = 0] \\ 2A_1 + 5A_2 = \dfrac{29}{2} & [\because s_1 = 1] \end{cases}$$

On solving these equations we get $A_1 = \frac{8}{3}$ and $A_2 = \frac{11}{6}$. Therefore, the total solution to the difference equation is given as

$$s_n = \frac{8}{3}(2)^n + \frac{11}{6}(5)^n + \left(-\frac{9}{2}\right)3^n$$

Example 8 *Solve the difference equation* $s_n + 6s_{n-1} + 9s_{n-2} = 3$*, given that* $s_0 = 0,\ s_1 = 1$.

Solution: Given that the difference equation

$$s_n + 6s_{n-1} + 9s_{n-2} = 3 \qquad \text{...(i)}$$

The characteristic equation is given as

$$r^2 + 6r + 9 = 0 \quad i.e., \ r = -3, -3$$

The homogeneous solution is given as

$$s_n^h = (A_1 + A_2 n)(-3)^n$$

Let the particular solution be $s_n^p = P$. Therefore, from equation (*i*), we get

$$P + 6P + 9P = 3; \qquad i.e., \ P = \frac{3}{16}$$

Thus, the total solution to the difference equation is given as

$$s_n = s_n^h + s_n^p = (A_1 + A_2 n)(-3)^n + \frac{3}{16}$$

Using initial conditions $s_0 = 0,\ s_1 = 1$, we get

$$\begin{cases} A_1 + \dfrac{3}{16} = 0 & \text{for } n = 0 \\ (A_1 + A_2)(-3) + \dfrac{3}{16} = 1 & \text{for } n = 1 \end{cases}$$

On solving these equations we get $A_1 = -\frac{3}{16}$ and $A_2 = -\frac{1}{12}$. Therefore, the total solution to the difference equation is given as

$$s_n = -\left(\frac{3}{16} + \frac{1}{12}n\right)(-3)^n + \frac{3}{16}.$$

Example 9 *Solve the difference equation* $s_n - 2s_{n-1} + s_{n-2} = 3$*, given that* $s_0 = 0,\ s_1 = 1$.

Solution: Given that the difference equation

$$s_n - 2s_{n-1} + s_{n-2} = 3 \qquad \text{...(i)}$$

The characteristic equation is given as

$$r^2 - 2r + 1 = 0 \qquad i.e., \qquad r = 1 \text{ is a root of multiplicity 2.}$$

The homogeneous solution is given as

$$s_n^h = (A_1 + A_2 n)(1)^n$$

It is clear that 1 is a characteristic root of multiplicity 2 and 3 can be written as $3 \cdot 1^2$. Let the particular solution be $s_n^p = Pn^2$. So, equation (*i*) reduces to

$$Pn^2 - 2P(n-1)^2 + P(n-2)^2 = 3 \qquad i.e., \qquad P = \frac{3}{2}$$

Thus, the total solution to the difference equation is given as

$$s_n = s_n^h + s_n^p = (A_1 + A_2 n)(1)^n + \frac{3}{2}n^2$$

Using initial conditions $s_0 = 0,\ s_1 = 1$, we get

$$\begin{cases} A_1 = 0 & for\ n = 0 \\ A_1 + A_2 + \dfrac{3}{2} = 1 & for\ n = 1 \end{cases}$$

On solving these equations, we get $A_1 = 0$ and $A_2 = -\frac{1}{2}$. Therefore, the total solution to the difference equation is given as

$$s_n = -\frac{n}{2} + \frac{3n^2}{2}$$

Example 10 *Show that the total solution of the difference equation*

$$s_n + 5s_{n-1} + 6s_{n-2} = 24r^2 \text{ is } A_1(-2)^n + A_2(-3)^n + 2n^2 + \frac{17}{3}n + \frac{77}{12},$$

where A_1 and A_2 are constants.

Solution: Given that the difference equation

$$s_n + 5s_{n-1} + 6s_{n-2} = 24r^2 \qquad ...(i)$$

The characteristic equation is given as

$$r^2 + 5r + 6 = 0$$

On solving the above equation we get the characteristic roots $r = -2, -3$. Therefore, the homogeneous solution is given as

$$s_n^h = A_1(-2)^n + A_2(-3)^n$$

It is given that the right hand side of the difference equation is $24r^2$. This is a polynomial of degree 2. Therefore, consider the particular solution as $s_n^p = P_1 n^2 + P_2 n + P_3$. So, equation (*i*) reduces to

$$[P_1 n^2 + P_2 n + P_3] + 5[P_1(n-1)^2 + P_2(n-1) + P_3] + 6\,[P_1(n-2)^2 + P_2(n-2) + P_3] = 24r^2$$

i.e.,

$$12P_1 r^2 + (12P_2 - 34P_1)r + (29P_1 - 17P_2 + 12P_3) = 24r^2$$

On equating the coefficients, we have

$$12P_1 = 24 \qquad i.e., \qquad P_1 = 2$$

$$12P_2 - 34P_1 = 0 \qquad i.e., \qquad P_2 = \frac{17}{3}$$

$$(29P_1 - 17P_2 + 12P_3) = 0 \quad i.e., \qquad P_3 = \frac{77}{12}$$

So, the particular solution is given as $s_n^p = 2n^2 + \frac{17}{3}n + \frac{77}{12}$.

Thus, the total solution to the difference equation is given as

$$s_n = s_n^h + s_n^p = A_1(-2)^n + A_2(-3)^n + 2n^2 + \frac{17}{3}n + \frac{77}{12}$$

Example 11 *Find the total solution of the difference equation* $s_n - s_{n-1} = 5$ *, given that* $s_0 = 2$.

Solution: Given that the difference equation $s_n - s_{n-1} = 5$ ... (i)

The characteristic equation is given as

$$r - 1 = 0 \quad i.e., \quad r = 1 \text{ is a characteristic root.}$$

The homogeneous solution to the difference equation is given as

$$s_n^h = A_1(1)^n$$

It is clear that 1 is a characteristic root and 5 can be written as 5.1^n. Let the particular solution be $s_n^p = Pn$. Therefore, equation (i) reduces to

$$Pn - P(n-1) = 5 \qquad i.e., \qquad P = 5$$

Thus, the total solution to the difference equation is given as

$$s_n = s_n^h + s_n^p = A_1 + 5n$$

Using initial condition $s_0 = 2$ we get $A_1 = 2$. Therefore, the total solution to the difference equation is given as

$$s_n = 2 + 5n$$

Example 12 *Find the total solution of the difference equation* $s_n - 2s_{n-1} = 5.2^n$ *, given that* $s_0 = 7$.

Solution: Given that the difference equation $s_n - 2s_{n-1} = 5 \cdot 2^n$ ...(i)

The characteristic equation is given as

i.e., $\qquad r - 2 = 0 \qquad r = 2$ is a characteristic root.

The homogeneous solution is given as $s_n^h = A_1(2)^n$

Let the particular solution be $s_n^p = Pn2^n$. So, equation (i) reduces to

$$Pn2^n - 2P(n-1)2^{n-1} = 5 \cdot 2^n$$

i.e., $\qquad 2^n[Pn - P(n-1)] = 5 \cdot 2^n$

i.e., $\qquad P = 5$

Thus, the total solution to the difference equation is given as

$$s_n = s_n^h + s_n^p = A_1(2)^n + 5n2^n$$

Using initial condition $s_0 = 7$, we get $A_1 = 7$. Therefore, the total solution to the difference equation is given as

$$s_n = 7(2)^n + 5n2^n.$$

Example 13 *Solve the difference equation* $s_n - 4s_{n-1} + 4s_{n-2} = (n+1)3^n$, *given that* $s_0 = 0$ *and* $s_1 = 2$.

Solution: Given that the difference equation

$$s_n - 4s_{n-1} + 4s_{n-2} = (n+1)3^n \qquad \text{...}(i)$$

The characteristic equation is given as

$$r^2 - 4r + 4 = 0 \text{ i.e., } r = 2 \text{ is a characteristic root of multiplicity 2.}$$

The homogeneous solution is given as $s_n^h = (A_1 + A_2 n)2^n$

Let the particular solution be $s_n^p = (P_1 + P_2 n)3^n$. So, equation (*i*) becomes

$$(P_1 + P_2 n)3^n - 4(P_1 + P_2(n-1))3^{n-1} + 4(P_1 + P_2(n-2))3^{n-2} = (n+1)3^n$$

i.e., $$3^n[P_1 + P_2 n - \frac{4}{3}(P_1 + P_2 n - P_2) + \frac{4}{9}(P_1 + P_2 n - 2P_2)] = n3^n + 3^n$$

i.e., $$3^n\left[\frac{1}{9}(P_1 + 4P_2) + nP_2\left(\frac{1}{9}\right)\right] = n3^n + 3^n.$$

On equating the coefficients, we get $P_2 = 9$ and $P_1 = -27$.

Therefore, the total solution is given as

$$s_n = s_n^h + s_n^p = (A_1 + A_2 n)2^n + (9n - 27)3^n$$

Using initial conditions $s_0 = 0$ and $s_1 = 2$, we get $A_1 = 27$ and $A_2 = 1$. Therefore, the total solution to the difference equation is given as

$$s_n - (27 + n)2^n + (9n - 27)3^n.$$

Example 14 *Find the total solution to the recurrence relation* $s_n - 4s_{n-1} + 4s_{n-2} = (n^2 + 1)2^n$, *given that* $s_0 = 0$ *and* $s_1 = 2$.

Solution: Given that the difference equation

$$s_n - 4s_{n-1} + 4s_{n-2} = (n^2 + 1)2^n \qquad \text{...}(i)$$

The characteristic equation is given as

$$r^2 - 4r + 4 = 0 \text{ i.e., } r = 2 \text{ is a characteristic root of multiplicity 2.}$$

The homogeneous solution is given as $s_n^h = (A_1 + A_2 n)2^n$

As the right hand side of the given difference equation is $(n^2 + 1)2^n$ and 2 is a characteristic root of multiplicity 2, the general particular solution to the difference equation is $s_n^p = (P_1 + P_2n + P_3n^2)n^2 2^n$. So, equation (*i*) reduces to

$$(P_1 + P_2n + P_3n^2)n^2 2^n - 4(P_1 + P_2(n-1) + P_3(n-1)^2)(n-1)^2 2^{n-1} +$$
$$4(P_1 + P_2(n-2) + P_3(n-2)^2)(n-2)^2 2^{n-2} = (n^2+1)2^n$$

i.e.,
$$2^n[P_1n^2 + P_2n^3 + P_3n^4 - 2P_1(n-1)^2 - 2P_2(n-1)^3 - 2P_3(n-1)^4$$
$$+P_1(n-2)^2 + P_2(n-2)^3 + (n-2)^4] = (n^2+1)2^n$$

i.e.,
$$2^n[2P_1 + (6n-6)P_2 + (12n^2 - 24n + 14)P_3] = (n^2+1)2^n$$

i.e.,
$$2^n[(2P_1 - 6P_2 + 14P_3) + n(6P_2 - 24P_3) + 12n^2P_3] = (n^2+1)2^n$$

On equating the coefficients, we get

$$12P_3 = 1;\ 6P_2 - 24P_3 = 0 \quad \text{and} \quad 2P_1 - 6P_2 + 14P_3 = 1$$

On solving these equations, we get, $P_1 = \frac{11}{12}$, $P_2 = \frac{1}{3}$ and $P_3 = \frac{1}{12}$.

Therefore, the total solution is given as

$$s_n = s_n^h + s_n^p = (A_1 + A_2n)2^n + \left(\frac{11}{12} + \frac{1}{3}n + \frac{1}{12}n^2\right)n^2 2^n$$

Using initial conditions $s_0=0$ and $s_1=2$, we get $A_1 = 0$ and $A_2 = \frac{-1}{3}$. Therefore, the total solution to the difference equation is given as

$$s_n = -\frac{1}{3}n2^n + \left(\frac{11}{12} + \frac{1}{3}n + \frac{1}{12}n^2\right)n^2 2^n$$

Example 15 *Solve the difference equation*

$$s_n = \sqrt{s_{n-1} + \sqrt{s_{n-2} + \sqrt{s_{n-3} + \sqrt{\cdots}}}}$$

given that $s_0 = 4$.

Solution: Given that the difference equation

$$s_n = \sqrt{s_{n-1} + \sqrt{s_{n-2} + \sqrt{s_{n-3} + \sqrt{\cdots}}}} \quad \text{...(i)}$$

On taking $n = n - 1$, we get

$$s_{n-1} = \sqrt{s_{n-2} + \sqrt{s_{n-3} + \sqrt{s_{n-4} + \sqrt{\cdots}}}} \quad \text{...(ii)}$$

On squaring the difference equation (*i*) we get

$$s_n^2 = s_{n-1} + \sqrt{s_{n-2} + \sqrt{s_{n-3} + \sqrt{\cdots}}}$$

i.e.,
$$s_n^2 = s_{n-1} + s_{n-1}$$

i.e.,
$$s_n^2 - 2s_{n-1} = 0 \quad \text{...(iii)}$$

Let us take $b_n = \lg s_n$. It implies that $s_n = 2^{b_n}$. So, equation (*iii*) reduces to

$$(2^{b_n})^2 - 2(2^{b_{n-1}}) = 0$$

i.e., $$2^{2b_n} - 2^{b_{n-1}+1} = 0$$

This is possible only when $2b_n - b_{n-1} - 1 = 0$ ...(*iv*)

Now, we have to solve this first order, linear and non-homogeneous difference equation (*iv*). So, the characteristic equation is given as

$$2r - 1 = 0 \quad i.e., \quad r = \frac{1}{2}.$$

Therefore, the homogeneous solution is given as $b_n^h = A\left(\frac{1}{2}\right)^n$.

Let the general particular solution to the equation (*iv*) is $b_n^p = P$. So, equation (*iv*) reduces to

$$2P - P - 1 = 0 \qquad i.e., \qquad P = 1.$$

Thus, the total solution is given as

$$b_n = b_n^h + b_n^p = A(\tfrac{1}{2})^n + 1 \qquad ...(v)$$

Again, we have $b_0 = \lg s_0 = \lg 4 = 2$. On imposing this condition we get $A + 1 = b_0 = 2$, *i.e.,* A = 1. Therefore, the total solution is given as

$$b_n = \left(\frac{1}{2}\right)^n + 1$$

i.e., $$\lg s_n = \left(\frac{1}{2}\right)^n + 1$$

i.e., $$s_n = 2^{\left(\frac{1}{2}\right)^n + 1}$$

Example 16 *Solve the difference equation* $s_n^2 - 2s_{n-1} = 0$, *given that the initial condition* $s_0 = 4$.

Solution: Given that $s_n^2 - 2s_{n-1} = 0$ ...(*i*)

Let us take $b_n = \lg s_n$. It implies that $s_n = 2^{b_n}$. So, equation (*i*) reduces to

$$(2^{b_n})^2 - 2(2^{b_{n-1}}) = 0$$

i.e., $$2^{2b_n} - 2^{b_{n-1}+1} = 0$$

This is possible only when $2b_n - b_{n-1} - 1 = 0$ ...(*ii*)

The above equation (*ii*) is of first order, linear and non-homogeneous and hence can be solved. So, the characteristic equation is given as

$$2r - 1 = 0 \quad i.e., \quad r = \frac{1}{2}.$$

Therefore, the homogeneous solution is given as $b_n^h = A\left(\frac{1}{2}\right)^n$.

Let the general particular solution to the equation (*ii*) is $b_n^p = \text{P}$. So, equation (*ii*) reduces to

$$2\text{P} - \text{P} - 1 = 0 \qquad \textit{i.e.,} \qquad \text{P} = 1.$$

Thus, the total solution is given as

$$b_n = b_n^h + b_n^p = \text{A}\left(\frac{1}{2}\right)^n + 1 \qquad \text{...(iii)}$$

Again, we have $b_0 = \lg s_0 = \lg 4 = 2$. On imposing this condition we get, $\text{A} + 1 = b_0 = 2$ *i.e.*, A = 1. Therefore, the total solution is given as

$$b_n = \left(\frac{1}{2}\right)^n + 1 \quad \textit{i.e.,} \quad \lg s_n = \left(\frac{1}{2}\right)^n + 1$$

i.e.,

$$s_n = 2^{\left(\frac{1}{2}\right)^n + 1}$$

Example 17 *Find the total solution of the difference equation* $ns_n + ns_{n-1} - s_{n-1} = 2^n$.

Solution: Given that the difference equation

$$ns_n + ns_{n-1} - s_{n-1} = 2^n$$

i.e.,

$$ns_n + (n-1)s_{n-1} = 2^n \qquad \text{...(i)}$$

Let us take $b_n = ns_n$. Therefore, $(n-1)s_{n-1} = b_{n-1}$. Thus, the difference equation (*i*) reduces to

$$b_n + b_{n-1} = 2^n \qquad \text{... (ii)}$$

The above equation (*ii*) is of first order, linear and non-homogeneous and hence can be solved. So, the characteristic equation is given as

$$r + 1 = 0 \qquad \textit{i.e.,} \qquad r = -1$$

The homogeneous solution is given as $b_n^h = \text{A}(-1)^n$.

Let the general particular solution to the equation (*ii*) is $b_n^p = \text{P}2^n$. So, equation (*ii*) reduces to

$$\text{P}2^n + \text{P}2^{n-1} = 2^n$$

On equating the coefficient of 2^n, we get $\text{P} = \frac{2}{3}$. Thus, the total solution to the difference equation (*ii*) is given as

$$b_n = b_n^h + b_n^p = \text{A}(-1)^n + \left(\frac{2}{3}\right)2^n$$

$$ns_n = \text{A}(-1)^n + \left(\frac{2}{3}\right)2^n \qquad \text{...(iii)}$$

The above equation (*iii*) is the required total solution.

Example 18 *Solve the difference equation* $s_n - ns_{n-1} = n!$; $n \geq 1$ *with initial condition* $s_0 = 2$.

Solution: Given that $s_n - ns_{n-1} = n!$; $n \geq 1$ ...(*i*)

Let us take $b_n = s_n / n!$. Therefore, $b_{n-1} = s_{n-1} /(n-1)!$. Thus, equation (*i*) reduces to

$$n!\ b_n - n(n-1)!\ b_{n-1} = n!$$

i.e.,

$$b_n - b_{n-1} = 1 \qquad \text{...(ii)}$$

The above equation (*ii*) is of first order, linear and non-homogeneous and hence can be solved. So, the characteristic equation is given as

$$r - 1 = 0 \ \ i.e., \ r = 1$$

Therefore, homogeneous solution is given as $b_n^h = A(1)^n$.

Since, the characteristic root of the difference equation is 1 and 1 can be written as 1^n, the general form of particular solution is $b_n^p = Pn$. So, equation (*ii*) reduces to

$$Pn - P(n-1) = 1 \qquad i.e., \qquad P = 1$$

Thus, the total solution to the difference equation (*ii*) is given as

$$b_n = b_n^h + b_n^p = A(1)^n + n \qquad \text{...(iii)}$$

Again, we have $b_0 = s_0 / 0! = 2$. On imposing the condition in equation (*iii*), we get A = 2. Thus, the total solution is given as

$$b_n = 2(1)^n + n$$

i.e., $$s_n / n! = 2(1)^n + n$$

i.e., $$s_n = n!\ \ (2 * 1^n + n)$$

Example 19 *Compute the product AB by using Strassen's matrix multiplication algorithm where*

$$A = \begin{bmatrix} 2 & 4 \\ 3 & 5 \end{bmatrix} \quad and \quad B = \begin{bmatrix} 1 & 6 \\ 3 & 2 \end{bmatrix}$$

Solution: Given that the matrices

$$A = \begin{bmatrix} 2 & \vdots & 4 \\ \cdots & \vdots & \cdots \\ 3 & \vdots & 5 \end{bmatrix} \quad \text{and} \quad B = \begin{bmatrix} 1 & \vdots & 6 \\ \cdots & \vdots & \cdots \\ 3 & \vdots & 2 \end{bmatrix}$$

Partition both the given matrices A and B into submatrices of order (1×1). Therefore, we get the submatrices A_{11}, A_{12}, A_{21}, A_{22}, B_{11}, B_{12}, B_{21}, B_{22} as below.

$$A_{11} = 2,\ A_{12} = 4,\ A_{21} = 3,\ A_{22} = 5$$
$$B_{11} = 1,\ B_{12} = 6,\ B_{21} = 3,\ B_{22} = 2$$

Now, compute fourteen matrices $A_1, A_2, \cdots, A_7, B_1, B_2, \cdots, B_7$ as below

$$\begin{array}{ll} A_1 = (A_{11} + A_{22}) = 7; & B_1 = (B_{11} + B_{22}) = 3; \\ A_2 = (A_{12} - A_{22}) = -1; & B_2 = (B_{21} + B_{22}) = 5; \\ A_3 = (A_{11} - A_{21}) = -1; & B_3 = (B_{11} + B_{12}) = 7; \\ A_4 = (A_{11} + A_{12}) = 6; & B_4 = B_{22} = 2; \\ A_5 = (A_{21} + A_{22}) = 8; & B_5 = B_{11} = 1; \\ A_6 = A_{11} = 2; & B_6 = (B_{12} - B_{22}) = 4; \\ A_7 = A_{22} = 5; & B_7 = (B_{21} - B_{11}) = 2 \end{array}$$

Compute seven matrices $M_1, M_2, \cdots, M_7$ of order (1×1) such that $M_i = A_iB_i$, $i = 1,2,..., 7$. Therefore, we get

$$M_1 = 21,\ M_2 = -5,\ M_3 = -7,\ M_4 = 12,\ M_5 = 8,\ M_6 = 8,\ M_7 = 10$$

Now, compute the submatrices $P_{11}, P_{12}, P_{21}, P_{22}$ as below:

$$P_{11} = M_1 + M_2 - M_4 + M_7 = 14$$
$$P_{12} = M_4 + M_6 = 20$$
$$P_{21} = M_5 + M_7 = 18$$
$$P_{22} = M_1 - M_3 - M_5 + M_6 = 28$$

Therefore, the product AB is given as

$$P = AB = \begin{bmatrix} P_{11} & \vdots & P_{12} \\ \cdots\cdots & \vdots & \cdots\cdots \\ P_{21} & \vdots & P_{22} \end{bmatrix} = \begin{bmatrix} 14 & \vdots & 20 \\ \cdots\cdots & \vdots & \cdots\cdots \\ 18 & \vdots & 28 \end{bmatrix}$$

Example 20 *Solve the difference equation* $s_n^2 - 2s_{n-1}^2 = 1$; *with initial condition* $s_0 = 2$.

Solution: Given that $s_n^2 - 2s_{n-1}^2 = 1$ ...(*i*)

Let us take $b_n = s_n^2$. Therefore, $b_{n-1} = s_{n-1}^2$. Thus, equation (*i*) reduces to

$$b_n - 2b_{n-1} = 1 \qquad ...(ii)$$

This is a first order, linear and non-homogeneous difference equation and hence can be solved. So, the characteristic equation is given as

i.e., $r - 2 = 0$ $\qquad r = 2$

Therefore, homogeneous solution is given as $b_n^h = A(2)^n$.

Let the general form of particular solution is $b_n^p = P$. So, equation (*ii*) reduces to

i.e., $P - 2P = 1$ $\qquad P = -1$

Thus, the total solution to the difference equation (*ii*) is given as

$$b_n = b_n^h + b_n^p = A(2)^n - 1 \qquad ...(iii)$$

Again, we have $b_0 = s_0^2 = 4$. On imposing the condition in equation (*iii*), we get A = 5. Thus, the total solution is given as

$$b_n = 5 * 2^n - 1 \quad i.e., \quad s_n^2 = 5 * 2^n - 1$$

i.e.,
$$s_n = \sqrt{5 * 2^n - 1}$$

Example 21 *Determine s_n if $s_0 = 0$, $s_1 = 1$, $s_2 = 4$, and $s_3 = 12$ satisfy the difference equation $s_n + c_1 s_{n-1} + c_2 s_{n-2} = 0$.*

Solution: Given that $s_n + c_1 s_{n-1} + c_2 s_{n-2} = 0$ ...(*i*)

and $s_0 = 0,\ s_1 = 1,\ s_2 = 4,\ s_3 = 12.$

On putting $n = 2$ in equation (*i*), we get

$$s_2 + c_1 s_1 + c_2 s_0 = 0; \quad i.e., \quad c_1 = -4$$

On putting $n = 3$ in equation (*i*) we get

$$s_3 + c_1 s_2 + c_2 s_1 = 0; \quad i.e., \quad 4c_1 + c_2 = -12$$

$$\Rightarrow \quad c_2 = 4$$

Therefore, equation (*i*) reduces to $s_n - 4s_{n-1} + 4s_{n-2} = 0$. The characteristic equation is given as

$$r^2 - 4r + 4 = 0 \quad i.e., \quad r = 2,\ 2$$

Thus, the total solution is given as $s_n = (An + B)2^n$. On imposing the initial condition $s_0 = 0$ and $s_1 = 1$, we get B = 0 and $A = \frac{1}{2}$. Therefore, the total solution to the difference equation is given as

$$s_n = n2^{n-1}.$$

Example 22 *Solve the difference equation $s_n + s_{n-1} + s_{n-2} = 0$, given that $s_0 = 0$ and $s_1 = 2$.*

Solution: Given that $s_n + s_{n-1} + s_{n-2} = 0$ ...(*i*)

The characteristic equation to the above equation is given as

$$r^2 + r + 1 = 0 \quad i.e., \quad r = -\frac{1}{2} \pm \frac{i\sqrt{3}}{2}$$

Thus, the homogeneous solution is given as

$$s_n = A_1\left(-\frac{1}{2} + \frac{i\sqrt{3}}{2}\right)^n + A_2\left(-\frac{1}{2} - \frac{i\sqrt{3}}{2}\right)^n$$

On using the initial conditions $s_0 = 0$ and $s_1 = 2$, we get

$$\begin{cases} A_1 + A_2 = 0; & n = 0 \\ A_1 - A_2 = \dfrac{4}{i\sqrt{3}}; & n = 1 \end{cases}$$

On solving above two equations we get $A_1 = \frac{2}{i\sqrt{3}}$ and $A_2 = \frac{-2}{i\sqrt{3}}$. Thus, the total solution is given as

$$s_n = \frac{2}{i\sqrt{3}}\left(-\frac{1}{2}+\frac{i\sqrt{3}}{2}\right)^n - \frac{2}{i\sqrt{3}}\left(-\frac{1}{2}-\frac{i\sqrt{3}}{2}\right)^n$$

Example 23 *Find the total solution of*

$$s_n + 5s_{n-1} + 6s_{n-2} = 42 \cdot 4^n$$

Solution: Given that $s_n + 5s_{n-1} + 6s_{n-2} = 42 \cdot 4^n$ ...(*i*)

The characteristic equation of the above difference equation is given as

i.e., $r^2 + 5r + 6 = 0;\ r = -2, -3$

Thus, the homogeneous solution is given as

$$s_n^h = A_1(-2)^n + A_2(-3)^n$$

Let the general form of particular solution is $s_n^p = P4^n$. So, from equation (*i*) we get

$$P4^n + 5P4^{n-1} + 6P4^{n-2} = 42{\cdot}4^n$$

i.e., $4^n(P + \frac{5}{4}P + \frac{6}{16}P) = 42 \cdot 4^n$

On equating the coefficient of 4^n both sides, we get P = 16 and hence $s_n^p = 16 \cdot 4^n$. Therefore, the total solution is given as

$$s_n = s_n^h + s_n^p = A_1(-2)^n + A_2(-3)^n + 16 \cdot 4^n$$

Example 24 *Find the total solution of the following recurrence relation*

$$s_n + s_{n-1} = 3n \cdot 2^n$$

Solution: Given that $s_n + s_{n-1} = 3n \cdot 2^n$ ...(*i*)

The characteristic equation of the above difference equation is given as

i.e., $r + 1 = 0;\quad r = -1$

Thus, the homogeneous solution is given as $s_n^h = A_1(-1)^n$. Let the general form of particular solution is $s_n^p = (P_1 n + P_2)2^n$. Therefore, from equation (*i*), we get

$$(P_1 n + P_2)2^n + (P_1(n-1) + P_2)2^{n-1} = 3n \cdot 2^n$$

i.e., $n2^n\left(\frac{3}{2}P_1\right) + 2^n\left(\frac{3}{2}P_2 - \frac{1}{2}P_1\right) = 3n \cdot 2^n$

On equating the coefficients and on solving we get $P_1 = 2$ and $P_2 = \frac{2}{3}$. Hence, the particular solution is given as $s_n^p = \left(2n + \frac{2}{3}\right)2^n$. Therefore, the total solution is given as

$$s_n = s_n^h + s_n^p = A_1(-1)^n + \left(2n + \frac{2}{3}\right)2^n$$

Example 25 *Find the general solution of the difference equation*

$$4s_n - 20s_{n-1} + 17s_{n-2} - 4s_{n-3} = 0$$

Solution: Given that the difference equation

$$4s_n - 20s_{n-1} + 17s_{n-2} - 4s_{n-3} = 0 \qquad ...(i)$$

The characteristic equation of the above difference equation is given as

$$4r^3 - 20r^2 + 17r - 4 = 0$$

i.e.,
$$(r-4)(2r-1)^2 = 0$$

i.e.,
$$r = 4,\ r = \frac{1}{2}, \frac{1}{2}$$

Therefore, the general solution to the difference equation is given as

$$s_n = A_1 4^n + (A_2 n + A_3)\left(\frac{1}{2}\right)^n$$

Example 26 *Find the general solution of the recurrence relation*

$$s_n + 6s_{n-1} + 12s_{n-2} + 8s_{n-3} = 0$$

Solution: Given that the difference equation

$$s_n + 6s_{n-1} + 12s_{n-2} + 8s_{n-3} = 0 \qquad ...(i)$$

The characteristic equation of the above difference equation is given as

$$r^3 + 6r^2 + 12r + 8 = 0$$

i.e.,
$$(r+2)^3 = 0$$

i.e.,
$$r = -2,\ -2,\ -2$$

It implies that $r = -2$ is a characteristic root of multiplicity 3. Therefore, the general solution to the difference equation is given as

$$s_n = (A_1 n^2 + A_2 n + A_3)(-2)^n$$

Example 27 *Obtain the partial decomposition and identify the sequence having in the following expressions as generating function.*

$$(a)\ G(x) = \frac{5+2x}{1\ \ 4x^2} \qquad (b)\ G(x) = \frac{6-29x}{1-11x+30x^2}$$

Solution: (*a*) Given that $G(x) = \dfrac{5+2x}{1-4x^2}$

$$\Rightarrow \qquad G(x) = \frac{5+2x}{(1-2x)(1+2x)} = \frac{3}{(1-2x)} + \frac{2}{(1+2x)}$$

$$\Rightarrow \qquad G(x) = 3(1-2x)^{-1} + 2(1+2x)^{-1} \qquad ...(i)$$

Note that $G(x) = \sum\limits_{n=0}^{\infty} s_n x^n$. Therefore, equation (*i*) reduces to

$$\sum_{n=0}^{\infty} s_n x^n = 3\sum_{n=0}^{\infty}(2x)^n + 2\sum_{n=0}^{\infty}(-2x)^n$$

On equating the coefficients of both sides we get

$$s_n = 3 \cdot 2^n + 2(-2)^n; \; n \geq 0$$

i.e.,
$$s_n = \begin{cases} 5 \cdot 2^n & \text{if } n \text{ is even} \\ 2^n & \text{if } n \text{ is odd} \end{cases}$$

(*b*) Given that $G(x) = \dfrac{6-29x}{1-11x+30x^2}$

$$\Rightarrow \quad G(x) = \frac{6-29x}{(1-6x)(1-5x)} = \frac{7}{(1-6x)} - \frac{1}{(1-5x)}$$

$$\Rightarrow \quad G(x) = 7(1-6x)^{-1} - (1-5x)^{-1} \qquad \ldots(i)$$

Note that $G(x) = \sum_{n=0}^{\infty} s_n x^n$. Therefore, equation (*i*) reduces to

$$\sum_{n=0}^{\infty} s_n x^n = 7\sum_{n=0}^{\infty} (6x)^n - \sum_{n=0}^{\infty} (5x)^n$$

On equating the coefficients of x^n both sides, we get

$$s_n = 7 \cdot 6^n - 5^n; \; n \geq 0$$

Example 28 *Obtain the partial decomposition and identify the sequence having in the following expressions as generating function.*

(*a*) $G(x) = \dfrac{32-22x}{2-3x+x^2}$ (*b*) $G(x) = \dfrac{1}{5-6x+x^2}$

Solution: (*a*) Given that $G(x) = \dfrac{32-22x}{2-3x+x^2}$

$$\Rightarrow \quad G(x) = \frac{32-22x}{(2-x)(1-x)} = \frac{10}{(1-x)} + \frac{12}{(2-x)}$$

$$\Rightarrow \quad G(x) = 10(1-x)^{-1} + 6\left(1-\frac{x}{2}\right)^{-1} \qquad \ldots(i)$$

Note that $G(x) = \sum_{n=0}^{\infty} s_n x^n$. Therefore, equation (*i*) reduces to

$$\sum_{n=0}^{\infty} s_n x^n = 10\sum_{n=0}^{\infty} x^n + 6\sum_{n=0}^{\infty} \left(\frac{x}{2}\right)^n$$

On equating the coefficients of x^n both sides, we get

$$s_n = 10 + 6\left(\frac{1}{2}\right)^n; \; n \geq 0$$

(*b*) Given that $G(x) = \dfrac{1}{5-6x+x^2}$

$$\Rightarrow \qquad G(x) = \frac{1}{(5-x)(1-x)} = \frac{1}{4(1-x)} - \frac{1}{4(5-x)}$$

$$\Rightarrow \qquad G(x) = \frac{1}{4}(1-x)^{-1} + \frac{1}{20}\left(1-\frac{x}{5}\right)^{-1} \qquad ...(i)$$

Note that $G(x) = \sum_{n=0}^{\infty} s_n x^n$. Therefore, equation ($i$) reduces to

$$\sum_{n=0}^{\infty} s_n x^n = \frac{1}{4}\sum_{n=0}^{\infty} x^n + \frac{1}{20}\sum_{n=0}^{\infty}\left(\frac{x}{5}\right)^n$$

On equating the coefficients of x^n both sides, we get

$$s_n = \frac{1}{4} + \frac{1}{20}\left(\frac{1}{5}\right)^n; \; n \geq 0$$

Example 29 *Solve the difference equation by using generating function.*

$$s_n - 6s_{n-1} + 5s_{n-2} = 0, \; n \geq 2; \; s_0 = 2, \; s_1 = 2$$

Solution: Multiplying both sides of difference equation by x^n, we get

$$s_n x^n - 6s_{n-1}x^n + 5s_{n-2}x^n = 0 \qquad ...(i)$$

Summing equation (ii) for all n, $n \geq 2$, we get

$$\sum_{n=2}^{\infty} s_n x^n - \sum_{n=2}^{\infty} 6s_{n-1}x^n + \sum_{n=2}^{\infty} 5s_{n-2}x^n = 0 \qquad ...(ii)$$

Note that $G(x) = \sum_{n=0}^{\infty} s_n x^n$. Therefore, equation ($iii$) reduces to

$$(G(x) - s_0 - s_1 x) - 6x\sum_{n=2}^{\infty} s_{n-1}x^{n-1} + 5x^2\sum_{n=2}^{\infty} s_{n-2}x^{n-2} = 0$$

i.e., $\qquad (G(x) - s_0 - s_1 x) - 6x(G(x) - s_0) + 5x^2 G(x) = 0$

i.e., $\qquad G(x)(1 - 6x + 5x^2) = s_0 + s_1 x - 6s_0 x = 2 - 10x$

i.e., $\qquad G(x) = \dfrac{2-10x}{1-6x+5x^2} = \dfrac{2}{(1-x)} = 2(1-x)^{-1}$

i.e., $\qquad \sum_{n=0}^{\infty} s_n x^n = 2\sum_{n=0}^{\infty} x^n$

On equating the coefficient of x^n both sides, we get $s_n = 2, \; n \geq 0$.

Example 30 *Solve the difference equation by using generating function.*

$$s_n - 5s_{n-1} + 6s_{n-2} = 0, \; n \geq 2; \; s_0 = 2, \; s_1 = 4$$

Solution: Multiplying both sides of difference equation by x^n, we get

$$s_n x^n - 5s_{n-1}x^n + 6s_{n-2}x^n = 0 \qquad ...(i)$$

Summing equation (i) for all n, $n \geq 2$, we get

$$\sum_{n=2}^{\infty} s_n x^n - \sum_{n=2}^{\infty} 5s_{n-1}x^n + \sum_{n=2}^{\infty} 6s_{n-2}x^n = 0 \qquad \ldots(ii)$$

Note that $G(x) = \sum_{n=0}^{\infty} s_n x^n$. Therefore, equation (ii) reduces to

$$(G(x) - s_0 - s_1x) - 5x\sum_{n=2}^{\infty} s_{n-1}x^{n-1} + 6x^2\sum_{n=2}^{\infty} s_{n-2}x^{n-2} = 0$$

i.e., $$(G(x) - s_0 - s_1x) - 5x(G(x) - s_0) + 6x^2G(x) = 0$$

i.e., $$G(x)(1 - 5x + 6x^2) = s_0 + s_1x - 5s_0x = 2 - 6x$$

i.e., $$G(x) = \frac{2-6x}{1-5x+6x^2} = \frac{2}{(1-2x)} = 2(1-2x)^{-1}$$

i.e., $$\sum_{n=0}^{\infty} s_n x^n = 2\sum_{n=0}^{\infty}(2x)^n$$

On equating the coefficient of both x^n sides, we get $s_n = 2^{n+1}$, $n \geq 0$.

Example 31 *Solve the difference equation by using generating function.*

$$4s_n + 3s_{n-1} - s_{n-2} = 5,\ n \geq 2;\ s_0 = 1,\ s_1 = 4$$

Solution: Multiplying both sides of difference equation by x^n, we get

$$4s_nx^n + 3s_{n-1}x^n - s_{n-2}x^n = 5x^n \qquad \ldots(i)$$

Summing equation (i) for all n, $n \geq 2$, we get

$$\sum_{n=2}^{\infty} 4s_nx^n + \sum_{n=2}^{\infty} 3s_{n-1}x^n - \sum_{n=2}^{\infty} s_{n-2}x^n = \sum_{n=2}^{\infty} 5x^n$$

i.e., $$4\sum_{n=2}^{\infty} s_nx^n + 3\sum_{n=2}^{\infty} s_{n-1}x^n - \sum_{n=2}^{\infty} s_{n-2}x^n = 5(x^2 + x^3 + x^4 + \cdots)$$

i.e., $$4(G(x) - s_0 - s_1x) + 3x(G(x) - s_0) - x^2G(x) = 5x^2(1 + x + x^2 + \cdots)$$

i.e., $$G(x)(4 + 3x - x^2) = 4s_0 + 4s_1x + 3s_0x + 5x^2(1-x)^{-1}$$

i.e., $$G(x)(4 + 3x - x^2) = 4 + 19x + \frac{5x^2}{(1-x)} = \frac{4 + 15x - 15x^2}{(1-x)}$$

i.e., $$G(x) = \frac{4 + 15x - 15x^2}{(1-x)(4+3x-x^2)} = \frac{4 + 15x - 15x^2}{(1-x)(4-x)(1+x)}$$

i.e., $$G(x) = \frac{2}{3(1-x)} + \frac{176}{15(4-x)} - \frac{13}{5(1+x)}$$

i.e., $$G(x) = \frac{2}{3}(1-x)^{-1} + \frac{176}{60}\left(1 - \frac{x}{4}\right)^{-1} - \frac{13}{5}(1+x)^{-1}$$

Since, $G(x) = \sum_{n=0}^{\infty} s_nx^n$. Therefore, above equation reduces to

$$\sum_{n=0}^{\infty} s_n x^n = \frac{2}{3}\sum_{n=0}^{\infty} x^n + \frac{44}{15}\sum_{n=0}^{\infty}\left(\frac{x}{4}\right)^n - \frac{13}{5}\sum_{n=0}^{\infty}(-x)^n$$

On equating the coefficient of x^n both sides, we get

$$s_n = \frac{2}{3} + \frac{44}{15}\left(\frac{1}{4}\right)^n - \frac{13}{5}(-1)^n,\ n \geq 0.$$

Example 32 *Solve the difference equation by using generating function.*

$$s_n - 5s_{n-1} + 6s_{n-2} = n + 2^n,\ n \geq 2;\ s_0 = 1,\ s_1 = 1$$

Solution: Multiplying both sides of difference equation by x^n, we get

$$s_n x^n - 5s_{n-1}x^n + 6s_{n-2}x^n = nx^n + (2x)^n \qquad \ldots(i)$$

with $s_0 = 1,\ s_1 = 1$. Summing above equation (i) for all $n, n \geq 2$, we get

$$\sum_{n=2}^{\infty} s_n x^n - \sum_{n=2}^{\infty} 5s_{n-1}x^n + \sum_{n=2}^{\infty} 6s_{n-2}x^n = \sum_{n=2}^{\infty} nx^n + \sum_{n=2}^{\infty}(2x)^n$$
$$= x\left[(1-x)^{-2} - 1\right] + (2x)^2(1-2x)^{-1}$$

Note that $G(x) = \sum_{n\geq 0} s_n x^n$. Therefore, above equation reduces to

$$(G(x) - s_0 - s_1 x) - 5x(G(x) - s_0) + 6x^2 G(x) = \frac{x}{(1-x)^2} - x + \frac{(2x)^2}{(1-2x)}$$

i.e., $$G(x)(1 - 5x + 6x^2) = \frac{x}{(1-x)^2} - x + \frac{(2x)^2}{(1-2x)} + s_0 + s_1 x - 5s_0 x$$

i.e., $$G(x)(1 - 5x + 6x^2) = \frac{x}{(1-x)^2} - x + \frac{(2x)^2}{(1-2x)} + 1 + x - 5x$$

i.e., $$G(x) = \frac{2 - 8x + 27x^2 - 35x^3 + 14x^4}{(1-x)^2(1-2x)(1-5x+6x^2)} = \frac{2 - 8x + 27x^2 - 35x^3 + 14x^4}{(1-x)^2(1-2x)^2(1-3x)}$$

i.e., $$G(x) = \frac{5}{4(1-x)} + \frac{1}{2(1-x)^2} - \frac{3}{(1-2x)} - \frac{2}{(1-2x)^2} + \frac{17}{4(1-3x)}$$

i.e., $$\sum_{n\geq 0} s_n x^n = \frac{5}{4}\sum_{n\geq 0} x^n + \frac{1}{2}\sum_{n\geq 0}(n+1)x^n - 3\sum_{n\geq 0}(2x)^n - 2\sum_{n\geq 0}(n+1)(2x)^n + \frac{17}{4}\sum_{n\geq 0}(3x)^n$$

On equating the coefficient of x^n both sides, we get

$$s_n = \frac{5}{4} + \frac{(n+1)}{2} - 3\cdot 2^n - 2(n+1)2^n + \frac{17}{4}\cdot 3^n,\ n \geq 0$$

i.e., $$s_n = \frac{7}{4} + \frac{n}{2} - (2n+5)2^n + \frac{17}{4}\cdot 3^n,\ n \geq 0$$

Example 33 *Solve the difference equation by using generating function.*

$$s_n + s_{n-1} = 4n - 3,\ n \geq 1;\ s_0 = 1$$

Solution: Multiplying both sides of difference equation by x^n, we get

$$s_n x^n + s_{n-1} x^n = 4nx^n - 3 \cdot x^n \qquad \text{...}(i)$$

with $s_0 = 1$. Summing above equation (i) for all n, $n \geq 1$, we get

$$\sum_{n\geq1} s_n x^n + \sum_{n\geq1} s_{n-1}x^n = \sum_{n\geq1} 4nx^n - \sum_{n\geq1} 3x^n = 4x(1-x)^{-2} - 3x(1-x)^{-1}$$

Note that $G(x) = \sum_{n\geq0} s_n x^n$. Therefore, above equation reduces to

$$(G(x) - s_0) + xG(x) = \frac{4x}{(1-x)^2} - \frac{3x}{(1-x)}$$

i.e., $$G(x)(1+x) = \frac{4x}{(1-x)^2} - \frac{3x}{(1-x)} + s_0 = \frac{4x}{(1-x)^2} - \frac{3x}{(1-x)} + 1$$

i.e., $$G(x) = \frac{1 - x + 4x^2}{(1-x)^2(1+x)} = -\frac{5}{2(1-x)} + \frac{2}{(1-x)^2} + \frac{3}{2(1+x)}$$

i.e., $$\sum_{n\geq0} s_n x^n = \frac{-5}{2}\sum_{n\geq0} x^n + 2\sum_{n\geq0}(n+1)x^n + \frac{3}{2}\sum_{n\geq0}(-x)^n$$

On equating coefficient of x^n both sides, we get

$$s_n = \frac{-5}{2} + 2(n+1) + \frac{3}{2}\cdot(-1)^n,\ n \geq 0.$$

Example 34 *Solve the recurrence relation by using generating function.*

$$s_n - 2s_{n-1} + s_{n-2} = 7,\ n \geq 2;\ s_0 = 2,\ s_1 = 4$$

Solution: Multiplying both sides of difference equation by x^n, we get

$$s_n x^n - 2s_{n-1}x^n + s_{n-2}x^n = 7x^n \qquad \text{...}(i)$$

with $s_0 = 2$ and $s_1 = 4$. Summing above equation (i) for all n, $n \geq 2$, we get the following equation.

$$\sum_{n\geq2} s_n x^n - \sum_{n\geq2} 2s_{n-1}x^n + \sum_{n\geq2} s_{n-2}x^n = \sum_{n\geq2} 7x^n \qquad \text{...}(ii)$$

Note that $G(x) = \sum_{n\geq0} s_n x^n$. Therefore, above equation (ii) reduces to

$$(G(x) - s_0 - s_1 x) - 2x(G(x) - s_0) + x^2 G(x) = 7x^2(1-x)^{-1}$$

i.e., $$G(x)(1 - 2x + x^2) = \frac{7x^2}{(1-x)} + s_0 + s_1 x - 2xs_0 = \frac{7x^2 - 2x + 2}{(1-x)}$$

i.e., $$G(x) = \frac{7x^2 - 2x + 2}{(1-x)^3} = \frac{7}{(1-x)} - \frac{12}{(1-x)^2} + \frac{7}{(1-x)^3}$$

i.e., $$\sum_{n\geq 0} s_n x^n = 7\sum_{n\geq 0} x^n - 12\sum_{n\geq 0}(n+1)x^n + 7\sum_{n\geq 0}\frac{(-3)(-4)\cdots(-3-n+1)(-x)^n}{n!}$$

i.e., $$\sum_{n\geq 0} s_n x^n = 7\sum_{n\geq 0} x^n - 12\sum_{n\geq 0}(n+1)x^n + 7\sum_{n\geq 0}\frac{(n+2)!\ x^n}{2(n)!}$$

On equating coefficient of x^n both sides, we get

$$s_n = 7 - 12(n+1) + 7\cdot\frac{(n+2)\ !}{2(n)\ !},\ n\geq 0$$

Example 35 *Solve the recurrence relation by using generating function.*

$$s_n - 7s_{n-1} + 10s_{n-2} = 3^n,\ n\geq 2;\ s_0 = 0,\ s_1 = 6$$

Solution: Multiplying both sides of difference equation by x^n, we get

$$s_n x^n - 7s_{n-1}x^n + 10s_{n-2}x^n = (3x)^n \quad \ldots(i)$$

with $s_0 = 0$ and $s_1 = 6$. Summing above equation (*i*) for all n, $n\geq 2$, we get the following equation.

$$\sum_{n\geq 2} s_n x^n - \sum_{n\geq 2} 7s_{n-1}x^n + \sum_{n\geq 2} 10s_{n-2}x^n = \sum_{n\geq 2}(3x)^n \quad \ldots(ii)$$

Note that $G(x) = \sum_{n\geq 0} s_n x^n$. Therefore, above equation (*ii*) reduces to

$$(G(x) - s_0 - s_1 x) - 7x(G(x) - s_0) + 10x^2 G(x) = (3x)^2(1-3x)^{-1}$$

i.e., $$G(x)(1-7x+10x^2) = \frac{9x^2}{(1-3x)} + s_0 + s_1 x - 7xs_0 = \frac{6x-9x^2}{(1-3x)}$$

i.e., $$G(x) = \frac{6x-9x^2}{(1-3x)(1-5x)(1-2x)} = \frac{-9}{2(1-3x)} + \frac{7}{2(1-5x)} + \frac{1}{(1-2x)}$$

i.e., $$\sum_{n\geq 0} s_n x^n = \frac{-9}{2}\sum_{n\geq 0}(3x)^n + \frac{7}{2}\sum_{n\geq 0}(5x)^n + \sum_{n\geq 0}(2x)^n$$

On equating coefficient of x^n both sides, we get

$$s_n = \frac{-9}{2}(3)^n + \frac{7}{2}(5)^n + (2)^n;\ n\geq 0$$

Example 36 *Find the generating function for the sequence 1, 3, 9, 27, ...*

Solution: Let $G(x)$ be the generating function for the given sequence. Therefore, $G(x)$ can be written as

$$\begin{aligned} G(x) &= 1 + 3x + 9x^2 + 27x^3 + \cdots \\ &= 1 + 3x + (3x)^2 + (3x)^3 + \cdots \\ &= (1-3x)^{-1} \end{aligned}$$

Example 37 *Find the exponential generating function for the sequence 1, 3, 9, 27, ...*

Solution: Let E(x) be the generating function for the given sequence.
Therefore, E(x) can be written as

$$E(x) = 1 + 3x + \frac{9x^2}{2!} + \frac{27x^3}{3!} + \cdots$$
$$= 1 + 3x + \frac{(3x)^2}{2!} + \frac{(3x)^3}{3!} + \cdots$$
$$= e^{3x}$$

Example 38 *Find the closed form of the generating function for the following sequence $\{s_n\}$, where $s_n = n + 3$.*

Solution: Given that $s_n = n+3$. On multiplying both sides by x^n and then taking summation over $n \geq 0$, we get

$$\sum_{n\geq 0} s_n x^n = \sum_{n\geq 0} nx^n + \sum_{n\geq 0} 3x^n$$

Note that, $G(x) = \sum_{n \geq o} s_n x^n$. Therefore, above equation reduces to

$$G(x) = x\sum_{n\geq 1} nx^{n-1} + 3\sum_{n\geq 0} x^n = x(1-x)^{-2} + 3(1-x)^{-1}.$$

Example 39 *Find the sequences corresponding to the following generating functions.*

(*a*) $(2 + x)^3$ (*b*) $(1 + x)^2 + e^x$

Solution: (*a*) $(2 + x)^3 = 8 + 12x + 6x^2 + x^3$

The sequence corresponding to the above generating function is given as (8, 12, 6, 1, 0, 0, ...).

(*b*) $(1+x)^2 + e^x = (1 + 2x + x^2) + (1 + x + \frac{x^2}{2!} + \frac{x^3}{3!} + \cdots)$

$$= 2 + 3x + \left(1 + \frac{1}{2!}\right)x^2 + \left(\frac{1}{3!}\right)x^3 + \cdots$$

The sequence corresponding to the above generating function is given as $\left(2,\ 3,\ \left(1+\frac{1}{2!}\right),\ \frac{1}{3!},\ \cdots\right)$.

Example 40 *Obtain the solution to the following first order linear recurrence relation by using iterative method.*

$$s_n = 2s_{n-1} + 3 \text{ with } s_0 = 5$$

Solution: Given that $s_n = 2s_{n-1} + 3$ and $s_0 = 5$. Therefore, by applying the recurrence relation repeatedly, we have

$$s_n = 2s_{n-1} + 3$$
$$= 2(2s_{n-2} + 3) + 3 = 2^2 s_{n-2} + 3(1 + 2)$$

$$= 2^2(2s_{n-3} + 3) + 2 \times 3 + 3 = 2^3 s_{n-3} + 3(1 + 2 + 2^2)$$
$$= 2^3(2s_{n-4} + 3) + 3(1 + 2 + 2^2) = 2^4 s_{n-4} + 3(1 + 2 + 2^2 + 2^3)$$
$$= \cdots \ \cdots \ \cdots \ \cdots$$
$$= 2^n s_0 + 3(1 + 2 + 2^2 + 2^3 + \cdots + 2^{n-1})$$
$$= 2^n \times 5 + 3\left(\frac{2^n - 1}{2 - 1}\right) = 2^n \times 5 + 3 \times (2^n - 1)$$
$$= 8 \times 2^n - 3$$

Example 41 *Compute $u_n = s_n * t_n$. It is given that $s_n = 3^n$, $n \geq 0$ and $t_n = 5^n$, $n \geq 0$.*

Solution: Given that $s_n = 3^n$, $n \geq 0$. On multiplying both sides by x^n and then taking summation over $n \geq 0$, we get

$$\sum_{n\geq 0} s_n x^n = \sum_{n\geq 0} 3^n x^n = \sum_{n\geq 0} (3x)^n = \frac{1}{(1-3x)}$$

Note that, $S(x) = \sum_{n\geq 0} s_n x^n$. So, above equation reduces to $S(x) = \dfrac{1}{(1-3x)}$.

Similarly, $t_n = 5^n$, $n \geq 0$. On multiplying both sides by x^n and then taking summation over $n \geq 0$, we get

$$\sum_{n\geq 0} t_n x^n = \sum_{n\geq 0} 5^n x^n = \sum_{n\geq 0} (5x)^n = \frac{1}{(1-5x)}$$

Note that, $T(x) = \sum_{n\geq 0} t_n x^n$. So, above equation reduces to $T(x) = \dfrac{1}{(1-5x)}$.

Now, we have

$$U(x) = S(x) * T(x) = \frac{1}{(1-3x)} * \frac{1}{(1-5x)} = \frac{5}{2(1-5x)} - \frac{3}{2(1-3x)}$$

Note that, $U(x) = \sum_{n\geq 0} u_n x^n$. So, above equation reduces to.

$$\sum_{n\geq 0} u_n x^n = \frac{5}{2}\sum_{n\geq 0} (5x)^n - \frac{3}{2}\sum_{n\geq 0} (3x)^n$$

On equating the coefficient of x^n in the above equation, we get

$$u_n = \frac{5}{2}(5)^n - \frac{3}{2}(3)^n = \frac{1}{2}\left(5^{n+1} - 3^{n+1}\right)$$

In the above solution, we have taken the generating functions S(x), T(x) and U(x) for the sequences $\{s_n\}$, $\{t_n\}$ and $\{u_n\}$ respectively.

EXERCISES

1. Find the total solution of the following difference equations.
   (a) $s_n - 6s_{n-1} - 7s_{n-2} = 0$, given that $s_0 = 0$ and $s_1 = 2$.
   (b) $s_n - 8s_{n-1} + 16s_{n-2} = 0$, given that $s_0 = 1$ and $s_1 = 3$.
   (c) $s_n - 7s_{n-1} + 5s_{n-2} = 0$, given that $s_0 = 1$ and $s_1 = 3$.
   (d) $s_n - 8s_{n-1} + 16s_{n-2} = 0$, given that $s_0 = 1$ and $s_1 = 3$.
   (e) $s_n - 3s_{n-1} + 3s_{n-2} - s_{n-3} = 0$, given that $s_0 = 0$, $s_3 = 3$ and $s_5 = 10$.

2. Find the total solution of the following difference equations.
   (a) $s_n - 2s_{n-1} + s_{n-2} = 7$, given that $s_0 = 2$ and $s_1 = 5$.
   (b) $s_n - 4s_{n-1} + 4s_{n-2} = (n+1)2^n$, given that $s_0 = 0$ and $s_1 = 4$.
   (c) $s_n - s_{n-1} - 6s_{n-2} = 3^n$, given that $s_0 = 1$ and $s_1 = 1$.
   (d) $s_n - 7s_{n-1} + 10s_{n-2} = 6n + 10$, given that $s_0 = 0$ and $s_1 = 3$.
   (e) $s_n - 7s_{n-1} + 5s_{n-2} = n^2 + 2$, given that $s_0 = 2$ and $s_1 = 6$.

3. Solve the following difference equations.
   (a) $s_n - 2s_{n-1} + 2s_{n-2} - s_{n-3} = 0$, with $s_0 = 2$, $s_1 = 1$
   (b) $s_n + 6s_{n-1} + 9s_{n-2} = 3 + 2^n$, with $s_0 = 0$ and $s_1 = 3$.
   (c) $s_n - 4s_{n-1} + 3s_{n-2} = 5$, with $s_0 = 1$ and $s_1 = 10$.
   (d) $s_n - 5s_{n-1} + 7s_{n-2} - 3s_{n-3} = 7$, with $s_0 = 0$, $s_1 = 3$ and $s_2 = 6$.
   (e) $s_n - 10s_{n-1} + 25s_{n-2} = 5^n$, with $s_0 = 1$ and $s_1 = 6$.

4. Solve the following recurrence equations with the given initial conditions.
   (a) $s_n - 9s_{n-1} + 27s_{n-2} - 81s_{n-3} = 2^n$; $s_0 = 2$, $s_1 = 5$ and $s_2 = 8$
   (b) $s_n - 9s_{n-1} + 20s_{n-2} = n^2 + 2n + 1$; $s_0 = 1$ and $s_2 = 7$.
   (c) $s_n + s_{n-1} - 6s_{n-2} = (n+1)5^n$; $s_0 = 3$ and $s_1 = 7$.
   (d) $s_n - 8s_{n-1} + 21s_{n-2} - 18s_{n-3} = (n^2+1)3^n$; $s_0 = 3$, $s_1 = 5$ and $s_2 = 7$.
   (e) $s_n + 7s_{n-1} + 10s_{n-2} = 3^n + 4^n$; $s_0 = 2$ and $s_1 = 5$.

5. Determine the particular solution of the following difference equations.
   (a) $s_n - 5s_{n-1} + 6s_{n-2} = 1$
   (b) $s_n - 2s_{n-1} + s_{n-2} = 2n + 3$
   (c) $2s_n - 7s_{n-1} + 3s_{n-2} = 3^n$
   (d) $s_n + 4s_{n-1} + 4s_{n-2} = n^2 - 2n + 3$
   (e) $s_n - 6s_{n-1} + 8s_{n-2} = 3$

6. Obtain an explicit formula for the sequence defined by the difference equation and initial condition by using backtracking method.
   (a) $s_n - s_{n-1} = 3$ with $s_1 = 5$
   (b) $s_n + s_{n-1} = 2$ with $s_1 = 3$
   (c) $s_n = 5s_{n-1} + 7$ with $s_1 = 2$

7. Obtain an explicit formula for the sequence defined by the difference equation and initial condition by using forward chaining method.

(*a*) $s_n = s_{n-1} + n$ with $s_1 = 4$

(*b*) $s_n - ns_{n-1} = 0$ with $s_1 = 7$

8. Use generating function to solve the following difference equations.

(*a*) $s_n + 8s_{n-1} + s_{n-2} = 0,\ \ n \geq 2$ with $s_0 = 0$ and $s_1 = 3$.

(*b*) $s_n - 6s_{n-1} + 9s_{n-2} = 0,\ \ n \geq 2$ with $s_0 = 2$ and $s_1 = 5$.

(*c*) $s_n - 4s_{n-1} + 4s_{n-2} = 0,\ \ n \geq 2$ with $s_1 = 1$ and $s_2 = 7$.

(*d*) $s_n - 2s_{n-1} + 2s_{n-2} = 0,\ \ n \geq 2$ with $s_0 = 1$ and $s_1 = 4$.

9. Solve each of the following recurrence relations by using generating functions.

(*a*) $s_n - 5s_{n-1} + 2s_{n-2} = 3n^2 + 1,\ n \geq 2$ with $s_0 = 0,\ s_1 = 3$.

(*b*) $s_n - 5s_{n-1} + 6s_{n-2} = 5n,\ n \geq 2$ with $s_0 = 0,\ s_1 = 12$.

(*c*) $s_n + 8s_{n-1} + 16s_{n-2} = 8,\ n \geq 2$ with $s_0 = 3,\ s_1 = 15$.

(*d*) $s_n + 6s_{n-1} - 7s_{n-2} = 2^n,\ n \geq 2$ with $s_0 = 5,\ s_1 = 8$.

(*e*) $s_n - 7s_{n-1} + 10s_{n-2} = (n+1)2^n,\ n \geq 2$ with $s_0 = 3,\ s_1 = 5$.

10. Obtain the closed form of the generating function for each of the following sequences.

(*a*) 2, 4, 8, 16, ... (*b*) –2, 2, –2, 2, –2, ...

(*c*) 0, 0, 1, 3, 5, 7, ... (*d*) 1, 2, 4, 8, 16, 32, ...

11. Show that the solution of the homogeneous linear recurrence relation

$s_{n+2} - 5s_{n+1} + 6s_n = 2^n$ is $s_n = \left(\frac{1}{6}\right)5^n$.

12. Show that $s_n = \mathrm{A}2^n + \mathrm{B}4^n$ is a solution of the recurrence relation

$s_n - 6s_{n-1} + 8s_{n-2} = 0.$

13. Show that $s_n = (\mathrm{A} + \mathrm{B}n)3^n$ is a solution of the recurrence relation

$s_n - 6s_{n-1} + 9s_{n-2} = 0\,.$

14. Obtain the solution to each of the following recurrence relation by using iterative method.

(*a*) $s_n = s_{n-1} + 3,\ s_0 = 5$ (*b*) $s_n = s_{n-1} + 3n,\ s_0 = 1$

15. Find the closed form of the exponential generating function for each of the following sequences.

(*a*) 1, 1, 1, 1, ... (*b*) 0, 0, 1, 3, 5, 7, ...

(*c*) 2, 4, 8, 16, ... (*d*) –1, 1, –1, 1, ...

16. Find the closed form of the generating function for each of the following sequence, $\{s_n\}$ where

(*a*) $s_n = 2n + 3$ (*b*) $s_n = 5^n$

(*c*) $s_n = n(7 + 3n)$ (*d*) $s_n = 7$

17. Compute the product AB using Strassen's matrix multiplication.

(a) $A = \begin{pmatrix} 2 & 5 & 1 \\ 0 & 7 & 3 \\ 7 & 5 & 0 \end{pmatrix}$ $B = \begin{pmatrix} 5 & 6 & 1 \\ 0 & 8 & 2 \\ 7 & 10 & 0 \end{pmatrix}$

(b) $A = \begin{pmatrix} 7 & 3 \\ 2 & 5 \end{pmatrix}$ $B = \begin{pmatrix} 5 & 2 \\ 0 & 8 \end{pmatrix}$

(c) $A = \begin{pmatrix} 4 & 3 \\ 2 & 1 \end{pmatrix}$ $B = \begin{pmatrix} 5 & 3 \\ 1 & 8 \end{pmatrix}$

18. Compute the product AB by using 7 multiplications and 15 additions as defined below:

$A_1 = A_{21} + A_{22}$ $A_5 = B_{12} - B_{11}$ $M_1 = A_2A_6$

$A_2 = A_1 - A_{11}$ $A_6 = B_{22} - A_5$ $M_2 = A_{11}B_{11}$

$A_3 = A_{11} - A_{21}$ $A_7 = B_{22} - B_{12}$ $M_3 = A_{12}B_{21}$

$A_4 = A_{12} - A_2$ $A_8 = A_6 - B_{21}$ $M_4 = A_3A_7$

$M_5 = A_1A_5$ $M_7 = A_{22}A_8$ $T_2 = T_1 + M_4$

$M_6 = A_4B_{22}$ $T_1 = M_1 + M_2$

The elements of the product matrix C can be computed as below:

$C_{11} = M_2 + M_3$ $C_{21} = T_2 - M_7$

$C_{12} = T_1 + M_5 + M_6$ $C_{22} = T_2 + M_5$

Show that these equations yield the correct result for the following matrices.

(a) $A = \begin{pmatrix} 4 & 3 \\ 2 & 1 \end{pmatrix}$ $B = \begin{pmatrix} 5 & 3 \\ 1 & 8 \end{pmatrix}$

(b) $A = \begin{pmatrix} 7 & 3 \\ 2 & 5 \end{pmatrix}$ $B = \begin{pmatrix} 5 & 2 \\ 0 & 8 \end{pmatrix}$

19. Determine the generating function of the numeric function s_n, where

$$s_n = \begin{cases} 3^n; & \textit{if } n \textit{ is even} \\ -3^n; & \textit{if } n \textit{ is odd} \end{cases}$$

20. Compute the partial decompositions and identify the sequence having in the following expressions as generating functions.

(a) $G(x) = \dfrac{3+2x}{1-9x^2}$ (b) $G(x) = \dfrac{10+5x}{1-6x+9x^2}$

(c) $G(x) = \dfrac{5+3x}{1-6x+5x^2}$ (d) $G(x) = \dfrac{15-3x}{2-3x+x^2}$

(e) $G(x) = \dfrac{1}{1-6x+x^2}$ (f) $G(x) = \dfrac{7x^2}{(1-2x)(1+3x)}$

21. Obtain the sequences corresponding to the generating function given below:

(*a*) $(2+x)^4$ (*b*) $(5+x)^2+(3+x)^4$

(*c*) $3x+2x^2+e^{3x}$ (*d*) $e^{3x}+e^{2x}$

22. The following algorithm computes x^n recursively, where x is a real number and n is a positive integer.

```
1. Algorithm Exp (x, n)
2. if n = 1, then
3.       return(x)
4. m = ⌊n/2⌋
5. return ( Exp(x, m) · Exp(x, n − m) )
6. end Exp
```

Let s_n be the number of multiplications required at line 5 to compute x^n. Find a recurrence relation and initial condition for the sequence $\{s_n\}$. Solve the obtained recurrence relation if n is a power of 2.

23. The following algorithm computes the maximum and minimum elements in a sequence.

```
1. Algorithm max_min (a, i, j, max, min)
2. if (i = j)
3.       max = a_i
4.       min = a_i
5. m = ⌊(i + j)/2⌋
6. max_min (a, i, m, max_left, min_left)
7. max_min (a, m + 1, j, max_right, min_right)
8. if (max_left > max_right), then
9.       max = max_left
10. else
11.       max = max_right
12. if (min_left > min_right), then
13.       min = min_right
14. else
15.       min = min_left
16. end max_min
```

Let s_n be the number of comparisons required at lines 8 and 12 for an input of size n. Find a recurrence relation and initial condition for the sequence $\{s_n\}$. Solve the obtained recurrence relation if n is a power of 2.

24. Determine t_n if $u_n = s_n + t_n$. It is given that

$$s_n = \begin{cases} 1 & n = 0 \\ 2 & n = 1 \\ 0 & n \geq 2 \end{cases} \quad \text{and} \quad u_n = \begin{cases} 1 & n = 0 \\ \\ 0 & n \geq 1 \end{cases}$$

25. Find $u_n = s_n * t_n$ in the following cases, where s_n and t_n are given below:

$$(a)\ s_n = \begin{cases} 1 & if\ 0 \leq n \leq 2 \\ 0 & if\ n \geq 3 \end{cases} \quad \text{and} \quad t_n = \begin{cases} 1 & if\ 0 \leq n \leq 2 \\ 0 & if\ n \geq 3 \end{cases}$$

$$(b)\ s_n = 2^n;\ n \geq 0 \quad \text{and} \quad t_n = \begin{cases} 1 & if\ 0 \leq n \leq 2 \\ 3^n & if\ n \geq 3 \end{cases}$$

26. Determine $u_n = s_n + t_n$ in the following cases, where s_n and t_n are given below:

$$(a)\ s_n = \begin{cases} 1 & if\ 0 \leq n \leq 2 \\ 0 & if\ n \geq 3 \end{cases} \quad \text{and} \quad t_n = \begin{cases} 1 & if\ 0 \leq n \leq 2 \\ 0 & if\ n \geq 3 \end{cases}$$

$$(b)\ s_n = 2^n;\ n \geq 0 \quad \text{and} \quad t_n = \begin{cases} 1 & if\ 0 \leq n \leq 2 \\ 3^n & if\ n \geq 3 \end{cases}$$

27. Compute the product AB using Strassen's matrix multiplication.

$$(a)\ A = \begin{pmatrix} 1 & 4 & 6 \\ 3 & 0 & 5 \end{pmatrix} \qquad B = \begin{pmatrix} 3 & 5 & 1 \\ 0 & 7 & 2 \\ 5 & 9 & 1 \end{pmatrix}$$

$$(b)\ A = \begin{pmatrix} 1 & 7 \\ 2 & 5 \end{pmatrix} \qquad B = \begin{pmatrix} 5 & 9 & 0 & 1 \\ 3 & 7 & 8 & 1 \end{pmatrix}$$

6

Combinatorics

6.0 INTRODUCTION

In one form or the other counting originated with the primitive man. He did not find any problem as long as it was within the limits of his fingers. As it crossed the limits of his fingers, he found out other devices and invented the numbers. But counting becomes laborious and impossible even with numbers. Hence new concepts and ideas are developed which gave rise to permutations and combinations. In discrete mathematics we frequently encounter the problem of counting. The branch of Discrete Mathematics dealing with counting problem is called combinatorics. The different techniques of counting problem are important in computer science, especially in algorithm analysis and design. In this chapter we will be dealing with two important aspects, *i.e.*, permutation and combination. Permutation is defined as the arrangement of elements whereas combination is defined as selection of elements.

6.1 FUNDAMENTAL PRINCIPLE OF COUNTING

Basically fundamental principle of counting is of two types. These are addition principle and multiplication principle.

Addition Principle: If one event can occur in n_1 different ways and another event can occur in n_2 different ways, then exactly one of these events takes place in $(n_1 + n_2)$ ways. The above addition principle can be extended to finite number of events. This addition principle is otherwise known as principle of disjunctive counting.

For example assume that a student have to choose an elective paper in computer science from one of the four lists containing 5, 6, 5 and 4 elective papers respectively.

It indicates that the student can choose an elective from the first list in 5 ways, from the second list in 6 ways, from the third list in 5 ways whereas from the fourth list in 4 ways. Therefore, the total number of ways that the student can choose an elective paper from the four lists is equal to $(5 + 6 + 5 + 4) = 20$.

Multiplication Principle: If one event can occur in n_1 different ways and another event can occur in n_2 different ways then both these events take place in $(n_1 n_2)$ ways. The above principle can be extended to finite number of events. This multiplication principle is otherwise known as principle of sequential counting.

For example consider a cinema hall with 3 entrances and 5 exits. Therefore, the number of ways that a person can enter and exit from the cinema hall is $(3 \times 5) = 15$.

6.2 FACTORIAL NOTATION

The continued product of first n natural numbers is denoted as $n!$ and read as factorial n. Mathematically,

$$n! = 1.\ 2.\ 3.\ 4.\ \ldots\ldots(n-1).\ n$$
$$= \{1.\ 2.\ 3.\ldots.\ (n-1)\}.\ n$$
$$= (n-1)!\ n$$

For example $5! = 5 \times 4 \times 3 \times 2 \times 1 = 120$ and $4! = 1 \times 2 \times 3 \times 4 = 24$. Therefore, it is clear that $5! = 4! \times 5 = 24 \times 5 = 120$.

6.3 PERMUTATION

Assume that, we have 3 digits 1, 3 and 5. Then the two digit numbers that can be formed out of the given digits if the digits are not repeated as: 13, 15, 35, 31, 51, 53. Therefore, there are 6 possible ways of getting 2 digit numbers out of three digits. This is nothing but the arrangement of 2 digits out of three digits. Similarly, if we make an arrangement of all the three digits, then we have 135, 153, 351, 315, 513 and 531 as possible arrangements. The number of different arrangement or ordering that can be made out of the given number of objects by taking some or all at a time is called the permutation. If r-objects are arranged out of n-distinct objects, then we call it as r-permutation.

Theorem The number of arrangements of r different objects out of n distinct objects ($r \le n$) is denoted by $P(n, r)$ and is call r-permutation of n objects. It is defined as

$$P(n, r) = n(n-1)(n-2) \ldots. (n-r+1)$$
$$= \frac{n!}{(n-r)!}$$

Proof: Given that there are n distinct objects and we have to arrange r objects. This is same as the number of ways in which r places can be filled up by n distinct objects.

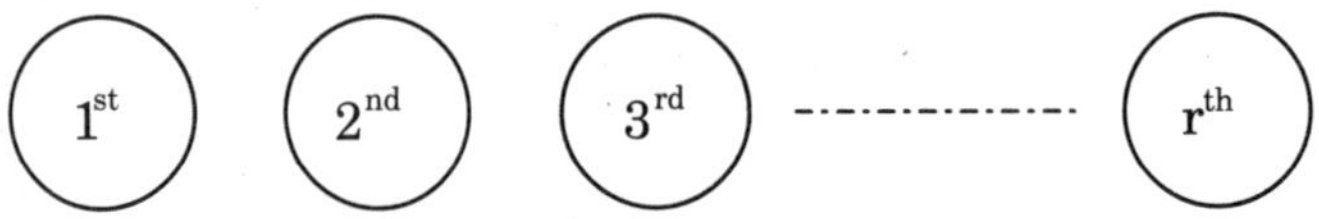

From the figure it is clear that any one of the n object can be put in 1st place *i.e.*, the 1st place can be filled up in n distinct ways. Therefore, we are left out with $(n-1)$ objects, so the 2nd place can be filled up in $(n-1)$ distinct ways. Since each way of filling up the first place can be associated with each way of filling up the second place, so by multiplication principle the first two places can be filled up in $n(n-1)$ ways. Similarly, for the 3rd place we have left out with $(n-2)$ objects, so the 3rd place can be filled up by any one of these $(n-2)$ objects. Therefore, by multiplication principle the first three places can be filled up in $n(n-1)(n-2)$ ways. Proceeding in this manner the total number of ways in which r places can be filled up is given as

$$P(n,r) = n(n-1)(n-2)\ldots r \text{ factors}$$
$$= n(n-1)(n-2) \ldots. (n-r+1)$$

$$= \frac{n(n-1)(n-2)(n-3)\cdots(n-r+1)(n-r)(n-r-1)\cdots 1}{1.2.3.4.\cdots\ (n-r-1)(n-r)}$$

$$= \frac{n!}{(n-r)!}$$

6.3.1 Permutation with Repetition

In this case there is no restriction on the number of times a particular object may occur in r-permutation of n objects. It implies that a repetition is allowed all the r places to be filled up by any of the n objects. Therefore, by multiplication principle the total number of ways in which r places can be filled up is given as

$$n \times n \times n \times \ldots\ldots \times n \quad (r\text{-factors})$$
$$= n^r$$

For example consider 3 prizes to be given to 4 students. Hence, it is clear that the 1st prize can be taken by any of the 4 students and similarly the 2nd prize can be taken by any of the 4 students and so on. Therefore, by using multiplication principle the total number of possible ways

$$= 4 \times 4 \times 4 = 64$$

6.3.2 Permutation of *n* Things not all Different

Assume that there n objects, among which K_1 are alike say A, K_2 are of second type say B. Let the elements are $A_1, A_2, A_3, \cdots, A_{k_1}$ and $B_1, B_2, B_3, \cdots, B_{k_2}$. But we know that, if the n objects are dissimilar, then it can be arranged in $n!$ different ways.

Now, consider the total number of permutations be X. For one of this permutation $A_1, A_2, A_3, \cdots, A_{k_1}$, k_1 objects can be arranged among themselves in $k_1!$ ways. Therefore, for X permutations the total number of possible ways are $Xk_1!$.

Again for one of these $Xk_1!$ permutations, the k_2 objects $B_1, B_2, B_3, \cdots, B_{k_2}$ can be arranged in $k_2!$ ways. Therefore, for $Xk_1!$ permutations, the total number of possible ways are equal to $(Xk_1!\, k_2!)$. Therefore, we get

$$Xk_1!\, k_2! = n!$$

This implies that $$X = \frac{n!}{k_1!\, k_2!}$$

Note: The above concept can be generalized to any number of objects. Permutation of n objects out of which k_1 objects are of 1st kind say A, k_2 objects are of 2nd kind say B, k_3 objects are of 3rd kind say C and so on is given by $X = \dfrac{n!}{k_1!\, k_2!\, k_3! \cdots}$, where $k_1 + k_2 + k_3 + \ldots = n$.

6.3.3 Circular Permutation

Arrangement of objects in a circle is called circular permutation. In this case of arrangement of objects, there is no distinction between the 1st place and the last place. Hence, in circular permutation we fix one of the n objects in the first place and arrange remaining $(n-1)$ objects among themselves. This can be done in $P(n-1, n-1)$ ways *i.e.*, $(n-1)!$.

For example, consider a problem of arranging 5 members for seating on a round table. It is a problem of circular permutation. Therefore, the total number of ways that the members can be seated in the round table is $(5-1)! = 4! = 24$.

Notes: 1. If the anticlockwise and clockwise arrangements are different, then the total number of arrangements is equal to $(n-1)!$.

2. If both clockwise and anticlockwise arrangements are same, then the total number of arrangements is equal to $\frac{1}{2}(n-1)!$.

6.3.4 Restricted Permutations

Sometimes in arrangement, we restrict some particular objects to be either always occur or will never occur. Such type of permutations are known as restricted permutations and this leads to two cases.

***Case* –1** Permutation of n-objects in r-places when 'x' particular objects will always occur. In such case, first the 'x' particular objects can be arranged in r-places and then remaining $(r-x)$ places will be filled up by the remaining $(n-x)$ objects. Therefore, the total number of arrangements is equal to $P(r, x) \times P(n-x, r-x)$.

***Case* – 2** Permutation of n-objects in r-places when 'x' particular objects will never occur. In this case $(n-x)$ objects will fill up the r-places. Therefore, the total number of arrangements is equal to $P(n-x, r)$.

6.3.5 Interpretation of Factorial 0

The value of zero (0) factorial can be determined with the help of permutation. We know that

$$P(n, r) = n(n-1)(n-2)\ldots(n-r+1)$$

On taking $r = n$ we get

$$\begin{aligned} P(n, n) &= n(n-1)(n-2)\ldots(n-n+1) \\ &= n(n-1)(n-2)\ldots 1 \\ &= n! \end{aligned}$$

This implies that $\frac{n!}{n!\ 0!} = n!$ and hence $0! = 1$.

■ 6.4 COMBINATION

We have seen that permutation is an ordered arrangement of objects. However, it is observed that order is not significant in some cases. For example, consider an examination and a student has to answer four questions out of seven questions. In this case, a selection is to be made irrespective of order. We call it as combination. Assume that, we have 3 objects A, B and C. If we consider the groups of two objects without taking into account the order then the different groups are AB, BC and CA. Therefore, the total number of groups is 3. This is nothing but the selection of 2 objects out of three objects. The number of different selections that can be made out of given number of objects by taking some or all of them at a time is called the combination. In combination we do not give importance to the order of arrangement.

Theorem The number of selections or combinations of r-different objects out of n-objects ($r \le n$) is denoted by $C(n, r)$ and we call it as r-combination of n-objects. This is defined as

$$C(n, r) = \frac{n!}{r!\ (n-r)!}.$$

Proof: We know that $P(n, r) = \dfrac{n!}{(n-r)!};\ r \le n$

As discussed earlier the order of arrangement is not significant in combination. In every combination there are r-objects that can be arranged among them in $r!$ number of ways. Therefore, $C(n, r)$ combinations will lead to $r! \times C(n, r)$ number of permutations. We call it as a permutation of r-objects out of n-objects *i.e.,* $P(n, r)$. Hence we get,

$$r!\ C(n, r) = P(n, r)$$

i.e.,

$$C(n,r) = \frac{P(n,r)}{r!}$$

$$= \frac{n!}{r!\,(n-r)!}.$$

6.4.1 Important Properties

The important properties of combination are discussed below.

(*a*) $C(n, r) = C(n, n-r)$

(*b*) $C(n, r) : C(n, r-1) = (n-r+1) : r$

Proof: (*a*) We know that $C(n,r) = \dfrac{n!}{r!(n-r)!}$

On taking $r = (n-r)$, we will get

$$C(n, n-r) = \frac{n!}{(n-r)!(n-\ (n-r))!}$$

$$= \frac{n!}{(n-r)!\ r!} = C(n, r)$$

i.e.,

$$C(n, r) = C(n, n-r)$$

(*b*) In order to prove $C(n, r) : C(n, r-1) = (n-r+1) : r$, consider the L.H.S.

$$\frac{C(n, r)}{C(n, r-1)} = \frac{\dfrac{n!}{r!(n-r)!}}{\dfrac{n!}{(r-1)!(n-r+1)!}}$$

$$= \frac{n!}{r!(n\ \ r)!} \times \frac{(r-1)!\ (n-r+1)!}{n!}$$

$$= \frac{n!}{(r-1)!\ r\ (n-r)!} \times \frac{(r-1)!\ (n-r)!\ (n-r+1)}{n!}$$

$$= \frac{n-r+1}{r}$$

6.4.2 Pascal's Identity

Let n and r be positive integers with $n \geq r$. Then $C(n, r) + C(n, r-1) = C(n+1, r)$

Proof:

$$C(n, r) + C(n, r-1) = \frac{n!}{r!\ (n-r)!} + \frac{n!}{(r-1)!\ (n-r+1)!}$$

$$= \frac{n!}{(r-1)!\ r(n-r)!} + \frac{n!}{(r-1)!\ (n-r)!\ (n-r+1)}$$

$$= \frac{n!}{(r-1)!\ (n-r)!}\left(\frac{1}{r} + \frac{1}{(n-r+1)}\right)$$

$$= \frac{n!}{(r-1)!\ (n-r)!}\left(\frac{n+1}{r(n-r+1)}\right)$$

$$= \frac{(n+1)!}{r\ !\ (n-r+1)!} = C(n+1, r)$$

This proves the Pascal's identity.

6.4.3 Restricted Combination

Sometimes we impose some restrictions that some particular objects to be either always occur or will never occur. Such type of combination is known as restricted combination and this leads to two cases.

***Case* – 1** Combination of n-objects taken 'r' at a time when 'x' particular objects will always occur. In such case, we keep aside the 'x' particular objects which always occur and then we choose the $(r-x)$ objects from the remaining $(n-x)$ objects. Therefore, the total number of combinations is equal to $C(n-x, r-x)$.

***Case* – 2** Combination of 'n' objects taken 'r' at a time when 'x' particular objects will never occur. In such cases, after removing 'x' objects from 'n' objects 'r' objects are to be selected from $(n-x)$ objects. Therefore, the total number of combinations is equal to $C(n-x, r)$.

Notes: 1. On taking $r = n$, we will get

$$C(n, n) = \frac{n!}{n!\ (n-n)!} = \frac{1}{0!} = \frac{1}{1} = 1$$

Therefore, $C(n, n) = 1$

2. On taking $r = 0$, we will get

$$C(n, n) = \frac{n!}{n!\ (n-n)!} = \frac{1}{0!} = \frac{1}{1} = 1$$

Therefore, $C\ (n, 0) = 1$

3. On taking $r = 1$, we will get

$$C(n, 1) = \frac{n!}{1!\ (n-1)!} = \frac{(n-1)!\ n}{(n-1)!} = n$$

Therefore, $C(n, 1) = n$

■ 6.5 THE BINOMIAL THEOREM

The sum of two distinct terms, say $(x + y)$, is called a binomial. Binomial theorem is defined as a formula for the power of a binomial, *i.e.*, $(x + y)^n$, $n \in \mathrm{N}$. The binomial expansion for the case $n = 2$ was used by the Greek mathematician Euclid. However, Omar Khayyam the Arab mathematician is credited with the binomial expansion for higher natural numbers. Later the great British scientist Sir Isaac Newton generalized the binomial theorem for negative integral and fractional indices.

Theorem Let n is a positive integer. Then for all x and y,

$$(x + y)^n = C(n, 0)x^n + C(n, 1)x^{n-1}y + C(n, 2)x^{n-2}y^2 + \cdots + C(n, n)y^n$$
$$= \sum_{r=0}^{n} C(n, r)x^{n-r}y^r$$

Proof: We will prove this by the method of mathematical induction on n.

For $n = 1$ we have $(x + y)^1 = x + y = C(1, 0)x + C(1, 1)y$

For $n = 2$ we have $(x + y)^2 = x^2 + 2xy + y^2$

$$= C(2, 0)x^2 + C(2, 1)x^{2-1}y + C(2, 2)y^2$$

So, the statement is true for $n = 1$ and 2. Assume that the statement is true for $n = k$. Therefore, we have,

$$(x+y)^k = C(k, 0)x^k + C(k, 1)x^{k-1}y + C(k, 2)x^{k-2}y^2 + \cdots + C(k, k)y^k$$
$$= \sum_{r=0}^{k} C(k, r)x^{k-r}y^r$$

Now, for $n = k + 1$, we have

$$(x + y)^{k+1} = (x + y)(x + y)^k = x(x + y)^k + y(x + y)^k$$
$$= x\sum_{r=0}^{k} C(k, r)x^{k-r}y^r + y\sum_{j=0}^{k} C(k, j)x^{k-j}y^j$$
$$= \sum_{r=0}^{k} C(k, r)x^{k+1-r}y^r + \sum_{j=0}^{k} C(k, j)x^{k-j}y^{j+1}$$
$$= x^{k+1} + \sum_{r=1}^{k} C(k, r)x^{k+1-r}y^r + \sum_{j=0}^{k} C(k, j)x^{k-j}y^{j+1}$$
$$= x^{k+1} + \sum_{r=1}^{k} C(k, r)x^{k+1-r}y^r + \sum_{r=1}^{k+1} C(k, r-1)x^{k-r+1}y^r \quad ; \text{[on taking } j = r - 1]$$
$$= x^{k+1} + \sum_{r=1}^{k} C(k, r)x^{k+1-r}y^r + \sum_{r=1}^{k} C(k, r-1)x^{k+1-r}y^r + y^{k+1}$$
$$= x^{k+1} + \sum_{r=1}^{k} \big(C(k, r) + C(k, r-1)\big)x^{k+1-r}y^r + y^{k+1}$$
$$= x^{k+1} + \sum_{r=0}^{k} C(k+1, r)x^{(k+1)-r}y^r + y^{k+1} \quad ; \quad \text{[on using Pascal's identity]}$$
$$= \sum_{r=0}^{k+1} C(k+1, r)x^{(k+1)-r}y^r$$

Therefore, the statement is true for $n = k + 1$ and hence the statement is true for all non-negative integer values of n.

6.5.1 Binomial Coefficient

The natural numbers $C(n, r)$, $r = 0, 1, 2, \ldots, n$ are called binomial coefficients. This is defined as below.

$$C(n,\ r) = \frac{n!}{r!\ (n-r)!};\ r = 0,\ 1,\ 2,\ 3,\ \cdots,\ n$$

6.5.2 General Term

If $T_0, T_1, T_2, \cdots, T_n$ are the terms in the binomial expansion, then the $(r+1)^{\text{th}}$ term T_{r+1} is called the general term. This is defined as below.

$$\begin{aligned} T_{r+1} &= C(n,\ r)x^{n-r}y^r \\ &= \frac{n!}{r!\ (n-r)!}x^{n-r}y^r \end{aligned}$$

6.5.3 Pascal Triangle

The geometric arrangement of the binomial coefficients in a triangle is known as Pascal's triangle named after Blaise Pascal (1623–1662). It can be constructed in this manner. Write only the number 1 on the first row. Any coefficient in a row in the triangle is obtained by adding the coefficients to its immediate left and the coefficient to the immediate right in the preceding row. If either the one to the right or left is not present, then substitute a zero in its place. For example, in the figure given below the numbers 1 and 2 in the third row are added to produce 3 in the fourth row. Generally, this construction uses Pascal's identity that we have already discussed.

$$\begin{array}{lccccccccccc}
n = 0 & & & & & & 1 & & & & & \\
n = 1 & & & & & 1 & & 1 & & & & \\
n = 2 & & & & 1 & & 2 & & 1 & & & \\
n = 3 & & & 1 & & 3 & & 3 & & 1 & & \\
n = 4 & & 1 & & 4 & & 6 & & 4 & & 1 & \\
n = 5 & 1 & & 5 & & 10 & & 10 & & 5 & & 1
\end{array}$$

6.5.4 Properties of Binomial Expansion

Some simple patterns are immediately apparent in binomial expansion $(x + y)^n$ where, n is a non-negative integer.

(*a*) The binomial expansion contains exactly $(n + 1)$ terms.

(*b*) In each term of the expansion, sum of powers of x and y is always n.

(*c*) All the terms of the expansion are obtained from the general term by taking $r = 0, 1, \ldots, n$.

6.5.5 Vander Monde's Identity

Let m, n and r be non-negative integers with r not exceeding either m or n. Then

$$C(m+n,\ r)=\sum_{j=0}^{r} C(m,\ r-j)C(n,\ j)$$

Proof: Let us consider two binomial expansions $(x+y)^m$ and $(x+y)^n$ where m and n both are non-negative integers. Therefore, by using binomial theorem we get

$$(x+y)^m=\sum_{r=0}^{m} C(m,\ r)x^{m-r}y^r \quad \text{...(i)}$$

$$(x+y)^n=\sum_{j=0}^{n} C(n,\ j)x^{n-j}y^j \quad \text{...(ii)}$$

On multiplying above two equations, we get

$$(x+y)^{m+n}=\left(\sum_{r=0}^{m} C(m,\ r)x^{m-r}y^r\right)\left(\sum_{j=0}^{n} C(n,\ j)x^{n-j}y^j\right)$$

On equating the binomial coefficient of $x^{m+n-r}y^r$ on both sides we will get

$$C(m+n,\ r)=C(m,\ 0)\,C(n,\ r)+C(m,\ 1)\,C(n,\ r-1)+\cdots+C(m,\ r)\,C(n,\ 0)$$

i.e.,
$$C(m+n,\ r)=\sum_{j=0}^{r} C(m,\ r-j)C(n,\ j)$$

6.5.6 Identities Involving Binomial Coefficients

In this section, we will discuss important identities involving binomial coefficients. Through out this section we shall use C_0, C_1, C_2, $C_3, \cdots, C_n$ respectively in place of $C(n,\ 0)$, $C(n,\ 1)$, $C(n,\ 2)$, $C(n,\ 3)$, $\cdots$, $C(n,\ n)$ for convenience of notation.

1. $C_0+\ C_1+C_2+\ C_3+\cdots+C_n=2^n$

Proof: We know from binomial theorem that

$$(1+x)^n=C_0+C_1x+C_2x^2+C_3x^3+\quad +C_nx^n$$

On putting $x = 1$ in the above identity, we get

$$(1+1)^n=C_0+C_1 1+C_2 1^2+C_3 1^3+\ \cdots\ +C_n 1^n$$

i.e.,
$$C_0+C_1+C_2+C_3+\ \cdots\ +C_n=2^n$$

2. $C_0+\ C_2+C_4+\ C_6+\cdots=2^{n-1}=C_1+\ C_3+C_5+\ C_7+\cdots$

Proof: We know from binomial theorem that

$$(1+x)^n=C_0+C_1x+C_2x^2+C_3x^3+\ \cdots\ +C_nx^n$$

On putting $x = 1$ and $x = -1$ in the above identity, we get respectively

$$C_0+C_1+C_2+C_3+C_4+C_5+\ \cdots\ +C_n=2^n \quad \text{...(i)}$$

$$C_0-C_1+C_2-C_3+C_4-C_5+\ \cdots\ +(-1)^nC_n=0 \quad \text{...(ii)}$$

On adding above two equations (1) and (2), we get

$$2C_0 + 2C_2 + 2C_4 + \cdots = 2^n$$

i.e., $$2(C_0 + C_2 + C_4 + \cdots) = 2^n$$

i.e., $$C_0 + C_2 + C_4 + \cdots = 2^{n-1}$$

Therefore, equation (2) reduces to

$$C_1 + C_3 + C_5 + C_7 + \cdots = 2^{n-1}$$

3. $C_0^2 + C_1^2 + C_2^2 + C_3^2 + \cdots + C_n^2 = C(2n, n)$

Proof: We know from Vander Monde's identity that

$$C(m+n, r) = \sum_{j=0}^{r} C(m, r-j)C(n, j)$$

On putting $m = r = n$ in the above identity, we get

$$C(n+n, n) = \sum_{j=0}^{r} C(n, n-j)C(n, j) = \sum_{j=0}^{r} (C(n, j))^2; \qquad [\because C(n, n-j) = C(n, j)]$$

i.e., $$C_0^2 + C_1^2 + C_2^2 + C_3^2 + \cdots + C_n^2 = C(2n, n)$$

Alternative Proof: We apply a special trick to get the identity. Since we are to sum the squares of binomial coefficients, by considering the coefficients of $(1+x)^n$ alone we may not get the result. We know from binomial theorem that

$$(1+x)^n = C_0 + C_1x + C_2x^2 + C_3x^3 + \cdots + C_nx^n$$

and $$(x+1)^n = C_0x^n + C_1x^{n-1} + C_2x^{n-2} + C_3x^{n-3} + \cdots + C_n$$

On multiplying both sides of above equations we get

$$(1+x)^{2n} = (C_0 + C_1x + \cdots + C_nx^n)(C_0x^n + C_1x^{n-1} + \cdots + C_n)$$

Therefore, the coefficient of x^n on both sides of the above equation must be equal. Thus, we get,

$$C(2n, n) = C_0^2 + C_1^2 + C_2^2 + \cdots + C_n^2$$

i.e., $$C_0^2 + C_1^2 + C_2^2 + \cdots + C_n^2 = \frac{(2n)!}{(n!)^2}.$$

6.6 BINOMIAL THEOREM FOR RATIONAL INDEX

Any binomial $(x + y)$ can be written in the form of $x\left(1+\frac{y}{x}\right)$ or $y\left(1+\frac{x}{y}\right)$. Therefore, it is clear that $(x+y)^n$ can be reduced to the form $(1+z)^n$. Now we state, the binomial theorem of $(1+z)^n$ when n is not a positive integer or a negative or positive fraction.

When n is a positive or negative fraction or negative integer and $|z|<1$

$$(1+z)^n = 1 + nz + \frac{n(n-1)}{2!}z^2 + \cdots + \frac{n(n-1)(n-2)\cdots(n-r+1)}{r!}z^r + \cdots$$

It is to be noted that the number of terms in the right hand side is infinite as the binomial coefficient $\frac{n(n-1)(n-2)\cdots(n-r+1)}{r!}$ never vanishes since $(n-r+1)$ can never be zero. The proof is beyond the scope of this book. However, we shall discuss the application of this theorem to approximation problems. For example, we can apply this theorem to compute the nth root of a positive integer.

The $(r+1)^{\text{th}}$ term T_{r+1} is considered as the general term. It is given as

$$T_{r+1} = \frac{n(n-1)(n-2)\cdots(n-r+1)}{r\ !} z^r.$$

6.7 THE CATALAN NUMBERS

In combinatorics, the Catalan numbers figure a series of natural numbers that generally occur in various counting problems. It is named after Belgian mathematician Eugene Charles Catalan.

Theorem The nth Catalan number C_n is defined in terms of binomial coefficient as

$$C_n = \frac{1}{n+1}\binom{2n}{n} = \frac{1}{n+1}\left(\frac{(2n)!}{n!\ n!}\right) = \frac{(2n)!}{(n+1)!\ n!} \quad \text{for } n \geq 0$$

Proof: There are many ways to proof the above relation. Here, we will be using generating function to proof the relation. It can be easily seen that the above relation satisfy the recurrence relation

$$C_0 = 1 \quad \text{and} \quad C_{n+1} = \sum_{i=0}^{n} C_i C_{n-i} \qquad \text{for } n \geq 0. \qquad \ldots(1)$$

The generating function for the Catalan numbers is defined by

$$c(x) = \sum_{n=0}^{\infty} C_n x^n$$

It is clear from the equation (1) that the sum on the right side of the above recurrence relation is the coefficient of x^n in the product $\left(\sum_{i=0}^{\infty} C_i x^i\right)^2$. Therefore, we have

$$\left(\sum_{i=0}^{\infty} C_i x^i\right)^2 = \sum_{n=0}^{\infty} C_{n+1} x^n$$

On multiplying both sides by x, we get

$$x\left(\sum_{i=0}^{\infty} C_i x^i\right)^2 = \sum_{n=0}^{\infty} C_{n+1} x^{n+1} = \sum_{n=1}^{\infty} C_n x^n = -1 + \sum_{n=0}^{\infty} C_n x^n$$

i.e., $\quad x(c(x))^2 = -1 + c(x)$

i.e., $\quad x(c(x))^2 - c(x) + 1 = 0$

i.e., $\quad c(x) = \dfrac{1-\sqrt{1-4x}}{2x}$

Now, we have to expand the square root as a power series. This can be done by using binomial theorem. Therefore, in general, we get

$$\sqrt{1-4x} = 1 - 2\sum_{n=1}^{\infty}\binom{2n-2}{n-1}\left(-\frac{1}{4}\right)^{n}\frac{(-4x)^{n}}{n}$$

On using the above relation in $c(x)$ we get

$$c(x) = \frac{1-1+2\sum_{n=1}^{\infty}\binom{2n-2}{n-1}\left(-\frac{1}{4}\right)^{n}\frac{(-4x)^{n}}{n}}{2x} = \frac{1}{x}\sum_{n=1}^{\infty}\binom{2n-2}{n-1}\frac{x^{n}}{n}$$

$$\Rightarrow \qquad c(x) = \sum_{n=1}^{\infty}\binom{2n-2}{n-1}\frac{x^{n-1}}{n} = \sum_{n=0}^{\infty}\binom{2n}{n}\frac{x^{n}}{n+1}$$

$$\Rightarrow \qquad \sum_{n=0}^{\infty} C_n x^n = \sum_{n=0}^{\infty}\binom{2n}{n}\frac{x^{n}}{n+1}$$

On equating coefficients of x^n both sides in the above equation we get the desired formula for C_n. Therefore,

$$C_n = \frac{1}{n+1}\binom{2n}{n}$$

Note: An alternative expression for n^{th} Catalan number C_n that indicates C_n is a natural number is given as

$$C_n = \binom{2n}{n} - \binom{2n}{n-1} \quad \text{for } n \geq 1$$

6.7.1 Applications of Catalan Numbers

It is observed that in many cases Catalan numbers provides solutions to the counting problems in combinatorics. Here we will discuss some examples, with illustrations of the case $C_3 = 5$.

1. Successive applications of a binary operator can be represented in terms of a binary tree. It indicates that C_n is the number of rooted ordered binary trees with $(n + 1)$ leaves (L). Here we have taken $n = 3$.

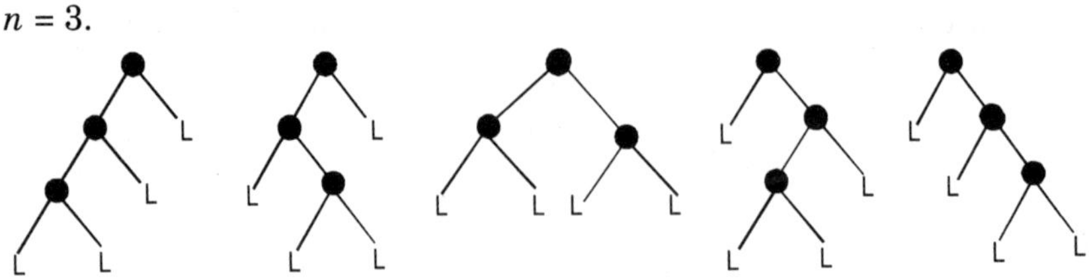

2. The Catalan number C_n is the number of different ways a convex polygon with $(n + 2)$ sides can be cut into triangles by connecting vertices with straight lines. The following pentagons illustrate the case $n = 3$.

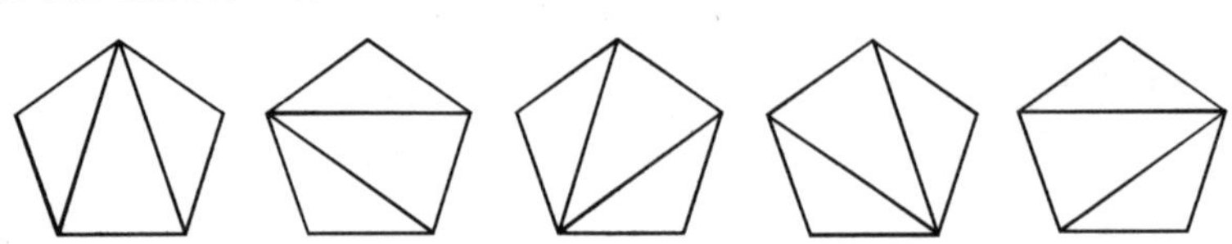

3. The Catalan number C_n is the number of Dyck words of length $2n$. A Dyck word is a string consisting of n A's and n B's such that no initial segment of the string has more B's than A's. The following Dyck words illustrate the case $n = 3$, *i.e.*, Dyck words of length 6.

AAABBB ABAABB ABABAB AABBAB AABABB

4. On considering the symbol A as an open parentheses and B as a close parentheses. The Catalan number C_n counts the number of expressions containing n pairs of parentheses that are correctly matched.

((( ))) ( )(( )) ( )( )( ) (( ))( ) (( )( ))

6.8 RAMSEY NUMBER

In combinatorics, we came across many interesting problems and a party problem is one among them. Before we discuss Ramsey number first we state the party problem. Prove that in any party with people there are three people who all know each other or that there are three people who do not all know each other. The simplest way to solve this problem is using graph theory, but it is one of the important illustrations in the area of logic that was developed by F. P. Ramsey. For two colours, Ramsey's theorem states that for any two positive integers m and n, there exists a smallest integer $R(m, n) = p$ such that if we arbitrarily colour the edges of the complete graph (that is a simple graph where each vertex 'u' is adjacent to every other vertex 'v') with 'p' vertices blue or red, then we must get a blue complete graph with 'm' vertices or a red complete graph with 'n' vertices as subgraph. Thus, it is clear that $R(m, n) = p$ signifies an integer that depends on both m and n. The above idea can be extended to finite number of colours. Therefore, the theorem states that for any given number of colours k, and any given integers $n_1, n_2, n_3,..., n_k$, there exists a number $R(n_1, n_2, n_3,..., n_k) = q$ such that if the edges of a complete graph with 'q' vertices are coloured with 'k' different colours, then for some i between 1 and k, it must contain a complete subgraph with n_i vertices whose edges are all colour i. The numbers $R(m, n)$ in Ramsey's theorem are known as Ramsey numbers. Now our aim is to compute the value of $R(m, n)$ for small values of m and n.

First of all, we have to think does it make any sense $R(m, 1)$ or $R(1, n)$. Let $R(m, 1) = p$. Since complete graph with 1 vertex does not have any edges, colouring the edges of complete graph with p vertices blue or red does not help us to get a complete graph with 1 vertex with blue or red edges. Therefore, we assume that both m and n are at least 2. Thus the smallest value of $R(m, n)$ must be $R(2, 2)$. Now the question arises, what complete graph with 2 vertices can wc colour to be sure that we get either a blue edge or a red edge? If the single edge is coloured blue, then we have a blue edge at the same time if it is coloured red, then we have a red edge. Therefore, it is cleared that $R(2, 2) = 2$.

Theorem If m and n are two positive integers both greater than 2, then

$$R(m, n) \le R(m-1, n) + R(m, n-1)$$

Proof: Let $p = R(m-1, n)$, $q = R(m, n-1)$ and $r = (p + q)$.

Consider a group $\{1, 2, 3, ..., r\}$ of r people. Let us assume that L_1 be the set of people known to person 1 whereas L_2 be the set of people not known to person 1. It indicates that the two sets together have $(r - 1)$ people; so either L_1 has at least p people or L_2 has at least q people. It leads to two cases.

(*a*) If L_1 has $p = R(m-1, n)$ people, then by definition, it contains a subset of $(m - 1)$ people known to each other or it has a subset of n people unknown to one another. In the earlier case the $(m - 1)$ people and person 1 together constitute m people known to each other.

Thus, in their case, a group of $R(m-1, n)+R(m, n-1)$ people necessarily includes m persons who all know each other or n people who do not know each other. Therefore,

$$R(m, n) \leq R(m-1, n)+R(m, n-1)$$

(*b*)The same argument follows by symmetry when L_2 contains q people.

Theorem If m and n are two positive integers both greater than 1, then

$$R(m, n) \leq \binom{m+n-2}{m-1}$$

Proof: When $m = 2$ and $n = 2$ we have $R(2, 2) = 2 = \binom{2}{1}$.

Therefore, the given relation holds with equality. We will proof this by the method induction on $k = m + n$. As we have just seen, the given relation is true when $k = 4$. Assume that the relation is true for $k - 1$. Therefore,

$$R(m-1, n) \leq \binom{m+n-3}{m-2} \text{ and } R(m, n-1) \leq \binom{m+n-3}{m-1}$$

On adding the above inequalities, we get

$$R(m-1, n) + R(m, n-1) \leq \binom{m+n-3}{m-2} + \binom{m+n-3}{m-1}$$

By using Pascal's identity, we get the following:

$$R(m-1, n) + R(m, n-1) \leq \binom{m+n-2}{m-1} \quad ...(i)$$

Again, we know from the previous theorem that

$$R(m, n) \leq R(m-1, n)+R(m, n-1) \quad ...(ii)$$

Therefore, from equations (*i*) and (*ii*), we get

$$R(m, n) \leq R(m-1, n) + R(m, n-1) \leq \binom{m+n-2}{m-1}$$

i.e.,

$$R(m, n) \leq \binom{m+n-2}{m-1}.$$

6.8.1 Ramsey Numbers

All known Ramsey numbers $R(m, n)$ for values of m and n up to 7 are shown in the following table. If the exact value is unknown, the table lists the best known bounds. $R(m, n)$ for values of m and n less than 3 are given by $R(1, n) = 1$ and $R(2, n) = n$ for all values of n. The development of Ramsey number research has been written by Stanislaw Radziszowski.

m \ n	1	2	3	4	5	6	7
1	**1**						
2	**1**	**2**					
3	**1**	**3**	**6**				
4	**1**	**4**	**9**	**18**			
5	**1**	**5**	**14**	**25**	**43–49**		
6	**1**	**6**	**18**	**35–41**	**58–87**	**102–165**	
7	1	7	23	49–61	80–143	113–298	205–540

The upper entries across the main diagonal are omitted because of trivial symmetry $R(m, n) = R(n, m)$.

SOLVED EXAMPLES

Example 1 *Find the value of* $\frac{7!}{3!}$ *and check whether* $\frac{7!}{3!} = 3!$

Solution: $\frac{7!}{3!} = \frac{3!4\cdot5\cdot6\cdot7}{3!} = 840$ and $3! = 3 \times 2 \times 1 = 6$

Therefore, $\frac{7!}{3!} \neq 3!$

Example 2 *Evaluate the following for n = 8 and r = 3.*

(a) $(n-r)!$ (b) $\frac{n!}{r!}$ (c) $\frac{n!}{r!(n-r)!}$

Solution: Given that $n = 8$ and $r = 3$. Therefore,

(a) $(n-r)! = (8-3)! = 5! = 5 \times 4 \times 3 \times 2 \times 1 = 120$

(b) $\frac{n!}{r!} = \frac{8!}{3!} = \frac{8\cdot7\cdot6\cdot5\cdot4\cdot3!}{3!} = 8\cdot7\cdot6\cdot5\cdot4 = 6720$

(c) $\frac{n!}{r!(n-r)!} = \frac{8!}{3!(8-3)!} = \frac{8!}{3!\ 5!}$

$$= \frac{8.7\cdot6\cdot5!}{3!\cdot5!} = \frac{8.7\cdot6}{3\cdot2\cdot1} = 56.$$

Example 3 *It is observed that in a computer system each user has a four to five character password, where each character is a lowercase letter or a digit. It is mandatory that each password must contain atleast one digit. Compute the total number of possible passwords.*

Solution: Assume that T be the total number of possible passwords. It is given that each password contains four to five characters. Let P_4 and P_5 denote the number of possible passwords of length 4 and 5 characters respectively. Therefore, by addition principle

$$T = P_4 + P_5$$

Now, our aim is to calculate P_4 and P_5. In order to calculate P_4, first of all we have to compute the number of passwords of lowercase letters and digits that are 4 characters long and then we have to subtract the number of passwords with no digits. Therefore, by using multiplication principle

$$P_4 = 36^4 - 26^4 = 1222640$$

Similarly, $P_5 = 36^5 - 26^5 = 48584800$

Therefore, $T = P_4 + P_5 = 1222640 + 48584800 = 49807440.$

Example 4 *If two dice are thrown once, then find the number of ways in which we can get an even sum.*

Solution: Given that two dice are thrown once. It is clear that the minimum sum we will get is 2 whereas the maximum sum is 12. Our aim is to calculate the number of ways that we will get the even sum *i.e.*, we need the sum 2, 4, 6, 8, 10 and 12.

Now, we will get the sum 2 in 1 way *i.e.*, (1, 1). Similarly, we will get the sum 4 in 3 ways *i.e.*, (1, 3), (2, 2) and (3, 1). Likewise we will get the sum 6, 8, 10 and 12 in 5, 5, 3 and 1 ways respectively. Therefore, the total number of ways in which we will get an even sum is equal to

$$1 + 3 + 5 + 5 + 3 + 1 = 18.$$

Example 5 *Show that* $(2n)! = 2^n (n!)\, 1.3.5\ldots(2n - 1)$

Solution: $(2n)! = 2n\,(2n - 1)\,(2n - 2)\,(2n - 3)\,(2n - 4) \ldots .\, 4.\, 3.\, 2.\, 1$

$$= [2n\,(2n - 2)\,(2n - 4)\ldots 4.2][(2n - 1)(2n - 3)\ldots .3.1]$$

$$= 2\,n\,2\,(n - 1)\,2\,(n - 2) \ldots ..\,(2.2)\,2\,(2n - 1)(2n - 3)\ldots .3.1$$

$$= 2n\;\; n\,(n - 1)\,(n - 2) \ldots ..\,2\,1\,[(2n - 1)(2n - 3) \ldots ..3.1]$$

$$= 2n\;\; (n!)\,1.3.5\ldots(2n - 1)(2n - 3)$$

Example 6 *Find the value of the followings.*

(*a*) *P*(8, 5) (*b*) $\dfrac{P(10,\ 6)}{P(12,\ 8)}$

Solution: (*a*) $P(8,5) = \dfrac{8!}{(8-5)!} = \dfrac{8!}{3!} = \dfrac{3!\ 4{\cdot}5{\cdot}6{\cdot}7{\cdot}8}{3!}$

$$= 6720$$

(*b*) $$\frac{P(10,6)}{P(12,8)} = \frac{\dfrac{10!}{(10-6)!}}{\dfrac{12!}{(12-8)!}} = \frac{10!}{4!} \times \frac{4!}{12!} = \frac{10!}{12!} = \frac{10!}{10! \times 11 \times 12} = \frac{1}{132}$$

Example 7 *How many bit code words can be generated of length 6?*

Solution: Here we have to construct bit code words of length 6. A code word is said to be a bit code word if all the digits are either 0 or 1. Consider the code word as below.

1st	2nd	3rd	4th	5th	6th

6 bit code word

From the code word it is clear that each place can be filled in two ways. Therefore, the total code words = 2 × 2 × 2 × 2 × 2 × 2 = 64.

Example 8 *In how many ways 25 answer papers are arranged so that a particular pair of answer papers shall not come together?*

Solution: Given that there are 25 answer papers. The restriction is that two answer papers cannot be put together.

Let us consider this particular pair of answer papers as one answer paper. So we have 24 answer papers instead of 25 answer papers. Now these 24 answer papers can be arranged in $P(24, 24) = 24!$ ways.

Now the 2 particular answer papers can arranged in 2! ways, on assuming that they are one. So the total number of ways when particular pair of answer papers is always together is equal to

$$24! \times 2! = 2 \times 24!$$

But the 25 answer papers can be arranged in $P(25, 25) = 25!$ ways.

Therefore, the number of ways in which a particular pair of answer papers is never together is equal to

$$= 25! - 2 \times 24! = 25 \times 24! - 2 \times 24!$$

$$= 24! \times (25 - 2) = 23 \times 24!$$

Example 9 *Find the number of ways in which 8 men and 4 women stand in a circle such that no two women are next to each other.*

Solution: According to circular arrangement 8 men can stand in a circle in $(8 - 1)! = 7!$ ways. It indicates that there are 8 locations available in the circle for 4 women.

So, first women can stand in 8 ways; second women can stand in 7 ways; third women can stand in 6 ways whereas fourth women can stand in 5 ways.

Therefore, the total number of ways in which no two women are next to each other

$$= 7! \times (8 \times 7 \times 6 \times 5)$$

$$= 7! \times P(8, 4) = 8467200.$$

Example 10 *Find the value of n if* $P(n, 7) = 12\,P(n, 5)$

Solution: Given that $P(n, 7) = 12\,P(n, 5)$

$$\Rightarrow \quad \frac{n!}{(n-7)!} = 12.\frac{n!}{(n-5)!}$$

$$\Rightarrow \quad \frac{(n-5)!}{(n-7)!} = \frac{12}{1}$$

$$\Rightarrow \quad \frac{(n-7)!\ (n-6)(n-5)}{(n-7)!} = 12$$

$$\Rightarrow \quad n^2 - 11n + 30 = 12$$

$$\rightarrow \quad n^2 - 11n + 18 = 0$$

$$\Rightarrow \quad (n-9)(n-2) = 0$$

$$\Rightarrow \quad n = 9 \text{ or } 2$$

From the equation it is clear that n must be greater than 7. Thus, $n = 9$.

Example 11 *Find the value of r if* $\dfrac{P(10,r-1)}{P(11,r-2)} = \dfrac{30}{11}$.

Solution: Given that $\dfrac{P(10,r-1)}{P(11,r-2)} = \dfrac{30}{11}$

$$\Rightarrow \quad 11 \times P(10,r-1) = 30 \times P(11,r-2)$$

$$\Rightarrow \quad \frac{10!}{(11-r)!}.\frac{(13-r)!}{11!} = \frac{30}{11}$$

$$\Rightarrow \quad \frac{10!}{(11-r)!}.\frac{(13-r)(12-r)(11-r)!}{11 \cdot 10!} = \frac{30}{11}$$

$$\Rightarrow \quad \frac{1}{11}.(13-r)(12-r) = \frac{30}{11}$$

$$\Rightarrow \quad (13-r)(12-r) = 30$$

$$\Rightarrow \quad 156 - 25r + r^2 = 30$$

$$\Rightarrow \quad r^2 - 25r + 126 = 0$$

$$\Rightarrow \quad (r-7)(r-18) = 0$$

$$\Rightarrow \quad r = 7 \text{ or } 18.$$

But if r =18, then both $P(10, 18 - 1)$ and $P(11, 18 - 2)$ are not defined as 17 and 16 both are greater than 10. Therefore, we take $r = 7$.

Example 12 *How many words can be formed from the letters of the word 'DAUGHTER' if the vowels always coming together?*

Solution: Given that the word 'DAUGHTER'. The alphabets present in the word are D, A, U, G, H, T, E, R. The vowels are A, U and E.

We have to form words in such a manner that the vowels shall be together. Now consider all the vowels as one letter say X.

These 3 letters can be arranged among themselves in 3! ways. Now, we have 6 letters, *i.e.*, D, G, H, T, R and X. These 6 letters can be arranged in $P(6, 6) = 6!$ ways.

Therefore, the number of words can be formed with under 'DAUGHTER' with vowels together is equal to

$$3! \times 6! = 6 \times 720 = 4320.$$

Example 13 *Find the number of ways in which 4 boys and 4 girls can sit around a round table if there is no restriction.*

Solution: Given that there are 4 boys and 4 girls. Total number of boys and girls is equal to 8. It is given that they have to sit around a round table. Therefore, the total number of ways $= (8 - 1)! = 5040$.

Example 14 *How many different 4 digit numbers may be formed out of the digits 1, 3, 5, 6, 8 and 9 when no digit is repeated in the same number?*

Solution: Given that there are six digits 1, 3, 5, 6, 8 and 9. We have to form 4-digit numbers in which no digits are to be repeated.

1st	2nd	3rd	4th

The 1st place can be filled up in 6 ways, as one of the six digits can be used, after this 5 digit is left out. So the 2nd place can be filled up in 5 ways. Similarly the 3rd place can be filled up in 4 ways and the 4th place can be filled up in 3 ways.

Therefore, the total number of four digit numbers that can be formed is equal to

$6 \times 5 \times 4 \times 3 = 360$

Example 15 *Compute the number of ways in which 3 boys and 2 girls are to be seated for a dinner such that no 2 girls and no 2 boys sit together.*

Solution: Let us consider 5 chairs for five students. Let the chairs be given below:

1st	2nd	3rd	4th	5th

The 2 girls can be placed in 2 even places. This can be done in $P(2, 2)$ ways. In between and at the extremities of 2 girls there are 3 odd places in which 3 boys can be seated. This can be done in $P(3, 3)$ ways. Therefore, by multiplication principle the total number of arrangements is equal to

$$P(2, 2) \times P(3, 3) = 2! \times 3! = 2 \times 6 = 12.$$

Example 16 *Find the total distinct numbers of six digits that can be formed with 0, 1, 3, 5, 7 and 9 and how many of them is divisible by 10?*

Solution: Given that the digits are 0, 1, 3, 5, 7 and 9. We have to form six digit numbers.

1st	2nd	3rd	4th	5th	6th

The 1st place cannot be filled up by 0 as it will remit in 5 digit number. So the 1st place can be filled up in 5 ways. For the 2nd place we are left with 5 digits, so the 2nd place can be filled up in 5 ways. Similarly 3rd place can be filled up in 4 ways, 4th place in 3 ways, 5th place in 2 ways and 6th place in 1 way.

Therefore, the total number of different numbers that can be formed is equal to $5 \times 5 \times 4 \times 3 \times 2 = 600$

Now let us find out the numbers that are divisible by 10. A number will be divisible by 10 if the unit place has 0. Therefore, the 6th place can be arranged by 1 way. The remaining 5 places can be arranged with 5 digits as $P(5, 5)$ ways. Therefore, the total six digit numbers that can be divisible by 10 is equal to

$$P(5, 5) \times 1 = \frac{5!}{(5-5)!} = \frac{5!}{0!} = 120.$$

Example 17 *How many 5 digit numbers can be formed using the digits 5, 0, 6, 5 and 6?*

Solution: Given that there are 5 digits 5, 0, 6, 5 and 6. In which 5 appears 2 times, 6 appear 2 times whereas 0 appears once.

So, the total number of permutation $= \dfrac{5!}{2!\ 2!} = \dfrac{5 \times 4 \times 3 \times 2!}{2! \times 2 \times 1} = 30.$

But these arrangements also include 0 in the first place. This leads to 4 digit numbers. So, we have to exclude such cases. Therefore, the remaining 4 digits can be arranged in

$$= \frac{4!}{2!\ 2!} = \frac{4 \times 3 \times 2!}{2! \times 2 \times 1} = 6 \text{ ways.}$$

Hence, the total 5 digit numbers = 30 – 6 = 24

Example 18 *Find the number of permutations in the letter 'MALAYALAM'.*

Solution: Given that the word MALAYALAM. This contains 9 letters consisting of two M's, two L's, four A's and one Y. So the no. of permutation is equal to

$$\frac{9!}{2!\ 2!\ 4!\ 1!} = 3780.$$

Example 19 *Adams has 5 friends. In how many ways can he invite two or more of them to a tea party?*

Solution: Adams has 5 friends and he has to invite two or more of them to a tea party, *i.e.*, he can invite 2 friends, 3 friends, 4 friends or 5 friends. Therefore, the total number of ways is equal to

$$C(5, 2) + C(5, 3) + C(5, 4) + C(5, 5) = 26.$$

Example 20 *Find the number of ways in which 'n' things of which 'm' are alike can be arranged in circular order.*

Solution: Let the number of arrangements = X

Let us consider one of these X arrangement. Now '*m*' objects are alike. If these '*m*' similar objects are replaced by dissimilar objects, than it can be arranged among them in *m*! ways.

Therefore, X permutation will give (X × *m*!) arrangements. But if '*n*' things are different then the number of circular arrangements = $(n - 1)!$.

Hence, $$X \times m! = (n - 1)!$$

$$\Rightarrow \qquad X = \frac{(n-1)!}{m!}$$

Example 21 *How many even numbers that can be formed by using all the digits 1, 2, 4, 5, 6, 7 and 9?*

Solution: Given that the digits are 1, 2, 4, 5, 6, 7 and 9. We have to form even numbers using all the seven digits. But we know that a number is said to be an even number if the unit place is divisible by 2 *i.e.*, the 1st place has an even number. So, the units place can be arranged by using the digits 2, 4 or 6.

Therefore, the total ways of arranging units place is $P(3, 1) = 3$.

Now the rest 6 places can be filled up by 6 numbers in $P(6, 6) = 6!$ ways. Therefore, the total even numbers = 3 × 6!

= 3 × 720 = 2160

Example 22 *There are 3 toys to be distributed among 7 children. In how many ways can it be done such that*

(a) No child gets more than one toy.

(b) There is no restriction as to the number of toys any child gets.

(c) No child gets all toys.

Solution: Given that there are 4 toys and 7 children.

(*a*) The first toy can be distributed in 7 ways as it can be given to any one of 7 children. The 2nd toy can be given to any one of the remaining 6 children. Similarly, the 3rd toy can be given in 5 ways. Therefore, by multiplication principle the total number of ways possible

$$= 7 \times 6 \times 5 = 210$$

(*b*) Given that there is no restriction to the number of toys any child gets. This implies that repetition is allowed. So, all the 3 toys can be given in 7 ways. Therefore, the required number of ways possible

$$= 7 \times 7 \times 7 = 343$$

(*c*) The number of ways in which a child gets all the 3 toys is 7 as one among 7 children gets all the 3 toys.

If there is no restriction on the number of toys any child gets, then total number of ways possible is 343. Therefore, the required number of ways in which no child gets all the toys

$$= 343 - 7 = 336.$$

Example 23 *How many code signals can be formed by hoisting 4 flags of different colours?*

Solution: Given that the 4 flags are of different colours. We can form signals by hoisting one, two, three or four flags.

By hoisting one flag, the number of code signals $= P(4, 1) = 4$

By hoisting 2 flags, the number of code signals $= P(4, 2) = 12$

By hoisting 3 flags, the number of code signals $= P(4, 3) = 24$

By hoisting 4 flags, the number of code signals $= P(4, 4) = 24$

So, by addition principle, the total number of code signals is equal to $4 + 12 + 24 + 24 = 64$.

Example 24 *An urn contains 9 balls including 5 red, 3 blue and 1 green. They are thrown one by one and arranged in a row. Assume that all the 9 balls are thrown, then find the number of different arrangements.*

Solution: Given that the urn contains 9 balls. *i.e.*, $n = 9$.

Number of red balls $= 5$ *i.e.*, $p = 5$

Number of blue balls $= 3$ *i.e.*, $q = 3$

Number of green balls $= 1$ *i.e.*, $r = 1$

As the objects are repeated, the total number of permutations are equal to

$$\frac{n!}{p!\ q!\ r!} = \frac{9!}{5!\ 3!\ 1!} = \frac{5!\ 6 \cdot 7 \cdot 8 \cdot 9}{5!\ 1 \cdot 2 \cdot 3} = 504$$

Example 25 *How many numbers greater than 7000 can be formed with the digits 2, 4, 7, 8 and 9 if no digits are repeated?*

Solution: We have to form numbers greater than 7000. So it shall be minimum four digit number. Given 5 digits are 2, 4, 7, 8 and 9. Let the four digit number as given below.

1st	2nd	3rd	4th

As the required number is greater than 7000, so the first place digit must be greater than 7. Therefore,

The 1st place can be filled up by 7, 8 or 9 *i.e.*, in $P(3, 1) = 3$ ways.
The rest three places can be filled up in $P(4, 3) = 24$ ways.
Therefore, the total number of four digit numbers greater than 7000 is equal to $24 \times 3 = 72$.

Again it is clear that any five digit number will be greater than 7000. Therefore, the total five digit numbers is equal to $P(5, 5) = 5! = 120$.

Hence, the total numbers greater than 7000 = 72 + 120 =192.

Example 26 *In a meeting there are 12 people. Each shakes hands with each other exactly once. Determine the total number of shake hands.*

Solution: Total number of people in the meeting = 12
Each shake hand involves two people.

$$\text{Total number of shake hands} = C(12, 2) = \frac{12!}{2!\,(12-2)!} = \frac{12!}{2!\;10!} = 66.$$

Example 27 *In how many ways a cricket team of eleven is chosen from a batch of 18 player? How many of them will*

(a) *Include a particular player*

(b) *Exclude a particular player*

Solution: We have to select 11 players from 18 players which can be done in

$$P(18, 11) = \frac{18!}{7!} \text{ ways.}$$

(a) Now, we have to select the team including a particular player. In this case one player is fixed. So, rest 10 players is to be selected from 17 players and it can be done in P(17, 10) ways.

i.e.,
$$P(17, 10) = \frac{17!}{7!} \text{ ways.}$$

(b) Now we have to select the team by excluding a particular player. In this case we have to exclude a particular player. So 11 players has to be chosen from 17 players, and this can be done in $P(17,11)$ ways.

i.e.
$$P(17, 11) = \frac{17!}{6!}, \text{ ways.}$$

Example 28 *If $C(n, r) = 56$ and $P(n, r) = 336$, find the value of r.*

Solution: Given that $C(n, r) = 56$ and $P(n, r) = 336$.

$$\Rightarrow \quad \frac{n!}{r!(n-r)!} = 56 \text{ and } \frac{n!}{(n-r)!} = 336$$

$$\Rightarrow \quad \frac{\dfrac{n!}{(n-r)!}}{\dfrac{n!}{r!(n-r)!}} = \frac{336}{56}$$

$$\Rightarrow \quad \frac{n!}{(n-r)!} \times \frac{r!(n-r)!}{n!} = 6$$

$$\Rightarrow \quad r! = 6 = 3!$$

$$\Rightarrow \quad r = 3$$

Example 29 *In a office club there are 4 ladies and 10 gentlemen, a committee has to be formed, with 2 ladies and 2 gentleman, in how many ways can this be done if*

(*a*) *Any lady and gentleman can be included?*

(*b*) *One particular gentleman must be there in the committee?*

Solution: Given that 4 ladies and 10 gentlemen are there in the office club. A committee of 2 ladies and 2 gentlemen has to be formed.

(*a*) Consider the case of any lady and gentlemen can be included. Now 2 ladies can be selected from 4 ladies in $C(4, 2)$ ways and 2 gentlemen can be selected from 10 gentlemen in $C(10, 2)$ ways.

Therefore, the total number of ways = $C(4, 2) \times C(10, 2)$

$$= \frac{4!}{2!(4-2)!} \times \frac{10!}{2!(10-2)!} = \frac{4!}{2! \cdot 2!} \times \frac{10!}{2! \cdot 8!} = 270$$

(*b*) Now, we have to select the committee with one particular gentleman. So, rest one gentleman can be selected from remaining 9 gentleman in $C(9, 1)$ ways. The 2 ladies can be selected from 4 ladies in $C(4, 2)$ ways.

Therefore, the total number of ways = $C(4, 2) \times C(9, 1)$

$$= \frac{4!}{2!(4-2)!} \times 9 = \frac{4!}{2! \cdot 2!} \times 9 = 54.$$

Example 30 *In an examination a minimum is to be secured in each of 4 subjects to pass. In how many ways can a student fail?*

Solution: A student can fail in any of the following manner

(*a*) He may fail in any one of the paper which is possible in $C(4, 1)$ ways.

(*b*) He may fail in any two of the papers which is possible in $C(4, 2)$ ways.

(*c*) He may fail in any three of the papers which is possible in $C(4, 3)$ways.

(*d*) He may fail in all the four papers which is possible in $C(4, 4)$ ways.

Thus, the number of ways in which a student can fail

$$= C(4, 1) + C(4, 2) + C(4, 3) + C(4, 4)$$
$$= C(5, 2) + C(5, 4) = 10 + 5 = 15.$$

Example 31 *Compute the total number of different words that can be generated out of 3 capitals, 3 consonants and 4 vowels if each word contains 2 consonants, 2 vowels and beginning with capital.*

Solution: Given 5 capitals, 3 consonants and 4 vowels. We have to form words with 2 consonants, 2 vowels and beginning with capital.

Now 2 consonants can be selected from 3 consonants in $C(3, 2)$ ways. Again 2 vowels can be selected from 4 vowels in $C(4, 2)$ ways. This group of 4 letters can be arranged among themselves in 4! ways.

Therefore, total number of words = $4! \times C(3, 2) \times C(4, 2)$

$$= 24 \times 3 \times 6 = 432.$$

Now each word shall begin with capital and there are 3 capitals. Therefore, total number of words = $3 \times 432 = 1296$

Example 32 *Find the middle term in the expansion of* $\left(x^3 + \frac{2}{x^4}\right)^{14}$.

Solution: In the above expansion $n = 14$. Therefore, the total number of terms is equal to 15. It indicates that there exists only one middle term *i.e.*, $\left(\frac{14}{2}+1\right) = 8^{\text{th}}$ term. Therefore,

$$T_8 = C(14, 7)(x^3)^{14-7}\left(\frac{2}{x^4}\right)^7$$
$$= 3432 \times 2^7 x^{-7}.$$

Example 33 *Compute the 10th term in the expansion of* $(3x^2 - 2y)^{13}$.

Solution: In the above expansion $n = 13$. Our aim is to compute the 10th term. Therefore, we get

$$T_{10} = C(13, 9)(3x^2)^{13-9}(-2y)^9 = 715 \times 3^4 \times (-2)^9 x^8 y^9.$$

Example 34 *A cricket team of 11 players is to be chosen from 16 players including 5 bowlers and 2 wicket keepers. In how many different ways can a team be formed so that the team consists of at least 3 bowlers and at least one wicket keeper?*

Solution: Given that number of bowlers = 5

Number of wicket keepers = 2

and number of other players = 16 – 7 = 9

The team of 11 has to be formed with at least 3 bowlers and at least one wicket keeper.

Bowler	*Wicket keeper*	*Others*	*No. of ways*
3	1	7	$C(5, 3)\times C(2,1)\times C(9,7)=720$
3	2	6	$C(5, 3)\times C(2, 2)\times C(9, 6)=840$
4	1	6	$C(5, 4)\times C(2,1)\times C(9,6)=840$
4	2	5	$C(5, 4)\times C(2,2)\times C(9,5)=630$
5	1	5	$C(5, 5)\times C(2,1)\times C(9,5)=252$
5	2	4	$C(5, 5)\times C(2,2)\times C(9, 4)=126$

Therefore, total number ways is given as

$$= 720 + 840 + 840 + 630 + 252 + 126$$
$$= 3408 \text{ ways.}$$

Example 35 *An examination paper consists of questions divided into groups A and B. Group A contains 7 questions and Group B contains 5 questions. A student is required to attempt 8* *questions selecting at least 3 from each Group. In how many ways can the candidate select question?*

Solution: Given that there are two groups in the question paper, Group A and Group B. Group A contains 7 questions and Group B contains 5 questions. A student has to answer 8 questions selecting at least 3 from each group. This can be done in the following ways:

Group A	*Group B*	*Number of ways*
3	5	$C(7,3)\times C(5,5)$
4	4	$C(7,4)\times C(5,4)$
5	3	$C(7,5)\times C(5,3)$

Thus, the total number of selection is given as

$$C(7, 3) \times C(5, 5) + C(7, 4) \times C(5, 4) + C(7, 5) \times C(5, 3)$$

$$= \frac{7\times 6\times 5}{6} + \frac{7\times 6\times 5}{6}\times 5 + \frac{7\times 6}{2}\times\frac{5\times 4}{2} = 420$$

A student can choose the question papers in 420 ways.

Example 36 *If n is a positive integer, then show that $5^{2n} - 24n - 1$ is divisible by 576.*

Solution: It is clear that $(5^{2n} - 24n - 1) = (25^n - 24n - 1)$. Therefore,

$$\begin{aligned}(25^n - 24n - 1) &= (1+24)^n - 24n - 1\\ &= (1 + 24n + C(n,\ 2)24^2 + \cdots + C(n,\ n)24^n) - 24n - 1\\ &= C(n,\ 2)24^2 + \cdots + C(n,\ n)24^n\\ &= 24^2(C(n,\ 2) + C(n,\ 3)24 + \cdots + C(n,\ n)24^{n-2})\\ &= 576 \times \text{a positive integer.}\end{aligned}$$

Therefore, $(5^{2n} - 24n - 1)$ is divisible by 576.

Example 37 *Compute the middle terms of the expansion* $\left(x^2 + \frac{1}{x^2}\right)^{13}$.

Solution: In the above expansion $n = 13$. Therefore, the total number of terms is equal to 14.

Hence, there exists two middle terms *i.e.*, $\frac{14}{2} = 7^{\text{th}}$ term and $\left(\frac{14}{2}+1\right) = 8^{\text{th}}$ term. Therefore, we get

$$T_7 = C(13,\ 6)(x^2)^{13-6}\left(\frac{1}{x^2}\right)^6 = 1716x^{12}$$

$$T_8 = C(13,\ 7)(x^2)^{13-7}\left(\frac{1}{x^2}\right)^7 = 1716x^{-12}$$

Example 38 *Find the sum of* $\frac{C_1}{C_0}+2\frac{C_2}{C_1}+3\frac{C_3}{C_2}+\cdots+n\frac{C_n}{C_{n-1}}$.

Solution: We know that $\frac{C_r}{C_{r-1}}=\frac{n+1-r}{r}$

On putting $r = 1, 2, 3, \ldots, n$, in the above equation and then adding, we get

$$\frac{C_1}{C_0}+2\frac{C_2}{C_1}+3\frac{C_3}{C_2}+\cdots+n\frac{C_n}{C_{n-1}}=n+2\frac{n-1}{2}+3\frac{n-2}{3}+\cdots+n\frac{1}{n}$$

$$=n+(n-1)+(n-2)+\cdots+1$$

$$=\frac{n(n+1)}{2}.$$

Example 39 *Is there any term independent of x in the expansion of* $\left(x^3-\frac{2}{x^2}\right)^{15}$ *? If yes, then compute the same.*

Solution: In the above expansion $n = 15$. Let the $(r+1)^{\text{th}}$ term is independent of x. Therefore,

$$T_{r+1}=C(15,\ r)(x^3)^{15-r}\left(-\frac{2}{x^2}\right)^r=C(15,\ r)(-2)^r x^{45-5r}$$

As $(r + 1)^{\text{th}}$ term is independent of x, so the power of x in T_{r+1} must be 0. Therefore, we get $(45 - 5r) = 0$. It implies that $r = 9$. It indicates that the 10th term is independent of x. Therefore, we have

$$T_{10}=C(15,\ 9)(-2)^9=-5005\times 2^9.$$

Example 40 *Calculate the value of* $(15)^{\frac{1}{4}}$ *up to third decimal places by using binomial theorem for any rational index.*

Solution: $(15)^{\frac{1}{4}}=(16-1)^{\frac{1}{4}}=16^{\frac{1}{4}}\left(1-\frac{1}{16}\right)^{\frac{1}{4}}$

$$=2\left(1-\frac{1}{16}\right)^{\frac{1}{4}}=2\left(1+\frac{1}{4}\left(-\frac{1}{16}\right)+\frac{\frac{1}{4}\left(-\frac{3}{4}\right)}{2!}\left(-\frac{1}{16}\right)^2+\cdots\right)$$

$$=2\left(1-\frac{1}{64}-\frac{3}{8192}-\cdots\right)$$

$$=2(1-0.0156-0.0003)$$

$$=1.968$$

In this case we neglect the higher order terms as their value will not affect the third place of decimal.

Example 41 *If* $R(m-1,\ n)$ *and* $R(m,\ n-1)$ *are both even and greater than 2, show that* $R(m,\ n) \le R(m-1,\ n) + R(m,\ n-1) - 1$.

Solution: Let , $p = R(m-1,\ n)$, $q = R(m,\ n-1)$ and $r = (p + q)$.

It suffices to establish that in any group of X = {1, 2, 3, ⋯, $r-1$} of $r-1$ people there is either a subgroup of m people all who know one another or a subgroup of n people all who do not know one another. Let k_i be the number of people known to person i for $i = 1,\ 2,\ 3,\ \cdots,\ r$. As knowing is mutual, $k_1 + k_2 + k_3 + \cdots + k_{r-1}$ is essentially even. Since $(r-1)$ is odd; so k_i is even for atleast one i. Let it be i =1.

Assume that L_1 be the set of people known to person 1 whereas L_2 be the set of people not known to person 1. As there are even numbers of people L_1, there must be even number of people in L_2. It indicates that either L_1 has at least $(p-1)$ people or L_2 has at least q people. But $(p-1)$ is odd. Therefore, either L_1 has at least p people or L_2 has at least q people.

(*a*) Suppose L_1 has at least p people. Again $p = R\ (m-1,\ n)$ indicates that L_1 must contain either $(m-1)$ people known to each other or it has n people unknown to one another. In the earlier case the $(m-1)$ people and person 1 together constitute m people known to each other.

(*b*) The case of q or more people in L_2 can be handled by symmetry.

Therefore, $R(m,\ n) \le R(m-1,\ n) + R(m,\ n-1) - 1$

Example 42 *Show that R (4, 3) = 9.*

Solution: We know that $R\ (3,\ 3) = 6$ and $R\ (4,\ 2) = 4$.

As R (3, 3) and R (4, 2) are both even, so by previous relation, we get

$$R(4,\ 3) \le R(3,\ 3) + R(4,\ 2) - 1 = 9$$

Now our aim is to prove that $R(4,3) = R(3,4) > 8$. In order to prove this relation we exhibit a group of 8 people which has no subgroup of 3 people all known to each other and no subgroup of 4 people all are not known to one another.

Example 43 *Show that R (5, 3) = 14.*

Solution: By theorem, we know that,

$$R(5,\ 3) \le R(4,\ 3) + R(5,\ 2) = 9 + 5 = 14$$

Now, we have to show that $R(5,\ 3) = R(3,\ 5) > 13$. In order to show this consider a group of 13 people sitting on a round table such that each person knows only the 5th person on his right and the 5th person on his left. In such a situation there is no subgroup of 3 people all known to one another or no subgroup of 5 people all unknown to one another.

Example 44 *Show that* $R(4,\ 4) = 18$.

Solution: By theorem, we know that,

$$R(4,\ 4) \le R(3,\ 4) + R(4,\ 3) = 9 + 9 = 18$$

Now, we have to show that $R(4,\ 4) > 17$. In order to show this, consider an arrangement of 17 people sitting on a round table such that each person knows exactly 6 people: the first, second and fourth persons on one's left and first, second and fourth persons on one's right. It can be verified that there is no subgroup of 4 people all known to one another or no subgroup of 4 people all unknown to one another.

Example 45 Show that $R(2, 3) = 3$.

Solution: By theorem, we know that,

$$R(2, 3) \leq R(1, 3) + R(2, 2) = 1 + 2 = 3$$

Now, we have show that $R(2, 3) \geq 3$. In order to show this, consider a complete graph with 3 vertices. Therefore, it is clear that we can not arbitrarily colour edges of a complete graph with 2 vertices either blue or red and force either a blue complete graph with 2 vertices or a red complete graph with 3 vertices. We might just have used red on the one edge.

Therefore, $R(2, 3) = 3$.

EXERCISES

1. In how many ways can 5 letters be posted in three letter boxes?
2. How many 3 digit even numbers can be formed from the digits 2, 5, 6, 7 and 8, if repetition of digits is not allowed?
3. If 5 candidates are contesting for the post of president of a college union and there are 200 voters, then in how many ways can the votes be given?
4. Evaluate the followings:

   (*i*) $P(15, 7)$ (*ii*) $P(7, 3)$ (*iii*) $P(6, 3)$

   (*iv*) $C(18, 12)$ (*v*) $C(11, 7)$ (*vi*) $C(15, 15)$
5. Determine the value of n, if

   (*i*) $C(n, 15) = C(n, 22)$

   (*ii*) $18\,C(n, 3) = 5\,C(n, 5)$

   (*iii*) $C(2n, 3): C(n, 2) = 12: 1$
6. In how many ways can the letters of the word MISSISSIPPI be rearranged?
7. Twelve boys compete in a race. In how many ways can the first three places be taken?
8. How many 2 digit odd numbers can be formed from the digits 1, 2, 3, 4, 6 and 9 if repetition of digits are not allowed?
9. There are 7 books in a shelf with different titles; 4 of these have green cover whereas others have red cover. In how many ways can these be arranged so that the green books are placed together?
10. A man has 7 friends. Find the number of ways that he can invite three or more to a dinner party.
11. In how many ways can a student of computer science choose 5 courses out of the courses $CS_1, CS_2, CS_3, \cdots, CS_7$ if CS_1, CS_2 are compulsory and CS_6, CS_7 cannot be taken together?
12. How many 5-digit numbers can be formed from the digits 0, 3, 5, 7, 8 and 9 if repetition of digits is not allowed?
13. Prove the followings

    (*i*) $P(n+1, r+1) = (n+1)\,P(n, r)$

    (*ii*) $P(n, n) = P(n, n-1)$

    (*iii*) $P(n-1, r) + r\,P(n-1, r-1) = P(n, r)$
14. A box contains 9 red and 6 black balls. In how many ways can 7 balls be drawn? Find out the number of cases in which one gets 4 red and 3 black balls.

15. Find out the number of factors of 210.

16. Find out the number of diagonals of an octagon.

17. Show that $R(2, 4) = 4$.

18. Forty points lie on a plane, out of which no three points are collinear. Find the number of straight lines that can be formed by joining pairs of points.

19. A committee of 4 ladies and 2 gentlemen is to be formed from 7 ladies and 6 gentlemen. In how many ways can the committee be formed?

20. If a set X has n number of elements and another set Y has m number of elements, then find out the number of relations from X to Y.

21. How many numbers can be formed between 3000 and 4000 by using the digits 1, 2, 4, 6, 9 without repeating any digit?

22. In how many ways can the letters of the word 'MATHEMATICS' be arranged if the vowels are to retain their places?

23. In how many ways 9 cards are drawn from a pack of 52 cards such that 4 aces are included?

24. An examination question paper consists of Group A and Group B. Group A contains 8 questions and Group B contains 6 questions. A student is required to attempt 8 questions selecting at least two questions from each Group. Find the number of ways that the candidate select question.

25. How many factors does 1155 have that are divisible by 3?

26. How many 4 letter words can be formulated from the word 'MATHEMATICS' if no letter is repeated?

27. In how many ways can a cricket team of 11 players be chosen from 9 batsman and 5 bowlers so as to include at least 5 batsmen?

28. An urn contains 5 red, 3 blue, 6 black and 2 green balls. In how many ways can the balls be rearranged?

29. A cricket team of 11 is to be chosen from 8 batsman and 6 bowlers. In how many ways the selection is made so as to include at least 4 batsman and 5 bowlers?

30. Find the value of r, if

(*a*) $C(n, r) = C(n, n-7)$

(*b*) $P(n, r) = C(n, r)$

(*c*) $C(n, r) + C(n, r-1) = C(17, 6)$

31. How many numbers can be formed from the digits 2, 4, 5, 8, 9 and 0 that are divisible by 5, if no digits are repeated in a number?

32. Find the value of r, if $P(20, r) = 13.\ P(20, r-1)$.

33. Find the value of the followings:

(*i*) $C(8, 2) + C(8, 1) + C(9, 3) + C(10, 4)$

(*ii*) $C(8, 5) + C(7, 4) + C(6, 3) + C(5, 2) + C(5, 1)$

34. In how many ways can four men and 4 women sit at a round table for a dinner so that no two men can occupy adjacent positions?

35. How many different words with two letters can be formed by using the letters of the word ENGINE, each containing one vowel and one consonant?

36. How many three letter words can be formed from the letters of the word SUCCESS such that

(*a*) they always contain the letter S

(*b*) they do not contain the letter C?

37. In an election there are 7 candidates for the post of secretary. How many ways can their names are printed on the ballot paper?

38. Find the value of n, if

(*a*) $P(n, 5) = 42\,P(n, 3)$ provided that $n > 4$

(*b*) $30\,P(n, 6) = P(n + 2, 7)$

(*c*) $3\,P(7, n) = 7\,P(6, n)$

(*d*) $P(6, n) = 360$

39. How many natural numbers below 1600 can be formed from the digits 0, 1, 2, 3, 4, 5, 7 and 9 if

(*a*) No digit is repeated.

(*b*) Repetition of a digit is allowed?

40. How many triangles can be formed by joining the angular points of a decagon?

41. Let there be three towns X, Y and Z. There are 5 buses running between towns X and Y whereas 8 buses between Y and Z. In how many ways can one person travel from X to Z?

42. Four letters are written and four envelops are addressed. In how many ways can all letters be placed in wrong envelops?

43. Let n is a positive integer and $(x+1)^n$ is expanded in decreasing powers of x. If the consecutive coefficients are in the ration 2:15:70, then find the value of n.

44. Find the 4th term in the expansion of $\left(3x + \dfrac{2}{x^3}\right)^{10}$ using rules of Pascal's triangle.

45. Expand the following by using binomial theorem.

(*a*) $(3x^2 + 2y^3)^5$ (*b*) $(4x^2 - 2x^3)^7$

(*c*) $\left(2x + \dfrac{3}{x^3}\right)^6$ (*d*) $\left(x^2 + \dfrac{a^2}{y^2}\right)^5$

46. Find the middle term(s) in the following expansions.

(*a*) $\left(\dfrac{a}{b} + \dfrac{b}{a}\right)^8$ (*b*) $\left(x + \dfrac{4}{x}\right)^{13}$ (*c*) $\left(x^{\frac{3}{2}} + y^{\frac{5}{2}}\right)^{10}$

(*d*) $(3x + 2y)^{15}$ (*e*) $\left(x^2 + \dfrac{b^2}{y^3}\right)^{20}$ (*f*) $(4a^2 + 2b^3)^9$

47. Find the 6th term in the expansion $\left(x^2 + \dfrac{3}{y^3}\right)^8$.

48. Is there any term independent of x in the following expansions? If yes, then compute the same.

(a) $\left(5x+\frac{a^3}{x}\right)^{10}$ (b) $\left(x^3-\frac{b^7}{x^5}\right)^{12}$ (c) $\left(\frac{4}{x^2}+\frac{x^2}{3}\right)^{8}$

49. Find the coefficient of x^4 in the expansion of $(1+2x+3x^2)\left(x+\frac{1}{x}\right)^{10}$.

50. Let there are 'n' different balls. Prove that the number of ways in which balls can be arranged so that two particular balls shall not be together is $(n-1)!(n-2)$.

51. Try to show that $R(3,\ 6)=18$ and $R(3,\ 5)=14$.

52. Prove that $R(m,\ n)=R(n,\ m)$.

53. Compute the square root of 2 and 3 up to three decimal places.

54. If $C(n, r) = C_r$, then prove the followings:

(a) $C_0+2C_1+3C_2+\cdots+(n+1)C_n=2^n+n2^{n-1}$

(b) $C_0C_r+C_1C_{r+1}+C_2C_{r+2}+\cdots+C_{n-r}C_n=C(2n,\ n-r)$

(c) $C_0C_1+C_1C_2+C_2C_3+\cdots+C_{n-1}C_n=C(2n,\ n-1)$

(d) $C_2+2C_3+3C_4+\cdots+(n-1)C_n=1+(n-2)\ 2^{n-1}$

(e) $C_0+3C_1+5C_2+\cdots+(2n+1)C_n=2^n+n2^n$

(f) $C_0-2C_1+3C_2-\cdots+(-1)^n(n+1)C_n=0$

(g) $\frac{C_1}{C_0}+\frac{2C_2}{C_1}+\frac{3C_3}{C_2}+\cdots+\frac{nC_n}{C_{n-1}}=\frac{n(n+1)}{2}$.

55. Show that $1^{99}+2^{99}+3^{99}+4^{99}+5^{99}$ is divisible by both 3 and 5. Use binomial theorem to prove this.

56. If three consecutive coefficients of $(1+x)^n$ are 35, 21 and 7, then compute the value of n.

7

Group Theory

■ 7.0 INTRODUCTION

In this chapter we will study the algebraic structure known as group which is the building block of "Abstract Algebra". Group is a one operational system. *i.e.*, it has one binary operation using which we can combine two elements of a set to get the third element. In this chapter we will discuss definition of group, subgroup, cyclic group, group homomorphism and etc. Group theory has also wide application in the areas of Computer Science specially in the field of binary coding.

■ 7.1 BINARY OPERATION ON A SET

Let A be a non-empty set. If f be a function from $(A \times A) \to A$, then f is said to be a binary operation on the set A. So the binary operation must satisfy the following two conditions, *i.e.*, f assigns an element $f(a, b)$ of A to every ordered pair (a, b) in $(A \times A)$ and only one element of A is assigned to each ordered pair; as the operation is a function.

Generally we use the symbols +, ×, *, o etc. for representing the binary operation on a set. So 'o' will be the binary operation in A if and only if

(*a*) $(a\ _0\ b) \in A \qquad \forall\, a, b \in A$

(*b*) $(a\ _0\ b)$ is unique.

We will use the symbol $(_0)$ to represent the binary operation in place of f and the element assigned to (a, b) by $(a\ _0\ b)$. It is clear that binary operation function is a special case of binary operation.

Let us consider the operation addition in the set of Natural numbers N.

Let $a, b \in N$; *i.e.*, a and b are two natural numbers. But we know that sum of any two natural numbers is again a natural number and is unique *i.e.*, $(a + b) \in N$ for all $a, b \in N$. Hence, + is a binary operation.

■ 7.2 ALGEBRAIC STRUCTURE

A non-empty set A along with one or more binary operations is called an algebraic structure. So if A is a non-empty set and o is a binary operation then (A, o) is a algebraic structure. Consider the examples of algebraic structures as (N, +), (R, ×) and (I, +).

7.2.1 Semi Group

An algebraic structure (G, o) is said to be a semi group if the binary operation (o) is associative in G.

i.e., $$a_0(b_0 c) = (a_0 b)_0 c\,;\ a, b, c \in \text{G}.$$

Let us consider the algebraic structure (N, o), where o is a usual product. We know that for any three natural numbers $a, b, c \in$ N, we have $a_0(b_0 c) = (a_0 b)_0 c$, as product is associative in N. This implies that (N, o) is a semi group.

7.2.2 Monoid

An algebraic structure (G, o) is said to be a monoid if the binary operation (o) is associative in G with an identity element e in G.

i.e., $a_0(b_0 c) = (a_0 b)_0 c$ and $a_0 e = e_0 a = a \;\forall\; a, b, c \in$ G, where e is the identity element of G.

Let us consider the algebraic structure (Z, +), where Z is the set of positive integers and the binary operation is an addition. It will become monoid if there exists an identity element e in Z such that

$$a + e = a \;\forall\; a \in \text{Z}$$

This implies that $e = 0$, but $0 \notin$ Z. Hence, (Z, +) is not a monoid.

■ 7.3 GROUP

A non-empty set G is said to be a group under the binary operation 'o' if the following conditions are satisfied. It is also to be noted that a group is a monoid with unit element e.

(*a*) Closure Law: For all $a, b \in$ G; $(a_0 b) \in$ G

(*b*) Associative Law: For all $a, b, c \in$ G, $a_0(b_0 c) = (a_0 b)_0 c$

(*c*) Identity: For all $a \in$ G, there exists an identity element $e \in$ G such that $(a_0 e) = a = (e_0 a)$, where e is called the identity element.

(*d*) Inverse: For all $a \in$ G, there exists an element $a^{-1} \in$ G such that $(a_0 a^{-1}) = e = (a^{-1}{}_0 a)$.

7.3.1 Commutative Group

A group G is said to be a commutative group or abelian group if the commutative law holds. *i.e.*,

$$(a_0 b) = (b_0 a) \;\forall\; a, b \in \text{G}$$

7.3.2 Finite and Infinite Group

If the number of elements in a group G is finite, then it is called a finite group. Otherwise it is called an infinite group.

7.3.3 Order of a Group

The number of elements in a finite group G is called the order of the group and is denoted by O(G).

Let us consider the group G ={a, e}, then O(G)= 2 *i.e.*, G is a group of order 2.

7.3.4 Order of an Element

Let G be a group and $a \in$ G, then the order of an element a is the least positive integer n such that $a^n = e$. If there exists no such n, then the order of a is infinity or zero.

Let us consider the set of integers G with the binary operation addition.

(*a*) *Closure Law:* We know that the sum of two integers is also an integer,

i.e., $(a + b) \in G \ \forall \ a, b \in G$.

(*b*) *Associative Law:* We know that the addition of integers is associative,

i.e., $a + (b + c) = (a + b) + c \ \forall \ a, b, c$ G.

(*c*) *Existence of Identity:* For every integer $a \in$ G there exists identity element $0 \in$ G such that

$$a + 0 = 0 + a = a$$

(*d*) *Existence of Inverse:* For every integer $a \in$ G, there exists inverse element $-a \in$ G such that $a + (-a) = (-a) + a = 0$. So every element of G has an additive inverse.

This implies that the set of integers G together with the binary operation addition (+) is a group.

Note

1. Since addition of Integers is Commutative *i.e.,* $(a + b) = (b + a)$ for all $a, b \in$ G, the group G is an abelian or commutative group.
2. Since G contains infinite element, S_o G is a commutative group of infinite order.

7.3.5 Theorem

If G be a group, then

(*a*) The identity element is unique.

(*b*) Every $a \in$ G has an unique inverse in G.

(*c*) For every $a \in$ G; $(a^{-1})^{-1} = a$

(*d*) For all $a, b \in$ G; $(a \,_o\, b)^{-1} = b^{-1} \,_o\, a^{-1}$

Proof: (*a*) If not and if possible let *e* and *f* be the two identity elements of group G. Thus we have

$$e \,_o\, f = f \text{ (Taking } e \text{ as identity)}$$

and $$e \,_o\, f = e \text{ (Taking } f \text{ as identity)}.$$

Now $(e \,_o\, f)$ is an unique element of G as G is a group. Therefore $f = e$, *i.e.,* Identity element is unique.

(*b*) Let $a \in$ G and $e \in$ G be the identity element of G. If not and if possible let $a_1 \in$ G and $a_2 \in$ G be two inverses of $a \in$ G. Therefore

$$(a \,_o\, a_1) = (a_1 \,_o\, a) = e \text{ and } (a \,_o\, a_2) = (a_2 \,_o\, a) = e.$$

Now $$a_2 \,_o\, (a \,_o\, a_1) = (a_2 \,_o\, e) = a_2 \quad \ldots (i)$$

and $$(a_2 \,_o\, a) \,_o\, a_1 = e \,_o\, a_1 = a_1 \quad \ldots (ii)$$

Again by associative property $a_2 \,_o\, (a \,_o\, a_1) = (a_2 \,_o\, a) \,_o\, a_1$. Therefore from equations (*i*) and (*ii*) we get $a_1 = a_2$. This implies that the inverse of an element is unique.

(*c*) Given that G is a group. Let $a \in$ G, this implies $a^{-1} \in$ G. Similarly $(a^{-1})^{-1} \in$ G. Let *e* be the identity element of G. Hence, we have $(a^{-1} \,_o\, a) = e$

$\Rightarrow$ $(a^{-1})^{-1} \,_o\, (a^{-1} \,_o\, a) = (a^{-1})^{-1} \,_o\, e$

$\Rightarrow$ $((a^{-1})^{-1} \,_o\, a^{-1}) \,_o\, a = (a^{-1})^{-1}$ [Using associative and identity law]

$\Rightarrow$ $e \,_o\, a = (a^{-1})^{-1}$ [Using identity law]

$\Rightarrow$ $a = (a^{-1})^{-1}$ [Using inverse law]

So, $(a^{-1})^{-1} = a$.

(*d*) Given that G is a group. Let $a, b \in G$

Now $(a \circ b)(b^{-1} \circ a^{-1}) = ((a \circ b) b^{-1}) \circ a^{-1}$ [Associative law]

$= (a \circ (b \circ b^{-1})) \circ a^{-1}$ [Associative law]

$= (a \circ e) \circ a^{-1}$ [Inverse law]

$= a \circ a^{-1}$ [Identity law]

$= e$ [Inverse law]

So, $(a \circ b)(b^{-1} \circ a^{-1}) = e$

Similarly it can be shown that $(b^{-1} \circ a^{-1})(a \circ b) = e$.

Thus it is clear that $(b^{-1} \circ a^{-1})$ is the inverse of $(a \circ b)$.

i.e., $(a \circ b)^{-1} = b^{-1} \circ a^{-1}$.

7.3.6 Theorem

Let G be a group and for all $a, b, c \in G$

(*i*) if $(a \circ b) = (a \circ c)$, then $b = c$ [Left cancellation law]

(*ii*) if $(b \circ a) = (c \circ a)$, then $b = c$ [Right cancellation law]

Proof: (*i*) Let G be a group and $a, b, c \in G$

Assume that $(a \circ b) = (a \circ c)$

$\Rightarrow$ $a^{-1} \circ (a \circ b) = a^{-1} \circ (a \circ c)$

$\Rightarrow$ $(a^{-1} \circ a) \circ b = (a^{-1} \circ a) \circ c$ [Associative law]

$\Rightarrow$ $e \circ b = e \circ c$ [Existence of inverse]

$\Rightarrow$ $b = c$ [Existence of identity]

So, if $(a \circ b) = (a \circ c)$, then $b = c$. This is called the left cancellation law.

(*ii*) Let G be a group and $a, b, c \in G$

Assume that $(b \circ a) = (c \circ a)$

$\Rightarrow$ $(b \circ a) \circ a^{-1} = (c \circ a) \circ a^{-1}$

$\Rightarrow$ $b \circ (a \circ a^{-1}) = c \circ (a \circ a^{-1})$ [Associative law]

$\Rightarrow$ $b \circ e = c \circ e$ [Existence of inverse]

$\Rightarrow$ $b = c$ [Existence of identity]

So, if $(b \circ a) = (c \circ a)$, then $b = c$. This is called the right cancellation law.

7.3.7 Theorem

Let G be a group and a, b be the elements of G, then

(*i*) The equation $ax = b$ has unique solution in G.

(*ii*) The equation $ya = b$ has unique solution in G.

Proof: (*i*) Let G be a group and let $a, b \in G$.

According to closure law $(a^{-1} b) \in G$, as $a^{-1} \in G$ and $b \in G$. Let $x = a^{-1} b$. Now

$$ax = a(a^{-1} b) = (a \circ a^{-1}) b = e \circ b = b$$

Therefore, $x = a^{-1} b$ is the solution to the equation $ax = b$.

Let us assume that x_1 and x_2 be two solutions to the equation $ax = b$. Hence, we have

$$ax_1 = b \text{ and } ax_2 = b.$$

This implies that $ax_1 = ax_2$. Therefore, by left cancellation law $x_1 = x_2$. Hence, the solution is unique.

(*ii*) Let G be a group and let $a, b \in G$.

According to closure law $(ba^{-1}) \in G$, as $a^{-1} \in G$ and $b \in G$. Let $y = ba^{-1}$. Now

$$ya = (ba^{-1})\, a = b\, (a^{-1} {}_o\, a) = b\, {}_o\, e = b$$

Therefore, $y = b\, a^{-1}$ is the solution to the equation $ya = b$.

Let us assume that y_1 and y_2 be two solutions to the equation $ya = b$. Hence, we have

$$y_1 a = b \text{ and } y_2 a = b.$$

This implies that $y_1 a = y_2 a$. Therefore by right cancellation law $y_1 = y_2$. Hence, the solution is unique.

7.3.8 Theorem

The order of all the elements of a finite group is finite and is less than or equal to the order of the group.

Proof: Let G be a finite group and the composition being multiplication.

Let $a \in G$.

This implies $\quad (a * a) = a^2 \in G.$

$\Rightarrow \quad (a * a^2) = a^3 \in G.$

$\Rightarrow \quad (a * a^3) = a^4 \in G$ and so on.

i.e., $a, a^2, a^3, a^4, a^5, \ldots$. are the elements of G. This implies that G has infinite order.

But it is given that G is of finite order. So there must exist two integers j and k such that

$$a^j = a^k \text{ for } j > k.$$

$\Rightarrow \quad a^j\, a^{-k} = a^k\, a^{-k} = e$

$\Rightarrow \quad a^{j-k} = e$

$\Rightarrow \quad a^l = e$, where $l = (j - k) \in I^+$ (Set of positive integers)

Now the set of all these positive integers l satisfying $a^l = e$ will have a least member say m. So, $a^m = e$.

Therefore, O(a) is finite. Let O(a) = n.

Now we have to show that O(a) ≤ O(G), *i.e.*, $n \le$ O(G). If not and if possible let us assume that $n >$ O(G).

Let $a \in G$. Therefore by closure property we have $a, a^2, a^3, a^4, a^5, \ldots, a^n \in G$. If they are not distinct then there exists two integers r and s such that $a^r = a^s$, $\quad 1 \le s < r \le n$

$\Rightarrow \quad a^{r-s} = e.$

Thus, O(a) = $(r - s)$ as $(r - s) < n$.

This contradicts to the fact that O(a) = n. Hence our supposition is wrong. Similarly this is not possible if $n >$ O(G).

Therefore, O(a) ≤ O(G).

7.3.9 Theorem

The order of any integral power of an element a can not exceed the order of a.

Proof: Let G be a group and $a \in$ G. Let us assume that the order of the element a is n *i.e.*, $O(a) = n$. This implies that $a^n = e$, where e is the identity element of G.

Suppose that a^m be the integral power of a.

Again $a^n = e$

$\Rightarrow$ $(a^n)^m = e^m = e$

$\Rightarrow$ $(a^{nm}) = e$

$\Rightarrow$ $(a^m)^n = e$

This implies that the order of the element a^m can not exceed the order of a, *i.e.*, $O(a^m) \leq n$.

7.3.10 Theorem

The order of an element a of a group G is the same as the order of its a^{-1}, *i.e.*, If G is a group and $a \in$ G, then $O(a) = O(a^{-1})$.

Proof: Let the order of an element a of a group G be m and that of a^{-1} be n. *i.e.*, $O(a) = m$ and $O(a^{-1}) = n$.

Now $O(a) = m$

This implies that $a^m = e$

$\Rightarrow$ $(a^m)^{-1} = e$

$\Rightarrow$ $(a^{-1})^m = e$

Therefore order of a^{-1} is less than or equal to m, *i.e.*, $O(a^{-1}) \leq m$. Thus, we have $n \leq m$ ...(*i*)

Again $O(a^{-1}) = n$

This implies that $(a^{-1})^n = e$

$\Rightarrow$ $(a^n)^{-1} = e$

$\Rightarrow$ $(a^n) = e$ [$\because$ $a^{-1} = e$ implies $a = e$]

Therefore, order of a is less than or equal to n *i.e.*, $O(a) \leq n$. Thus we have $m \leq n$. ...(*ii*)

Hence from equations (*i*) and (*ii*) it is clear that $m = n$. Therefore, the order of an element a of a group G is the same as the order of its inverse a^{-1}.

■ 7.4 SUBGROUP

A non-empty subset H of a group G is said to be subgroup of G if H forms the group under the binary operation defined on G. As every set is subset of itself, so G is subset of itself and hence G is subgroup of G.

7.4.1 Theorem

A non-empty subset H of a group G is said to be subgroup of G if and only if

(*i*) $a, b \in$ H implies $(a \,_0\, b) \in$ H [Closure law] and

(*ii*) $a \in$ H implies $a^{-1} \in$ H [Inverse law]

Proof: Let H be a subgroup of G. Therefore, H satisfies all properties of a group. Thus closure law and existence of inverse holds.

Conversely, suppose that

(*i*) $a, b \in$ H implies $(a \,_0\, b) \in$ H [Closure law] and

(*ii*) $a \in$ H implies $a^{-1} \in$ H [Inverse law]

Now we have to show that H is a subgroup of G *i.e.*, Existence of identity and associative law holds in H.

Associative Law: Let $a, b, c \in$ H. This implies that $a, b, c \in$ G as H $\subseteq$ G. Thus we get $a \,{}_0\, (b \,{}_0\, c) = (a \,{}_0\, b) \,{}_0\, c$. Therefore associative law holds in H.

Existence of Identity: Given that $a \in$ H implies $a^{-1} \in$ H. So, by closure law $(a \,{}_0\, a^{-1}) \in$ H. *i.e.*, $e \in$ H. Therefore, identity element exists in H. Thus H is a subgroup of G.

7.4.2 Theorem

A non-empty subset H of a group G is the subgroup of G if and only if $(a \,{}_0\, b^{-1}) \in$ H for $a, b \in$ H.

Proof: Let H be a subgroup of G. Therefore, H satisfies all properties of a group G. Thus closure law and existence of inverse holds.

Let $a, b \in$ H. Again by existence of inverse we have $b^{-1} \in$ H. Therefore by closure property $(a \,{}_0\, b^{-1}) \in$ H as $a \in$ H, $b^{-1} \in$ H.

Conversely, suppose that $(a \,{}_0\, b^{-1}) \in$ H for $a, b \in$ H.

Now we have to show that H is a subgroup of G *i.e.*, H satisfies all the four properties of the group G.

Given that $(a \,{}_0\, b^{-1}) \in$ H for $a, b \in$ H. Hence we have $(b \,{}_0\, b^{-1}) \in$ H. This implies that $e \in$ H. So, identity element exists in H. Again $e \in$ H and $a \in$ H implies that $(e \,{}_0\, a^{-1}) \in$ H. *i.e.*, $a^{-1} \in$ H. Thus inverse element for every element exist in H.

Also $a \in$ H, $b^{-1} \in$ H implies that $(a \,{}_0\, (b^{-1})^{-1}) \in$ H. *i.e.*, $(a \,{}_0\, b) \in$ H. So, closure law holds in H. As H is a subset of G and G is a group, so associative law also holds in H.

Therefore, H is a subgroup of G.

7.4.3 Theorem

Intersection of two subgroups of a group G is also a subgroup of G.

Proof: Let H and K be two subgroups of group G. So, H $\subseteq$ G and K $\subseteq$ G. This implies that (H $\cap$ K) is also a subset of G.

Now our claim is to show that (H $\cap$ K) is a subgroup of G *i.e.*, Closure law and existence of inverse holds in (H $\cap$ K).

Closure Law: Let $a, b \in$ (H $\cap$ K)

$\Rightarrow$ $a, b \in$ H and $a, b \in$ K

$\Rightarrow$ $(a \,{}_0\, b) \subset$ H and $(a \,{}_0\, b) \in$ K [$\because$ H and K are subgroups]

$\Rightarrow$ $(a \,{}_0\, b) \in$ (H $\cap$ K).

Existence of Inverse: Let $a \in$ (H $\cap$ K).

$\Rightarrow$ $a \in$ H and $a \in$ K.

$\Rightarrow$ $a^{-1} \in$ H and $a^{-1} \in$ K [$\because$ H and K are subgroups]

$\Rightarrow$ $a^{-1} \in$ (H $\cap$ K)

Therefore, (H $\cap$ K) satisfies both closure law and inverse axiom. Hence, (H $\cap$ K) is a subgroup of G.

7.4.4 Theorem

If H is a non-empty finite subset of a group G, then H is a subgroup of G if and only if H is closed under multiplication.

Proof: (Necessary Part). Let H be a multiplicative subgroup of the group G.

As H is a subgroup G so it satisfies all the properties of group, hence the closure properties.

i.e., $a, b \in \text{H}$ implies $(a \,_0\, b) \in \text{H}$.

(Sufficient Part) Let H be non-empty finite subset of group G and H is closed under multiplication.

i.e., $a, b \in \text{H}$ implies $(a \,_0\, b) \in \text{H}$.

Now we have to show that every element of H has an inverse element in H *i.e.,* $a \in \text{H}$ implies $a^{-1} \in \text{H}$.

Let $a \in \text{H}$. This implies that $(a \,_0\, a) = a^2 \in \text{H}$. Similarly $a^3 = (a^2 \,_0\, a) \in \text{H}$ and so on. Therefore we get

$$\text{H} = \{a, a^2, a^3, a^4, \ldots, a^m, \ldots\}$$

This indicates that H is an infinite set. But, it is given that H is a finite set. So, all the elements of H listed above are not distinct. Thus there exists two integers j and k such that $a^j = a^k$ for $j > k > 0$.

This implies that $a^j a^{-k} = a^k a^{-k} = e$

i.e., $a^{j-k} = e$ ...(*i*)

Now as $j > k > 0$ are two positive integers we have $(j - k) \geq 1$. *i.e.,* $(j - k - 1) \geq 0$.

So, $a^{(j-k-1)} \in \text{H}$, as H contains elements of type a^m.

Again $a \in \text{H}$ and $a^{(j-k-1)} \in \text{H}$ implies that

$(a \,_0\, a^{(j-k-1)}) \in \text{H}$ [By closure law]

i.e., $a^{j-k} = e \in \text{H}$.

Therefore, $(a \,_0\, a^{(j-k-1)}) = e$ ...(*ii*)

From equation it is clear that $a^{(j-k-1)}$ is the inverse element of a. *i.e.,* $a^{-1} = a^{(j-k-1)}$. So, inverse element exists in H. Therefore, H is a subgroup of G.

7.4.5 Definition

Let G be a group and H is subgroup of G. Now for $a, b \in \text{G}$ we say "a is congruent to b mod H" written as $a \equiv b \text{ mod H}$ if $a\, b^{-1} \in \text{H}$.

7.4.6 Theorem

Let G be a group and H is a subgroup of G. Then show that relation $a \equiv b$ mod H is an equivalence relation.

Proof: Given G be a group and H be a subgroup of G. We have to show that the relation $a \equiv b$ mod H is an equivalence relation.

Reflexive: Let $a \in \text{H}$. This implies that $a^{-1} \in \text{H}$.

Hence by closure axiom we have $(a \,_0\, a^{-1}) \in \text{H}$.

i.e., $a \equiv a$ mod H.

Symmetric: Suppose that $a \equiv b$ mod H.

This implies that $ab^{-1} \in \text{H}$

$\Rightarrow$ $(ab^{-1})^{-1} \in \text{H}$ [By existence of inverse]

$\Rightarrow$ $(b^{-1})^{-1} a^{-1} \in \text{H}$ [By theorem]

$\Rightarrow \qquad ba^{-1} \in H$

$\Rightarrow \qquad b \equiv a \bmod H.$

Transitive: Suppose that $a \equiv b \bmod H$ and $b \equiv c \bmod H$

This implies that $\quad ab^{-1} \in H$ and $b\,c^{-1} \in H$

$\Rightarrow \qquad (ab^{-1})(b\,c^{-1}) \in H$ [By closure law]

$\Rightarrow \qquad a(b^{-1}b)\,c^{-1} \in H$ [By associative law]

$\Rightarrow \qquad (ae)\,c^{-1} \in H$ [By existence of inverse]

$\Rightarrow \qquad ac^{-1} \in H$ [By existence of identity]

i.e., $\qquad a \equiv c \bmod H$

Therefore, the relation $a \equiv b \bmod H$ is an equivalence relation.

■ 7.5 CYCLIC GROUP

A group G is called the cyclic group if for any $a \in G$ all other elements are of type a^n, where n is any integer. 'a' is called the generator of G. A cyclic group may have more than one generator and the generator is denoted by (a). Therefore a cyclic group G is of the type

$$G = \{x \mid x = a^n \; ; n \text{ is any integer}\}$$

The elements of G is of the form $\ldots, a^{-3}, a^{-2}, a^{-1}, a^0 = e, a, a^2, a^3, \ldots$

7.5.1 Theorem

Every cyclic group is an abelian group.

Proof: Let G be a cyclic group with generator a. Let a^m and a^n be two elements of G, *i.e.*, a^m, $a^n \in G$.

Now $\qquad (a^m \,_0\, a^n) = a^{m+n} = a^{n+m} = (a^n \,_0\, a^m).$

Therefore, G is an abelian group.

7.5.2 Theorem

G be a cyclic group with generator a, then a^{-1} is also the generator.

Proof: Let G be a cyclic group with generator a. So, $a^n \in G$, where n is some integer.

Now $\qquad a^n = (a^{-1})^{-n} \; ; -n \in I$

This indicates that every element can be expressed as some powers of a^{-1}. Therefore, a^{-1} is also the generator of G.

7.5.3 Theorem

The order of cyclic group is same as the order of its generator.

Proof: Let G be a cyclic group with generator a and let the order of a be n, *i.e.*, $a^n = e$.

Now we have to show that the order of cyclic group G is n *i.e.*, G contains exactly n elements.

Let m be an integer; $m > n$ and $a^m \subset G$.

As $m > n$, we have by division algorithm $m = nk + r \; ; 0 \le r < n$.

Therefore, $a^m = a^{nk+r} = a^{nk}\, a^r = (a^n)^k\, a^r = e^k\, a^r = a^r$. This implies that $a^m = a^r$, *i.e.*, a^m is one of the element from $a, a^2, a^3, \ldots, a^n$. Therefore, G can not have more than n elements.

Now we have to show G contains exactly n elements *i.e.*, all the elements $a, a^2, a^3, \ldots, a^n$ are distinct.

If not and if possible, let there be repetition. *i.e.*, $a^m = a^r$; $0 < r < m$. Thus we have

$$a^m a^{-r} = a^r a^{-r} = e \text{ i.e., } a^{m-r} = e \text{ with } 0 < (m - r) < n.$$

This contradicts to the fact that the order of a is n. So, our supposition is wrong. Hence all elements are distinct. Therefore, G contains exactly n elements.

7.5.4 Theorem

A finite group G of order n containing an element of order n must be cyclic.

Proof: Let us consider G be a finite group of order n. Let a be an element of G with order n.

This implies that

$$a^n = e.$$

Let us construct an cyclic group G_1 with generator a. Thus we have $G_1 = \{a, a^2, a^3, \ldots, a^n = e\}$

But we know that order of group and order of its generator is same. This implies that

$$O(G_1) = O(a) = O(G).$$

i.e., $O(G_1) = O(G)$

Again let $a \in G_1$

$\Rightarrow$ $a^n \in G_1$

$\Rightarrow$ $a^n \in G$

This implies that $G_1 \subseteq G$; but $O(G_1) = O(G)$. Therefore, $G = G_1$. Hence G is a cyclic group.

7.5.5 Theorem

Subgroup of a cyclic group is itself a cyclic group.

Proof: Let G be a cyclic group with generator a and let H be the subgroup of G.

Now as H is contained in G, the elements of H are of the type a^k. Let m be the least positive integer such that

$$a^m \in H$$

Let $a^k \in H$, where k is an integer greater than m *i.e.*, $k > m$.

This implies that $k = mn + r; 0 \leq r \leq m\, a^{-mn}$ ....(*i*)

But we know that the elements of H are in the form of integral power of a. Therefore, $a^{mn} \in H$. This implies that $a^{-mn} \in H$.

Now $a^k \in H$ and $a^{-mn} \in H$. So by closure property we have $a^k \cdot a^{-mn} \in H$

$\Rightarrow$ $a^{k-mn} \in H$

$\Rightarrow$ $a^{mn+r-mn} \in H$

$\Rightarrow$ $a^r \in H$

This contradicts to the assumption that m is the least positive integer for which $a^m \in H$. So $a^r \in H$ is possible only if $r = 0$. Thus we have from equation (*i*) $k = mn$.

Thus $a^k = a^{mn} = (a^m)^n$.

Therefore, H is a cyclic group with generator a^m.

7.6 COSETS

Associated with any subgroup there are two cosets namely left coset and right coset. Let G be a group and H be any subgroup of G and let $a \in G$. The left coset of H in G is the set aH given by

$$aH = \{x \mid x = ah, \forall h \in H\}$$

The right coset of H in G is the set Ha given by

$$Ha = \{x \mid x = ha, \forall h \in H\}$$

The cosets are not necessarily subgroup of G. If G is an abelian group, then the left coset of H in G is equal to the right coset of H in G.

7.6.1 Theorem

If H is subgroup of a group G and $h \in H$, then $Hh = H = hH$.

Proof: Given that H is a subgroup of group G and $h \in H$.

Our claim is $Hh = H$ *i.e.*, $Hh \subseteq H$ and $H \subseteq Hh$.

Let $h_1 \in H$. Again $h_1 \in H$ and $h \in H$ implies that $(h_1 h) \in H$.

But we know that being the right coset $(h_1 h) \in Hh$.

Thus, $(h_1 h) \in Hh \Rightarrow (h_1 h) \in H.$

Therefore, $Hh \subseteq H$ ...(*i*)

Now $h_1 \in H$

and $h_1 = h_1 e$

$= h_1(h^{-1} h)$ [By existence of identity]

$= (h_1 h^{-1}) h$ [By associative law]

Therefore, $h_1 = (h_1 h^{-1}) h \in Hh$

Hence, we get $h_1 \in H \Rightarrow h_1 \in Hh$. Thus we get

$$H \subseteq Hh \qquad \text{...}(ii)$$

Therefore, on combining equations (*i*) and (*ii*), we get $Hh = H$.

Similarly it can be shown that $hH = H$.

Therefore, we have $Hh = H = hH$.

7.7 HOMOMORPHISM

A mapping ϕ defined from a group G_1 with binary operation (o) to the group G_2 with binary operation (*) is said to be homomorphism if

$$\phi(x \circ y) = \phi(x) * \phi(y) \; \forall x, y \in G_1$$

7.7.1 Theorem

If ϕ is a homomorphism defined from G_1 to G_2, then

(*i*) $\phi(e_1) = e_2$; e_1 is the identity element of G_1 and e_2 is the identity element of G_2.

(*ii*) $\phi(x^{-1}) = (\phi(x))^{-1}$; $\forall x \in G_1$

Proof: (*i*) Given that ϕ is a homomorphism from G_1 to G_2. Also given that e_1 is the identity element of G_1 and e_2 is the identity element of G_2.

Let $x \in G_1$. This implies that $\phi(x) \in G_2$. Now $e_1 \in G_1$ such that $x e_1 = x$.

Therefore, $\phi(x) = \phi(x e_1)$

$$= \phi(x)\,\phi(e_1) \qquad [\because\ \phi \text{ is a homomorphism}]$$

i.e., $$\phi(x) = \phi(x)\,\phi(e_1) \qquad ...(i)$$

Again as e_2 is the identity element of G_2, we have

$$\phi(x) = \phi(x)\,e_2 \qquad ...(ii)$$

Hence from equations (*i*) and (*ii*) it is clear that

$$\phi(x)\,\phi(e_1) = \phi(x) = \phi(x)\,e_2$$

$$\Rightarrow \quad \phi(e_1) = e_2 \qquad \text{[Left cancellation law]}$$

(*ii*) Given that ϕ is a homomorphism from G_1 to G_2. Also given that e_1 is the identity element of G_1 and e_2 is the identity element of G_2.

Let $x \in G_1$ such that $(x\,x^{-1}) = e_1$

$$\Rightarrow \quad \phi((x\,x^{-1})) = \phi(e_1)$$

$$\Rightarrow \quad \phi(x)\,\phi(x^{-1}) = \phi(e_1) \qquad [\phi \text{ is a homomorphism}]$$

$$\Rightarrow \quad \phi(x)\,\phi(x^{-1}) = e_2$$

Hence, it is clear that $\phi(x^{-1})$ is the inverse element of $\phi(x)$.

Thus we have $\phi(x^{-1}) = (\phi(x))^{-1}\,;\ \forall\, x \in G_1$

i.e., Inverse element corresponds to the inverse element.

7.7.2 Theorem

If $\phi : G_1 \to G_2$ is a homomorphism, then $\phi(G_1)$ is a subgroup of G_2.

Proof: Given $\phi : G_1 \to G_2$ is a homomorphism. Then for $x, y \in G_1$ we have $\phi(x\,y) = \phi(x)\,\phi(y)$.

Again $\phi(x) \in \phi(G_1)$, $\phi(y) \in \phi(G_1)$ such that

$$\phi(x)\,\phi(y) = \phi(x\,y) \in \phi(G_1).$$

Therefore the closure property is satisfied.

Also $y \in G_1$ implies $y^{-1} \in G_1$ such that $(y\,y^{-1}) = e_1$, where e_1 is the identity element of G_1. Thus we have

$$\phi(y\,y^{-1}) = \phi(e_1)$$

$$\Rightarrow \quad \phi(y)\,\phi(y^{-1}) = \phi(e_1) \qquad [\because\ \phi \text{ is a homomorphism}]$$

$$\Rightarrow \quad \phi(y)\,\phi(y^{-1}) = e_2\,; \text{ where } e_2 \text{ identity element of } \phi(G_1).$$

Therefore, $(\phi(y))^{-1} = \phi(y^{-1})$.

This indicates that for every element $\phi(y) \in \phi(G_1)$ there exist inverse element $\phi(y^{-1})$ in $\phi(G_1)$.

Hence, $\phi(G_1)$ is a subgroup of G_2.

SOLVED EXAMPLES

Example 1 *Show that the subtraction is not a binary operation on the set of natural numbers N.*

Solution: We know that 'o' will be a binary operation in N if and only if $(a\ {}_o\ b) \in N\ \forall\ a, b \in N$ and $(a\ {}_o\ b)$ is unique. Here the binary operation is subtraction (–). It is clear that for $a, b \in N$, $(a - b)$ may or may not belongs to N. Let us take $a = 5$ and $b = 10$, so $(a - b) = -5 \notin N$. Hence, the subtraction is not a binary operation.

Example 2 *The operation 'o' defined by the relation* $(a\ {}_o\ b) = \dfrac{a}{b}$ *is not a binary operation in the set of real number R.*

Solution: We know that 'o' will be a binary operation in R if and only if $(a \,_0\, b) \in R \;\forall\; a, b \in R$ and $(a \,_0\, b)$ is unique. Here the binary operation 'o' defined by the relation $(a \,_0\, b) = \frac{a}{b}$. It is clear that $\frac{a}{b}$ is not defined for $b = 0$. Let us take $a = 5$ and $b = 0$, but $\frac{5}{0}$ is not defined. Hence the operation 'o' defined by the relation $(a \,_0\, b) = \frac{a}{b}$ is not a binary operation on R.

Example 3 *Is the following a valid definition of binary operation.*

(a) $(a \,_0\, b) = a\,b + 2b$ *on R*
(b) $(a \,_0\, b) = a^b$ *on* I^+
(c) $(a \,_0\, b) = Min\,(a, b)$ *on R*

Solution: (*a*) Let $a, b \in R$

$\Rightarrow$ $(ab) \in R$ [Product of two real numbers is also a real number]

Again, $b \in R$ implies that $2b \in R$.

We know that addition of two real numbers is also a real number, so $(ab + 2b) \in R$ and it is unique. Hence, the operation $(a \,_0\, b) = ab + 2b$ on R satisfies the definition of binary operation.

(*b*) Let $a, b \in I^+$ (Set of positive integers)

Given that $(a \,_0\, b) = a^b$. We know that a positive integer raised to the power by a positive integer will always result on a positive integer and it is unique also.

This implies that a^b is a unique positive integer. Hence $(a \,_0\, b) = a^b$ is a binary operation.

(*c*) Let $a, b \in R$ (Set of real numbers)

Given that $(a \,_0\, b) = \text{Min}\,(a, b)$. Which is equal to either a or $b \in R$. This implies that $(a \,_0\, b) = \text{Min}\,(a, b)$ is a binary operation in R.

Example 4 *Let A = {0, 1}, then define the binary operations for and (∧) and or (∨).*

Solution: Given that A = {0, 1}. The binary operations for and (∧) and or (∨) is given as below.

∨	1	0
1	1	1
0	1	0

∧	1	0
1	1	0
0	0	0

Example 5 *Complete the following table so that the binary operation (o) is commutative.*

o	*a*	*b*	*c*
a	*b*		
b	*c*	*b*	*a*
c	*a*		*c*

Solution: A binary operation (o) in set G is said to be commutative if $(a \,_0\, b) = (b \,_0\, a)$. Since binary operation (o) is commutative we have the followings.

$$(a \,_0\, b) = (b \,_0\, a) = c$$
$$(a \,_0\, c) = (c \,_0\, a) = a$$
$$(c \,_0\, b) = (b \,_0\, c) = a$$

Thus the complete table is given below.

o	a	b	c
a	b	c	a
b	c	b	a
c	a	a	c

Example 6 *For the algebraic structure (G, o), defined by* $(a\ _0\ b) = a + b - ab$; $a, b \in G$. *Show that G is a semi group, monoid and also show that commutative property holds.*

Solution: Given that for all $a, b \in$ G, $(a\ _0\ b) = a + b - ab$.

Semi group: Let $a, b, c \in$ G

Now $\quad a\ _0\ (b\ _0\ c) = a\ _0\ (b + c - bc)$

$= a + (b + c - bc) - a(b + c - bc)$

$= a + b + c - bc - ab - ac + abc$

Again $\quad (a\ _0\ b)\ _0\ c = (a + b - ab)\ _0\ c$

$= (a + b - ab) + c - c\ (a + b - ab)$

$= a + b + c - ab - ca - bc + abc$

Comparing the above two we see $a\ _0\ (b\ _0\ c) = (a\ _0\ b)\ _0\ c$. This implies that (G, o) is a semi group.

Monoid : We know that the algebraic structure (G, o) is monoid if it is a semi group and has an identity element. We have already shown that (G, o) is a semi group.

Let us now try to find the unit element $e \in$ G such that $(a\ _0\ e) = a$

i.e., $\quad a + e - ae = a$

$\Rightarrow \quad a + e\ (1 - a) = a$

$\Rightarrow \quad e\ (1 - a) = a - a = 0$

$\Rightarrow \quad e = 0$

So, the unit element 0 (Zero) exist in G.

Commutative Law: G is said to be commutative if $(a\ _0\ b) = (b\ _0\ a)$.

Now $(a\ _0\ b) = a + b - ab$ and $(b\ _0\ a) = b + a - ba$. This implies that $(a\ _0\ b) = (b\ _0\ a)$, hence commutative.

Example 7 *A set G = {a, b, c, d}, the binary operation (o) on this set is defined by the following figure. Find the followings.*

(a) $(a\ _0\ b)$ *and* $(b\ _0\ a)$

(b) Is binary operation (o) commutative.

(c) Is binary operation (o) associative.

o	a	b	c	d
a	a	c	b	d
b	d	a	b	c
c	a	d	a	a
d	d	b	a	c

Solution: Given that G = $\{a, b, c, d\}$. The binary operation (o) on this set is defined by the following figure.

o	a	b	c	d
a	a	c	b	d
b	d	a	b	c
c	c	d	a	a
d	d	b	a	c

(*a*) $(a\ _0\ b) = c$ and $(b\ _0 a) = d$

(*b*) As $(a\ _0\ b) \neq (b\ _0\ a)$, so the binary operation as defined is not commutative.

(*c*) Now $a\ _0\ (b\ _0\ c) = a\ _0\ (b) = c$ and

$$(a\ _0\ b)\ _0\ c = c\ _0\ c = a$$

So, $$a\ _0\ (b\ _0\ c) \neq (a\ _0\ b)\ _0\ c.$$

This implies that the binary operation defined above is not associative.

Example 8 *Given the algebraic structure (G, o), defined by the following table. Show that G is a semi group, monoid and find the unit element.*

o	a	b	c
a	c	b	a
b	b	c	b
c	a	b	c

Solution: Given the algebraic structure (G, o), defined by the following table as

o	a	b	c
a	c	b	a
b	b	c	b
c	a	b	c

Semi group: Let $a, b, c \in$ G

Now
$$a\ _0\ (b\ _0\ c) = a\ _0\ b = b \text{ and } (a\ _0\ b)\ _0\ c = b\ _0\ c = b$$
$$b\ _0\ (a\ _0\ c) = b\ _0\ a = b \text{ and } (b\ _0\ a)\ _0\ c = b\ _0\ c = b$$
$$c\ _0\ (a\ _0\ b) = c\ _0\ b = b \text{ and } (c\ _0\ a)\ _0\ b = a\ _0\ b = b$$

Therefore, the algebraic structure (G, o) is a semi group.

Monoid: We know that the algebraic structure (G, o) is monoid if it is a semi group and has an identity element.

We have already shown that (G, o) is a semi group.

Let us now try to find the unit element $e \in$ G. It is very clear from the table that

$$(a\ _0\ c) = a = (c\ _0\ a)$$
$$(b\ _0\ c) = b = (c\ _0\ b) \text{ and}$$
$$(c\ _0\ c) = c = (c\ _0\ c)$$

Therefore, the identity element is given as $e = c$.

Thus the algebraic structure (G, o) is a monoid.

Example 9 *G contains real numbers 1, – 1 under the usual multiplication. Then show that G is a commutative group of order 2.*

Solution: Given G = $\{1, -1\}$ and the binary operation (o) is multiplication (*). Let us construct the table.

*	1	–1
1	1	–1
–1	–1	1

Closure Law: From the above table it is clear that the binary operation with any two element results either 1 or –1. So the closure property is satisfied.

Associative Law: From the above table it is clear that the associative property is also satisfied.

Existence of Identity: From the above table it is clear that the identity element is $1 \in G$.

Existence of Inverse: From the above table it is clear that $(1 * 1) = 1$ and $(-1 * -1) = 1$. Therefore, 1 and – 1 are their own inverses.

Commutative Law: From the table it is clear that $1 * (-1) = (-1) * 1$. So, commutative property is also satisfied.

Therefore, G is an abelian group of order 2.

Example 10 *Show that the set $G = \{1, \omega, \omega^2\}$ is a group with respect to binary operation multiplication, where ω is the cube root of unity.*

Solution: Given ω is the cube root of unity. Thus we have $\omega^3 = 1$ and $1 + \omega + \omega^2 = 0$. It is also given that the binary operation (o) is also a multiplication. Let us construct the table.

*	1	ω	ω^2
1	1	ω	ω^2
ω	ω	ω^2	1
ω^2	ω^2	1	ω

Closure Law: As seen from the table all the elements belongs to the Set G. So, Closure law is satisfied.

Associative Law: The elements of G are complex numbers and we know that complex number multiplication is associative. Thus associative law is satisfied.

Existence of Identity: From the table it is clear that $1 \in G$ is the identity element.

Existence of Inverse: From the table it is also clear that $1 * 1 = 1$; $\omega * \omega^2 = \omega^3 = 1$ and $\omega^2 * \omega = \omega^3 = 1$.

Hence, it is clear that ω is the inverse element of ω^2, ω^2 is the inverse element of ω and 1 is the inverse element of 1. So, every element of G has its inverse in G.

Therefore, the set $G = \{1, \omega, \omega^2\}$ is a group with respect to binary operation multiplication.

Example 11 *Give an example of a group of second order such that every element is its own inverse.*

Solution: Let us consider the set $G = \{1, -1\}$ and the binary operation (o) is multiplication (*). Let us construct the table.

*	1	–1
1	1	–1
–1	–1	1

Closure Law: From the above table it is clear that the binary operation with any two element results either 1 or – 1. So the closure property is satisfied.

Associative Law: From the above table it is clear that the associative property is also satisfied.

Existence of Identity: From the above table it is clear that the identity element is $1 \in G$.

Existence of Inverse: From the above table it is clear that $(1 * 1) = 1$ and $(-1 * -1) = 1$. Therefore, 1 and – 1 are their own inverses.

Therefore, G is a group of second order. *i.e.*, $O(G) = 2$.

Example 12 *G is a set of all non-zero real numbers and let* $(a\ _o\ b) = \frac{ab}{2}$. *Show that (G, o) is an abelian group.*

Solution: Given that G is a set of real numbers.

Closure Law: Let $a, b \in$ G. We know that product of any two real numbers is a real number. This implies that $(ab) \in$ G. Similarly $\frac{ab}{2} \in$ G. Therefore, $(a\ _o\ b) \in$ G.

Associative Law: Let $a, b, c \in$ G.

Now $a\ _o\ (b\ _o\ c) = a\ _o \left(\frac{bc}{2}\right) = \frac{abc}{4}$ and $(a\ _o\ b)\ _o\ c = \left(\frac{ab}{2}\right)\ _o\ c = \frac{abc}{4}$. Therefore, $a\ _o\ (b\ _o\ c) = (a\ _o\ b)\ _o\ c$

Existence of Identity: Let $a \in$ G and $e \in$ G be the identity element such that $(a\ _o\ e) = a$. *i.e.,* $\frac{ae}{2} = a$. This implies that $e = 2 \in$ G. So every element of G has 2 as the identity element.

Existence of Inverse: Let $a \in$ G and $a^{-1} \in$ G be the inverse element of a.

Thus we have $(a\ _o\ a^{-1}) = e = 2$. This implies that $a^{-1} = \frac{4}{a} \in$ G as $4 \in$ G, $a \in$ G and ratio of two non-zero real number is a real number.

Commutative Law: Now $(a\ _o\ b) = \frac{ab}{2} = \frac{ba}{2} = (b\ _o\ a)$. This implies that $(a\ _o\ b) = (b\ _o\ a)$. Therefore, commutative law also holds in G.

Thus (G, o) is an abelian group.

Example 13 *Let G be the set of all (2 ×2) real matrices* $\begin{pmatrix} a & b \\ c & d \end{pmatrix}$ *where* $(ad - bc) \neq 0$ *is a rational number. Prove that G forms a group under multiplication.*

Solution: Let us consider G be the set of all (2 × 2) real matrices $\begin{pmatrix} a & b \\ c & d \end{pmatrix}$ where $(ad - bc) \neq 0$ is a rational number.

Closure Law: Let A, B $\in$ G. *i.e.*, A and B are two matrices of order (2 × 2). This implies that (A × B) is also a real matrix of order (2 × 2).

From the definition it is clear that $|A| \neq 0$ and $|B| \neq 0$, hence, $|A \times B| = |A| \times |B| \neq 0$. Thus (A × B) is a matrix of order (2 × 2) and $|A \times B| \neq 0$. So, (A × B) $\in$ G. Therefore, closure law is satisfied.

Associative Law: We know that matrix multiplication is associative. So, for A, B, C $\in$ G we have A × (B × C) = (A × B) × C. Therefore, associative law is satisfied.

Existence of Inverse: The matrix $I = \begin{pmatrix} 1 & 0 \\ 0 & 1 \end{pmatrix} \in$ G, since $|I| = 1 \neq 0$, will act as the identity element since

$$\begin{pmatrix} a & b \\ c & d \end{pmatrix}\begin{pmatrix} 1 & 0 \\ 0 & 1 \end{pmatrix} = \begin{pmatrix} a & b \\ c & d \end{pmatrix}$$

Existence of Inverse: For every A $\in$ G, there exists inverse $A^{-1} \in$ G such that $(A \times A^{-1}) = I$, where

$$A^{-1} = \frac{\text{Adj}(A)}{|A|} = \frac{1}{|A|}\begin{pmatrix} c_{11} & c_{12} \\ c_{21} & c_{22} \end{pmatrix}; c_{ij} \text{ is the cofactor of } a_{ij}.$$

Therefore, G forms a group under multiplication.

Example 14 *Let G be the set $\{a_0, a_1, a_2, \ldots, a_6\}$. The binary operation (o) is defined as below. Check whether (G, o) is a group or not.*

$$(a_i \circ a_j) = \begin{cases} a_{i+j} & \text{if } (i+j) < 7 \\ a_{i+j-7} & \text{if } (i+j) \geq 7 \end{cases}$$

Solution: Let G = $\{a_0, a_1, a_2, \ldots, a_6\}$. The binary operation (o) is defined as

$$(a_i \circ a_j) = \begin{cases} a_{i+j} & \text{if } (i+j) < 7 \\ a_{i+j-7} & \text{if } (i+j) \geq 7 \end{cases}$$

Based on the binary operation defined above we have

$$(a_0 \circ a_0) = a_0;\ (a_0 \circ a_1) = a_1;\ (a_5 \circ a_1) = a_6;$$

$(a_5 \circ a_2) = a_{7-7} = a_0$; $(a_6 \circ a_3) = a_{9-7} = a_2$; and so on. Thus we have the following table.

0	a_0	a_1	a_2	a_3	a_4	a_5	a_6
a_0	a_0	a_1	a_2	a_3	a_4	a_5	a_6
a_1	a_1	a_2	a_3	a_4	a_5	a_6	a_0
a_2	a_2	a_3	a_4	a_5	a_6	a_0	a_1
a_3	a_3	a_4	a_5	a_6	a_0	a_1	a_2
a_4	a_4	a_5	a_6	a_0	a_1	a_2	a_3
a_5	a_5	a_6	a_0	a_1	a_2	a_3	a_4
a_6	a_6	a_0	a_1	a_2	a_3	a_4	a_5

Closure Law: From the table it is clear that all the elements are in G. So, closure property is satisfied.

Associative Law: Let $a_i, a_j, a_k \in$ G. Now it is evident from the table that

$$a_i \circ (a_j \circ a_k) = \begin{cases} a_{(i+j+k)} & \text{if } (i+j+k) < 7 \\ a_{(i+j+k-7)} & \text{if } 7 < (i+j+k) < 14 \\ a_{(i+j+k-14)} & \text{if } (i+j+k) \geq 14 \end{cases}$$

Same definition holds for $(a_i \circ a_j) \circ a_k$. So, associative property is satisfied.

Existence of Identity: From the first row of the table it is clear that $a_0 \in$ G is the identity element as $(a_0 \circ a_j) = a_j$ for all j = 0, 1, 2, 3, 4, 5, 6.

Existence of Inverse: From the table it is clear that a_0 is the inverse of a_0 and a_j is the inverse element of $a_{(7-j)}$ for J = 1, 2, 3, 4, 5, 6. So, every element has its inverse in G.

Therefore, G is a group under the binary operation defined above.

Example 15 *If G is a group of even order then there exists an element $a \neq e$ such that $a^2 = e$.*

Solution: Given G is a group of even order.

Let there be n elements and its n number of inverses. So altogether there are $2n$ number of elements. Again in this $2n$ number of elements there is an identity element $e \in$ G and $e^{-1} = e$.

This implies G contains $(2n - 1)$ number of elements, but it is given that G is of even order. So, there must exist at least one element which is its own inverse.

i.e., $a \in G \Rightarrow a^{-1} = a$

$\Rightarrow \quad (a \circ a^{-1}) = (a \circ a)$

$\Rightarrow \quad e = a^2$

i.e., $a^2 = e.$

Example 16 *Suppose that G is a group and $a^2 = a$, $a \in G$. Prove that $a = e$.*

Solution: Given that $a^2 = a$

$\Rightarrow \quad (a \circ a) = (a \circ e).$

So, by left cancellation law $a = e$.

Example 17 *If every element of the group G is its own inverse, then it is an abelian group.*

Solution: Given every element of group G is its own inverse.

Let $a, b \in G$ implies that $a^{-1} = a$ and $b^{-1} = b$.

Again by closure law $a, b \in G$ implies that $(a \circ b) \in G$ and $(a \circ b)^{-1} = (a \circ b)$.

$\Rightarrow \quad (b^{-1} \circ a^{-1}) = (a \circ b)$ [By theorem]

$\Rightarrow \quad (b \circ a) = (a \circ b)$ $[a^{-1} = a$ and $b^{-1} = b]$

Therefore, G is an abelian group.

Example 18 *If G is a group with $(a \circ b)^n = a^n b^n$ for three consecutive integers, then G is an abelian group.*

Solution: Given G is a group and $(a \circ b)^n = a^n b^n$ for three consecutive integers.

Let the three consecutive integers be n, $(n + 1)$ and $(n + 2)$. So, by definition we have

$$(a \circ b)^n = a^n b^n; \qquad (a \circ b)^{n+1} = a^{n+1} b^{n+1} \text{ and}$$
$$(a \circ b)^{n+2} = a^{n+2} b^{n+2}$$

Now $(a \circ b)^{n+2} = (a \circ b)^{n+1} (a \circ b)$

$\Rightarrow \quad a^{n+2} b^{n+2} = (a^{n+1} b^{n+1}) (a \circ b)$

$\Rightarrow \quad (a^{n+1} a) b^{n+2} = (a^{n+1}) (b^{n+1} a) \circ b$

$\Rightarrow \quad (a^{n+1}) (a b^{n+2}) = (a^{n+1}) (b^{n+1} a) \circ b$

$\Rightarrow \quad (ab^{n+2}) = (b^{n+1} a) \circ b$ [Left cancellation law]

$\Rightarrow \quad (ab^{n+1}) b = (b^{n+1} a) \circ b$

$\Rightarrow \quad (ab^{n+1}) = (b^{n+1} a)$ [Right cancellation law]

$\Rightarrow \quad a^n (ab^{n+1}) = a^n (b^{n+1} a)$

$\Rightarrow \quad (a^{n+1} b^{n+1}) = (a^n b^n) (b \circ a)$

$\Rightarrow \quad (a \circ b)^{n+1} = (a \circ b)^n (b \circ a)$

$\Rightarrow \quad (a \circ b)^n (a \circ b) = (a \circ b)^n (b \circ a)$

$\Rightarrow \quad (a \circ b) = (b \circ a)$ [Left cancellation law]

This implies that G is an abelian group.

Example 19 *G is the set of all integers and the binary operation (o) is defined by $(a \circ b) = a - b$. Test whether G is a group.*

Solution: Given that G is the set of all integers and the binary operation (o) is defined by $(a \circ b) = a - b$.

Closure Law : We know that difference of two integers is an integer. So, for $a, b \in G$, we have $(a - b) \in G$. Thus $(a \circ b) \in G$. Therefore closure law is satisfied.

Associative Law: Let $a, b, c \in G$.

Now
$$a \circ (b \circ c) = a \circ (b - c)$$
$$= a - (b - c)$$
$$= a - b + c \qquad \ldots(i)$$

Again $(a \circ b) \circ c = (a - b) \circ c$
$= (a - b) - c$
$= a - b - c$ ... (ii)

From the equations (i) and (ii) it is clear that $a \circ (b \circ c) \neq (a \circ b) \circ c$. Therefore, associative law is not satisfied. Hence G is not a group.

Example 20 *Let G = {1, – 1, i, – i } be a group under the binary operation multiplication. Find the order of elements.*

Solution: Given G = {1, – 1, i, – i } be a group under the binary operation multiplication. Therefore, the identity element is 1.

Now $O(1) = 1$ $[\because (1)^1 = 1]$
$O(-1) = 2$ $[\because (1)^2 = 1]$
$O(i) = 4$ $[\because (i)^4 = (i^2)^2 = (-1)^2 = 1]$
$O(-i) = 4$ $[\because (-i)^4 = (i)^4 = (i^2)^2 = (-1)^2 = 1]$

Example 21 *G is the set of positive integers and the binary operation (o) is defined by $(a \circ b) = ab$. Test whether G is a group.*

Solution: G is the set of positive integers and the binary operation (o) is defined by

$$(a \circ b) = (a\, b).$$

Closure Law: We know that product of two positive integers is a positive integer. So, for $a, b \in$ G, we have $(a\, b) \in$ G. Thus $(a \circ b) \in$ G.

Therefore, closure law is satisfied.

Associative Law: Let $a, b, c \in$ G.

Now $a \circ (b \circ c) = (a\, b\, c) = (a \circ b) \circ c$. Therefore associative law is satisfied.

Existence of Identity: G is the set of positive integer. This implies that $1 \in$ G and $(1 \circ a) = (a \circ 1) = a$ for all $a \in$ G. Therefore 1 is the identity element of G.

Existence of Inverse : Let $a \in$ G and let b be the inverse of a. Thus we have $(a \circ b) = 1$. This implies that $b = \frac{1}{a} \notin$ G. Since $\frac{1}{a}$ is not a positive integer. So, inverse element does not exist in G.

Therefore, G is not a group.

Example 22 *Let G = {a, a^2, a^3, a^4, a^5, a^6 = e } be a group under the binary operation multiplication. Find the order of elements.*

Solution: Let G = $\{a, a^2, a^3, a^4, a^5, a^6 = e\}$ be a group under the binary operation multiplication.

Now $O(a) = 6$ $[\because a^6 = e]$
$O(a^2) = 3$ $[\because (a^2)^3 = a^6 = e]$
$O(a^3) = 2$ $[\because (a^3)^2 = a^6 = e]$
$O(a^4) = 3$ $[\because (a^4)^3 = a^{12} = (a^6)^2 = e]$
$O(a^5) = 6$ $[\because (a^5)^6 = a^{30} = (a^6)^5 = e]$
$O(a^6) = 1$ $[\because (a^6)^1 = e]$

Example 23 *Let G = {0, 1, 2, 3, 4, 5} be a group under the binary operation addition modulo 6. Find the order of elements of the group.*

Solution: Given G = {0, 1, 2, 3, 4, 5} be a group under the binary operation addition modulo 6. Here the identity element is 0 (Zero). *i.e.*, O(0) = 1 as $0^1 = 0$. Let us now find out the order of 1.

$1^1 = 1$
$1^2 = 1 \oplus_6 1 = 2$
$1^3 = 1 \oplus_6 1^2 = 1 \oplus_6 2 = 3$

$$1^4 = 1 \oplus_6 1^3 = 1 \oplus_6 3 = 4$$
$$1^5 = 1 \oplus_6 1^4 = 1 \oplus_6 4 = 5$$
$$1^6 = 1 \oplus_6 1^5 = 1 \oplus_6 5 = 0$$

Therefore, $\quad O(1) = 6$

Similarly $\quad O(2) = 3; O(3) = 2; O(4) = 3; O(5) = 6.$

Example 24 *Let G is a group and order of every element $a \neq e$ of the group G is two. Show that G is an abelian group.*

Solution: Given G is a group and order of every element $a \neq e$ of the group G is two.

Let $a \in$ G. This implies that $O(a) = 2$. *i.e.*, $a^2 = e$

$\Rightarrow \quad (a \,_0\, a) = e$

$\Rightarrow \quad (a \,_0\, a) \,_0\, a^{-1} = e \,_0\, a^{-1} = a^{-1}$ [Existence of identity]

$\Rightarrow \quad a \,_0\, (a \,_0\, a^{-1}) = a^{-1}$ [Associative law]

$\Rightarrow \quad a \,_0\, e = a^{-1}$ [Existence of inverse]

$\Rightarrow \quad a = a^{-1}$ [Existence of identity]

So, every element of G is its own inverse. *i.e.*, For $a, b \in$ G we have $a = a^{-1}$ and $b = b^{-1}$.

Again by closure law $(a \,_0\, b) \in$ G and $(a \,_0\, b)^{-1} = (a \,_0\, b)$. This implies that $(b^{-1} \,_0\, a^{-1}) = (a \,_0\, b)$

i.e., $\quad (b \,_0\, a) = (a \,_0\, b)$.

Therefore, G is an abelian group.

Example 25 *Show that in a additive group of integers G the order of every element except 0 (zero) is infinite.*

Solution: Given G is the additive group of integers. The identity element in case of additive group of integers is 0 (zero). This implies that $O(0) = 1$ as $0^1 = 0$. Let us consider the next element 1.

Now
$$1^1 = 1$$
$$1^2 = 1 + 1 = 2$$
$$1^3 = 1 + 1 + 1 = 3$$

and so on. From this it is clear that there exists no such n for which $1^n = 0$. This implies that order of 1 is infinite. The same argument also holds for other integers.

Example 26 *Let G be a group and the order of a, b and $(a \,_0\, b)$ be two. Show that G is an abelian group.*

Solution: Given G be a group and the order of a, b and $(a \,_0\, b)$ be two. *i.e.*, $a^2 = e$; $b^2 = e$ and $(a \,_0\, b)^2 = e$.

Now $\quad (a \,_0\, b)^2 = e$

$\Rightarrow \quad (a \,_0\, b)\,(a \,_0\, b) = e$

$\Rightarrow \quad (a \,_0\, b)\,(a \,_0\, b) = e \,_0\, e = a^2 \,_0\, b^2$

$\Rightarrow \quad a \,_0\, (b \,_0\, a) \,_0\, b = (a \,_0\, a) \,_0\, (b \,_0\, b)$

$\Rightarrow \quad a \,_0\, (b \,_0\, a) \,_0\, b = a \,_0\, (a \,_0\, b) \,_0\, b$ [Associative law]

$\Rightarrow \quad (b \,_0\, a) = (a \,_0\, b)$ [Cancellation law]

Therefore, G is an abelian group.

Example 27 *Is union of two subgroups of a group G is a subgroup of G? If no then explain with the help of a counter example.*

Solution: The union of two subgroups of a group G is not a subgroup of G.

Let us consider the group G = {... −3, −2, −1, 0, 1, 2, 3, ...} with the binary operation addition. Let us define two subgroups H_1 and H_2 of G as

$H_1 = \{0, \pm 2, \pm 4, \pm 6, ...\}$ and $H_2 = \{0, \pm 3, \pm 6, \pm 9, ...\}$. Hence, $(H_1 \cup H_2) = \{0, \pm2, \pm3, \pm4, \pm6, ...\}$. From this it is clear that $2, 3 \in (H_1 \cup H_2)$ implies that $(2 + 3) = 5 \notin (H_1 \cup H_2)$. Therefore, closure law is not satisfied.

Hence, $(H_1 \cup H_2)$ is not a subgroup of G.

Example 28 *Suppose* $G = \{... , 2^{-3}, 2^{-2}, 2^{-1}, 1, 2, 2^2, 2^3, ...\}$ *is the multiplicative group. Let* $H = \{1, 2, 2^2, 2^3, ...\}$ *be the subset of G. Test whether H is a subgroup of G or not.*

Solution: Given $G = \{... , 2^{-3}, 2^{-2}, 2^{-1}, 1, 2, 2^2, 2^3, ...\}$ is the multiplicative group. Let $H = \{1, 2, 2^2, 2^3, ...\}$ be the non-empty subset of G and $a, b \in H$ implies that $(a\ _0\ b) \in H$. So, H is closed under multiplication.

Again $2 \in H$ implies $2^{-1} \notin H$. Thus inverse axiom is not satisfied. Therefore, H is not the subgroup of G.

Example 29 *Let G be a group of integers under addition and H is the subset of G consisting of all multiples of* $n \in N$. *Show that H is a subgroup for every value of n.*

Solution: Given G be a group of integers under addition and H is the subset of G consisting of all multiples of $n \in N$.

Let $a, b \in H$. This implies that a and b are both multiples of n. Therefore $(a + b)$ is also a multiple of n. So, $(a\ _0\ b) = (a + b) \in H$. Again $a \in H$ implies $-a \in H$. Therefore, H is a subgroup of G.

Example 30 *Suppose that G be the set of all ordered pairs (a, b) of Real numbers;* $a \neq 0$. *The binary operation (o) is defined by* $(a, b)\ _0\ (c, d) = (ac, bc + d)$. *Show that (G, o) is a non-abelian group. Let H be a subset of G containing elements of the form (1, b). Does H is a subgroup of G.*

Solution: Given G be the set of all ordered pairs (a, b); $a, b \in R, a \neq 0$. The binary operation is defined as $(a, b)\ _0\ (c, d) = (ac, bc + d)$.

We have to show that G is a group. *i.e.*, G satisfies all the four properties of the group.

Closure Law: Let $(a, b), (c, d) \in G$; This implies that $a \neq 0$ and $c \neq 0$.

Now $\quad (a, b)\ _0\ (c, d) = (ac, bc + d) \in G \qquad [ac \neq 0; ac, bc + d \in R]$

Associative Law: Let $(a, b), (c, d), (e, f) \in G$.

$$\text{Now} \quad (a, b)\ _0\ [(c, d)\ _0\ (e, f)] = (a, b)\ _0\ (ce, de + f) = (ace, bce + de + f) \qquad ...(i)$$

$$\text{Again} \quad [(a, b)\ _0\ (c, d)]\ _0\ (e, f) = (ac, bc + d)\ _0\ (e, f) = (ace, bce + de + f) \qquad ...(ii)$$

So, from equations (*i*) and (*ii*) we have

$$(a, b)\ _0\ [(c, d)\ _0\ (e, f)] = [(a, b)\ _0\ (c, d)]\ _0\ (e, f)$$

Existence of Identity: Let $(a, b) \in G$. Let (u, v) be the identity element. Thus we have

$$(a, b)\ _0\ (u, v) = (a, b)$$

i.e., $\quad (au, bu + v) = (a, b)$

This implies that $au = a$ and $bu + v = b$ *i.e.*, $u = 1$ and $v = 0$. So, the identity element

$$(u, v) = (1, 0) \in G.$$

Existence of Inverse: Let $(a, b) \in G$ and let (u, v) be the inverse element of (a, b). Thus, we have

$$(a, b)\ _0\ (u, v) = (1, 0) \qquad [\because\ (1, 0) \text{ is the identity element}]$$

i.e., $\quad (au, bu + v) = (1, 0)$

This implies that $au = 1$ and $bu + v = 0$ *i.e.*, $u = \frac{1}{a}$ and $v = \frac{-b}{a}$.

So, the inverse element of (a, b) is $\left(\frac{1}{a}, \frac{-b}{a}\right) \in G$. Thus G satisfies all the four properties of group and hence G is a group.

Commutative Law: Let $(a, b), (c, d) \in G$.

Thus we have $(a, b)\,_o\,(c, d) = (ac, bc + d)$ and

$(c, d)\,_o\,(a, b) = (ca, da + b)$.

Hence it is clear that $(a, b)\,_o\,(c, d) \neq (c, d)\,_o\,(a, b)$. Therefore, G is not an abelian group.

Let H be a subset of G containing elements of the form $(1, b)$. Now we have check whether H is subgroup or not.

Let $(1, b), (1, c) \in H$. Such that $(1, b)\,_o\,(1, c) = (1, b + c) \in H$. Hence closure law holds in H.

Let $(1, b) \in H$. The inverse of $(1, b)$ is $(1, -b) \in H$. Hence every element of H has an inverse element.

Therefore, H is a subgroup of G.

Example 31 *The set of all integers under addition is a cyclic group with generator 1.*

Solution: Let G be the set of all integers.

Now $1^0 = 0$ [$\because$ $1^0 = e = 0$] $1^1 = 1$; $1^2 = 1 + 1 = 2$; $1^3 = 1 + 1 + 1 = 3$ and so on ..... $1^{-1} = -1$; $1^{-2} = (1^2)^{-1} = -2$; $1^{-3} = (1^3)^{-1} = -3$ and so on..... So, all the elements of G can be expressed as some powers of 1.

Example 32 *Let G = {0, 1, 2, 3, 4, 5}. Show that G is the cyclic group with generator 1 under addition modulo 6.*

Solution: Given that G = {0, 1, 2, 3, 4, 5}. Here the generator is 1. Again $1^1 = 1$

$$1^2 = 1 \oplus_6 1 = 2$$
$$1^3 = 1 \oplus_6 1^2 = 1 \oplus_6 2 = 3$$
$$1^4 = 1 \oplus_6 1^3 = 1 \oplus_6 3 = 4$$
$$1^5 = 1 \oplus_6 1^4 = 1 \oplus_6 4 = 5$$
$$1^6 = 1 \oplus_6 1^5 = 1 \oplus_6 5 = 0$$
$$1^7 = 1 \oplus_6 1^6 = 1 \oplus_6 0 = 1$$

Therefore, we get, $G = \{1, 1^2, 1^3, 1^4, 1^5, 1^6 = 0\}$. This indicates that G is the cyclic group with generator 1.

Example 33 *Prove that any group of order 3 is cyclic.*

Solution: Given G is a group of order 3. So G contains 3 elements and one of this element is e whereas the other two are distinct elements. Let the distinct elements of G be a and b.

i.e., $G = \{a, b, e\}$

Now by closure property $a \in G, b \in G$ implies $(a \cdot b) \in G$. As G has only three elements, we have the following possibilities.

(*i*) $(a \cdot b) = a$; (*ii*) $(a \cdot b) = b$ or (*iii*) $(a \cdot b) = e$.

Suppose that $(a \cdot b) = a$

$\Rightarrow$ $(a \cdot b) = a \cdot e$

$\Rightarrow$ $b = e$ [Left cancellation law]

Suppose that $(a \cdot b) = b$

$\Rightarrow$ $(a \cdot b) = b \cdot e = e \cdot b$ [Existence of identity]

$\Rightarrow$ $a = e$ [Right cancellation law]

We have taken that a and b are two distinct elements other than e. Hence, both $(a \cdot b) = a$ and $(a \cdot b) = b$ are not possible. Thus we must have $(a \cdot b) = e$.

Similarly $a \in G$ implies $a^2 \in G$. Hence there arises three cases. *i.e.,* (*i*) $a^2 = e$; (*ii*) $a^2 = a$ or (*iii*) $a^2 = b$.

Suppose that $a^2 = e$

$\Rightarrow \qquad a^2 = (a \cdot b) \qquad [(a \cdot b) = e]$

This implies that $a = b$. This is not possible as a and b are distinct.

Suppose that $\qquad a^2 = a$

$\Rightarrow \qquad (a \cdot a) = a \cdot e$

This implies that $a = e$. This is also not possible as a is other than e. Hence we must have $a^2 = b$. Thus we get

$$G = \{e, a, b\} = \{e, a, a^2\}$$

Therefore, G is a cyclic group with generator a.

Example 34 *Let G be the additive group of integers. i.e., G = {... , –3, – 2, – 1, 0, 1, 2, 3, ...}. Let H be the subgroup of G given by H = { ... , – 9, – 6, –3, 0, 3, 6, 9, ... }. Form the right cosets and left cosets.*

Solution: Given that G = {... , –3, –2, –1, 0, 1, 2, 3, ...}.

H = { ... , –9, –6, –3, 0, 3, 6, 9, ... }

Let us now form the right cosets.

Now $0 \in$ G, so

$$H = H + 0 = \{ \ldots, -9+0, -6+0, -3+0, 0+0, 3+0, 6+0, \ldots \}$$
$$= \{ \ldots, -9, -6, -3, 0, 3, 6, 9, \ldots \}$$

Again $1 \in$ G, so

$$H + 1 = \{ \ldots, -9+1, -6+1, -3+1, 0+1, 3+1, 6+1, \ldots \}$$
$$= \{ \ldots, -8, -5, -2, 1, 4, 7, 10, \ldots \}$$

Similarly $2 \in$ G, so

$$H + 2 = \{ \ldots, -9+2, -6+2, -3+2, 0+2, 3+2, 6+2, \ldots \}$$
$$= \{ \ldots, -7, -4, -1, 2, 5, 8, 11, \ldots \}$$

Example 35 *Let $\phi : G \to G$ defined by $\phi(x) = e$, for all $x \in G$, where e is the identity element. Show that ϕ is a homomorphism.*

Solution: Given that $\phi : G \to G$ defined by $\phi(x) = e \ \forall \ x \in G$, where e is the identity element.

Let $x, y \in G$

$\Rightarrow \qquad \phi(x) = e$ and $\phi(y) = e$

Now $x, y \in G$ implies that $(x\,y) \in G$

Therefore, $\qquad \phi(x\,y) = e$

$$= e \cdot e = \phi(x)\,\phi(y)$$

i.e., $\qquad \phi(x\,y) = \phi(x)\,\phi(y)$

Hence ϕ is a homomorphism.

In this way we can form the right cosets and the left cosets.

Example 36 *Let $\phi : G_1 \to G_2$ defined by $\phi(x) = 2^x$, where G_1 is a group of Real numbers under addition and G_2 is a group of non-zero Real numbers under multiplication. Show that ϕ is a homomorphism.*

Proof: Given that G_1 is a group of Real numbers under addition and G_2 is a group of non-zero Real numbers under multiplication. Let $x, y \in G_1$

This implies $\qquad \phi(x) = 2^x \in G_2$ and $\phi(y) = 2^y \in G_2$

Now $x, y \in G_1$ implies that$(x + y) \in G_1$

Therefore, $\qquad \phi(x + y) = 2^{x+y}$

$$= 2^x\,2^y = \phi(x)\,\phi(y)$$

i.e., $\qquad \phi(x + y) = \phi(x)\,\phi(y)$

Hence, ϕ is a homomorphism.

EXERCISES

1. Define the binary operation. Show that the binary operation multiplication is closed on the set A = {1, – 1}.
2. Show that the addition and multiplication are associative binary operation in the set of rational numbers.
3. Show that (I, +) and (R, +) are semi group.
4. Show that the set of integers and real numbers are abelian group under ordinary addition but is not a group under ordinary multiplication.
5. Show that $G = \{\ldots, 3^{-3}, 3^{-2}, 3^{-1}, 1, 3, 3^2, 3^3, \ldots\}$forms an infinite abelian group under ordinary multiplication.
6. Is the set of all even natural numbers forms a group
   (*i*) under addition (*ii*) under multiplication.
7. Distinguish between abelian and non-abelian group. Explain with the help of examples.
8. Let G is a semi-group and for any $a, b \in G$; $a^2 b = b = b a^2$. Show that G is an abelian group.
9. Distinguish between the order of an element of a group and order of the group.
10. Show that every finite group of order less than six (6) must be abelian.
11. If G be a cyclic group of order 10, then find out how many generators are there in G.
12. Show that a cyclic group is abelian. Show by an example that the converse is not true.
13. G is a group of all real numbers under addition and H is the set of all integers. Then show that H is a subgroup of G.
14. Can an abelian group have non-abelian subgroup?
15. Can a non-abelian group have an abelian subgroup?
16. Can a non-abelian group have a non-abelian subgroup?
17. Show that homomorphic image of an abelian group is abelian.
18. Show that $G = \{1, \omega, \omega^2\}$ is the cyclic group under multiplication , where ω is the cube root of unity.
19. Show that $G = \{1, -1, i, -i\}$ is the cyclic group under multiplication, where i is the imaginary quantity such that $i^2 = -1$.
20. Form two cyclic subgroups of a cyclic group $G = \{a, a^2, a^3, \ldots, a^8 = e\}$and how many generators are there in G.
21. Let G_1 be a group of non zero Real numbers under multiplication and $G_2 = \{-1, 1\}$be a group under multiplication. Let $\phi : G_1 \to G_2$ defined by
$$\phi(x) = \begin{cases} 1 & \text{if } x > 0 \\ -1 & \text{if } x < 0 \end{cases}$$
   Show that ϕ is a homomorphism.
22. Let $\phi: G_1 \to G_2$ defined by $\phi(x) = \log_{10}(x)$, where G_1 is a group of positive Real numbers under multiplication and G_2 is a group of all Real numbers under addition. Show that ϕ is homomorphism.
23. Let $\phi: G_1 \to G_2$ defined by $\phi\begin{pmatrix} a & b \\ c & d \end{pmatrix} = ad - bc$, where G_1 be a group of all (2×2) matrix $\begin{pmatrix} a & b \\ c & d \end{pmatrix}$: $ad - bc \neq 0$ under matrix multiplication and G_2 be a group of non zero Real numbers under multiplication. Show that ϕ is a homomorphism.
24. Show that homomorphic image of an abelian group is abelian.

8

Codes and Group Codes

■ 8.0 INTRODUCTION

When we want to send a message to someone, we send it through some communication channel. This transmission of message over a channel entails some chances of undesirable interference in the channel sometimes deliberate and sometimes due to random defects in the channel. This leads to coding problem. The coding problem is to represent distinct messages by distinct sequence of letters from a given alphabet set.

For example, in a Morse code we represent a message by dots and dashes. Similarly over alphabet can be {0, 1} *i.e.*, binary alphabet.

When a message is to be transmitted, then the message is first given by the source to the encoder, the encoder converts the message into the code word. The encoded message is then sent through the channel, where noise may occur and change the message. When this message arrives at the decoder at the receivers end, it is equated to most likely code word.

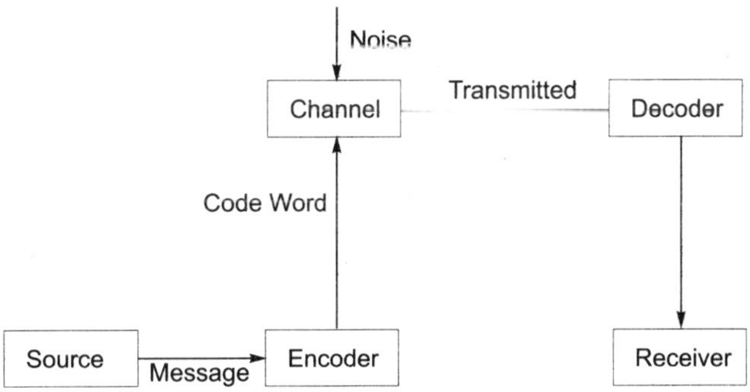

Communication Channel with Noise

■ 8.1 TERMINOLOGIES

We will use the following terms in our discussion.

Word: A word is the sequence of letters drawn from the alphabet set.

Code: Code is the collection of words to represent a distinct message.

Code word: A word represented by a code is called the code word.

Block Code: A code consisting of words that are of same length is called Block code. One of the advantages of using the Block Code is its ability to correct errors.

8.2 ERROR CORRECTION

When we transmit a message from the source to the destination, due to the presence of noise in the communication channel the message may get altered, *i.e.*, some of the 1's transmitted may be received as 0's and some of the 0's may be received as 1's. So the received message is no more is the transmitted message. Now we would want to recover the transmitted message from the received message. This is called error correction.

8.3 GROUP CODES

Let A be the set of all binary sequence of length n. Let us define a binary operation ⊕ in A such that X, Y ∈ A implies (X ⊕ Y) ∈ A *i.e.*, a sequence of length n. Where

$$(X \oplus Y) = \begin{cases} 1 & \text{if X, Y differs in position} \\ 0 & \text{if X, Y are same in position} \end{cases}$$

The set A together with the binary operation ⊕, *i.e.*, (A, ⊕) forms a group and a subset G of A is called the group code if (G, ⊕) is a subgroup of (A, ⊕).

Let us consider X = 1 0 0 1 0 0 1 and Y = 0 1 0 1 0 0 1. Therefore, we have

$$(X \oplus Y) = 1\ 1\ 0\ 0\ 0\ 0\ 0.$$

8.4 WEIGHT OF CODE WORD

Let A be the set of all binary sequence of length n. Let X be a code word in A, the weight of X denoted by ω(X) is the number of 1's in X.

Let us consider the code words X = 10101 and Y = 00011. The number of 1's present in X are three whereas the number of 1's present in Y are two. So, the weight of X is 3 and the weight of Y is 2.

i.e., $$\omega(X) = 3 \quad \text{and} \quad \omega(Y) = 2.$$

8.5 DISTANCE BETWEEN THE CODE WORDS

Let A be the set of all binary sequence of length n. Let X and Y be two code words in A, the distance between X and Y denoted by d (X, Y) and is defined as the weight of ω (X ⊕ Y).

i.e., $$d(X, Y) = \omega(X \oplus Y)$$

The distance between the two code words gives the number of positions in which they differ.

Let us consider code words X = 01011 and Y = 10101. Now the distance between X and Y is defined as ω (X ⊕ Y). Now

$$X = 01011$$
$$Y = 10101$$
$$\overline{\quad (X \oplus Y) = 11110 \quad}$$

Therefore, $$d(X, Y) = \omega(X \oplus Y) = 4$$

8.5.1 Theorem

Let A be the set of all binary sequence of length n. The distance between two code words X and Y satisfies the following properties.

(*a*) Commutative law *i.e.*, $d(X, Y) = d(Y, X)$

(*b*) Triangle's inequality *i.e.*, $d(X, Y) \le d(X, Z) + d(Z, Y)$.

Proof: (*a*) Let A be the set of all binary sequence of length n. Let X and Y be two code words in A.

Therefore, $X \oplus Y = Y \oplus X$

This implies that $\omega(X \oplus Y) = \omega(Y \oplus X)$

Thus, $d(X, Y) = d(Y, X)$

(*b*) Let A be the set of all binary sequence of length n. Let X ,Y and Z be three code words in A.

We know that $\omega(X)$ is the number of 1's in X and $(X \oplus X) = 0$. This implies that $\omega(U \oplus V) \le \omega(U) + \omega(V)$ ... (1)

Now, $\omega(X \oplus Y) = \omega(X \oplus Z \oplus Z \oplus Y)$ $[\because (Z \oplus Z) = 0]$

$\le \omega(X \oplus Z) + \omega(Z \oplus Y)$ [By equation (1)]

Therefore, $d(X, Y) \le d(X, Z) + d(Z, Y)$.

■ 8.6 ERROR CORRECTION FOR BLOCK CODE

We know that block code is a code consisting of words that are of same length. The advantage of using block code is its ability to correct the errors.

Let G be a Block code, the distance of G is defined as the minimum distance between any pair of distinct code words in G. The ability of Block codes to correct the errors depends on its distance.

Let a word has been transmitted and we received a word Y (say). Now there is a likelihood of received word containing an error. Now we will like to have the transmitted word corresponding to the received word Y.

We can use two methods. *i.e.*, Maximum likelihood decoding criterion and Minimum distance decoding criterion.

8.6.1 Maximum Likelihood Criterion

Let $X_1, X_2, \ldots\ldots, X_n$ be the code words in G. One of this is transmitted and we have received the code word Y. The received word may contain error and we are interested to find the word transmitted. Maximum likelihood criterion says that compute the conditional probabilities $P(X_1 \mid Y), P(X_2 \mid Y), \ldots P(X_n \mid Y)$. Where $P(X_i \mid Y)$ means the probability that X_i is transmitted when the received word is Y. Let

$$P(X_k \mid Y) = \operatorname*{Max}_i \{P(X_i|Y)\};\ i = 1, 2, \ldots\ldots, n$$

Then X_k is the transmitted word.

8.6.2 Minimum Distance Decoding Criterion

In the minimum distance decoding criterion we compute d (X_1, Y), d (X_2, Y), d (X_3, Y), ..., d (X_n ,Y). Let us define

$$d(X_k, Y) = \underset{i}{\text{Min}}\{d(X_i, Y)\};\ i = 1, 2, 3, \dots\dots\dots, n$$

Then, X_k is taken as the transmitted word when the received word is y.

■ 8.7 COSETS

Let (G, ⊕) be a group code. Let a word y is received. Then the coset with respect to y denoted by (G ⊕ y) is defined as

$$(G \oplus y) = \{X_i \oplus y \mid X_i \in G, i \in N\}$$

Again $d(X_i, Y) = \omega(X_i \oplus y)$. So the weights of the words in the coset (G ⊕ y) are the distances between the code words in G and y.

The decoding procedure includes the followings:

1. Determine all cosets of G.
2. For each coset, choose the coset leader, *i.e.*, the word of smallest weight.
3. For the received word y, ($e \oplus y$) is the transmitted word.

SOLVED EXAMPLES

Example 1 *Let X = 0101011 and Y = 1010101. Find (X ⊕Y).*

Solution: Given that X = 0101011 and Y = 1010101

Now

$$\begin{aligned} X &= 0101011 \\ Y &= 1010101 \\ \hline (X \oplus Y) &= 1111110 \end{aligned}$$

Therefore, (X ⊕ Y) = 1111110.

Example 2 *A is a set of all binary sequence of length n. Show that (A, ⊕) forms a group.*

Solution: Given that is aw set of all binary sequence of length n, say for our convenience we take the length to be 5.

Closure Law: Let X = 01011 and Y = 10101

Now X ⊕ Y = 01011 ⊕ 10101 = 11110

This is again a code word of length 5.

Therefore, X, Y ∈ A implies (X ⊕ Y) ∈ A. So, closure law holds.

Associative Law: Let X = 10101, Y = 10000 and Z = 01010

Now (Y ⊕ Z) = 10000 ⊕ 01010 = 11010

Therefore, X ⊕ (Y ⊕ Z) = 10101 ⊕ 11010 = 01111

So, X ⊕ (Y ⊕ Z) = 01111 ...(*i*)

Again (X ⊕ Y) = 10101 ⊕ 10000 = 00101

Therefore, (X ⊕ Y) ⊕ Z = 00101 ⊕ 01010 = 01111

So, $(X \oplus Y) \oplus Z = 01111$ ...(*ii*)

Therefore from equations (*i*) and (*ii*) we get $(X \oplus Y) \oplus Z = X \oplus (Y \oplus Z)$. So, associative law holds.

Existence of Identity: A code word with all zeros of specified length will act as the identity element.

Let $X = 10101$ and $Y = e = 00000$ such that

$$X \oplus Y = 10101 \oplus 00000 = 10101 = X$$

Therefore, $(X \oplus Y) = X$

So, $Y = 00000 \in A$ acts as an identity element.

Existence of Inverse: A code word itself is inverse of its own.

Let $X = 10101$ such that $(X \oplus X) = 10101 \oplus 10101 = 00000 = e$. Therefore, $(X \oplus X) = e$. This implies that every code word is its own inverse. So, $(A, \oplus)$ satisfies all the properties of group and hence called group codes.

Example 3 *Illustrate by example distance function satisfies the commutative and triangle's inequality.*

Solution: Commutative Law: Let $X = 101010$ and $Y = 010101$

So, $(X \oplus Y) = 101010 \oplus 010101 = 111111$

Therefore, $d(X, Y) = \omega(X \oplus Y) = 6$ ....(*i*)

Again, $(Y \oplus X) = 010101 \oplus 101010 = 111111$

Therefore, $d(Y, X) = \omega(Y \oplus X) = 6$ ...(*ii*)

So, from equations (*i*) and (*ii*) it is clear that $d(X, Y) = d(Y, X)$.

Triangle's inequality: Let us take $X = 101010$, $Y = 100010$ and $Z = 101000$. Now

$$(X \oplus Y) = 101010 \oplus 100010 = 001000$$
$$(X \oplus Z) = 101010 \oplus 101000 = 000010$$
$$(Z \oplus Y) = 101000 \oplus 100010 = 001010$$

Therefore,

$$d(X, Y) = \omega(X \oplus Y) = 1$$
$$d(X, Z) = \omega(X \oplus Z) = 1$$
$$d(Z, Y) = \omega(Z \oplus Y) = 2$$

Thus, we have $d(X, Z) + d(Z, Y) = 1 + 2 = 3 \geq d(X, Y) = 1$

i.e., $d(X, Y) \leq d(X, Z) + d(Z, Y)$

Example 4 *In the minimum distance criterion, a code of distance (2t + 1) can correct t or fewer transmission errors.*

Solution: Let X be the transmitted word and Y be the received word.

Now if t or less number of errors has occurred during the transmission we will have

$$d(X, Y) \leq t \qquad \text{...}(i)$$

Now since the distance is $(2t + 1)$, so for any code word X_1 we have

$$d(X, X_1) \geq 2t + 1 \qquad \text{...}(ii)$$

Since the distance means the minimum distance between any pairs of distinct code words. Again from triangle's inequality we have

$$d(X, X_1) \leq d(X, Y) + d(Y, X_1)$$

$$\Rightarrow \quad 2t + 1 \leq d(X, X_1) \leq t + d(Y, X_1)$$

$\Rightarrow \qquad 2t + 1 \leq t + d\,(Y, X_1)$

$\Rightarrow \qquad d(Y, X_1) \geq (t + 1) \qquad \ldots (iii)$

So, we have $\quad d(X, Y) \leq t$ and $d(Y, X_1) \geq (t + 1)$

From the minimum distance decoding criterion X will be selected as the transmitted word.

EXERCISES

1. Let X and Y be two code words. Find $(X \oplus Y)$ in each of the following cases.

(*a*) X = 11111 and Y = 11111
(*b*) X = 11111 and Y = 00000
(*c*) X = 1010101 and Y = 0101010
(*d*) X = 00101101 and Y = 11101100
(*e*) X = 1100110 and Y = 0011101

2. Find the weight of the following code words.

(*a*) X = 11111
(*b*) X = 11111
(*c*) X = 1010101
(*d*) X = 00101101
(*e*) X = 1100110

3. Find the distance between the code words in each of the following cases.

(*a*) X = 10011 and Y = 00000
(*b*) X = 01101 and Y = 10110
(*c*) X = 0011101 and Y = 0101010
(*d*) X = 11101101 and Y = 10011101
(*e*) X = 1100110 and Y = 1110110

4. Illustrate by example distance function satisfies the associative law and triangle's inequality.

5. Illustrate by example $(A, \oplus)$ forms a group, where A is a set of all binary sequence of length 7.

9

Ring Theory

■ 9.0 INTRODUCTION

As we have discussed group as an algebraic structure, in this chapter we will discuss about "Ring" which is quite different from the group in a way that it is two operational systems viz; addition and multiplication whereas the group is a one operational system. In ring theory many of the notions of group theory will be extended to the system with two operations.

■ 9.1 RING

A non-empty set R with two binary operations addition (+) and multiplication (.) defined in it is said to be associative ring if it satisfies the following properties.

Under Addition

(*a*) Closure Axiom: For $a, b \in R$; $a + b \in R$

(*b*) Associative Axiom: For $a, b, c \in R$; $(a + b) + c = a + (b + c)$

(*c*) Existence of Identity: For every element $a \in R$, there exist an identity element $0 \in R$ such that

$$a + 0 = 0 + a = a \ \forall \ a \in R$$

(*d*) Existence of Inverse: For every element $a \in R$ there exist an inverse element $-a \in R$ such that

$$a + (-a) = 0$$

(*e*) Commutative Axiom: For $a, b \in R$; $(a + b) = (b + a)$

Under Multiplication

(*a*) Closure Axiom : For $a, b \in R$; $(a \,.\, b) \in R$

(*b*) Associative Axiom : For $a, b, c \in R$; $(a \,.\, b) \,.\, c = a \,.\, (b \,.\, c)$

(*c*) Distributive Axiom : For $a, b, c \in R$ we have

(*i*) $a \,.\, (b + c) = a \,.\, b + a \,.\, c$ [Left distributive law]

(*ii*) $(b + c) \,.\, a = b \,.\, a + c \,.\, a$ [Right distributive law]

9.1.1 Theorem

If R is a ring, then for all $a, b, c \in R$

(*i*) $a \cdot 0 = 0 \cdot a = 0$

(*ii*) $a(-b) = (-a)b = -(ab)$

(*iii*) $(-a)(-b) = ab$

(*iv*) $a \cdot (b - c) = a \cdot b - a \cdot c$

(*v*) $(b - c) \cdot a = b \cdot a - c \cdot a$

Proof: (*i*) We know that $0 = 0 + 0$

This implies that $a \cdot 0 = a \cdot (0 + 0)$

$= a \cdot 0 + a \cdot 0$ [Distributive law]

So, $a \cdot 0 = a \cdot 0 + a \cdot 0$ ... (1)

Again, $a \cdot 0 = a \cdot 0 + 0$ ... (2) [Additive identity law]

Therefore from equations (1) and (2), we have

$$a \cdot 0 + 0 = a \cdot 0 + a \cdot 0$$

$\Rightarrow$ $0 = a \cdot 0$ [Left cancellation law]

Thus, $a \cdot 0 = 0$ ....(3)

Similarly, $0 = 0 + 0$

This implies that $0 \cdot a = (0 + 0) \cdot a$

$= 0 \cdot a + 0 \cdot a$ [Distributive law]

So, $0 \cdot a = 0 \cdot a + 0 \cdot a$ ...(4)

Again, $0 \cdot a + 0 = 0 \cdot a$ ...(5) [Additive identity law]

Therefore from equations (4) and (5), we have

$$0 \cdot a + 0 = 0 \cdot a + 0 \cdot a$$

$\Rightarrow$ $0 = 0 \cdot a$ ...(6) [Left cancellation law]

Combining equations (3) and (6) we get

$$a \cdot 0 = 0 \cdot a = 0$$

(*ii*) Given $b \in R$, so by existence of additive inverse $(-b) \in R$ such that $b + (-b) = 0$

$\Rightarrow$ $a \cdot (b + (-b)) = a \cdot 0$

$\Rightarrow$ $a \cdot b + a \cdot (-b) = 0$ [By previous (*i*)]

$\Rightarrow$ $a \cdot (-b) = -(ab)$ ...(1)

Again $a \in R$ implies that $-a \in R$ such that

$$a + (-a) = 0$$

$\Rightarrow$ $(a + (-a)) \cdot b = 0 \cdot b$

$\Rightarrow$ $a \cdot b + (-a) \cdot b = 0$ [By previous (*i*)]

$\Rightarrow$ $(-a) \cdot b = -(ab)$ ...(2)

Combining equations (1) and (2), we get

$$a \cdot (-b) = -(a \cdot b) = (-a) \cdot b$$

(*iii*) Given $a, b \in R$ implies that $-a, -b \in R$.

Now, $(-a)(-b) = (-a) \cdot x; \quad x = -b$

$= -(a \cdot x)$ [By previous (*ii*)]

$= -(a \cdot (-b))$

$= -(-(a \cdot b))$ [By previous (*ii*)]

$= a \cdot b$

Therefore, $(-a).(-b) = a.b$

(*iv*) $a.(b-c) = a.(b+(-c))$

$= a.b + a.(-c)$ [Left distributive law]

$= a.b + (-(a.c))$

$= a.b - a.c$

Therefore, $a.(b-c) = a.b - a.c$

(*v*) $(b-c).a = (b+(-c)).a$

$= b.a + (-c).a$ [Right distributive law]

$= b.a + (-(c.a))$

$= b.a - c.a$

Therefore, $(b-c).a = b.a - c.a$

9.1.2 Theorem

If R is a ring with unit element then

(*i*) $(-1).a = -a$

(*ii*) $(-1)(-1) = 1$

Proof: (*i*) Given R is a ring with unit element *i.e.,* $1 \in R$.

Now $0 = 0.a$

$= (1+(-1)).a$

$= 1.a + (-1).a$

$= a + (-1).a$

i.e., $a + (-1).a = 0$

Therefore, $(-1).a = -a$

(*ii*) In the previous we have proved that $(-1).a = -a$.

Let $a = -1$

So, $(-1)(-1) = -(-1) = 1$

Therefore, $(-1)(-1) = 1$

9.2 SPECIAL TYPES OF RING

In this section we will discuss special types of ring. These are mostly used in algebraic structure. Here, we discuss the basic definitions in order to get a clear idea.

9.2.1 Commutative Ring

A ring R is said to be commutative ring if under multiplication

$$(a.b) = (b.a) \ \forall \ a, b \in R.$$

9.2.2 Ring with Unit Element

A ring R is said to be ring with unit element if there exist an element $1 \in R$ such that

$$(1.a) = (a.1) = a \ \forall \ a \in R$$

9.2.3 Null Ring

The singleton set {0} with binary operation + and . defined as

$$0 + 0 = 0 \quad \text{and} \quad 0 . 0 = 0$$

is called a null ring or zero ring.

9.2.4 Boolean Ring

A ring R is said to be Boolean ring if $a^2 = a$ for all $a \in$ R.

9.2.5 Division Ring

A ring R is said to be division ring if the non-zero elements of R forms a group under multiplication.

9.2.6 Zero Divisor

Let R be a commutative ring, an element $a \neq 0 \in$ R is said to be zero divisor if there exists $b \neq 0$ such that

$$(a . b) = 0; \quad a, b \in \text{R}$$

Let us consider a set R of integers. From the discussion given below it is clear that R is a commutative ring with unit element.

Under Addition

(*i*) Closure Axiom: We know that the addition of two integer is again an integer.

i.e., $$a, b \in \text{R} \Rightarrow (a + b) \in \text{R}$$

(*ii*) Associative Axiom: We know that addition of integers is associative.

i.e., $$a + (b + c) = (a + b) + c\ ; \quad \forall \quad a, b, c \in \text{R}$$

(*iii*) Existence of Identity: For all $a \in$ R, there exists $0 \in$ R such that

$$(a + 0) = (0 + a) = a$$

(*iv*) Existence of Inverse: For every $a \in$ R there exists $-a \in$ R such that

$$a + (-a) = (-a) + a = 0.$$

(*v*) Commutative Axiom: For any $a, b \in$ R we know that the addition of integers is commutative.

i.e., $$(a + b) = (b + a)$$

Under Multiplication

(*i*) Closure Axiom: We know that multiplication of two integers is again an integer.

i.e., $$(a . b) \in \text{R} \quad \forall \quad a, b \in \text{R}.$$

(*ii*) Associative Axiom: We know that integer multiplication is associative.

i.e., $$a . (b . c) = (a . b) . c \quad \forall \quad a, b, c \in \text{R}.$$

(*iii*) Distributive Laws: Set of integers follow both left distributive and right distributive property

i.e., $$a . (b + c) = (a . b) + (a . c)$$

and $$(b + c) . a = (b . a) + (c . a) \quad \forall\, a, b, c \in \text{R}$$

(*iv*) Commutative Law: We know that multiplication of integers is commutative,

i.e., $(a \,.\, b) = (b \,.\, a)$ for all $a, b \in$ R.

(*v*) Unit Element: As R contains integers, so $1 \in$ R.

Again $(a \,.\, 1) = (1 \,.\, a) = a \;\forall a \in$ R

Therefore, R is a commutative ring with unit element.

9.3 RING WITHOUT ZERO DIVISOR

A commutative ring R is said to be without zero divisor if for $a, b \in$ R

$a \,.\, b = 0$ implies $a = 0$ or $b = 0$ or both a and b are zero.

Set of integers I is a ring without zero divisor as product of integers is zero only if any one of them is zero.

9.3.1 Theorem

A commutative ring R is without zero divisor if and only if the cancellation law holds.

Proof: (*Necessary part*) Let the commutative ring R does not have zero divisor.

Let $a, b, c \in$ R, $a \neq 0$ and $ab = ac$

$\Rightarrow$ $ab - ac = 0$

$\Rightarrow$ $a(b - c) = 0$

As $a \neq 0$ and R does not have zero divisor, so we must have $(b - c) = 0$. This implies that $b = c$.

Hence left cancellation law holds.

Similarly it can be shown that right cancellation also holds.

(*Sufficient part*) Let the cancellation law holds in the ring R. We have to show that R has no zero divisor.

If possible, let $(a \,.\, b) = 0$ with $a \neq 0$ and $b \neq 0$

$\Rightarrow$ $(a \,.\, b) = (a \,.\, 0)$ $[\because \; a \,.\, 0 = 0]$

Hence by left cancellation $b = 0$. This contradicts to the fact that $b \neq 0$.

Therefore, R is a ring without zero divisor.

9.4 INTEGRAL DOMAIN

A commutative ring without zero divisors is an integral domain. Set of integers is an integral domain since it forms a commutative ring but does not have zero divisors.

9.5 DIVISION RING

If the non-zero elements of a ring R form a group under multiplication then the ring R is said to be a division ring.

9.6 FIELD

A ring F is said to be a field if the non-zero elements form a multiplicative abelian group. It is defined also as a commutative division ring. Besides this if, every element $a \neq 0$ of an integral domain has a multiplicative inverse a^{-1}, then the integral domain is called as a field.

9.6.1 Theorem

Every field is an integral domain.

Proof: Suppose that F be a field. This indicates that F is a commutative division ring. *i.e.*, The non-zero elements of F forms a group under multiplication.

Our claim is to show F is an integral domain. *i.e.*, F does not have the zero divisor.

Let $a, b \in$ F and $a \neq 0$ such that

$$a \,.\, b = 0$$

$$\Rightarrow \quad a^{-1}.(a\,.\,b) = a^{-1}.\,0 \qquad [\because a \neq 0 \Rightarrow a^{-1} \in F]$$

$$\Rightarrow \quad (a^{-1}.\,a)\,.\,b = 0 \qquad \text{[Associative law]}$$

$$\Rightarrow \quad 1\,.\,b = 0$$

$$\Rightarrow \quad b = 0$$

Thus, we get $(a\,.\,b) = 0, a \neq 0 \Rightarrow b = 0$

Similarly, we can show that $(a\,.\,b) = 0, b \neq 0 \Rightarrow a = 0$

So, the field F does not have a zero divisor. Hence, F is an integral domain.

9.6.2 Theorem

A finite integral domain is a field.

Proof: Let R be a finite integral domain.

Let $\quad R = \{x_1, x_2, \ldots\ldots, x_n\}$; where the elements of R are distinct.

i.e., $\quad x_i \neq x_j$ for all $i \neq j$; where $i, j = 1, 2, \ldots., n$.

Now since R is an integral domain, so R is a commutative ring without zero divisors.

Our claim is to prove R is a field. *i.e.*, it is sufficient to prove that R contains the unit element and every non-zero element has multiplicative inverse.

(*Existence of unity*) Let $a \neq 0 \in$ R

Now, $\quad ax_1, ax_2, ax_3, \ldots., ax_n \in$ R and all these elements are distinct. If not,

let $ax_i = ax_j$ for $i \neq j$

$$\Rightarrow \quad (a\,x_i - a\,x_j) = 0$$

$$\Rightarrow \quad a\,(x_i - x_j) = 0$$

$$\Rightarrow \quad (x_i - x_j) = 0 \qquad [\because a \neq 0 \text{ is the additive identity}]$$

$$\Rightarrow \quad x_i = x_j$$

This contradicts to the statement $x_1, x_2, \ldots, x_n$ are all distinct.

So, $\quad ax_1, ax_2, ax_3, \ldots, ax_n \in$ R and are distinct. Therefore one of these elements must be equal to 'a' since $a \in$ R.

Let $a = ax_k$

i.e., $\quad ax_k = a = x_k a \qquad$ [R is commutative]

Let us take any element $x_m \in$ R. Now x_m must be equal to $(a\,x_r)$ for some value of r. $1 \leq r \leq n$.

i.e., $\quad a\,x_r = x_m = x_r\,.{\times}a$

Now, $\quad x_k\,.\,x_m = x_k\,(a\,x_r)$

$$= (x_k\,.\,a)\,.\,x_r \qquad \text{[Associative law]}$$

$$= a\,.\,x_r = x_m$$

So, $x_k \cdot x_m = x_m$. This implies that 'x_k' is the identity element and is denoted by 1. Thus, we have the unit element in R.

(*Multiplicative Inverse*)

We have proved that $1 \in$ R. So 1 must be equal to 'ax_i' for some i.

i.e., $ax_i = 1$. Therefore there is some $b \in$ R such that

$$a \cdot b = 1 = b \cdot a$$

Hence, b is the multiplicative inverse of the non-zero element a.

9.6.3 Theorem

The commutative ring $Z_p = \{0, 1, 2, \ldots, p-1\}$ under the operation $\oplus_p$ and $\otimes_p$ is a field if and only if p is a prime number.

Proof: Given $Z_p = \{0, 1, 2, \ldots, p-1\}$ is a commutative ring under addition and multiplication modulo p.

Suppose that p is a prime number.

Let $a, b \in Z_p$ and $a \neq 0, b \neq 0$ and let $(a \cdot b) \equiv 0 \bmod p$.

This implies $p \mid ab$. *i.e.*, $p \mid a$ or $p \mid b$. Therefore, we get

$$a \equiv 0 \bmod p \quad \text{or } b \equiv 0 \bmod p.$$

This contradicts to the fact p is a prime number. Hence, Z_p does not have zero divisors.

Therefore, Z_p is a field.

Conversely, suppose that Z_p is a field. We have to show that p is a prime number.

Suppose that p is not a prime number.

$$\Rightarrow \quad p = m \cdot n \quad (1 < m, n < p)$$

$$\Rightarrow \quad m \cdot n \equiv 0 \bmod p \qquad \ldots (1)$$

Now
$$\begin{aligned} n &= 1 \cdot n \bmod p \\ &= (m^{-1} \cdot m)\, n \bmod p \qquad [\because m^{-1} \cdot m = 1] \\ &= m^{-1} (m\, n) \bmod p \\ &= m^{-1} \cdot 0 = 0 \end{aligned}$$

Thus, we get $n = 0$. This is a contradiction.

Therefore, p is a prime number.

■ 9.7 THE PIGEONHOLE PRINCIPLE

If n objects are distributed over m places and if $(n > m)$ then some places will receive at least two objects. So if n objects are distributed over m places in such a way that no place receives more than one object, then each place will receive exactly one object. This principle is known as Pigeonhole principle.

■ 9.8 CHARACTERISTICS OF A RING

Let (R, + , .) be a ring with 0 as zero element. If there exist a positive integer 'n' such that

$$n \cdot a = a + a + a + \ldots + a \ (n \text{ times}) = 0 \ \forall\, a \in \text{R}.$$

Then such smallest positive integer 'n' is called the characteristic of the ring. Thus, the characteristic of a ring R is defined as

$$\text{Ch(R)} = \begin{cases} \text{Smallest positive integer } n \text{ such that } n\,a = 0, \forall\, a \in \text{R} \\ 0 \quad \text{otherwise} \end{cases}$$

If no such 'n' exists then the ring R is said to have a characteristic zero or infinite.

Let us consider the ring $I_6 = \{0, 1, 2, 3, 4, 5\}$ with the binary operations $\oplus_6$, $\otimes_6$. Then the characteristic of this ring R will be 6 since $6 \,.\, a = 0$ for all $a \in I_6$.

9.8.1 Theorem

The characteristic of a ring with unity is 0 or $n > 0$ depending on whether unity element is regarded as the member of additive group has the order 0 or 'n' respectively.

Proof: Let R be a ring with unity 1. Hence, there arises two cases.

Case 1 : If the order of 1 is zero then obviously the characteristic of ring is zero.

Case 2 : If the order of 1 is n (finite), then

$$\underbrace{1 + 1 + 1 \ldots\ldots + 1}_{n\text{ – times}} = 0 \quad \forall\, a \in \text{R}$$

$\Rightarrow$ $n\,.1 = 0$

Now for any $a \in \text{R}$ we have

$$\begin{aligned} na &= a + a + \ldots\ldots\ldots\ldots + a \;(n \text{ terms}) \\ &= 1 \cdot a + 1 \cdot a + \ldots\ldots\ldots\ldots + 1 \cdot a \qquad [\because \;1 \text{ is the unity}] \\ &= (1 + 1 + \ldots\ldots\ldots\ldots +1) \cdot a \\ &= (n\,.\,1)\,.\,a = (0\,.\,a) = 0 \end{aligned}$$

i.e., $na = 0$

Therefore, the characteristic of R is n.

9.8.2 Theorem

The characteristic of an integral domain is either 0 or a prime number.

Proof: Let R be an integral domain. We have to show that the characteristic of R *i.e.,* Ch (R) is either 0 or a prime number.

Let Ch (R) = n

Let if possible assume that $n \neq 0$ and not a prime number. Therefore, $n = n_1 \,.\, n_2$ with n_1, n_2 less than n.

Now as the characteristic of R is n, we have the order of the unit element e is 'n'. *i.e.,* $O(e) = n$

$\Rightarrow$ $n\,.\,e = 0$

$\Rightarrow$ $(n_1 \,.\, n_2)\, e = 0$

$\Rightarrow$ $n_1 \,.\, (n_2 \,.\, e) = 0$ [Associative law]

$\Rightarrow$ $(n_1 \,.\, e)\,(n_2 \,.\, e) = 0$ $[\because (n_1 \,.\, e) = n_1]$

As R does not have zero divisor so $n_1 \,.\, e = 0$ or $n_2 \,.\, e = 0$. This indicates that the characteristic of R is either 'n_1' or 'n_2'.

This is a contradiction to the assumption that characteristic of R is 'n'. So, our assumption was wrong. Therefore, 'n' is zero or a prime number.

■ 9.9 SUB RING

For a ring (R, +, .), a non-empty subset S of R is called a sub ring of R if (S,+, .) forms a ring under the binary operations defined in R.

For the ring (I, +, .) the subset of even integers is a sub ring.

9.9.1 Theorem

The necessary and sufficient condition for (S, +, .) to be a sub ring of the ring (R, +, .) is

(*i*) $a - b \in S \;\forall\; a, b \in S$

(*ii*) $a \,.\, b \in S \;\forall\; a, b \in S$

Where S is the sub set of R.

Proof: (*Necessary part*) Suppose that (S, +, .) be the sub ring of the ring (R, +, .). This implies that S is a group with respect to addition.

Now, for $b \in S$ we have $(-b) \in S$.

Again since S is closed under addition so, $(a + (-b)) \in S$ for $a \in S$, and $(-b) \in S$. *i.e.*, $(a - b) \in S$.

Similarly since S is closed under multiplication we have

$$a \in S, b \in S$$

$$\Rightarrow \quad a \,.\, b \in S$$

(*Sufficient part*) Suppose that

(*i*) $a - b \in S \;\forall\; a, b \in S$ and

(*ii*) $a \,.\, b \in S \;\forall\; a, b \in S$

Now $\quad a \in S, a \in S$

$\Rightarrow \quad (a - a) \in S$

$\Rightarrow \quad 0 \in S \quad \ldots(i)$

Again, $\quad 0 \in S, a \in S$

$\Rightarrow \quad (0 - a) \in S$

i.e., $\quad -a \in S$

Again, $\quad a \in S, -b \in S$

$\Rightarrow \quad a - (-b) \in S$

i.e., $\quad (a + b) \in S$

The addition and commutative axiom under addition holds in R so it will hold in S. Therefore, (S, +, .) is an abelian group. The remaining postulates will hold in S as they hold in R.

■ 9.10 HOMOMORPHISM

Let R_1 and R_2 be two rings, then the mapping

$\phi: R_1 \to R_2$ is said to be homomorphism if it satisfies the following conditions.

(*a*) $\phi(a + b) = \phi(a) + \phi(b)$

(*b*) $\phi(a \,.\, b) = \phi(a) \,.\, \phi(b) \;\forall\; a, b \in R$

9.10.1 Theorem

If ϕ is homomorphism from ring R_1 into Ring R_2, then

(*i*) $\phi(0) = 0$

(*ii*) $\phi(-a) = -\phi(a)$

Proof: (*i*) Let $a \in R_1$. Then there exists an identity element $0 \in R_1$ such that $(a + 0) = a$

$\Rightarrow \quad \phi(a+0) = \phi(a)$

$\Rightarrow \quad \phi(a)+\phi(0) = \phi(a)$ [ϕ is a homomorphism]

$\Rightarrow \quad \phi(a)+\phi(0) = \phi(a)+0$ [0 is additive identity of R_2]

$\Rightarrow \quad \phi(0) = 0$ [Left cancellation law]

(*ii*) For the ring R_1, $a \in R_1$ implies $-a \in R_1$

Now, $\quad a + (-a) = 0$

$\Rightarrow \quad \phi(a+(-a)) = \phi(0)$

$\Rightarrow \quad \phi(a)+\phi(-a) = 0$ [$\because \phi(0) = 0$]

This indicates that $\phi(-a)$ is the additive inverse of $\phi(a)$ in R_2.

Therefore, $\phi(-a) = -\phi(a)$.

9.10.2 Theorem

Let R_1 is a ring with unit element 1 and ϕ is a homomorphism of R_1 into R_2, then $\phi(1)$ is the unit element of R_2.

Proof: Given that the mapping ϕ is homomorphism from ring R_1 into R_2.

i.e., $\quad \phi : R_1 \rightarrow R_2$ is homomorphism.

Let $1 \in R_1$, this implies that $\phi(1) \in R_2$.

Now for any $\quad a_1 \in R_2$, we have $a_1 = \phi(a)$ for some $a \in R_1$.

Therefore, $\quad \phi(1) . a_1 = \phi(1) . \phi(a)$

$= \phi(1 . a)$ [ϕ is a homomorphism]

$= \phi(a)$ [Existence of identity]

$= a_1$

Therefore, $\phi(1) . a_1 = a_1$. Hence, $\phi(1)$ is the unit element of R_2.

9.10.3 Theorem

Every homomorphic image of a commutative ring is commutative.

Proof: Let R be a commutative ring and ϕ is a homomorphic mapping from R into R′. *i.e.*, R′ is the homomorphic image of the commutative ring R.

Our claim is R′ is commutative.

Let $a', b' \in R'$. Hence there exists $a, b \in R$ such that

$$a' = \phi(a), \quad \text{and} \quad b' = \phi(b),$$

Now, $\quad a' . b' = \phi(a) . \phi(b)$

$= \phi(a . b)$ [ϕ is a homomorphism]

$= \phi(b . a)$ [R is commutative]

$= \phi(b) . \phi(a)$ [ϕ is a homomorphism]

Therefore, $\quad a' . b' = \phi(b) . \phi(a) = b' . a'$

Hence the homomorphic image R′ is commutative.

9.11 KERNEL OF HOMOMORPHISM OF RING

If ϕ is a homomorphism from ring R into R′, then the kernel of homomorphism is a set denoted by I (ϕ) containing elements of R which are mapped to the additive identity element of R′.

i.e., $$I(\phi) = \{x \in R \mid \phi(x) = 0; 0 \in R'\}$$

9.11.1 Theorem

If ϕ is homomorphism from R into R′ with kernel I(ϕ) then

(*i*) I (ϕ) is a subgroup of R under addition

(*ii*) If $a \in I(\phi)$ and $x \in R$, then $(x \,.\, a)$ and $(a \,.\, x) \in I(\phi)$

Proof: (*i*) Given ϕ is homomorphism from R into R′ with kernel I(ϕ)

Our claim is I(ϕ) is a subgroup of R under addition. *i.e.,* I(ϕ) satisfies the closure and inverse axiom.

Let $a, b \in I(\phi)$

This implies that $\phi(a) = 0$ and $\phi(b) = 0$

Now, $\phi(a + b) = \phi(a) + \phi(b) = 0 + 0 = 0$

Hence, $\phi(a + b) = 0$

Therefore, $(a + b) \in I(\phi)$

Again $\phi(-a) = -\phi(a) = 0$

Hence, $\phi(-a) = 0$

Therefore, $-a \in I(\phi)$

This implies that I (ϕ) is subgroup under addition.

(*ii*) Suppose that $a \in I(\phi)$ and $x \in R$.

Now, $\phi(a \,.\, x) = \phi(a) \,.\, \phi(x)$ [ϕ is a homomorphism]

$= 0 \,.\, \phi(x)$ [$a \in I(\phi) \Rightarrow \phi(a) = 0$]

$= 0$

So, $\phi(a \,.\, x) = 0$

This implies that $(a \,.\, x) \in I(\phi)$

Similarly, it can be shown that $(x \,.\, a) \in I(\phi)$.

9.12 ISOMORPHISM

A mapping ϕ from ring R into R′ is said to be isomorphism if

(*i*) ϕ is homomorphism

(*ii*) ϕ is one-one

i.e., A homomorphism ϕ of R into R′ is said to be isomorphism if it is one-to-one mapping.

9.12.1 Theorem

The homomorphism ϕ defined from the ring R into R′ is an isomorphism if and only if I(ϕ) = (0).

Proof: (*Necessary part*) Let ϕ: R $\to$ R′ is an isomorphism.

This implies that ϕ is a homomorphism and one-one.

Let $a \in I(\phi) \Rightarrow \phi(a) = 0$

$\Rightarrow \quad \phi(a) = \phi(0)$ [ϕ is homomorphism; $\phi(0) = 0$]

$\Rightarrow \quad a = 0$ [ϕ is one-one]

So, $\quad a \in I(\phi) \Rightarrow a = 0 \quad \forall\, a \in R$

Therefore, $\quad I(\phi) = (0)$

(*Sufficient part*) Let $\quad I(\phi) = (0)$

Let $x, y \in R$ and $\phi(x) = \phi(y)$

Now, $\quad \phi(x) = \phi(y)$

$\Rightarrow \quad \phi(x) - \phi(y) = 0$

$\Rightarrow \quad \phi(x - y) = 0$

$\Rightarrow \quad (x - y) \in I(\phi) = (0)$

Therefore, $\quad x - y = 0$, hence $x = y$.

So, $\quad \phi(x) = \phi(y) \Rightarrow x = y$

This implies that ϕ is one-one and hence ϕ is isomorphism.

SOLVED EXAMPLES

Example 1 *Show that the set of all square matrix of order $(m \times m)$ under the binary operations addition and multiplication is a non-commutative ring.*

Solution: Let R be a set of all square matrices of order $(m \times m)$.

We have to show that R is a ring, *i.e.*, R satisfies all the eight properties of ring.

Under Addition

Closure Law: Let A and B be two square matrices of order $(m \times m)$.

So, (A + B) will be a square matrix of order $(m \times m)$

This implies $\quad (A + B) \in R$

i.e., $\quad A, B \in R \quad \Rightarrow \quad (A + B) \in R$

Associative Law: We know that matrix addition is associative. *i.e.*, A, B, C $\in$ R implies that

$$A + (B + C) = (A + B) + C$$

Existence of Identity: For every square matrix A $\in$ R, there exists null matrix $[0]_{m \times m} \in R$ such that

$$A + 0 = 0 + A = A$$

Existence of Inverse: For every A $\in$ R there exist inverse element $(-A) \in R$ such that

$$A + (-A) = 0$$

Commutative Law: We know that matrix addition is commutative, *i.e.*, For A, B $\in$ R we have

$$(A + B) = (B + A)$$

Under Multiplication

Closure Law: Let A, B $\in$ R. *i.e.*, A and B are two square matrices of order $(m \times m)$. Now multiplying A and B we will get a matrix of order $(m \times m)$.

i.e., $\quad A \,.\, B \in R$

Associative Law: We know that matrix multiplication is associative.

i.e., $$A . (B . C) = (A . B) . C \quad \forall \quad A, B, C \in R$$

Distributive Law: Let A, B, C ∈ R. *i.e.,* A, B and C are three square matrices of order $(m \times m)$. Also we know that

$$A . (B + C) = A . B + A . C$$

Therefore, R satisfies all the properties of Ring. Hence, R is a ring.

Example 2 *If R is a Boolean ring, then prove that*

(i) $a + a = 0 \ \forall \quad a \in R$

(ii) $a + b = 0$ *implies* $a = b \ \forall\, a, b \in R$

(iii) *R is a commutative ring.*

Solution: Given that R is Boolean ring.

i.e., $$a^2 = a \ \forall \ a \in R$$

(i) Let $a \in R$, this implies that $(a + a) \in R$

$\Rightarrow \quad (a + a)^2 = (a + a) \qquad [\because a^2 = a]$

$\Rightarrow \quad (a + a) . (a + a) = a + a$

$\Rightarrow \quad a . (a + a) + a . (a + a) = a + a \qquad$ [Distributive law]

$\Rightarrow \quad (a . a + a . a) + (a . a + a . a) = a + a \qquad$ [Distributive law]

$\Rightarrow \quad (a^2 + a^2 + a^2 + a^2) = a + a$

$\Rightarrow \quad a + a + a + a = a + a$

$\Rightarrow \quad a + a = 0 \qquad$ [Cancellation law]

(ii) Suppose that $\quad a + b = 0 \ \forall\, a, b \in R$

Again, we have proved that $a + a = 0$

Thus we have $\quad a + b = a + a$

This implies that $\quad b = a \qquad$ [Cancellation law]

(iii) Let $a, b \in R$, this implies that $(a + b) \in R$. As R is a Boolean ring, so we have $(a + b)^2 = a + b$

$\Rightarrow \quad (a + b) . (a + b) = a + b$

$\Rightarrow \quad a . (a + b) + b . (a + b) = a + b \qquad$ [Distributive law]

$\Rightarrow \quad (a . a + a . b) + (b . a + b . b) = a + b \qquad$ [Distributive law]

$\Rightarrow \quad a^2 + a . b + b . a + b^2 = a + b$

$\Rightarrow \quad a + a . b + b . a + b = (a + b)$

$\Rightarrow \quad a . b + b . a = 0$

$\Rightarrow \quad a . b = b . a \qquad [a + b = 0$ implies $a = b]$

Therefore, R is a commutative ring.

Example 3 *If R is a ring with unity 1 = 0; then show that R is a singleton set.*

Solution: Given R is a ring with unity 1 = 0 and let $a \in R$

Now $\quad a = 1 . a = 0 . a = 0$

The above argument is true for all $a \in R$

Therefore, $\quad R = \{0\}$

Hence R is a singleton set with 0 as its element.

Example 4 *Let R is a set satisfying all the properties of ring except the commutative axiom under addition. If R has the unit element, then prove that R is a ring.*

Solution: Given that R is a set satisfying all the properties of ring except the commutative axiom under addition.

i.e., $a + b = b + a$

It is also given that R contains unit element, *i.e.,* $1 \in R$

Let $a, b \in R$ implies that $(a + b) \in R$ [Closure law]

Again, $1 \in R \Rightarrow (1 + 1) \in R$

Now, $(a + b) . (1 + 1) = a . (1 + 1) + b . (1 + 1)$ [Distributive law]

$= a . 1 + a . 1 + b . 1 + b . 1$ [Distributive law]

$= (a + a) + (b + b)$ ......(*i*)

Again, $(a + b) . (1 + 1) = (a + b) . 1 + (a + b) . 1$ [Distributive law]

$= (a + b) + (a + b)$ ......(*ii*)

Combining equations (*i*) and (*ii*) we get

$(a + a) + (b + b) = (a + b) + (a + b)$

$\Rightarrow a + \{a + (b + b)\} = a + \{b + (a + b)\}$ [Associative law]

$\Rightarrow a + (b + b) = b + (a + b)$ [Cancellation law]

$\Rightarrow (a + b) + b = (b + a) + b$ [Associative law]

$\Rightarrow (a + b) = (b + a)$ [Cancellation law]

Therefore, R is a ring.

Example 5 *Let R be a ring of all square matrices of order (2 ×2). Show that R has zero divisor.*

Solution: Let us consider two square matrices A and B of the ring R as

$$A = \begin{bmatrix} 5 & 0 \\ 0 & 0 \end{bmatrix} \text{ and } B = \begin{bmatrix} 0 & 0 \\ 5 & 0 \end{bmatrix}$$

Here $A \neq 0$ and $B \neq 0$, but

$$(A . B) = \begin{bmatrix} 5 & 0 \\ 0 & 0 \end{bmatrix} \begin{bmatrix} 0 & 0 \\ 5 & 0 \end{bmatrix} = \begin{bmatrix} 0 & 0 \\ 0 & 0 \end{bmatrix} \in R$$

Therefore, R is a ring with zero divisor.

Example 6 *Let R is a ring with unity and $(x . y)^2 = x^2 . y^2 \ \forall x, y \in R$. Show that R is a commutative ring.*

Solution: Given R is a ring with unity, *i.e.,* $1 \in R$. Also given that $(x . y)^2 = x^2 . y^2 \ \forall x, y \in R$.

Again, $(y + 1) \in R$ as $y \in R$ and $1 \in R$

Therefore, $(x . (y + 1))^2 = x^2 . (y + 1)^2$

$\Rightarrow (x y + x)^2 = x^2 . (y^2 + 2y + 1)$

$\Rightarrow (x y + x) . (x y + x) = x^2 . (y^2 + 2y + 1)$

$\Rightarrow x y . x y + x y . x + x . x y + x^2 = x^2 y^2 + 2x^2 y + x^2$

$\Rightarrow (x y)^2 + x y . x + x . x y + x^2 = x^2 y^2 + 2x^2 y + x^2$

$\Rightarrow x^2 y^2 + (x y . x + x . x y) + x^2 = x^2 y^2 + 2x^2 y + x^2$

$\Rightarrow x y x + x x y = 2x^2 y$ [Left and right cancellation law]

$\Rightarrow x y x + x^2 y = x^2 y + x^2 y$

$\Rightarrow x y x = x^2 y$ [Cancellation law]

Now on replacing x by $(x+1)$ we have

$$(x+1)\,y\,(x+1) = (x+1)^2\,y$$

$\Rightarrow$ $$(xy+y)(x+1) = (x^2+2x+1)\,y$$

$\Rightarrow$ $$xyx + xy + yx + y = x^2y + 2xy + y$$

$\Rightarrow$ $$x^2y + xy + yx + y = x^2y + 2xy + y \qquad [\because xyx = x^2y]$$

$\Rightarrow$ $$xy + yx = 2xy \qquad \text{[Left and right cancellation law]}$$

$\Rightarrow$ $$yx = xy$$

Therefore, R is commutative ring.

Example 7 *Let R = {0, 1, 2, 3, 4, 5} be a ring under binary operations $\oplus_6$ and $\otimes_6$. Show that R is a ring with zero divisor.*

Solution: Given that R = {0, 1, 2, 3, 4, 5} be a ring under binary operations $\oplus_6$ and $\otimes_6$.

Here $2 \in$ R and $3 \in$ R are two non zero elements such that

$$2 \cdot 3 = 0$$

Therefore, R is a ring with zero divisor.

Example 8 *R is the set of integer mod 7 under addition and multiplication mod 7. Show that R is a commutative ring with unit element.*

Solution: Given R is the set of integer mod 7 under addition and multiplication mod 7. The operation is defined as

(*i*) $a + b = c$ where c is the remainder of $a + b$ when divided by 7.

(*ii*) $a \cdot b = c$ where c is the remainder of $a \cdot b$ when divided by 7.

So, it is clear that R contains 7 elements. *i.e.*, R = {0, 1, 2, 3, 4, 5, 6}.

Table for addition modulo 7

+	0	1	2	3	4	5	6
0	0	1	2	3	4	5	6
1	1	2	3	4	5	6	0
2	2	3	4	5	6	0	1
3	3	4	5	6	0	1	2
4	4	5	6	0	1	2	3
5	5	6	0	1	2	3	4
6	6	0	1	2	3	4	5

Table for multiplication modulo 7

	0	1	2	3	4	5	6
0	0	0	0	0	0	0	0
1	0	1	2	3	4	5	6
2	0	2	4	6	1	3	5
3	0	3	6	2	5	1	4
4	0	4	1	5	2	6	3
5	0	5	3	1	6	4	2
6	0	6	5	4	3	2	1

Under Addition

Closure Law: From the table for addition modulo 7 it is clear that for any

$$a, b \in R \Rightarrow (a + b) \in R$$

Associative Law: From the table for addition modulo 7 it is clear that for any

$$a + (b + c) = (a + b) + c \quad \forall \quad a, b, c \in R.$$

Let $a = 1, b = 3, c = 5$.

Therefore, we have

$$a + (b + c) = 1 + (3 + 5) = 1 + 1 = 2$$

and $$(a + b) + c = (1 + 3) + 5 = 4 + 5 = 2$$

Therefore, $$a + (b + c) = (a + b) + c$$

Existence of Identity: From first row of the table for addition modulo 7 it is clear that $0 \in R$ is the identity element.

i.e., $$0 + a = a \quad \forall \quad a \in R$$

Existence of Inverse: From the table for addition modulo 7 it is clear that the inverse elements of 0, 1, 2, 3, 4, 5, 6 are 0, 6, 5, 4, 3, 2, $1 \in R$ respectively. The inverse element of 3 is 4 because $3 + 4 = 0$.

i.e., For every $a \in R$ there exists an $(-a) \in R$ such that $a + (-a) = 0$.

Commutative Law: From the table for addition modulo 7 it is clear that

$$a + b = b + a \quad \forall \quad a, b \in R$$

Under Multiplication

Closure Law: From the table for multiplication modulo 7 it is clear that for all

$$a, b \in R \Rightarrow a \,.\, b \in R$$

Associative Law: From the table for multiplication modulo 7 it is clear that

$$a \,.\, (b . c) = (a \,.\, b) \,.\, c \quad \forall \quad a, b, c \in R$$

Let $a = 3, b = 4, c = 6$.

Therefore we have

$$a \,.\, (b \,.\, c) = 3 \,.\, (4 \,.\, 6) = 3 \,.\, 3 = 2 \text{ and}$$

$$(a \,.\, b) \,.\, c = (3 \,.\, 4) \,.\, 6 = 5 \,.\, 6 = 2$$

Therefore, we get $a \,.\, (b \,.\, c) = (a \,.\, b) \,.\, c$

Distributive Law:

Let $a = 1, b = 4, c = 2$.

Hence, we have

$$a \,.\, (b + c) = 1 \,.\, (4 + 2) = 1 \,.\, 6 = 6 \text{ and}$$

$$(a \,.\, b) + (a \,.\, c) = (1 \,.\, 4) + (1 \,.\, 2) = 6$$

Therefore, we get $a \,.\, (b + c) = a \,.\, b + a \,.\, c$

Commutative Law: From the table for multiplication modulo 7 it is clear that

$$a \,.\, b = b \,.\, a \quad \forall \quad a, b \in R$$

Unit Element: From the table for multiplication modulo 7, the 2nd row or column indicates that $1 \in R$ is the identity element for every element $a \in R$. Hence for every $a \in R$ there exists unit element $1 \in R$ such that

$$(1 \,.\, a) = a \quad \forall \quad a \in R$$

Therefore, R is commutative ring with unit element.

Example 9 *Show that if R is a ring with unity, then any non-zero element with multiplicative inverse in R cannot be the zero divisor.*

Solution: Given that R is a ring with unity.

Let $a \in R$ and $a \neq 0$.

Again $a \neq 0$ implies $a^{-1} \in R$

Suppose that $(a \,.\, b) = 0$ with $b \neq 0 \in R$

$$\Rightarrow \quad a^{-1} . (a . b) = a^{-1} . 0$$

$$\Rightarrow \quad (a^{-1} . a)\, b = 0$$

$$\Rightarrow \quad b = 0 \quad [(a^{-1} . a) = 1]$$

This is a contradiction. This contradicts to the fact that $b \neq 0$. This indicates that, a is not the zero divisor.

Example 10 *For the ring $R = M_{2\times 2}(I)$, show that the subset S defined as*

$$S = \left\{ \begin{pmatrix} x & x+y \\ x+y & x \end{pmatrix} : x, y \in I \right\} \text{ is a sub ring.}$$

Solution: Given $R = M_{2\times 2}(I)$ be a ring and

$$S = \left\{ \begin{pmatrix} x & x+y \\ x+y & x \end{pmatrix} : x, y \in I \right\}$$

Putting $x = y = 0$, we have

$$\begin{pmatrix} 0 & 0 \\ 0 & 0 \end{pmatrix} \in S \Rightarrow S \text{ is non-empty. } i.e., S \neq \varphi.$$

To prove S is a sub ring we have to show that S satisfies two axioms

(*i*) $A - B \in S$; $\quad A, B \in S$

(*ii*) $A . B \in S$; $A, B \in S$

Let $A = \begin{pmatrix} x & x+y \\ x+y & x \end{pmatrix}$ and $B = \begin{pmatrix} u & u+v \\ u+v & u \end{pmatrix}$; $\quad u, v, x, y \in I$

Now,
$$A - B = \begin{pmatrix} x-u & x+y-u-v \\ x+y-u-v & x-u \end{pmatrix} \in S$$

and
$$A . B = \begin{pmatrix} x & x+y \\ x+y & x \end{pmatrix} \begin{pmatrix} u & u+w \\ u+w & u \end{pmatrix}$$

$$= \begin{pmatrix} xu+(x+y)(u+w) & (x)(u+w)+(x+y)u \\ (x+y)u+x(u+w) & (x+y)(u+w)+xu \end{pmatrix} \in S$$

This implies that all the entries of the matrices (A – B) and A.B are integers. Therefore, S is a sub ring.

Example 11 *Let R = {0, 1, 2, 3, 4} be a commutative ring with respect to the binary operations $\oplus_5$ and $\otimes_5$. Show that R is an integral domain.*

Solution: Given that R = {0, 1, 2, 3, 4} be a commutative ring with respect to the binary operations $\oplus_5$ and $\otimes_5$.

It is also clear that there is no such nonzero element in R for which $(a \,.\, b) = 0$. Hence R is an integral domain.

Example 12 *Let R = {0, 1, 2, 3, 4, 5, 6, 7} be a commutative ring under the binary operations $\oplus_8$ and $\otimes_8$ · Show that R is not an integral domain.*

Solution: Given that R = {0, 1, 2, 3, 4, 5, 6, 7} be a commutative ring under the binary operations $\oplus_8$ and $\otimes_8$

Now $2 \in R$ and $4 \in R$ such that $2 \otimes_8 4 = 0$

Therefore, R is not an integral domain.

Example 13 *R = {u, v, w, t} define the operations + and . in such a way that R will be a ring.*

+	*u*	*v*	*w*	*t*
u	*u*	*v*	*w*	*t*
v	*v*	*u*	*t*	*w*
w	*w*	*t*	*u*	*v*
t	*t*	*w*	*v*	*u*

.	*u*	*v*	*w*	*t*
u	*u*	*u*	*u*	*u*
v	*u*	*v*		
w	*u*	*v*		*t*
t	*u*		*u*	

(*a*) *Using the associative and distributive law determine the entries in the blank space.*
(*b*) *Is it a commutative ring?*
(*c*) *Does it have unity? If yes, find the unit element.*
(*d*) *Is the ring an integral domain or a field.*

Solution: (*a*) Now, $(w \,.\, w) = w \,.\, (v + t)$ $[\because (v + t) = w]$

$= w \,.\, v + w \,.\, t$ [Distributive law]

$= v + t = w$

Therefore, $w \,.\, w = w$

Again, $(t \,.\, v) = (w + v) \,.\, v$ $[\because (w + v) = t]$

$= w \,.\, v + v \,.\, v$ [Distributive law]

$= v + v = u$

Therefore, $t \,.\, v = u$

Again, $(t \,.\, t) = t \,.\, (v + w)$ $[\because (w + v) = t]$

$= t \,.\, v + t \,.\, w$ [Distributive law]

$= u + u = u$

Therefore, $(t \,.\, t) = u$

Similarly, $(v \,.\, w) = (t + w) \,.\, w$ $[\because (t + w) = v]$

$= t \,.\, w + w \,.\, w$ [Distributive law]

$= u + w = w$

Therefore, $(v \,.\, w) = w$

And $v \,.\, t = (t + w) \,.\, t$ $[\because (w + t) = v]$

$= t \,.\, t + w \,.\, t$ [Distributive law]

$= u + t = t$

So, the complete table is given as

.	u	v	w	t
u	u	u	u	u
v	u	v	w	t
w	u	v	w	t
t	u	v	u	u

(*b*) From the above table it is clear that $(v \,.\, w) = w$ and $(w \,.\, v) = v$. This implies $(v \,.\, w) \neq (w \,.\, v)$. Thus, R is not a commutative ring.

(*c*) As it is clear from the table, the ring does not contain unity and hence does not have unit element.

(*d*) Since R is not commutative, so it is neither integral domain nor field.

Example 14 *If $a, b, c, d \in R$ and R is a ring then evaluate $(a + b).(c + d)$.*

Solution: Given R is a ring and $a, b, c, d \in$ R

Now,
$$\begin{aligned}(a + b) \,.\, (c + d) &= u \,.\, (c + d); && \text{[Let } u = a + b\text{]}\\ &= u \,.\, c + u \,.\, d && \text{[Distributive law]}\\ &= (a + b) \,.\, c + (a + b) \,.\, d\\ &= a \,.\, c + b \,.\, c + a \,.\, d + b \,.\, d && \text{[Distributive law]}\end{aligned}$$

Example 15 *If R is a ring and $(x + y)^2 = x^2 + 2xy + y^2$ then prove that R is commutative for all $x, y \in R$.*

Solution: Given R is a ring and $(x + y)^2 = x^2 + 2xy + y^2$ for $x, y \in$ R.

$$\begin{aligned}\Rightarrow\quad (x + y)(x + y) &= x^2 + 2xy + y^2\\ \Rightarrow\quad x(x + y) + y(x + y) &= x^2 + 2xy + y^2\\ \Rightarrow\quad xx + xy + yx + yy &= x^2 + 2xy + y^2\\ \Rightarrow\quad x^2 + xy + yx + y^2 &= x^2 + 2xy + y^2\\ \Rightarrow\quad xy + yx &= 2xy && \text{[Cancellation law]}\\ \Rightarrow\quad xy + yx &= xy + xy\\ \Rightarrow\quad yx &= xy && \text{[Cancellation law]}\end{aligned}$$

Therefore, R is commutative ring.

Example 16 *For a commutative ring R with characteristic 2 show that*

$$(a + b)^2 = a^2 + b^2 = (a - b)^2 \quad \forall \quad a, b \in \text{R}$$

Solution: Let R be a commutative ring with characteristic 2. This indicates that $(2 \,.\, a) = 0$ for all $a \in$ R.

Now,
$$\begin{aligned}(a + b)^2 &= (a + b) \,.\, (a + b)\\ &= a\,a + a\,b + b\,a + b\,b\\ &= a^2 + a\,b + a\,b + b^2 && \text{[R is commutative]}\\ &= a^2 + 2ab + b^2\\ &= a^2 + 0 + b^2 && [2a = 0]\\ &= a^2 + b^2\end{aligned}$$

$\Rightarrow \qquad (a+b)^2 = a^2 + b^2$

Similarly, $\qquad (a-b)^2 = a^2 - 2ab + b^2$

$\qquad = a^2 + 0 + b^2 = a^2 + b^2$

$\Rightarrow \qquad (a-b)^2 = a^2 + b^2$

Therefore, $\qquad (a+b)^2 = a^2 + b^2 = (a-b)^2$

Example 17 *Show that a Boolean ring R is a commutative ring with characteristic 2.*

Solution: Given R is a Boolean ring. This implies $a^2 = a$ for all $a \in$ R.

Let $a, b \in$ R

$\Rightarrow \qquad (a+b) \in R$

$\Rightarrow \qquad (a+b)^2 = (a+b)$ $\qquad [\because a^2 = a]$

$\Rightarrow \qquad (a+b).(a+b) = (a+b)$

$\Rightarrow \qquad a\,a + a\,b + b\,a + b\,b = a + b$

$\Rightarrow \qquad a^2 + a\,b + b\,a + b^2 = (a+b) + 0$

$\Rightarrow \qquad a + a\,b + b\,a + b = (a+b) + 0$

$\Rightarrow \qquad (a+b) + (a\,b + b\,a) = (a+b) + 0$ $\qquad$ [Associative law]

$\Rightarrow \qquad a\,b + b\,a = 0$

$\Rightarrow \qquad a\,b = b\,a$ $\qquad [\because$ In Boolean ring $(a+b) = 0 \Rightarrow a = b]$

Therefore, R is a commutative ring.

Again, $\qquad (a+a)^2 = (a+a)$

$\Rightarrow \qquad a^2 + a^2 + a^2 + a^2 = a + a$

$\Rightarrow \qquad (a+a) + (a+a) = (a+a) + 0$

$\Rightarrow \qquad a + a = 0$ $\qquad$ [Left cancellation law]

$\Rightarrow \qquad 2a = 0$

Hence, R is a commutative ring with characteristic 2.

Example 18 *For the ring (I, ⊕, ⊙) with binary operation defined as $x \oplus y = x + y - 1$ and $x \odot y = x + y - xy$, show that the subset S of all odd integers is a sub ring.*

Solution: Suppose that S be the set of all odd integers. Let $a, b \in$ S. This implies a and b are odd integers.

Now, $\qquad a \oplus b = a + b - 1$

Again $(a + b)$ is even as sum of odds is even.

$\Rightarrow \qquad a + b - 1$ is odd

$\Rightarrow \qquad a + b - 1 \in S$

$\Rightarrow \qquad a \oplus b \in S$

Similarly, $a \odot b = a + b - ab$. However, we know that $(a + b)$ is even and $(a\,b)$ is odd. Therefore $(a + b - a\,b)$ is odd.

$\Rightarrow \qquad a \odot b \in S$

Let $a \in$ S, then the additive inverse is $-a$ which is odd hence belongs to S, *i.e.*, $-a \in$ S.

Therefore, S is a sub ring.

Example 19 *Show that isomorphic image of a division ring is division ring.*

Solution: Let R be a division ring. Therefore, the non zero elements of R forms a group under multiplication.

Let ϕ be an isomorphism defined from R into R′. *i.e.*, ϕ: R $\to$ R′.

Let $a \neq 0 \in R \Rightarrow a^{-1} \neq 0 \in R$.

As ϕ is isomorphism, so $\phi(a) \neq 0$. We have to show that $\phi(a^{-1}) = \phi(a)^{-1}$.

Again, $\qquad \phi(a) . \phi(a^{-1}) = \phi(a . a^{-1}) = \phi(1) = 1' \in R'$

$\Rightarrow \qquad \phi(a) . \phi(a^{-1}) = 1'$

Therefore, we get $\qquad \phi(a^{-1}) = \phi(a)^{-1}$.

This indicates that every non-zero element of R′ has an inverse. Thus R′ is a division ring.

Example 20 *Show that the isomorphic image of an integral domain is an integral domain.*

Solution: Let R be an integral domain and ϕ be *a* isomorphism from R into R′, *i.e.*, ϕ: R $\to$ R′.

Since R is an integral domain, so it is a commutative ring without zero divisors.

Let $a, b \in R$, $a \neq 0$ and $b \neq 0$ such that $(a . b) \neq 0$.

$\Rightarrow \qquad \phi(a . b) \neq 0$

$\Rightarrow \qquad \phi(a) . \phi(b) \neq \phi(0)$

$\Rightarrow \qquad \phi(a) . \phi(b) \neq 0$

Since ϕ is an isomorphism, $a \neq 0$, $b \neq 0$ implies that $\phi(a) \neq 0$, $\phi(b) \neq 0$.

Therefore, we get $\phi(a) \neq 0$ and $\phi(b) \neq 0$ implies $\phi(a) . \phi(b) \neq 0$. Hence, R′ is without zero divisor.

Again we know that isomorphic image of a commutative ring is a commutative ring.

This indicates that R′ is a commutative ring without zero divisor, thus is an integral domain.

EXERCISES

1. Prove that the set of Real numbers R forms a ring under ordinary addition multiplication.
2. Show that the set of Rational numbers Q forms a commutative ring with unit element under ordinary addition and multiplication.
3. Let S = $\{a + b\sqrt{2} \mid a$ and b are integers$\}$ forms a ring under addition and multiplication.
4. The set R = {0, 1, 2, 3, 4, 5} is a commutative ring with unit element under $\oplus_6$ and $\otimes_6$.
5. The operations $a \oplus b = (a + b + 1)$ and $a \otimes b = (a + b + ab)$ are defined on the set of integers. Show that I forms a commutative ring under the operations defined. Does it have unit element?
6. Show that set of Real numbers of the type $(a + b\sqrt{2})$; $a, b \in R$ is an integral domain.
7. Show that ring of integers (I, +) is an integral domain but not field.
8. $R_P = \{0, 1, 2, \ldots, P - 1\}$, where P is a prime. Show that R_P is an integral domain.
9. Show that the set of numbers given by $\{a + b\sqrt{2}; a, b \in I\}$ is a ring under ordinary addition and multiplication.
10. Let R = $\{a, b, c, d, e\}$. The operations + and . on R is defined as

+	a	b	c	d	e
a	a	b	c	d	e
b	b	c	d	e	a
c	c	d	e	a	b
d	d	e	a	b	c
e	e	a	b	c	d

.	a	b	c	d	e
a	a	a	a	a	a
b	a	b	c	d	e
c	a	c	e	b	d
d	a	d	b	e	e
e	a	e	d	c	b

(*a*) Show that (R, +, .) is a commutative ring.
(*b*) What is the additive identity and unity ?
(*c*) What are the inverse elements of *a*, *b*, *c*, *d* ?

11. Show that set of all rational numbers is a commutative ring with unity under ordinary addition and multiplication.

12. Do the following sets forms an integral domain with respect to ordinary addition and multiplication? If yes, then test whether they are field.

(*a*) $I\left(\sqrt{2}\right) = \{x \mid x = b\sqrt{2} : b \text{ is rational}\}$
(*b*) Set of even integers
(*c*) Set of positive integers.

13. Show that (I, +, .) is a sub ring of (Q, +, .) which is a sub ring of (R, +, .) which is a sub ring of (C, +, .). Where

I : Set of integers
Q : Set of rational numbers
R : Set of real numbers
C : Set of complex numbers

14. Give an example of each of the followings.

(*a*) A non-commutative ring
(*b*) Ring without zero divisor
(*c*) Division ring
(*d*) A ring which is not an integral domain.

15. Show that set of all square matrix of order $(n \times n)$ is a non-commutative ring with unity under the matrix addition and multiplication.

16. Show that set of even integers under ordinary addition and multiplication is a commutative ring without unit element.

17. The set of rational numbers under usual addition and multiplication is a field.

18. Let R be a ring. Prove that $(n\,a)\,(m\,b) = (n\,m)\,(a\,b)$ for all $a, b \in R$ and $m, n \in I$.

19. Give an example of a ring which contains an element $a \neq 0$ such that $a^3 = 0$. Is it an integral domain?

20. Given *a*, *b* be two elements of a field F with characteristic 3. Show that $(a + b)^3 = a^3 + b^3$.

21. Prove that for a field

(*a*) $\dfrac{a}{b} = \dfrac{c}{d} \Leftrightarrow ad = bc$
(*b*) $(-a)^{-1} = -(a^{-1})$
(*c*) $\dfrac{a}{b} - \dfrac{c}{d} = \dfrac{ad - bc}{bd}$
(*d*) $\dfrac{-a}{-b} = \dfrac{a}{b}$

22. R is a ring with unit element 1 and ϕ is a homomorphism of R onto R_1. Then prove that $\phi(1)$ is the unit element of R_1.

10

Boolean Algebra

■ 10.0 INTRODUCTION

For centuries mathematicians felt there was a connection between mathematics and logic, but no one could find this missing link before George Boole. In 1854 he introduced symbolic logic known as Boolean Algebra, Boolean function, Boolean expression, Boolean ring and many more honour the nineteenth century mathematician George Boole. Each variable in Boolean algebra has either of two values: true or false. The purpose of this two - state algebra was to solve logic problems.

Almost after a century of Boole's work, it was observed by C.E. Shannon in 1938, that Boolean algebra could be used to analyze electrical circuits. This was developed by Shannon while he analyzed telephone switching circuits. Because of Shannon's work, engineers realized that Boolean algebra could be applied to Computer electronics.

This chapter introduces the Gate, Combinatorial Circuits, Boolean Expression, Boolean Algebra, Boolean Functions and Various Normal Forms.

■ 10.1 GATES

In logic we have discussed about the logical connectives $\neg$, $\wedge$ and $\vee$. The connectives $\wedge$ and $\vee$ can be considered as circuits connected in series and parallel respectively. A circuit with one or more input signals but only one output signal is known as a gate. Gates are digital circuits because of input and output signals, which are either low or high. Gates are also called logical circuits because they can be analyzed with Boolean algebra. In gates, the connectives $\neg$, $\wedge$ and $\vee$ are usually denoted by the symbols $'$, . and + respectively. The block diagrams for different gates are discussed below.

10.1.1 A NOT Gate

A NOT gate receives input x, where x is a bit (binary digit) and produces output x' where

$$x' = \begin{cases} 1 & \text{if } x = 0 \\ 0 & \text{if } x = 1 \end{cases}$$

The output state is always the opposite of the input state. The output is sometimes called the complement of the input. A NOT gate is drawn as shown in the following figure.

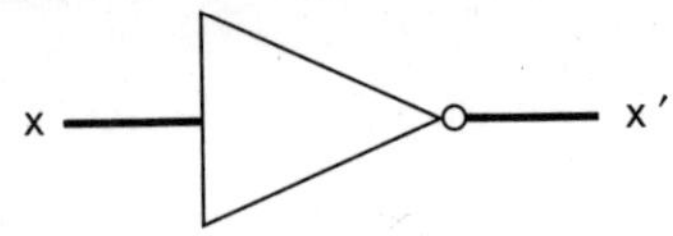

10.1.2 An AND Gate

An AND gate receives inputs x_1 and x_2, where x_1 and x_2 are bits, and produces output $(x_1 \wedge x_2)$, where

$$(x_1 \wedge x_2) = \begin{cases} 1 & \text{if } x_1 = x_2 = 1 \\ 0 & \text{otherwise} \end{cases}$$

An AND gate may have more inputs also but the output is always one. An AND gate is drawn as shown in the following figure.

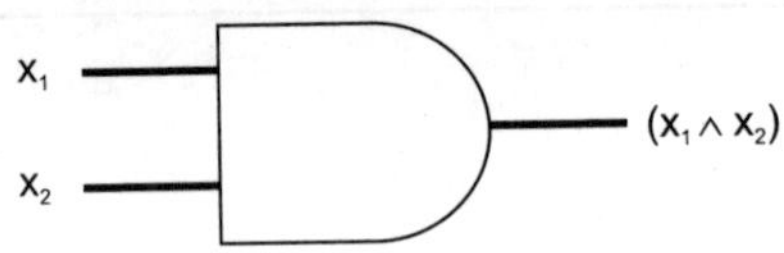

(2 input AND gate)

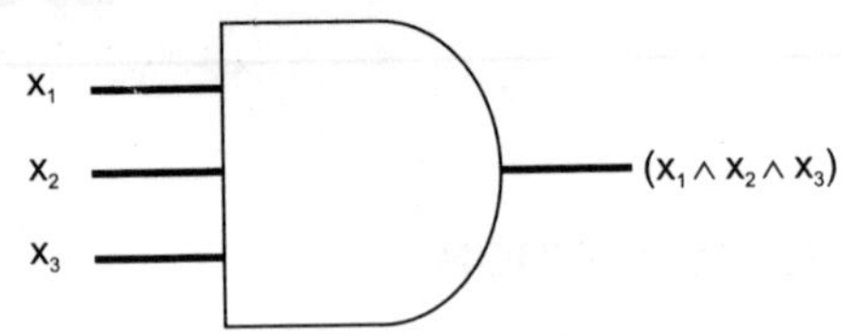

(3 input AND gate)

10.1.3 An OR Gate

An OR gate receives inputs x_1 and x_2, where x_1 and x_2 are bits, and produces output $(x_1 \vee x_2)$, where

$$(x_1 \vee x_2) = \begin{cases} 1 & \text{if } x_1 = 1 \text{ or } x_2 = 1 \\ 0 & \text{otherwise} \end{cases}$$

An OR gate may have more inputs also but the output is always one. An OR gate is drawn as shown in the following figure.

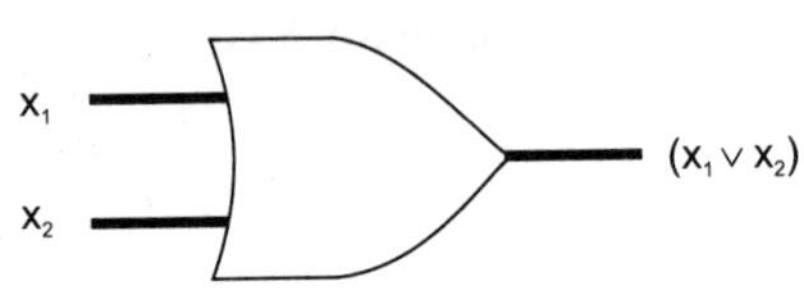

(2 input OR gate)

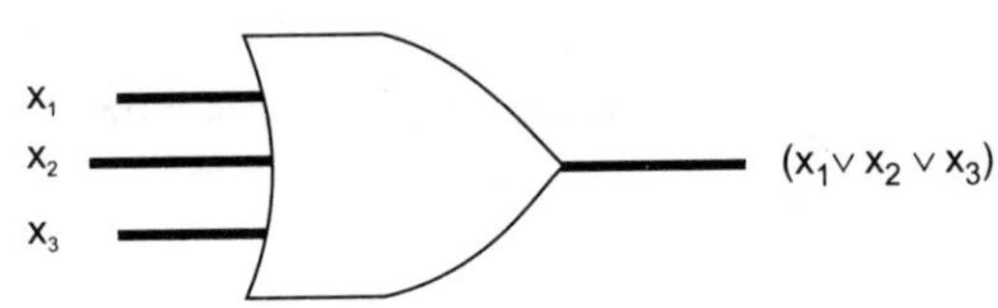

(3 input OR gate)

The logic tables for the basic AND, OR and NOT gates are given below.

x_1	x_2	$(x_1 \wedge x_2)$
1	1	1
1	0	0
1	0	0
0	0	0

x_1	x_2	$(x_1 \vee x_2)$
1	1	1
1	0	1
0	1	1
0	0	0

x	x'
1	0
0	1

10.2 MORE LOGIC GATES

There are some other types of gates which are useful and frequently used in computer science. These are called NAND, NOR, XOR and XNOR gates. The block diagrams for these different gates are given below.

10.2.1 NOR Gate

A NOR gate receives inputs x_1 and x_2 where x_1 and x_2 are bits, and produces output $(x_1 \vee x_2)'$, where

$$(x_1 \vee x_2)' = \begin{cases} 1 & \text{if } x_1 = x_2 = 0 \\ 0 & \text{otherwise} \end{cases}$$

A NOR gate may have more inputs also, but the output is always one. A NOR gate is drawn as shown in the following figure.

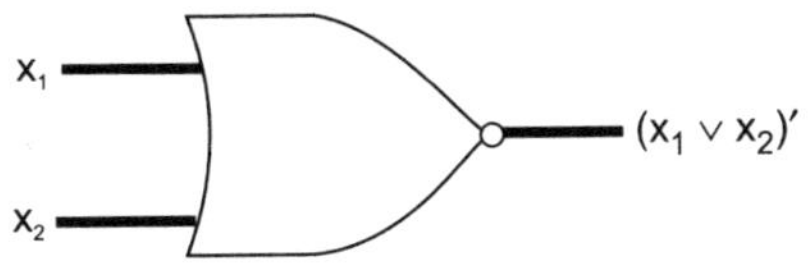

(2 input NOR gate)

According to de Morgan's first theorem, we have

$$(x_1 \vee x_2)' = x_1' \wedge x_2' \quad i.e., \quad (x_1 + x_2)' = x_1' . x_2'$$

10.2.2 NAND Gate

A NAND gate receives inputs x_1 and x_2, where x_1 and x_2 are bits, and produces output $(x_1 \wedge x_2)'$, where

$$(x_1 \wedge x_2)' = \begin{cases} 1 & \text{if } x_1 = 0 \text{ or } x_2 = 0 \\ 0 & \text{otherwise} \end{cases}$$

A NAND gate may have more inputs also, but the output is always one. A NAND gate is drawn as shown in the following figure.

According to the de Morgan's second theorem we have

$$(x_1 \wedge x_2)' = x_1' + x_2' \quad i.e., \quad (x_1 . x_2)' = x_1' + x_2'$$

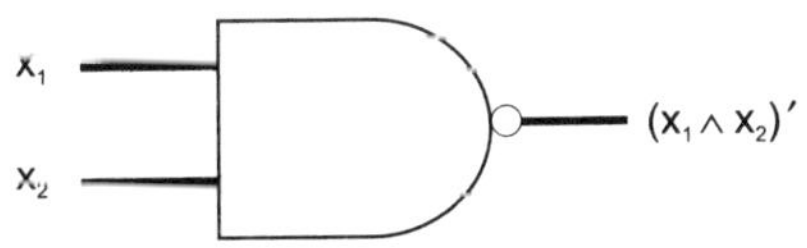

10.2.3 XOR Gate (Exclusive OR Gate)

A XOR gate receives inputs x_1 and x_2, where x_1 and x_2 are bits, and produces output $(x_1 \overline{\vee} x_2)$ or $(x_1 \oplus x_2)$, where

$$(x_1 \oplus x_2) = \begin{cases} 1 & \text{if } x_1 = 1 \text{ or } x_2 = 1 \text{ but not both} \\ 0 & \text{otherwise} \end{cases}$$

From the definition, it is clear that, the Exclusive OR gate, *i.e.* XOR gate produces 1 that have an odd number of 1's. A XOR gate may have more inputs also, but the output is always one. A XOR gate is drawn as shown in the following figure.

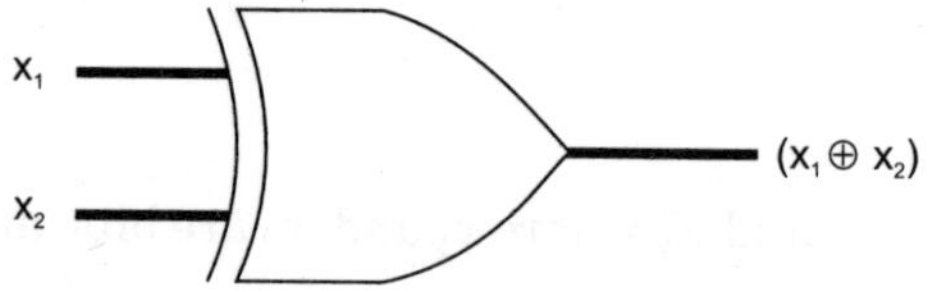

10.2.4 XNOR Gate (Exclusive NOR Gate)

A XNOR gate receives inputs x_1 and x_2, where x_1 and x_2 are bits, and produces output x_1 XNOR x_2 where

$$x_1 \text{ XNOR } x_2 = \begin{cases} 1 \text{ if } x_1 \text{and } x_2 \text{ are same bits} \\ 0 \text{ otherwise} \end{cases}$$

XNOR gate may have more inputs also, but the output is always one. In this case it recognizes even-parity words. Even parity means a word has an even number of 1's. For example 11100111 has even parity because it contains six 1's. Odd parity means a word has an odd number of 1's. For example 1101 has odd parity because it contains three 1's.

A XNOR gate is drawn as shown in the following figure.

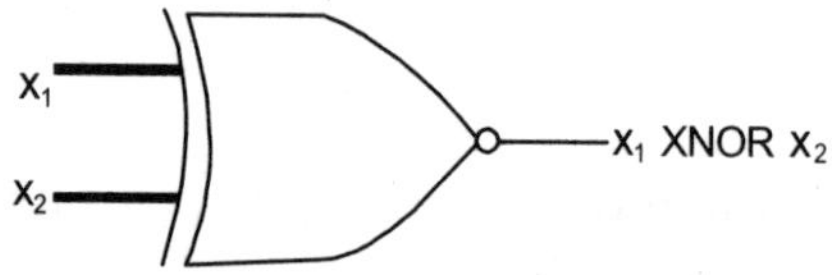

The logic tables for the above NOR, NAND, XOR and XNOR gates are given below.

x_1	x_2	$(x_1 \wedge x_2)'$
1	1	0
1	0	1
0	1	1
0	0	1

x_1	x_2	$(x_1 \vee x_2)'$
1	1	0
1	0	0
0	1	0
0	0	1

x_1	x_2	$(x_1 \oplus x_2)$
1	1	0
1	0	1
0	1	1
0	0	0

x_1	x_2	$(x_1 \text{ XNOR } x_2)$
1	1	1
1	0	0
0	1	0
0	0	1

10.3 COMBINATORIAL CIRCUIT

In digital computer electronics, there are only two possibilities, *i.e.,* 0 and 1, for the smallest, indivisible object. These 0 and 1 are known as binary digits (bit). A bit in one part of a circuit is transmitted to another part of the circuit as a voltage. Thus two voltage levels are needed. *i.e.*, high voltage level and low voltage level. A high voltage level communicates 1 whereas a low voltage level communicates 0.

A combinatorial circuit is a circuit which produces an unique output for every combination of inputs. A combinatorial circuit has no memory, previous inputs and the state of the system do not affect the output of a combinatorial circuit. These circuits can be constructed using gates which we have already discussed.

Let us consider the circuit

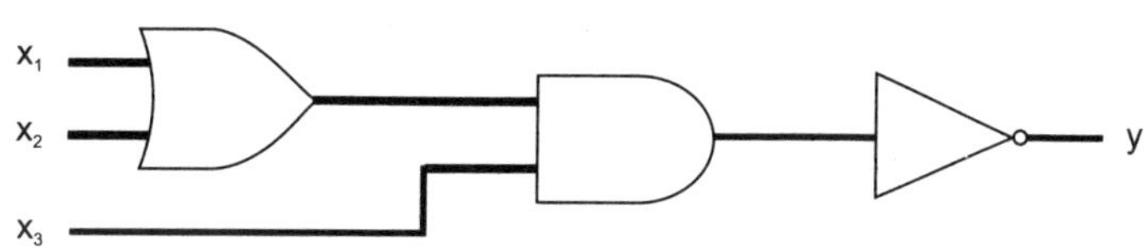

x_1	x_2	x_3	y
1	1	1	0
1	1	0	1
0	1	1	0
1	0	1	0
0	0	1	1
0	1	0	1
1	0	0	1
0	0	0	1

The logic table for the above circuit is given in the above table. From the table it is clear that the output y is uniquely defined for each combination of inputs x_1, x_2 and x_3. Therefore, the circuit is a combinatorial circuit.

If $x_1 = 1$ and $x_2 = 1$, then the output of OR gate is 1. Now the input for AND gate is 1 and 0, so the output of AND gate is 0. Since the input to the NOT gate is 0, the output $y = 1$.

Consider another circuit as

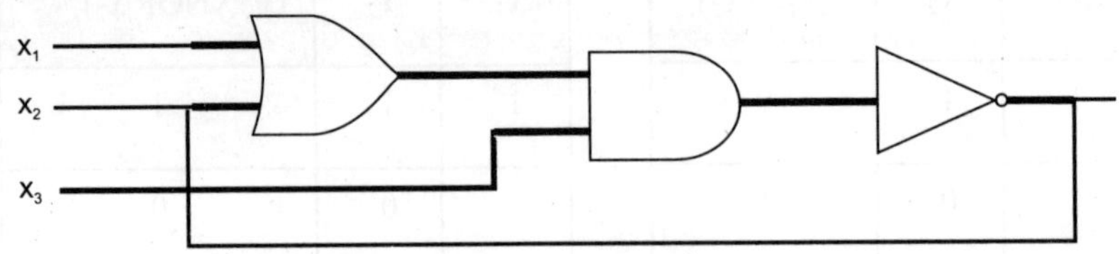

The above circuit is not a combinatorial circuit, as the output y is not defined uniquely for every combination of inputs x_1, x_2 and x_3.

■ 10.4 BOOLEAN EXPRESSION

Any expression built up from the variables $x_1, y_1, z_1, x_2, y_2, z_2, \ldots$ by applying the operations $\wedge, \vee$ and $'$ a finite number of times. If X_1 and X_2 are Boolean expressions, then (X_1), X_2', $(X_1 \wedge X_2)$ and $(X_1 \vee X_2)$ are also Boolean expressions. The output of a combinatorial circuit is also a Boolean expression.

Let us consider the combinatorial circuit as

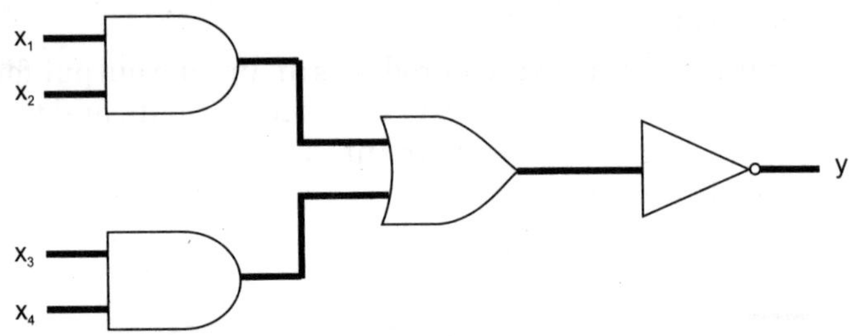

The Boolean expression to the above circuit is given as $((x_1 \wedge x_2) \vee (x_3 \wedge x_4))'$.

10.4.1 Theorem

If $\wedge$, $\vee$ and $'$ are connectives defined earlier, then the following properties hold.

(*i*) Associative Laws: For all $a, b, c \in \{0, 1\}$

$$(a \wedge b) \wedge c = a \wedge (b \wedge c) \text{ and}$$
$$(a \vee b) \vee c = a \vee (b \vee c)$$

(*ii*) Identity Laws: For all $a \in \{0, 1\}$

$$(a \wedge 1) = a \quad \text{and} \quad (a \vee 0) = a$$

(*iii*) Commutative Laws: For all $a, b \in \{0, 1\}$

$$(a \wedge b) = (b \wedge a) \quad \text{and}$$
$$(a \vee b) = (b \vee a)$$

(*iv*) Complement Laws: For all $a \in \{0, 1\}$

$$(a \wedge a') = 0 \quad \text{and}$$
$$(a \vee a') = 1$$

(*v*) Distributive Laws: For all $a, b, c \in \{0, 1\}$

$$a \vee (b \wedge c) = (a \vee b) \wedge (a \vee c) \quad \text{and}$$
$$a \wedge (b \vee c) = (a \wedge b) \vee (a \wedge c)$$

Proof: Proofs of (*i*), (*ii*), (*iii*) and (*iv*) are immediate consequences of the definitions. We prove only the first distributive law. Here we simply evaluate both sides of law for all possible values of $a, b, c \in \{0, 1\}$ and verify that in each case we obtain the same result.

We must show that $a \vee (b \wedge c) = (a \vee b) \wedge (a \vee c)$

a	b	c	$(b \wedge c)$	$a \vee (b \wedge c)$	$(a \vee b)$	$(a \vee c)$	$(a \vee b) \wedge (a \vee c)$
1	1	1	1	1	1	1	1
0	0	1	0	0	0	1	0
0	1	0	0	0	1	0	0
1	0	0	0	1	1	1	1
1	1	0	0	1	1	1	1
1	0	1	0	1	1	1	1
0	1	1	1	1	1	1	1
0	0	0	0	0	0	0	0

Therefore, $a \vee (b \wedge c) = (a \vee b) \wedge (a \vee c)$

10.4.2 de Morgan's Laws

If x_1, x_2 are bits, *i.e.*, $x_1, x_2 \in \{0, 1\}$, then

(*i*) $(x_1 \wedge x_2)' = x_1' \vee x_2'$

(*ii*) $(x_1 \vee x_2)' = x_1' \wedge x_2'$

Proof: We prove only the first de Morgan's Law.

i.e., $(x_1 \wedge x_2)' = x_1' \vee x_2'$

Construct the logical table.

x_1	x_2	$(x_1 \wedge x_2)'$	x_1'	x_2'	$x_1' \vee x_2'$
1	1	0	0	0	0
1	0	1	0	1	1
0	1	1	1	0	1
0	0	1	1	1	1

Therefore, $(x_1 \wedge x_2)' = x_1' \vee x_2'$.

■ 10.5 EQUIVALENT COMBINATORIAL CIRCUITS

Two combinatorial circuits, each having inputs $x_1, x_2, \ldots, x_n$ are said to be equivalent if they produce the same outputs for same inputs *i.e.*, the output for both the circuits remains same if the circuits receive same inputs.

Consider the following combinatorial circuits.

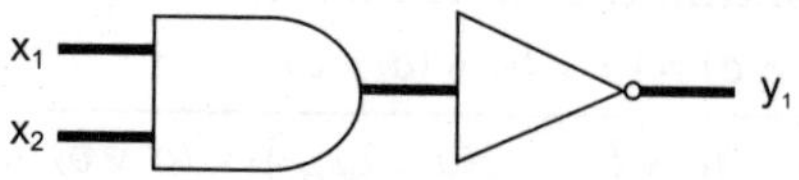

Figure 1

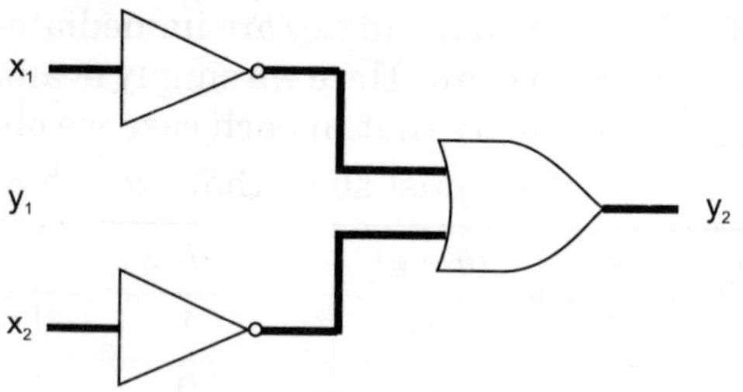

Figure 2

The logic tables for both the circuits are given below, which are identical.

x_1	x_2	y_1
1	1	0
1	0	1
0	1	1
0	0	1

x_1	x_2	y_2
1	1	0
1	0	1
0	1	1
0	0	1

From the logic tables it is clear that both the combinatorial circuits are equivalent.

10.6 BOOLEAN ALGEBRA

A Boolean algebra B consists of a set S together with two binary operations $\wedge$ and $\vee$ on S, a singular operation $'$ on S and two specific elements 0 and 1 of S such that the following laws hold. We write B = {S, $\wedge$, $\vee$, $'$, 0, 1}.

(*a*) Associative Laws: For all $a, b, c \in$ S

$$(a \wedge b) \wedge c = a \wedge (b \wedge c)$$

and $$(a \vee b) \vee c = a \vee (b \vee c)$$

(*b*) Commutative Laws: For all $a, b \in$ S

$$(a \wedge b) = (b \wedge a)$$

and $$(a \vee b) = (b \vee a)$$

(*c*) Distributive Laws: For all $a, b, c \in$ S

$$a \wedge (b \vee c) = (a \wedge b) \vee (a \wedge c)$$

and $$a \vee (b \wedge c) = (a \vee b) \wedge (a \vee c)$$

(*d*) Identity Laws: For all $a \in$ S

$$(a \wedge 1) = a \quad \text{and} \quad (a \vee 0) = a$$

(*e*) Complement Laws: For all $a \in$ S

$(a \wedge a') = 0$ and $(a \vee a') = 1$

10.6.1 Theorem

In a Boolean algebra; if $(a \vee b) = 1$ and $(a \wedge b) = 0$, then $b = a'$, *i.e.*, the complement is unique.

Proof: Suppose that $(a \vee b) = 1$ and $(a \wedge b) = 0$

Now
$$b = (b \vee 0) \quad \text{[Identity law]}$$
$$= b \vee (a \wedge a') \quad \text{[Complement law]}$$
$$= (b \vee a) \wedge (b \vee a') \quad \text{[Distributive law]}$$

$$= (a \vee b) \wedge (b \vee a') \quad \text{[Commutative law]}$$
$$= 1 \wedge (b \vee a') \quad \text{[Given condition]}$$
$$= (b \vee a') \quad \text{[Identity law]}$$

This implies $b = (b \vee a')$ ...(*i*)

Again a'

$$= (a' \vee 0) \quad \text{[Identity law]}$$
$$= a' \vee (a \wedge b) \quad \text{[Given condition]}$$
$$= (a' \vee a) \wedge (a' \vee b) \quad \text{[Distributive law]}$$
$$= 1 \wedge (a' \vee b) \quad \text{[Complement law]}$$
$$= (a' \vee b) \quad \text{[Identity law]}$$
$$= (b \vee a') \quad \text{[Commutative law]}$$

This implies $a' = (b \vee a') = b$ [Equation 1]

10.6.2 Theorem

In a Boolean algebra B = (S, ∨, ∧, ′ , 0, 1) ; the following properties hold.

(*a*) Idempotent Laws: For all $x \in S$

$$(x \vee x) = x \quad \text{and} \quad (x \wedge x) = x$$

(*b*) Bound Laws: For all $x \in S$

$$(x \vee 1) = 1 \quad \text{and} \quad (x \wedge 0) = 0$$

(*c*) Absorption Laws: For all $x, y \in S$

$$x \wedge (x \vee y) = x \quad \text{and} \quad x \vee (x \wedge y) = x$$

(*d*) Involution Laws: For all $x \in S$

$$(x')' = x$$

(*e*) 0 and 1 Laws: $0' = 1$ and $1' = 0$

(*f*) de Morgan's Laws: For all $x, y \in S$

$$(x \wedge y)' = x' \vee y'$$

and

$$(x \vee y)' = x' \wedge y'$$

Proof: (*a*)

$$x = x \vee 0 \quad \text{[Identity law]}$$
$$= x \vee (x \wedge x') \quad \text{[Complement law]}$$
$$= (x \vee x) \wedge (x \vee x') \quad \text{[Distributive law]}$$
$$= (x \vee x) \wedge 1 \quad \text{[Complement law]}$$
$$= (x \vee x) \quad \text{[Identity law]}$$

Therefore, $(x \vee x) = x$

Again

$$x = x \wedge 1 \quad \text{[Identity law]}$$
$$= x \wedge (x \vee x') \quad \text{[Complement law]}$$
$$= (x \wedge x) \vee (x \wedge x') \quad \text{[Distributive law]}$$
$$= (x \wedge x) \vee 0 \quad \text{[Complement law]}$$
$$= (x \wedge x) \quad \text{[Identity law]}$$

Therefore, $(x \wedge x) = x$

(*b*)

$$(x \vee 1) = (x \vee 1) \wedge 1 \quad \text{[Identity law]}$$
$$= (x \vee 1) \wedge (x \vee x') \quad \text{[Complement law]}$$
$$= ((x \vee 1) \wedge x) \vee ((x \vee 1) \wedge x') \quad \text{[Distributive law]}$$

$= ((x \wedge x) \vee (1 \wedge x)) \vee ((x \wedge x') \vee (1 \wedge x'))$

$= (x \vee (1 \wedge x)) \vee ((x \wedge x') \vee (1 \wedge x'))$ [Idempotent law]

$= (x \vee x) \vee ((x \wedge x') \vee x'))$ [Identity law]

$= (x \vee x) \vee (0 \vee x')$ [Complement law]

$= x \vee (0 \vee x')$ [Idempotent law]

$= (x \vee x')$ [Identity law]

$= 1$ [Complement law]

Therefore, $(x \vee 1) = 1$

Again, $(x \wedge 0) = (x \wedge 0) \vee 0$ [Identity law]

$= (x \wedge 0) \vee (x \wedge x')$ [Complement law]

$= ((x \wedge 0) \vee x) \wedge ((x \wedge 0) \vee x')$ [Distributive law]

$= ((x \vee x) \wedge (0 \vee x)) \wedge ((x \vee x') \wedge (0 \vee x'))$

$= ((x \vee x) \wedge x) \wedge ((x \vee x') \wedge x')$ [Identity law]

$= (x \wedge x) \wedge ((x \vee x') \wedge x')$ [Idempotent law]

$= (x \wedge x) \wedge ((x \wedge x') \vee (x' \wedge x'))$ [Distributive law]

$= x \wedge ((x \wedge x') \vee (x' \wedge x'))$ [Idempotent law]

$= x \wedge (0 \vee (x' \wedge x'))$ [Complement law]

$= x \wedge (0 \vee x')$ [Idempotent law]

$= x \wedge x'$ [Identity law]

$= 0$ [Complement law]

Therefore, $(x \wedge 0) = 0$

(*c*) $x \wedge (x \vee y) = (x \vee 0) \wedge (x \vee y)$ [Identity law]

$= x \vee (0 \wedge y)$ [Distributive law]

$= x \vee (y \wedge 0)$ [Commutative law]

$= x \vee 0$ [Bound law]

$= x$ [Identity law]

Therefore, $x \wedge (x \vee y) = x$

Again, $x \vee (x \wedge y) = (x \wedge 1) \vee (x \wedge y)$ [Identity law]

$= x \wedge (1 \vee y)$ [Distributive law]

$= x \wedge (y \vee 1)$ [Commutative law]

$= x \wedge 1$ [Bound law]

$= x$ [Identity law]

Therefore, $x \vee (x \wedge y) = x$

(*d*) $x' \vee x = x \vee x'$ [Commutative law]

$= 1$ [Complement law]

i.e., $x' \vee x = 1$

Also, $x' \wedge x = x \wedge x'$ [Commutative law]

$= 0$ [Complement law]

i.e., $x' \wedge x = 0$

Thus we have $x' \vee x = 1$ and $x' \wedge x = 0$

Therefore, $x = (x')'$ *i.e.*, $(x')' = x$

(*e*) We know that $(0 \vee 1) = (1 \vee 0) = 1$

i.e., $(0 \vee 1) = 1$

Again by Theorem $(0 \wedge 1) = (1 \wedge 0) = 0$

Thus we have $(0 \vee 1) = 1$ and $(0 \wedge 1) = 0$

Therefore, $1 = 0'$ and $0' = 1$

Similarly we also have $(1 \vee 0) = 1$ and $(1 \wedge 0) = 0$

Therefore, $0 = 1'$ and $1' = 0$

(*f*) Let $a = (x \wedge y)$ and $b = (x' \vee y')$

Now
$$
\begin{aligned}
(a \vee b) &= (x \wedge y) \vee b \\
&= (x \vee b) \wedge (y \vee b) && \text{[Distributive law]} \\
&= (x \vee (x' \vee y')) \wedge (y \vee (x' \vee y')) \\
&= ((x \vee x') \vee y') \wedge (y \vee (x' \vee y')) && \text{[Associative law]} \\
&= (1 \vee y') \wedge (y \vee (x' \vee y')) && \text{[Complement law]} \\
&= (1 \vee y') \wedge (y \vee (y' \vee x')) && \text{[Commutative law]} \\
&= (1 \vee y') \wedge ((y \vee y') \vee x') && \text{[Associative law]} \\
&= (1 \vee y') \wedge (1 \vee x') && \text{[Complement law]} \\
&= 1 \wedge 1 && \text{[Bound law]} \\
&= 1 && \text{[Idempotent law]}
\end{aligned}
$$

Again,
$$
\begin{aligned}
(a \wedge b) &= (x \wedge y) \wedge (x' \vee y') \\
&= ((x \wedge y) \wedge x') \vee ((x \wedge y) \wedge y') && \text{[Distributive law]} \\
&= ((y \wedge x) \wedge x') \vee ((x \wedge y) \wedge y') && \text{[Commutative law]} \\
&= (y \wedge (x \wedge x')) \vee (x \wedge (y \wedge y')) && \text{[Associative law]} \\
&= (y \wedge 0) \vee (x \wedge 0) && \text{[Complement law]} \\
&= 0 \vee 0 && \text{[Bound law]} \\
&= 0 && \text{[Idempotent law]}
\end{aligned}
$$

Therefore, $(a \vee b) = 1$ and $(a \wedge b) = 0$

This implies that $b = a'$ *i.e.*, $a' = b$

i.e., $(x \wedge y)' = (x' \vee y')$

Similarly the other de Morgan's law $(x \vee y)' = (x' \wedge y')$ can be proved.

■ 10.7 DUAL OF A STATEMENT

The dual of a statement involving Boolean expressions is obtained by replacing 0 by 1, 1 by 0, $\wedge$ by $\vee$, and $\vee$ by $\wedge$. Two Boolean expressions are said to be dual of each other if one expression is obtained from other by replacing 0 by 1, 1 by 0, $\wedge$ by $\vee$, and $\vee$ by $\wedge$.

Consider the statement $(x \wedge y)' = x' \vee y'$. The dual of above statement is $(x \vee y)' = x' \wedge y'$. Similarly the Boolean expressions $(x \wedge 1) = x$ and $(x \vee 0) = x$ are dual of each other.

10.7.1 Theorem

In Boolean algebra, the dual of a theorem is also a theorem.

Proof: Suppose that T is a theorem in Boolean algebra. Then there is a proof P of T involving definitions of a Boolean algebra. Let P_1 be the sequence of statements obtained by replacing 0 by 1, 1 by 0, $\wedge$ by $\vee$ and $\vee$ by $\wedge$. Then P_1 is a proof of the dual of T.

10.8 BOOLEAN FUNCTION

Let B = (S, $\vee$, $\wedge$, ′, 0, 1) be a Boolean algebra and let X $(x_1, x_2, x_3, \ldots, x_n)$ be a Boolean expression in '*n*' variables. A function $f: B^n \to B$ is called a Boolean function if f is of the form

$$f(x_1, x_2, x_3, \ldots, x_n) = X(x_1, x_2, x_3, \ldots, x_n)$$

Let us consider the example of a Boolean function f: $B^3 \to B$; B = {0, 1} defined by

$$f(x_1, x_2, x_3) = x_1 \wedge (x_2 \vee \bar{x}_3)$$

The inputs and outputs are given in the following table.

x_1	x_2	x_3	$f(x_1, x_2, x_3)$
1	1	1	1
1	1	0	1
1	0	1	0
0	1	1	0
1	0	0	1
0	1	0	0
0	0	1	0
0	0	0	0

10.8.1 Representations of Boolean Functions

We have seen that Boolean functions are nothing but the evaluation functions of Boolean expressions. It is also to be noted that two Boolean expressions give rise to the same evaluation function if and only if they are equivalent. Therefore, we identify a Boolean function with any of the equivalent Boolean expressions, whose evaluation function gives it.

This gives rise to the representation of a Boolean function. There are several ways for representing Boolean functions. These are

(*a*) Tabular Representation
(*b*) *n* Space Representation
(*c*) Cube Representation

Here we will discuss only tabular representation.

Tabular Representation: We know that, a Boolean function is completely determined by its evaluation over any Boolean algebra. In tabular representation, the procedure is very clear. We consider a row R of the table where the output is 1. We then form the combination $(x_1 \wedge x_2 \wedge x_3 \wedge \ldots \wedge x_n)$ and place a bar over each x_i whose value is 0 in row R. The combination formed is 1 if and only if x_i have the values given in row R. We thus OR the terms to obtain the Boolean expression.

To clear the procedure let us consider the Boolean function given by the following table.

x_1	x_2	x_3	$f(x_1, x_2, x_3)$
1	1	1	1 ← Row 1
1	1	0	0
1	0	1	1 ← Row 3
0	1	1	0
1	0	0	0
0	1	0	1 ← Row 6
0	0	1	0
0	0	0	0

From the table it is clear that, the output is 1 for the rows 1, 3 and 6. Consider the first row of the table and the combination is $(x_1 \wedge x_2 \wedge x_3)$ as $x_1 = x_2 = x_3 = 1$. Similarly for third row of the table we may construct the combination $(x_1 \wedge \overline{x_2} \wedge x_3)$ as $x_1 = 1, x_2 = 0, x_3 = 1$. Thus for sixth row the combination is $(\overline{x_1} \wedge x_2 \wedge \overline{x_3})$.

Therefore, the Boolean function $f(x_1, x_2, x_3)$ is given as

$$f(x_1, x_2, x_3) = (x_1 \wedge x_2 \wedge x_3) \vee (x_1 \wedge \overline{x_2} \wedge x_3) \vee (\overline{x_1} \wedge x_2 \wedge \overline{x_3}).$$

10.9 VARIOUS NORMAL FORMS

In this section we will discuss about two normal forms *i.e.*, disjunctive normal form and conjunctive normal form.

10.9.1 Disjunctive Normal Form

A Boolean function $f: B^n \to B$ which consists of a sum of elementary products is called the disjunctive normal form of the given function f.

Let $f: B^n \to B$ is a Boolean function. If f is not identically zero, let $A_1, A_2, A_3, \ldots, A_k$ denote the elements A_i of B_2^n, for which $f(A_i) = 1$,

where, $A_i = (a_1, a_2, \ldots, a_n)$.

For each A_i set $m_i = (y_1 \wedge y_2 \wedge y_3 \wedge \ldots \wedge y_n)$

where, $y_i = \begin{cases} x_i & \text{if } a_i - 1 \\ x_i{}' & \text{if } a_i = 0 \end{cases}$

Then, $f(x_1, x_2, x_3, \ldots, x_n) = m_1 \vee m_2 \vee m_3 \vee \ldots \vee m_k$. This representation of a Boolean function is called the disjunctive normal form.

Let us consider the Boolean function $(x_1 \oplus x_2)$. The truth table for this function is given below.

x_1	x_2	$(x_1 \oplus x_2)$
1	1	0
1	0	1
0	1	1
0	0	0

The disjunctive normal form of this function is given as

$$(x_1 \oplus x_2) = (x_1 \wedge \bar{x}_2) \vee (\bar{x}_1 \wedge x_2)$$

10.9.2 Conjunctive Normal Form

A Boolean function $f: B^n \to B$ which consists of a product of elementary sums is called the conjunctive normal form of the given function f.

Let $f: B^n \to B$ is a Boolean function. If f is not identically one, let $A_1, A_2, A_3, \ldots, A_k$ denote the elements A_i of B_2^n, for which $f(A_i) = 0$,

where, $$A_i = (a_1, a_2, a_3, \ldots, a_n).$$

For each A_i set

$$M_i = (y_1 \vee y_2 \vee y_3 \vee \ldots \vee y_n)$$

where, $$y_i = \begin{cases} x_i & \text{if } a_i = 0 \\ x_i' & \text{if } a_i = 1 \end{cases}$$

Then, $f(x_1, x_2, x_3, \ldots, x_n) = M_1 \wedge M_2 \wedge M_3 \wedge \ldots \wedge M_k$. This representation of a Boolean function is called the conjunctive normal form.

Let us consider the Boolean function $(x_1 \oplus x_2)$. The truth table for this function is given below.

x_1	x_2	$(x_1 \oplus x_2)$
1	1	0
1	0	1
0	1	1
0	0	0

From the table it is clear that, the output is 0 for the rows 1 and 4. Consider the first row of the table and the combination is $(\bar{x}_1 \vee \bar{x}_2)$. Similarly for the fourth row the combination is $(x_1 \vee x_2)$. So the conjunctive normal form for this function is given as

$$(x_1 \oplus x_2) = (\bar{x}_1 \vee \bar{x}_2) \wedge (x_1 \vee x_2)$$

Note: A term of the form $(y_1 \wedge y_2 \wedge y_3 \wedge \ldots \wedge y_n)$, where each y_i is either x_i or $\bar{x}_i$ is called a minterm where as a term of the form $(y_1 \vee y_2 \vee y_3 \vee \ldots \vee y_n)$, where each y_i is either x_i or $\bar{x}_i$ is called a maxterm.

SOLVED EXAMPLES

Example 1 *Construct an AND gate using three NOR gates.*

Solution: The output to an AND gate is $(x \wedge y)$, if the inputs are x and y. The output to a NOR gate is $\left(\overline{x \vee y}\right)$, if the inputs are x and y. The gating network is given further:

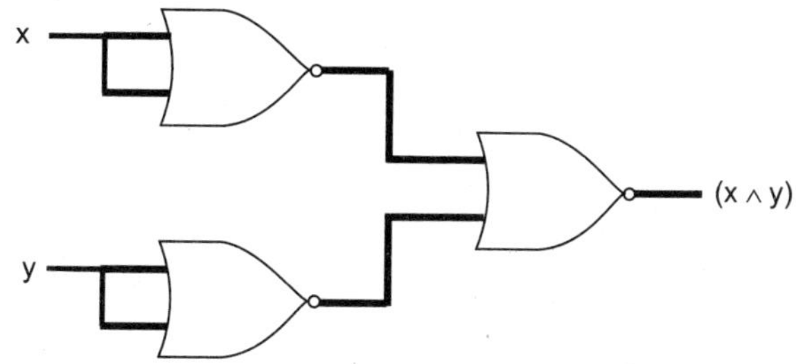

From the diagram given above it is clear that the output to the first NOR gate is $(\overline{x \vee x}) = \bar{x}$. Similarly the output to the second NOR gate is $(\overline{y \vee y}) = \bar{y}$. Therefore, the output to the final NOR gate is $(x \wedge y)$.

Example 2 *Construct an OR gate using three NAND gates.*

Solution: The output to an OR gate is $(x \vee y)$, if the inputs are x and y. The output to an NAND gate is $(\overline{x \wedge y})$, if the inputs are x and y. The gating network is given as below.

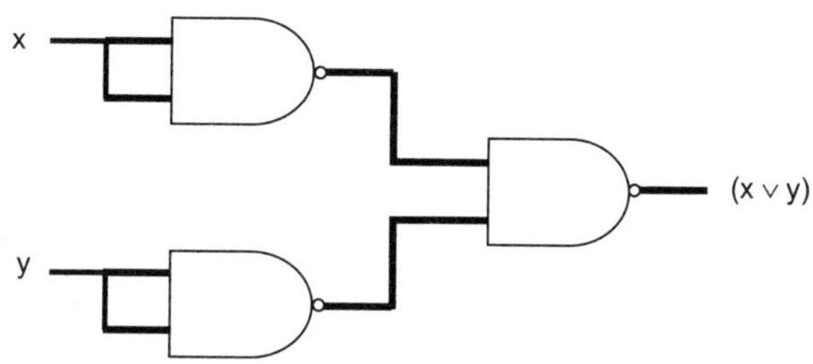

Example 3 *Describe a gating network corresponding to the statement (x . y) + (y . z) + (z . x).*

Solution: Given statement is $(x \cdot y) + (y \cdot z) + (z \cdot x)$. The gating network is given as

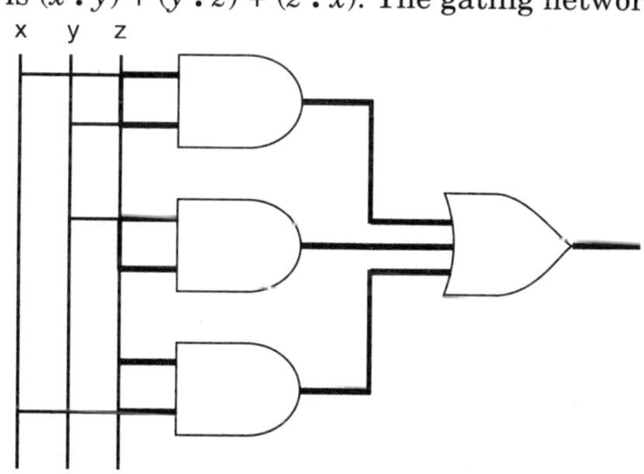

Example 4 *Describe a gating network corresponding to the statement*

$$(\overline{x+y})(z \cdot u)+(\overline{x \cdot y})(z+u)$$

Solution: Given statement is $(\overline{x+y})(z \cdot u)+(\overline{x \cdot y})(z+u)$. The gating network is given as below.

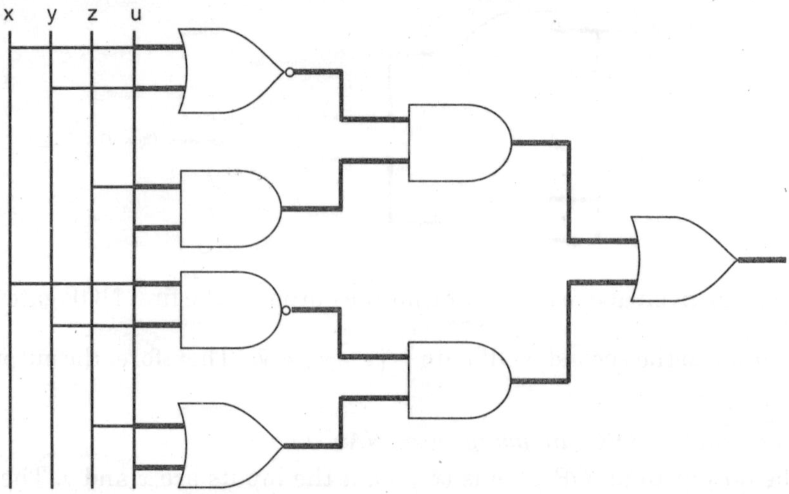

Example 5 *Describe the output of the following gating network.*

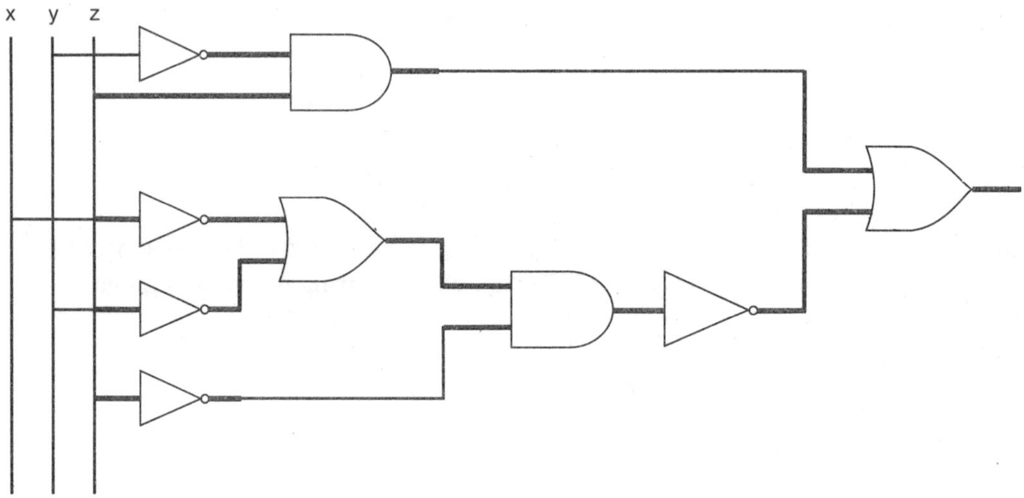

Solution: Consider the gating network given above. The output to the above gating network is given as

$$(\bar{y}.z) + (\overline{(\bar{x}+\bar{y}).\bar{z}}) = \bar{y}z + (\overline{\bar{x}+\bar{y}}) + \bar{\bar{z}} \qquad \text{[de Morgan's law]}$$

$$= \bar{y}z + \bar{\bar{x}}.\bar{\bar{y}} + z \qquad \text{[de Morgan's law]}$$

$$= \bar{y}z + xy + z$$

Example 6 *Construct a gating network using inverter and OR gate corresponding to the statement* $(x \,.\, y) + (y \,.\, z) + (z \,.\, x)$.

Solution: Given statement is $(x \,.\, y) + (y \,.\, z) + (z \,.\, x)$. The gating network is given futher.

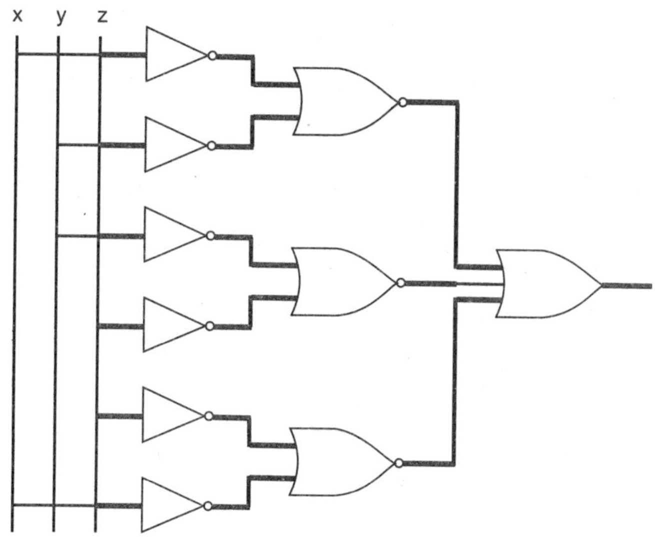

Example 7 *Find the value of the Boolean expression given below for $x = 1$, $y = 1$ and $z = 0$.*

$$(x \wedge (y \vee (x \wedge \overline{y}))) \vee ((x \wedge \overline{y}) \vee (\overline{x \wedge \overline{z}}))$$

Solution: Given that the value of the inputs are $x = 1$, $y = 1$ and $z = 0$. Now, the value of $(x \wedge \overline{y})$ is 0.

The value of $(y \vee (x \wedge \overline{y}))$ is 1

The value of $(x \wedge (y \vee (x \wedge \overline{y})))$ is 1

Similarly, the value of the $(\overline{x \wedge \overline{z}})$ is 0

The value of $((x \wedge \overline{y}) \vee (\overline{x \wedge \overline{z}}))$ is 0

So, the value of the Boolean expression

$(x \wedge (y \vee (x \wedge \overline{y}))) \vee ((x \wedge \overline{y}) \vee (\overline{x \wedge \overline{z}}))$ is 1.

Example 8 *Construct an AND gate using inverters and three NOR gates.*

Solution: Output to an AND gate is $(x \wedge y)$ or xy, if the inputs are x and y. The output to a NOR gate is $\overline{(x \vee y)}$, if the inputs are x and y. The gating network is given below.

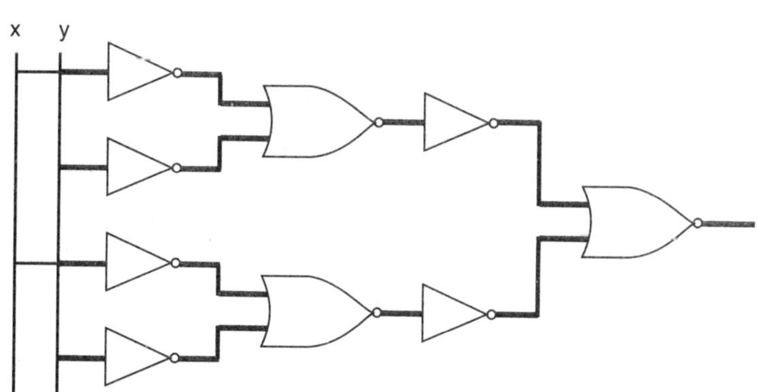

Example 9 *Write the Boolean expression that represents the combinatorial circuit, write the logic table and write the output of each gate symbolically.*

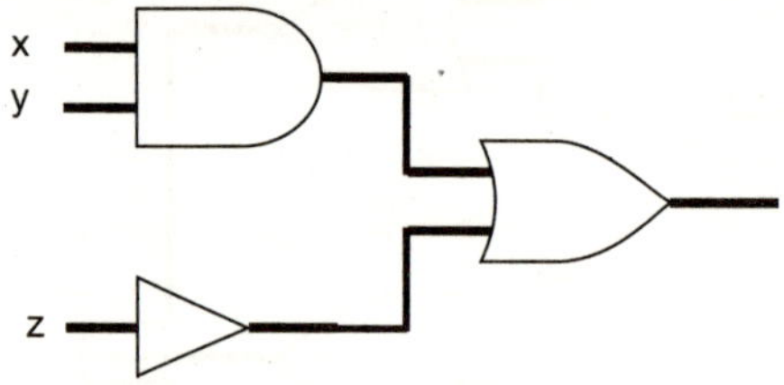

Solution: Given the gating network as below.

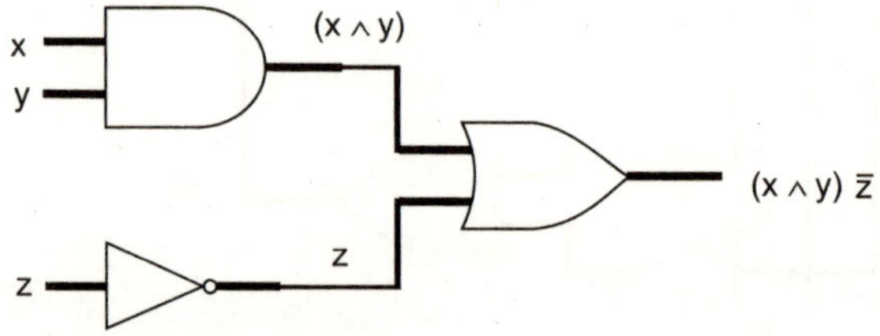

The Boolean expression that represents the combinatorial circuit is $((x \wedge y) \vee \bar{z})$. The logic table is given as below.

x	y	z	$(x \wedge y)$	$(x \wedge y) \vee \bar{z}$
1	1	1	1	1
1	1	0	1	1
1	0	1	0	0
0	1	1	0	0
1	0	0	0	1
0	1	0	0	1
0	0	1	0	0
0	0	0	0	1

Example 10 *If $(x + y) = (x + z)$ and $(x' + y) = (x' + z)$, then $y = z$.*

Solution: Given that $(x + y) = (x + z)$ *i.e.,* $(x \vee y) = (x \vee z)$

And $(x' + y) = (x' + z)$ *i.e.,* $(x' \vee y) = (x' \vee z)$

Now,

$$\begin{aligned}
y &= y \vee 0 && \text{[Identity law]}\\
&= y \vee (x \wedge x') && \text{[Complement law]}\\
&= (y \vee x) \wedge (y \vee x') && \text{[Distributive law]}\\
&= (x \vee y) \wedge (x' \vee y) && \text{[Commutative law]}\\
&= (x \vee z) \wedge (x' \vee z) && \text{[Given condition]}\\
&= (z \vee x) \wedge (z \vee x') && \text{[Commutative law]}\\
&= z \vee (x \wedge x') && \text{[Distributive law]}\\
&= z \vee 0 && \text{[Complement law]}\\
&= z && \text{[Identity law]}
\end{aligned}$$

Therefore, $y = z$.

Example 11 *Given the Boolean function f, write f in its disjunctive normal form.*

x	y	z	$f(x, y, z)$
1	1	1	1
1	1	0	1
1	0	1	0
0	1	1	0
1	0	0	0
0	1	0	1
0	0	1	0
0	0	0	1

Solution: From the table given below it is clear that, the output is 1 for the rows 1, 2, 6 and 8. For the first row the combination is $(x \wedge y \wedge z)$. Similarly for rows 2, 6 and 8 the combinations are $(x \wedge y \wedge \bar{z})$, $(\bar{x} \wedge y \wedge z)$ and $(\bar{x} \wedge \bar{y} \wedge \bar{z})$ respectively.

Thus, the disjunctive normal form to the above function f is given as

$$f(x, y, z) = (x \wedge y \wedge z) \vee (x \wedge y \wedge \bar{z}) \vee (\bar{x} \wedge y \wedge z) \vee (\bar{x} \wedge y \wedge z)$$

x	y	z	$f(x, y, z)$
1	1	1	1← Row 1
1	1	0	1 ← Row 2
1	0	1	0
0	1	1	0
1	0	0	0
0	1	0	1← Row 6
0	0	1	0
0	0	0	1← Row 8

Example 12 *Show that the combinatorial circuits (a) and (b) are equivalent.*

(*a*)

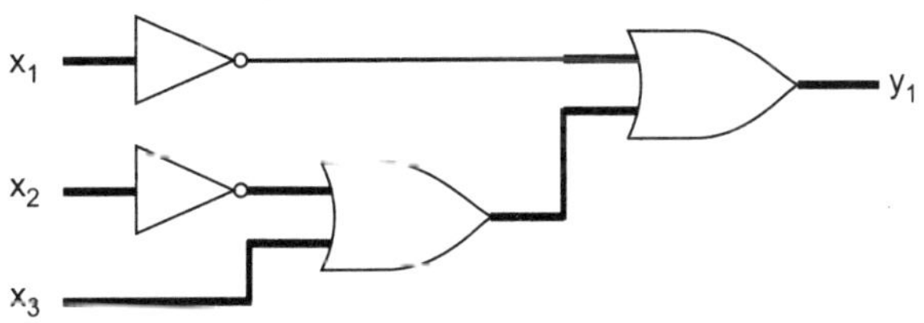

(*b*)

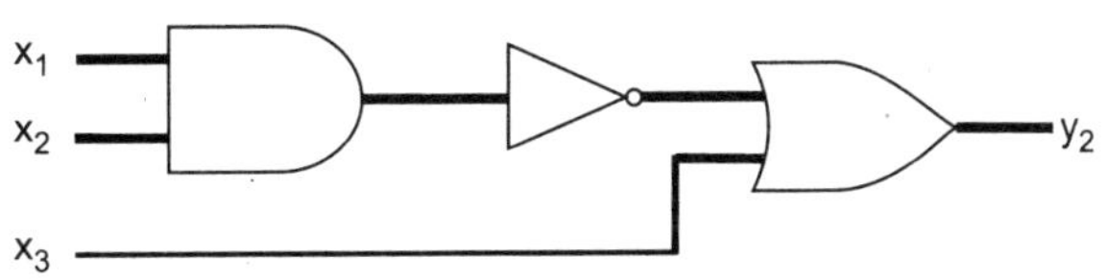

Solution: Given combinatorial circuits are

(*a*)

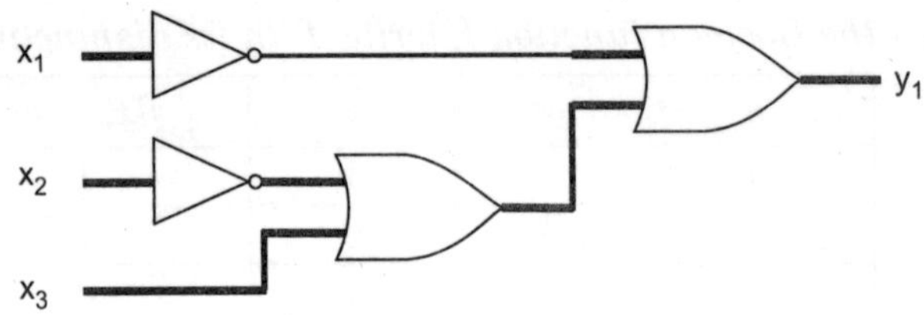

(*b*)

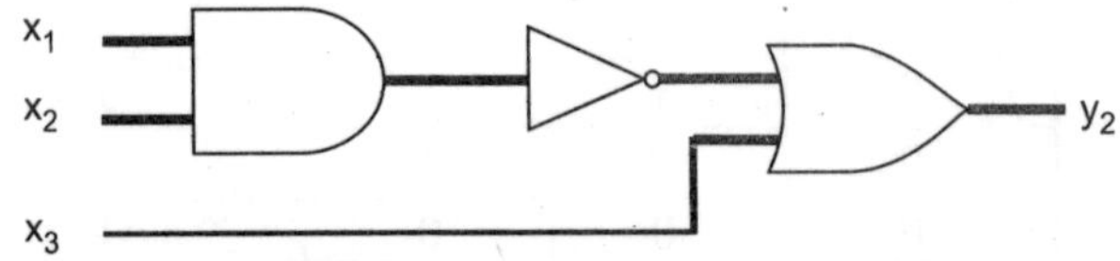

The output y_1 for combinatorial circuit (a) is given as

$$y_1 = \bar{x}_1 \vee (\bar{x}_2 \vee x_3) = (\bar{x}_1 \vee \bar{x}_2) \vee x_3 = \left(\overline{x_1 \wedge x_2}\right) \vee x_3$$

The output y_2 for combinatorial circuit (b) is given as $y_2 = \left(\overline{x_1 \wedge x_2}\right) \vee x_3$. Hence, the combinatorial circuits (a) and (b) are equivalent.

Example 13 *Reduce the following Boolean products to either 0 or a fundamental product.*

(a) $x\,y\,x'z$ (b) $x\,y\,z'y\,x$

Solution: (a)

$$\begin{aligned} x\,y\,x'z &= x\,x'\,y\,z && \text{[Commutative law]} \\ &= 0\,y\,z && \text{[Complement law]} \\ &= 0 && \text{[Bound law]} \end{aligned}$$

i.e., $x\,y\,x'z = 0$

(b)

$$\begin{aligned} x\,y\,z'y\,x &= x\,y\,y\,z'\,x && \text{[Commutative law]} \\ &= x\,y\,z'\,x && \text{[Idempotent law]} \\ &= x\,y\,x\,z' && \text{[Commutative law]} \\ &= x\,x\,y\,z' && \text{[Commutative law]} \\ &= x\,y\,z' && \text{[Idempotent law]} \end{aligned}$$

i.e., $x\,y\,z'y\,x = x\,y\,z'$

Example 14 *Given the Boolean function f, write f in its conjunctive normal form.*

x	y	z	$f(x, y, z)$
1	1	1	1
1	1	0	1
1	0	1	0
0	1	1	0
1	0	0	0
0	1	0	1
0	0	1	0
0	0	0	1

Solution: Given the Boolean function f as below.

x	y	z	$f(x, y, z)$
1	1	1	1
1	1	0	1
1	0	1	0 ← Row 3
0	1	1	0 ← Row 4

Contd...

1	0	0	0 ← Row 5
0	1	0	1
0	0	1	0 ← Row 7
0	0	0	1

From the table it is clear that, the output is 0 for the rows 3, 4, 5 and 7. For the third row the combination is $(\bar{x} \vee y \vee \bar{z})$. Similarly for rows 4, 5 and 7 the combinations are $(x \vee \bar{y} \vee \bar{z})$, $(\bar{x} \vee y \vee z)$ and $(x \vee y \vee \bar{z})$ respectively.

Thus, the conjunctive normal form to the above function is given as

$$f(x, y, z) = (\bar{x} \vee y \vee \bar{z}) \wedge (x \vee \bar{y} \vee \bar{z}) \wedge (\bar{x} \vee y \vee z) \wedge (x \vee y \vee \bar{z}).$$

Example 15 *Design a combinatorial circuit that computes exclusive OR; XOR of x and y.*

Solution: Let the inputs to the XOR gate be x and y. The logic table for XOR gate is given below.

x	y	$x \oplus y$
1	1	0
1	0	1
0	1	1
0	0	0

So, the disjunctive normal form of this function is given as

$$x \oplus y = (x \wedge y') \vee (x' \wedge y)$$

The combinatorial circuit corresponding to $(x \oplus y)$ is given below.

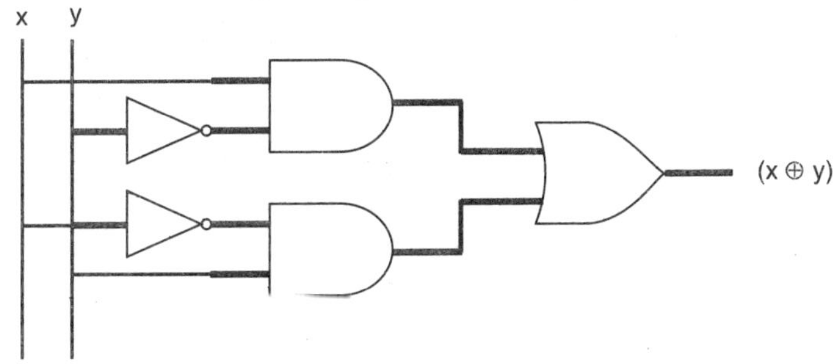

Example 16 *Find the disjunctive and conjunctive normal form of the given function and draw the combinatorial circuit corresponding to the disjunctive normal form.*

x	y	z	$f(x, y, z)$
1	1	1	0
1	1	0	0
1	0	1	0
0	1	1	1
1	0	0	1
0	1	0	1
0	0	1	1
0	0	0	0

Solution: Given Boolean function is

x	y	z	$f(x, y, z)$
1	1	1	0
1	1	0	0
1	0	1	0
1	0	0	1← Row 4
0	1	1	1← Row 5
0	1	0	1← Row 6
0	0	1	1← Row 7
0	0	0	0

From the table it is clear that the output is 1 for rows 4, 5, 6 and 7. For the fourth row the combination is $(x \wedge y' \wedge z')$. Similarly the combinations $(x' \wedge y \wedge z)$, $(x' \wedge y \wedge z')$, and $(x' \wedge y' \wedge z)$ are for rows 5, 6 and 7 respectively. So, the disjunctive normal form to the above function is given as

$$f(x, y, z) = (x \wedge y' \wedge z') \vee (x' \wedge y \wedge z) \vee (x' \wedge y \wedge z') \vee (x' \wedge y' \wedge z).$$

Similarly, corresponding to the output 0 for rows 1, 2, 3 and 8, the conjunctive normal form to the above function is given as

$$f(x, y, z) = (x' \vee y' \vee z') \wedge (x' \vee y' \vee z) \wedge (x' \vee y \vee z') \wedge (x \vee y \vee z).$$

The combinatorial circuit corresponding to the disjunctive normal form is given below.

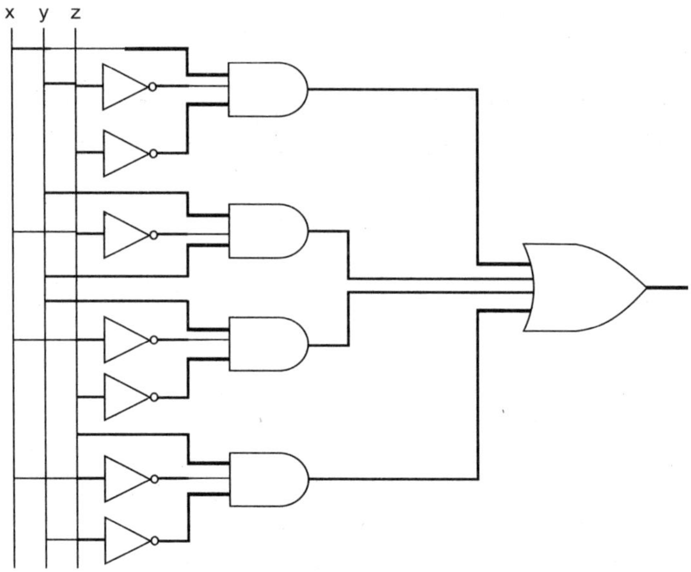

Example 17 *Find the disjunctive normal form of the function using algebraic technique.*

$$f(x, y) = (x \vee y) \wedge (x' \vee y')$$

Solution:

$$f(x, y) = (x \vee y) \wedge (x' \vee y')$$

$= (x \wedge (x' \vee y')) \vee (y \wedge (x' \vee y'))$ [Distributive law]

$= (x \wedge x') \vee (x \wedge y') \vee (y \wedge x') \vee (y \wedge y')$ [Distributive law]

$= 0 \vee (x \wedge y') \vee (y \wedge x') \vee 0$ [Complement law]

$= (x \wedge y') \vee (y \wedge x')$ [Identity law]

i.e., $f(x, y) = (x \wedge y') \vee (y \wedge x')$

Which is the disjunctive normal form of the function $f(x, y)$.

Example 18 *Find the disjunctive normal form for the following combinatorial circuit.*

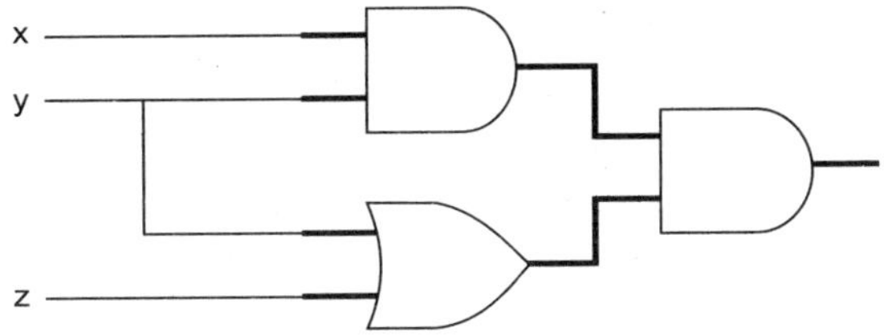

Solution: Given that the combinatorial circuit as

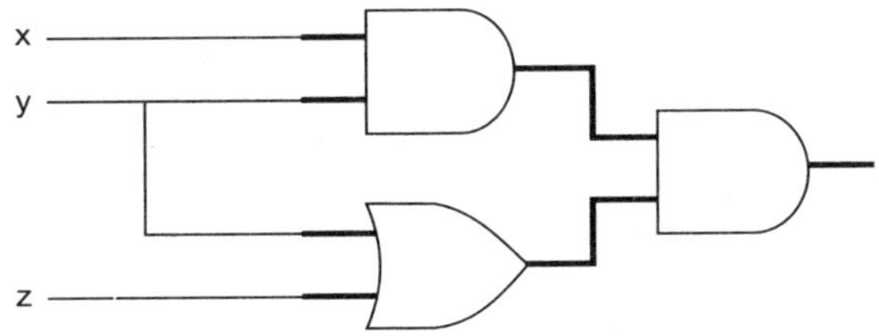

The output of the above combinatorial circuit is given as $f(x, y, z) = (x \wedge y) \wedge (y \vee z)$. The logic table for the above expression is given below. From the table it is clear that the function has output 1 for rows 1 and 2. For the first row the combination is $(x \wedge y \wedge z)$ whereas for second row the combination is $(x \wedge y \wedge z')$. Thus, the disjunctive normal form for the above function is given as

$$f(x, y, z) = (x \wedge y \wedge z) \vee (x \wedge y \wedge z')$$

x	y	z	$(x \wedge y)$	$(y \vee z)$	$(x \wedge y) \wedge (y \vee z)$
1	1	1	1	1	1
1	1	0	1	1	1
1	0	1	0	1	0
0	1	1	0	1	0
1	0	0	0	0	0
0	1	0	0	1	0
0	0	1	0	1	0
0	0	0	0	0	0

EXERCISES

1. Find the disjunctive normal form of each function using algebraic technique.

(*a*) $f(x, y) = x \vee (x \wedge y)$

(*b*) $f(x, y, z) = x \vee y \wedge (x \vee z')$

(*c*) $f(x, y, z) = x \vee (y' \vee (x\, y' \vee x\, z'))$

2. Reduce the following Boolean products to either 0 or a fundamental product.

(*a*) $x\, y\, z\, y$ (*b*) $x\, y\, z'\, y\, x'\, z'$ (*c*) $x\, y'\, z\, x\, y'$ (*d*) $x\, y\, z'\, t\, y'\, t$

(*e*) $x\, y'\, x\, z'\, t\, y'$

3. Write the logic table for the circuit given below.

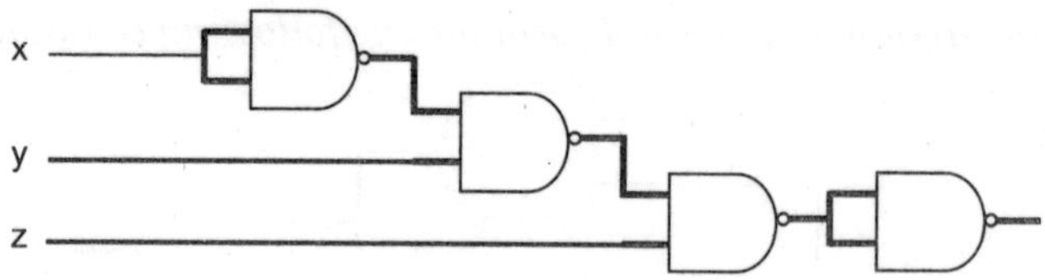

4. Find the disjunctive normal form of a Boolean expression having a logic table the same as the given table and draw the combinatorial circuit corresponding to the disjunctive normal form.

x	y	z	f
1	1	1	1
1	1	0	1
1	0	1	0
0	1	1	0
1	0	0	1
0	1	0	0
0	0	1	1
0	0	0	0

5. Are the combinatorial circuits equivalent? Explain.

(*i*) (*a*)

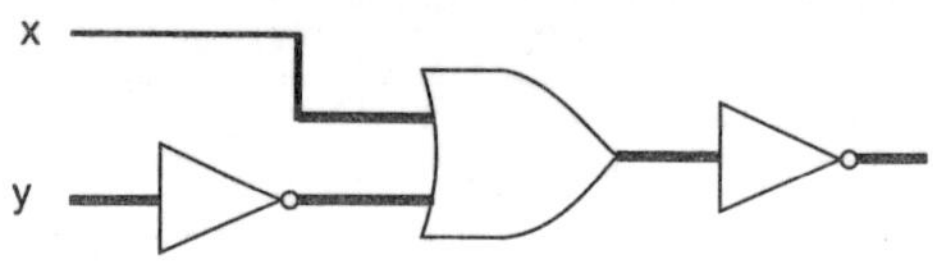

(*b*)

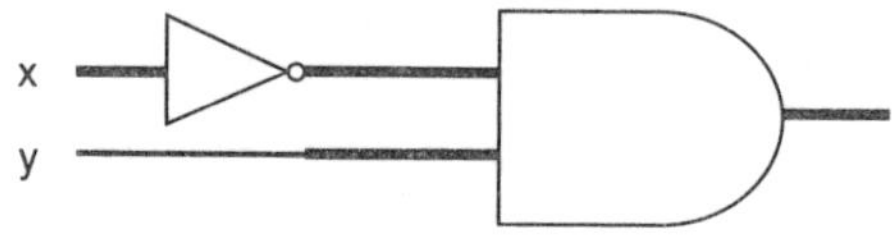

(*ii*) (*a*)

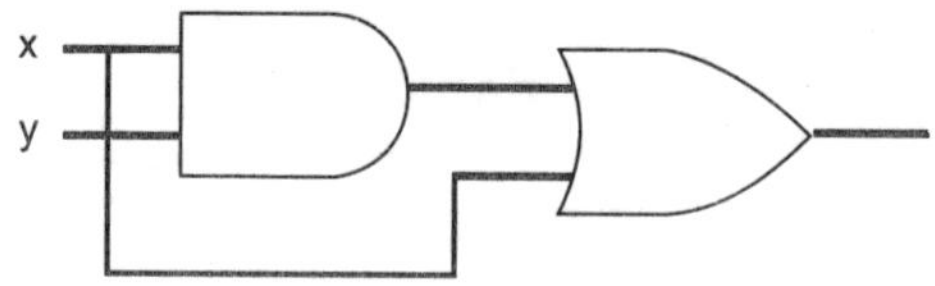

(*b*)

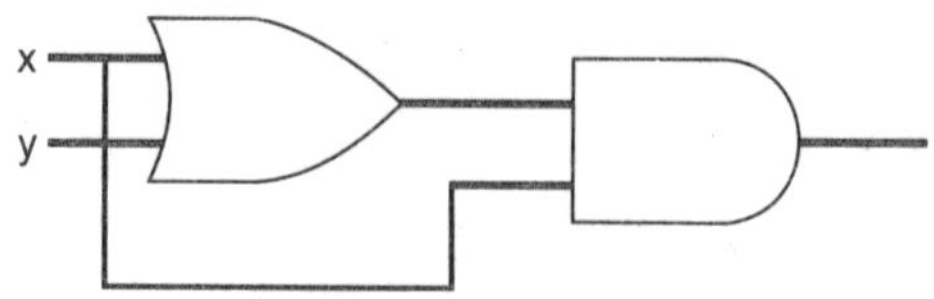

(*iii*) (*a*)

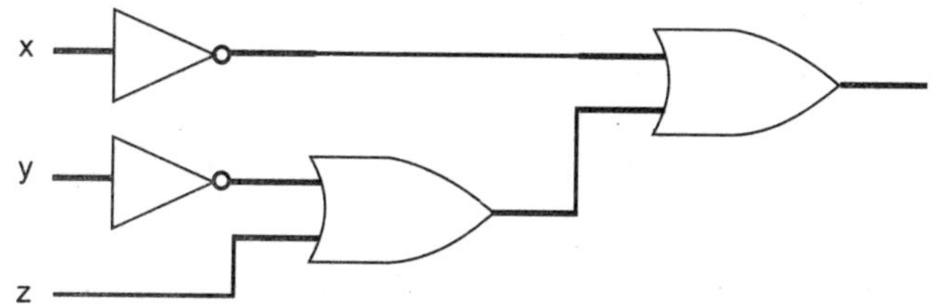

(*b*)

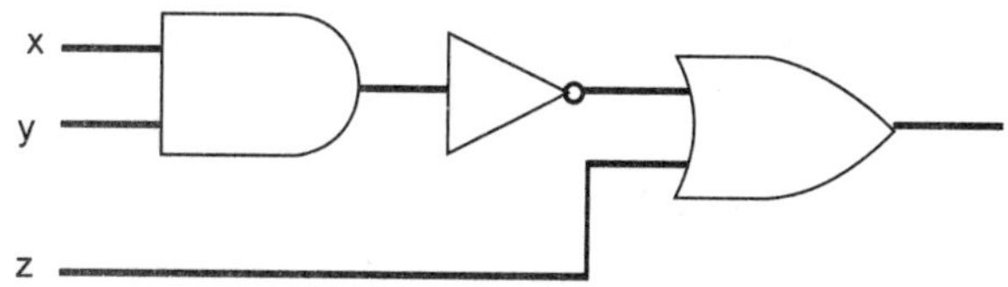

6. Find the Boolean expression in disjunctive normal form for the circuit given below.

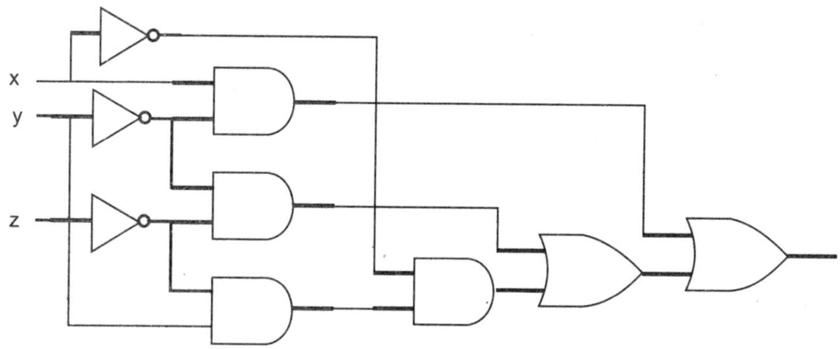

7. Find the disjunctive normal form of each function corresponding to the logic tables given below.

(*a*)

x	y	$f(x, y)$
1	1	1
1	0	0
0	1	1
0	0	1

(*b*)

x	y	$f(x, y)$
1	1	0
1	0	1
0	1	0
0	0	1

(c)

x	y	z	$f(x, y, z)$
1	1	1	1
1	1	0	0
1	0	1	1
0	1	1	1
1	0	0	0
0	1	0	0
0	0	1	0
0	0	0	1

(d)

x	y	z	$f(x, y, z)$
1	1	1	0
1	1	0	0
1	0	1	1
0	1	1	0
1	0	0	1
0	1	0	1
0	0	1	1
0	0	0	1

8. Find the conjunctive normal form of each function given in question 7.

9. Draw the logic circuit (Combinatorial circuit) with inputs x, y, z and output Y which corresponds to each Boolean expression.

(i) $Y = x'y\,z + x'y\,z' + x\,y\,z'$

(ii) $Y = x\,y'\,z + x\,z' + y'\,z$

10. Construct a combinatorial circuit that represents the following Boolean function.

x	y	z	$f(x, y, z)$
1	1	1	0
1	1	0	1
1	0	1	1
0	1	1	1
1	0	0	1
0	1	0	1
0	0	1	1
0	0	0	0

11. Write the dual of each Boolean equation.

(a) $(a \wedge 1) \vee (0 \vee a') = 0$

(b) $a \wedge (a' \vee b) = a \wedge b$

(c) $a \vee (a' \wedge b) = a \vee b$

(d) $(a \vee 1) \wedge (a \vee 0) = a$

(e) $(a \wedge a') \vee (a \wedge 0) = a$

(*f*) $(a \vee b) \wedge (b \vee c) = (a \wedge c) \vee b$

[**Hint:** To obtain the dual equation, interchange $\vee$ and $\wedge$, and interchange 0 and 1]

12. Discuss a XOR gate with four inputs x, y, z and t.

13. Express the following Boolean expression $f(x, y, z)$ as a sum of products and then in its complete sum- of- products form.

(*a*) $f(x, y, z) = x\,(x\,y' + x'y + y'\,z)$

(*b*) $f(x, y, z) = (x' + y)' + y'\,z$

(*c*) $f(x, y, z) = (x + y'\,z)\,(y + z')$

14. Express the output Y as a Boolean expression in the inputs x, y, z, t and u for the logic circuits given below.

(*a*)

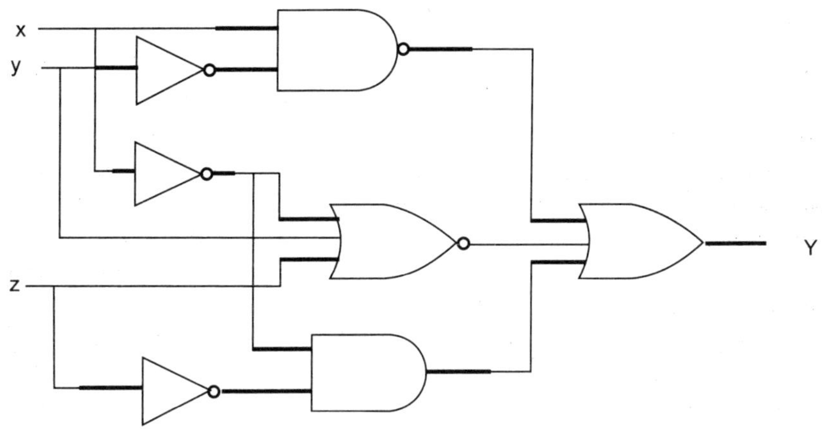

(*b*)

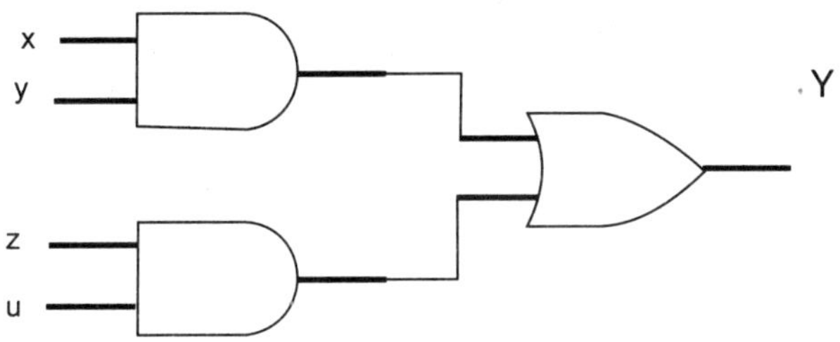

(*c*)

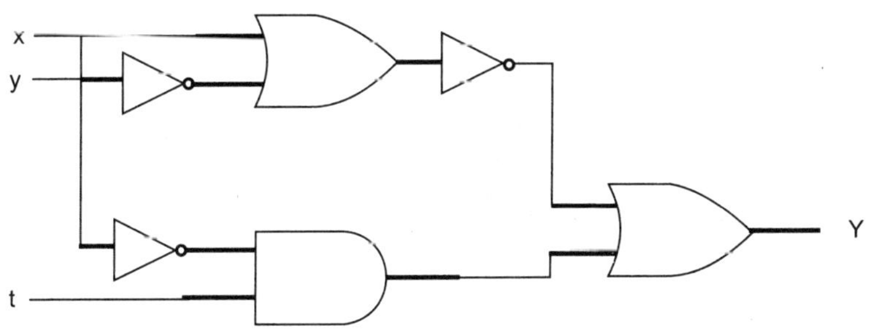

11

Introduction to Lattices

■ 11.0 INTRODUCTION

In this chapter, we shall introduce the fundamental concepts of Lattices. After defining them as particular kind of poset (partial ordered sets) we shall show that they could be introduced as algebraic systems possessing some specific properties. Here we will discuss Lattices, Duality principle, Distributed lattices, Bounded lattices, Complemented lattices and some special kind of lattices.

■ 11.1 LATTICES

Lattices is a partially ordered set (poset) in which every two elements have a unique least upper bound (L.U.B.) and a unique greatest lower bound (G.L.B.) *i.e.*, a lattice is a poset $\langle L, \leq \rangle$ in which every subset $\{a, b\}$ has a least upper bound and greatest lower bound. Where

$$\text{L.U.B}(\{a, b\}) = a \vee b \qquad \text{(join of } a \text{ and } b\text{)}$$

$$\text{G.L.B}(\{a, b\}) = a \wedge b \qquad \text{(meet of } a \text{ and } b\text{)}$$

Let us consider the poset $(N, \leq)$; where N is a set of natural numbers and $\leq$ is the ordinary less than or equal to relation. To show $(N, \leq)$ is a lattice, it is sufficient to define the L.U.B. and G.L.B. in N.

Now, Let $a, b \in N$

$$\text{L.U.B.}(\{a, b\}) = (a \vee b) = \text{Max}(a, b) \text{ and}$$

$$\text{G.L.B.}(\{a, b\}) = (a \wedge b) = \text{Min}(a, b)$$

Therefore, $(N, \leq)$ is a lattice.

11.1.1 Theorem

For a lattice $(B, \leq)$; $a, b \in B$

$$a \leq (a \vee b) \text{ and } (a \wedge b) \leq a.$$

Proof: Given (B, ≤) is a lattice and a , $b \in$ B

Now, $(a \vee b)$ = L.U.B. ({a, b})

i.e., $(a \vee b)$ is an upper bound of both a and b.

This implies that $a \leq (a \vee b)$.

Again, $(a \wedge b)$ = G.L.B. ({a, b})

i.e., $(a \wedge b)$ is the lower bound of both a and b.

This implies that $(a \wedge b) \leq a$.

■ 11.2 HASSE DIAGRAM

In principle, it is possible to draw a diagram which shows the order relation on a finite poset. Let (B, ≤) be a poset. Define the relation ≤ on B by x R y if and only if $x \leq y$ but $x \neq y$ for $x, y \in$ B. Given a partial order ≤ on B, 'y' is said to cover 'x' if $x < y$ and there is no element 'z' in B such that x R z and z R y.

A Hasse diagram of a poset (B, ≤) is a graphical representation consisting of points labeled by the members of B, with a line segment directed generally upward from x to wherever y covers x.

Let us consider B = D(12); set of all positive divisors of 12. Therefore, B = {1, 2, 3, 4, 6, 12}.

Let us define the relation x R y means x is a divisor of y for $x, y \in$ B. Thus we get

R = {(1, 2), (1, 3), (1, 4), (1, 6), (1, 12), (2, 4), (2, 6), (2, 12), (3, 6), (3, 12), (4, 12), (6, 12)}

From the above relation it is clear that 4 does not cover 1 because there exists 2 such that 1 R 2 and 2 R 4. Similarly, 6 does not cover 1, 12 does not cover 1, 12 does not cover 2 and 12 does not cover 3. Again it is also clear that 2 covers 1, 3 covers 1, 6 covers both 2 and 3 and 12 covers both 4 and 6. Therefore, the Hasse diagram is given below.

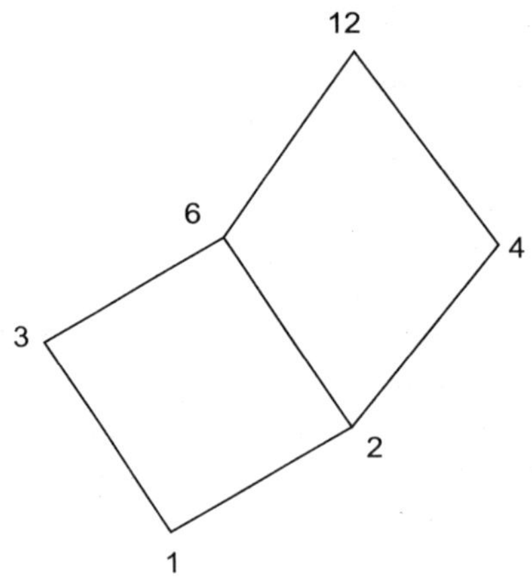

Note: We can distinguish lattices by looking at their Hasse diagrams. Because any two elements have a common predecessor and a common successor, the Hasse diagram of a lattice always is made up as a combination of closed polygons and thus its name lattice. A poset which has polygons open above or below is not a lattice because of lack of supremum or infimum.

Consider the following Hasse diagram.

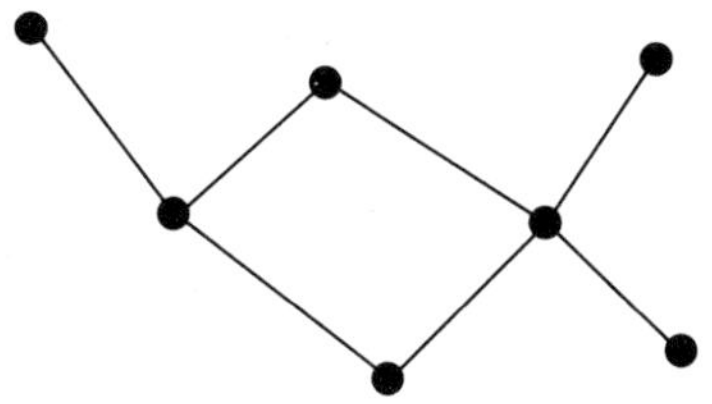

Figure 1

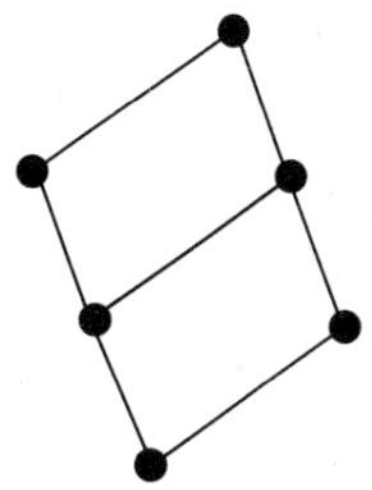

Figure 2

The above Hasse diagrams are posets, but figure 1 which is open above as well as open below is not a lattice whereas figure 2 is a lattice.

■ 11.3 PRINCIPLE OF DUALITY

Given a valid statement for a lattice we can obtain another valid statement by replacing the relation $\leq$ with $\geq$, join with meet and meet with join operation. This is known as the principles of duality in lattices.

11.3.1 Theorem

Let B be a lattice with $a, b, c \in$ B, then following properties holds.

1. **Idempotent Properties**
   (*a*) $(a \vee a) = a$
   (*b*) $(a \wedge a) = a$
2. **Commutative Properties**
   (*a*) $(a \vee b) = (b \vee a)$
   (*b*) $(a \wedge b) = (b \wedge a)$
3. **Associative Properties**
   (*a*) $a \vee (b \vee c) = (a \vee b) \vee c$
   (*b*) $a \wedge (b \wedge c) = (a \wedge b) \wedge c$
4. **Absorption Properties**
   (*a*) $a \vee (a \wedge b) = a$
   (*b*) $a \wedge (a \vee b) = a$

Proof:

1. **Idempotent Properties**
   (*a*) We know that $(a \vee b) =$ L.U.B. $(\{a, b\})$

This implies that $\quad a \leq (a \vee a) \quad$ ...(*i*)

Again, $\quad a \leq a$

This implies $\quad (a \vee a) \leq a \quad$ ...(*ii*)

Combining (*i*) and (*ii*) we will get

$$(a \vee a) = a$$

(*b*) Applying the principle of duality, we have,

$$(a \wedge a) = a$$

2. Commutative Properties

(*a*) We know that $(a \vee b) = \text{L.U.B.}(\{a, b\})$

and $(b \vee a) = \text{L.U.B.}(\{b, a\})$

$= \text{L.U.B.}(\{a, b\})$

$= (a \vee b)$

Therefore, $(a \vee b) = (b \vee a)$

(*b*) Applying the principle of duality, we have

$$(a \wedge b) = (b \wedge a)$$

3. Associative Properties

(*a*) Let $a \vee (b \vee c) = d$

and $(a \vee b) \vee c = e$

This implies that $a \le d$ and $(b \vee c) \le d$ $[\because a \le (a \vee b); b \le (a \vee b)]$

$\Rightarrow$ $a \le d, b \le d, c \le d$

$\Rightarrow$ $(a \vee b) \le d$ and $c \le d$

$\Rightarrow$ $(a \vee b) \vee c \le d$

i.e., $e \le d$ ... (*i*)

Again, $(a \vee b) \vee c = e$

$\Rightarrow$ $(a \vee b) \le e, c \le e$

$\Rightarrow$ $a \le e, b \le e, c \le e$

$\Rightarrow$ $a \le e, (b \vee c) \le e$

$\Rightarrow$ $a \vee (b \vee c) \le e$

$\Rightarrow$ $d \le e$ ... (*ii*)

Therefore from equations (*i*) and (*ii*) we have $d = e$

i.e., $a \vee (b \vee c) = (a \vee b) \vee c$

(*b*) Applying the principle of duality, we have

$$a \wedge (b \wedge c) = (a \wedge b) \wedge c$$

4. Absorption Properties

(*a*) We know that $a \vee (a \wedge b) = \text{L.U.B.}(\{a, a \wedge b\})$

This implies that $a \le a \vee (a \wedge b)$ ... (*i*)

Again, $a \le a$ and $(a \wedge b) \le a$ $[\because (a \wedge b) = \text{G.L.B.}(\{a, b\})]$

$\Rightarrow$ $a \vee (a \wedge b) \le (a \vee a) = a$

$\Rightarrow$ $a \vee (a \wedge b) \le a$ ... (*ii*)

Combining equations (*i*) and (*ii*) we get

$$a = a \vee (a \wedge b)$$

(*b*) Applying the principle of duality, we have

$$a \wedge (a \vee b) = a$$

11.3.2 Theorem

Let (B, $\leq$) be a lattice. For any a, b, c, d in Lattice B if $a \leq b$ and $c \leq d$, then $(a \vee c) \leq (b \vee d)$ and $(a \wedge c) \leq (b \wedge d)$.

Proof: Given (B, $\leq$) is a lattice and $a, b, c, d \in$ B.

Suppose that, $a \leq b$ and $c \leq d$

We know that $(b \vee d) =$ L.U.B. $(\{b, d\})$

This implies that $b \leq (b \vee d)$ and $d \leq (b \vee d)$

$\Rightarrow$ $a \leq (b \vee d)$ and $c \leq (b \vee d)$ $[\because\ a \leq b$ and $c \leq d]$

So, $(b \vee d)$ is an upper bound of a and c. Again $(a \vee c)$ is the least upper bound of a and c. Therefore,

$$(a \vee c) \leq (b \vee d)$$

Again, we know that $(a \wedge c) =$ G.L.B. $(\{a, c\})$

This implies that $(a \wedge c) \leq a$ and $(a \wedge c) \leq c$

$\Rightarrow$ $(a \wedge c) \leq b$ and $(a \wedge c) \leq d$ $[\because\ a \leq b$ and $c \leq d]$

Therefore, $(a \wedge c)$ is the lower bound of b and d. Again $(b \wedge d)$ is the greatest lower bound of b and d. Hence, we get

$$(a \wedge c) \leq (b \wedge d)$$

11.3.3 Theorem

Let (B, $\leq$) be a lattice. For any $a, b, c \in$ B we have

(*a*) If $a \leq b, a \leq c,$

then $a \leq (b \vee c)$ and $a \leq (b \wedge c)$

(*b*) If $a \geq b, a \geq c,$

then $a \geq (b \wedge c)$ and $a \geq (b \vee c)$

Proof: (*a*) Given that (B, $\leq$) be a lattice and $a, b, c \in$ B.

Suppose that $a \leq b, a \leq c$. This indicates that a is a lower bound of $\{b, c\}$.

Therefore, $a \leq$ G.L.B. $(\{b, c\}) = (b \wedge c)$

i.e., $a \leq (b \wedge c)$

Again, $(b \vee c) =$ L.U.B. $(\{b, c\})$

This implies that $b \leq (b \vee c)$

Also by hypothesis $a \leq b$

Therefore, we have $a \leq b \leq (b \vee c)$

i.e., $a \leq (b \vee c)$

i.e., $a \leq b, a \leq c,$

$\Rightarrow$ $a \leq (b \vee c)$ and $a \leq (b \wedge c)$

(*b*) On applying the principle of duality we can prove that if $a \geq b, a \geq c,$

then $a \geq (b \wedge c)$ and $a \geq (b \vee c)$

■ 11.4 DISTRIBUTIVE LATTICE

A lattice B is said to be distributive lattice if for $a, b, c \in$ B, it satisfies the following distributive laws.

(*a*) $a \wedge (b \vee c) = (a \wedge b) \vee (a \wedge c)$

(*b*) $a \vee (b \wedge c) = (a \vee b) \wedge (a \vee c)$

If a lattice is not distributive then it is called "non distributive" lattice.

11.4.1 Theorem

If the meet operation is distributive over the join operation in a Lattice, then the join operation is also distributive over the meet operation and vice versa.

Proof: Let (B, ≤) be a Lattice and the meet operation is distributive over the joint operation.

i.e., $a \wedge (b \vee c) = (a \wedge b) \vee (a \wedge c);\ a, b, c \in B$

Now, $(a \vee b) \wedge (a \vee c) = [(a \vee b) \wedge a] \vee [(a \vee b) \wedge c]$

$= a \vee [(a \vee b) \wedge c]$ [Absorption law]

$= a \vee [(a \wedge c) \vee (b \wedge c)]$

$= [a \vee (a \wedge c)] \vee (b \wedge c)$ [Associative law]

$= a \vee (b \wedge c)$ [Absorption law]

Therefore, $a \vee (b \wedge c) = (a \vee b) \wedge (a \vee c)$

11.4.2 Theorem

In any distributive lattice (B, ≤), the joint cancellation law holds.

i.e., If $(a \vee b) = (a \vee c)$ and $(a \wedge b) = (a \wedge c)$

then $b = c$

Proof: Suppose that $(a \vee b) = (a \vee c)$ and $(a \wedge b) = (a \wedge c)$

Now, $b = b \vee (a \wedge b)$ [Absorption law]

$= b \vee (a \wedge c)$ [Hypothesis]

$= (b \vee a) \wedge (b \vee c)$ [Distributive law]

$= (a \vee b) \wedge (b \vee c)$ [Commutative law]

$= (a \vee c) \wedge (b \vee c)$ [Hypothesis]

$= (a \wedge b) \vee c$ [Distributive law]

$= (a \wedge c) \vee c$ [Hypothesis]

$= c$ [Absorption law]

Therefore, $b = c$

11.4.3 Theorem

In a distributed lattice (B, ≤) the following equality holds for all $a, b, c \in B$

$$(a \wedge b) \vee (b \wedge c) \vee (c \wedge a) = (a \vee b) \wedge (b \vee c) \wedge (c \vee a)$$

Proof: Suppose that B be a distributive lattice with $a, b, c \in B$.

Now, $(a \wedge b) \vee (b \wedge c) \vee (c \wedge a) = [\{(a \wedge b) \vee b\} \wedge \{(a \wedge b) \vee c\}] \vee (c \wedge a)$

$= [\, b \wedge \{(a \wedge b) \vee c\}] \vee (c \wedge a)$ [Absorption law]

$= [\, b \wedge \{(a \vee c) \wedge (b \vee c)\}] \vee (c \wedge a)$ [Distributive law]

$= [(a \vee c) \wedge \{b \wedge (b \vee c)\}] \vee (c \wedge a)$ [Associative law]

$= [(a \vee c) \wedge b] \vee (c \wedge a)$ [Absorption law]

$= \{(a \vee c) \vee (c \wedge a)\} \wedge \{b \vee (c \wedge a)\}$ [Distributive law]

$= (c \vee a) \wedge (b \vee c) \wedge (b \vee a)$ [Distributive law]

$= (a \vee b) \wedge (b \vee c) \wedge (c \vee a)$

Therefore,

$$(a \wedge b) \vee (b \wedge c) \vee (c \wedge a) = (a \vee b) \wedge (b \vee c) \wedge (c \vee a).$$

■ 11.5 BOUNDED LATTICE

A lattice B is said to be bounded if it has a lower bound and an upper bound. The universal lower bound and the universal upper bound are denoted by 0 and 1 respectively in a bounded lattice.

11.5.1 Universal Lower Bound

Let (B, ≤) be a lattice. An element $a \in$ B is said to be universal lower bound if

$$a \leq b \;\forall\; b \in \text{B}.$$

11.5.2 Universal Upper Bound

Let (B, ≤) be a lattice. An element $a \in$ B is said to be universal upper bound if

$$b \leq a \;\forall\; b \in \text{B}.$$

11.5.3 Theorem

The universal lower bound and the universal upper bound are unique.

Proof: Let us first show that the universal lower bound is unique. Suppose two the contrary there exists two universal lower bound a and b of the lattice (B, ≤).

Therefore, $a, b \in \text{B}$

Now as 'a' is the universal lower bound we have

$$a \leq b \qquad ...(i)$$

Similarly, as 'b' is the universal lower bound we have

$$b \leq a \qquad ...(ii)$$

Hence, from equations (*i*) and (*ii*), we get

$$a = b$$

Therefore, our supposition is wrong. Thus, the universal lower bound is unique.

Similarly, it can be shown that the universal upper bound is also unique.

11.5.4 Theorem

In a bounded lattice (B, ≤), the universal upper and lower bounds 1 and 0 clearly satisfy the followings for any element $a \in$ B.

(*i*) $a \vee 1 = 1$ (*ii*) $a \wedge 1 = a$

(*iii*) $a \vee 0 = a$ (*iv*) $a \wedge 0 = 0$

Proof: (*i*) We know that for any lattice $(B, \leq)$

$$a \leq (a \vee b) \text{ for } a, b \in B$$

So, $$1 \leq (a \vee 1) \qquad \dots (i)$$

Again, since 1 is the universal upper bound

$$(a \vee 1) \leq 1 \qquad \dots (ii)$$

Combining (*i*) and (*ii*) we get $(a \vee 1) = 1$

(*ii*) We know that for any lattice $(B, \leq)$

$$(a \wedge b) \leq a \text{ for } a, b \in B$$

So, $$(a \wedge 1) \leq a \qquad \dots (i)$$

Again, since 1 is the universal upper bound we have $a \leq 1$. Also we know that $a \leq a$.

Therefore,

$$(a \wedge a) \leq (a \wedge 1)$$

i.e., $$a \leq (a \wedge 1) \qquad \dots (ii)$$

Combining (*i*) and (*ii*) we get $(a \wedge 1) = a$

Similarly, (*iii*) and (*iv*) can be proved.

■ 11.6 COMPLEMENTED LATTICE

A lattice $(B, \leq)$ is said to be complemented lattice if every element in the lattice has a complement. Let $(B, \leq)$ be a lattice with 0 and 1 as its universal lower bound and upper bound respectively. An element b is said to be complement of $a \in B$ if

$$(a \vee b) = 1 \text{ and } (a \wedge b) = 0$$

From the commutative property, if 'b' is complement of 'a' then 'a' is also complement of 'b'.

11.6.1 Theorem

In a bounded distributive lattice, if a complement exists then it is unique.

Proof: Let $(B, \leq)$ be a bounded distributive lattice.

Let $a \in B$ and a_1, a_2 are two complements of a. Hence by definition we have

$$a \vee a_1 = 1; \qquad a \vee a_2 = 1$$

$$a \wedge a_1 = 0; \qquad a \wedge a_2 = 0$$

Now,
$$a_1 = (a_1 \vee 0) = a_1 \vee (a \wedge a_2) \qquad [\because (a \wedge a_2) = 0]$$
$$= (a_1 \vee a) \wedge (a_1 \vee a_2) \qquad \text{[Distributive law]}$$
$$= 1 \wedge (a_1 \vee a_2)$$
$$= (a_1 \vee a_2)$$

So, $$a_1 = (a_1 \vee a_2) \qquad \dots (i)$$

Similarly,
$$a_2 = (a_2 \vee 0) = a_2 \vee (a \wedge a_1) \qquad [\because (a \wedge a_1) = 0]$$
$$= (a_2 \vee a) \wedge (a_2 \vee a_1) \qquad \text{[Distributive law]}$$
$$= (a \vee a_2) \wedge (a_1 \vee a_2)$$

$$= 1 \wedge (a_1 \vee a_2)$$

$$= (a_1 \vee a_2)$$

So, $$a_2 = (a_1 \vee a_2) \quad \ldots (ii)$$

Therefore, from equations (*i*) and (*ii*) we get

$$a_1 = a_2$$

Thus, in a bounded distributive lattice, if a complement exists then it is unique.

11.7 SOME SPECIAL LATTICES

Here we will discuss the basic definition of some special type of lattices such as Boolean lattice, sublattice and isomorphic lattices.

11.7.1 Boolean Lattice

A complemented and distributive lattice is called a Boolean lattice.

11.7.2 Sublattice

Let (B, ≤) be a lattice. Then any non-empty subset L of B is called a sublattice of B if

$$(a \vee b) \in \text{L and } (a \wedge b) \in \text{L}; \forall\, a, b \in \text{L}$$

In general if D(n) be a lattice and if m divides n, D(m) is a sublattice of D(n).

11.7.3 Isomorphic Lattices

Let (B_1, ≤) and (B_2, ≤) be two lattices, then $f : B_1 \to B_2$ is an isomorphism if

$$f(a \wedge b) = f(a) \wedge f(b) \text{ and } f(a \vee b) = f(a) \vee f(b) \text{ for all } a, b \in \text{A}$$

If two lattices are isomorphic as posets then they are said to be isomorphic lattices.

SOLVED EXAMPLES

Example 1 *Show that (I , |) is a lattice; where I is the set of positive integers and the relation | is defined as a | b if and only if a divides b.*

Solution: To show (I, |) is lattice, first of all we have to show that (I, |) is a poset. Here the relation is defined as

$$a \,\text{R}\, b \quad : \quad a \mid b\,; \qquad a, b \in \text{I}$$

i.e., a R b : a divides b

Reflexive: It is clear that for every $a \in$ I, a divides a, *i.e.*, $a \mid a$ for every $a \in$ I.

Anti Symmetric: Suppose that a R b and b R a.

i.e., a divides b and b divides a.

This implies that $a = b$.

i.e., a R b and b R a implies that $a = b$.

Transitive: Suppose that a R b and b R c

i.e., a divides b and b divides c.

This implies $b = a\,k_1$ and $c = b\,k_2$; k_1 and $k_2 \in$ I

Now, $c = b\,k_2 = a\,(k_1\,k_2)$.

This indicates that a divides c.

i.e., a R b and b R c implies a R c.

Therefore, (I, |) is a poset. To show it is a lattice, it is sufficient to define the L.U.B. and G.L.B. in I.

Now, let a, $b \in$ I

L.U.B. $(\{a, b\}) = (a \vee b) =$ L.C.M. (a, b) and

G.L.B. $(\{a, b\}) = (a \wedge b) =$ G.C.D. (a, b)

Therefore, (I, |) is a lattice.

Example 2 *For a lattice B ; a , b ∈ B prove the following*

(i) $(a \vee b) = b$ *if and only if* $a \leq b$

(ii) $(a \wedge b) = a$ *if and only if* $a \leq b$

(iii) $(a \wedge b) = a$ *if and only if* $(a \vee b) = b$

Solution: Given B is a lattice and $a, b \in$ B

(*i*) Suppose that $(a \vee b) = b$

Our claim is that $a \leq b$

Now, $(a \vee b) =$ L.U.B. $(\{a, b\})$

i.e., $a \leq (a \vee b)$

i.e., $a \leq b$ $[\because b = (a \vee b)]$

Conversely, suppose that $a \leq b$.

Our claim is that $(a \vee b) = b$

Given that $a \leq b$. Also we know that $b \leq b$. Hence it is clear that b is an upper bound of a and b. Again $(a \vee b)$ is the least upper bound, so

$$(a \vee b) \leq b \quad \ldots (i)$$

Again since $(a \vee b)$ is an upper bound of a and b. So,

$$b \leq (a \vee b) \quad \ldots (ii)$$

Hence, from equations (1) and (2) we get

$$(a \vee b) = b.$$

(*ii*) Suppose that $(a \wedge b) = a$

Our claim is $a \leq b$

We know that $(a \wedge b) =$ G.L.B. $(\{a, b\})$

i.e., $(a \wedge b) \leq b$

This implies that $a \leq b$ $[\because (a \wedge b) = a]$

Conversely, suppose that $a \leq b$

Our claim is $(a \wedge b) = a$.

Given $a \leq b$, also we know that $a \leq a$.

Hence it is clear that 'a' is the lower bound of both a and b. Again $(a \wedge b)$ is the G.L.B. of both a and b. Therefore,

$$a \leq (a \wedge b) \quad \ldots (iii)$$

Also, $(a \wedge b)$ is the lower bound of a and b. Therefore,

$$(a \wedge b) \leq a \qquad \ldots (iv)$$

Combining (iii) and (iv) we get $(a \wedge b) = a$.

(iii) On combining the proofs of (i) and (ii) we can get

$$(a \wedge b) = a \text{ if and only if } (a \vee b) = b$$

Example 3 *Let B be the power set of S = {1, 2, 3} and (B, ⊴) be a poset defined by X ≤Y if X ⊆ Y for X, Y ∈ B. Draw the Hasse diagram of the poset (B, ⊴).*

Solution: Given that B be the power set of S = {1, 2, 3}.

Therefore, B = {ϕ, {1}, {2}, {3}, {1, 2}, {1, 3}, {2, 3}, {1, 2, 3}}.

Given that (B, ≤) be a poset. Where the relation ≤ is defined as

XRY : X ⊆ Y for X, Y ∈ B

Therefore, the Hasse diagram is given as

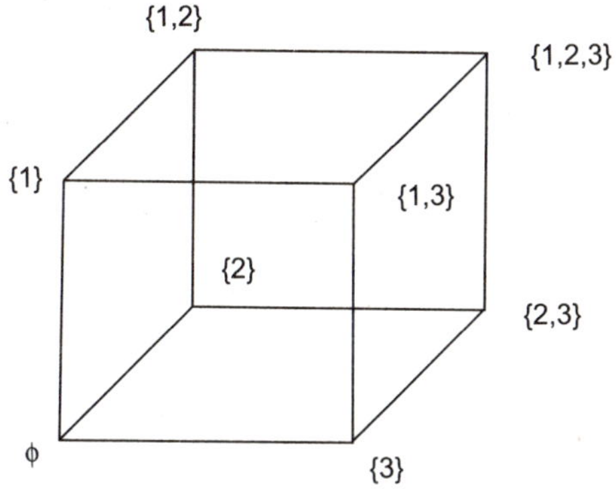

Example 4 *Set of all positive divisors of 30 i.e., D(30), forms a poset under the relation x ≤y means x divides y for x, y ∈ D(30). Draw the Hasse diagram.*

Solution: D(30) ={1, 2, 3, 5, 6, 10, 15, 30}.

Let us define the relation x R y means x is a divisor of y for $x, y \in$ D(30). Thus we get

R = {(1, 2), (1, 3), (1, 5), (1, 6), (1, 10), (1, 15), (1, 30), (2, 6), (2, 10), (2, 30), (3, 6), (3, 15), (3, 30), (5, 10), (5, 15), (5, 30), (6, 30), (10, 30), (15, 30)}

From the above relation it is clear that 6 does not cover 1 because there exists 2 such that 1 R 2 and 2 R 6. Similarly 10 does not cover 1, 15 does not cover 1 and 30 does not cover 1, 2, 3, 5. Again it is also clear that 2 covers 1, 3 covers 1, 5 covers 1 and so on.

Therefore, the Hasse diagram is given below.

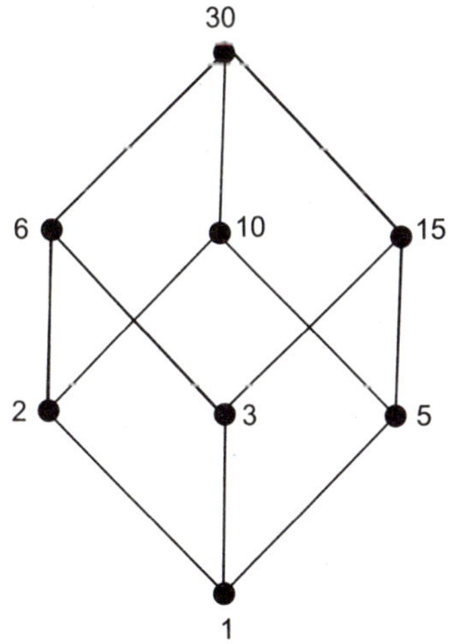

Example 5 *Draw Hasse diagrams of all lattices with four elements.*

Solution: Hasse diagrams of all lattices with four elements are given below.

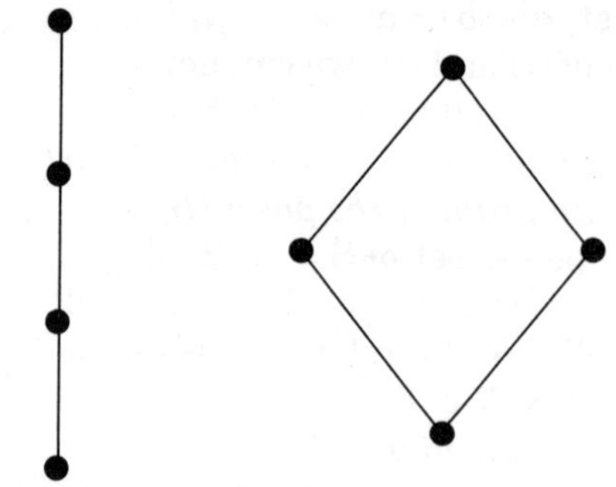

Example 6 *If B = D(24) be a lattice, then find all the sublattices of D(24). Also draw the Hasse diagram.*

Solution: Given that B = D(24) be a lattice. Where D(24) is the set of all positive divisors of 24. Therefore,

$$D(24) = \{1, 2, 3, 4, 6, 8, 12, 24\}$$

The Hasse diagram for the above lattice is given below.

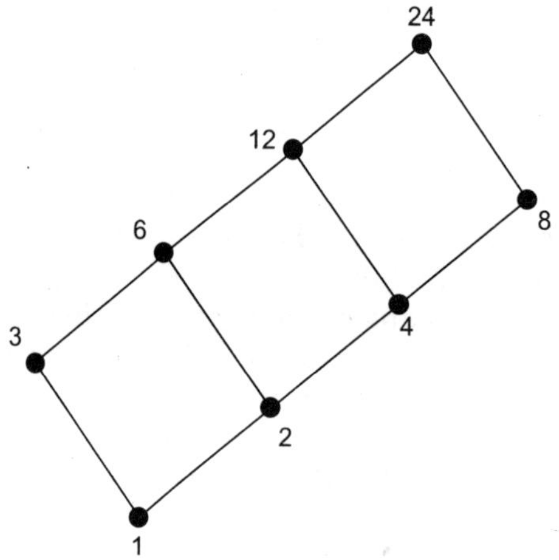

The sublattices of D(24) are D(6) = {1, 2, 3, 6}, {2, 4, 6, 12} and {4, 8, 12, 24}. Another sublattice is D(12) = {1, 2, 3, 4, 6, 12} as 12 divides 24.

EXERCISES

1. Let n be the positive integer and D(n) be the set of all positive divisor of n, then show that D(n) is a lattice under the relation of divisibility.
2. Let A = {1, 2, 3, 4, 5, 6}. We define x R y as $x \leq y$ if x divides y. Draw the Hasse diagram of the poset (A, ≤).
3. Draw the Hasse diagram of (P(A), ⊆). Where A = {1, 2} and P(A) is the power set of A.
4. Draw the Hasse diagram of (P(A), ⊆). Where A = {0, 1, 2, 3} and P(A) is the power set of A.
5. Draw the Hasse diagram of (D(n), |) for n = 6, 16, 24, 32, 100.
6. Draw Hasse diagrams of all lattices with five elements.

7. Show that set of all positive divisors of 105 *i.e.*, D(105), forms a poset under the relation $x \leq y$ means x divides y for $x, y \in$ D(105). Draw the Hasse diagram.
8. Show that the poset with the Hasse diagram given below is not a lattice.

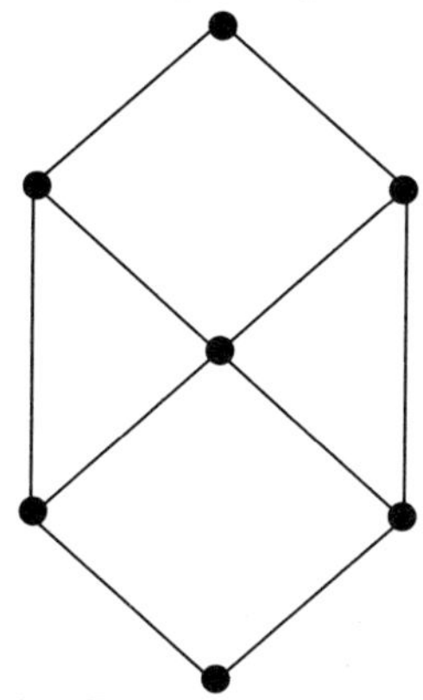

9. Prove that for all a, b, c in a lattice B,

   $[(a \wedge b) \quad \vee (a \wedge c)] \wedge [(a \wedge b) \vee (b \wedge c)]$
10. If B = D(30) be a lattice, then find all the sublattices of D(30). Also draw the Hasse diagram.

12

Graph Theory

■ 12.0 INTRODUCTION

Graph theory has applications in many areas like Mathematics, Computer Science, Engineering, Communication Science etc. Oystein Ore, the prominent graph theorist and author of the first graph theory book said in that "the theory of graphs is one of the few fields of mathematics with a definite birth date". Graph theory is considered to have begun in 1736 with the publication of Euler's solution of the Konigsberg Bridge problem. In 1936, Denes Konig wrote the first book on graph theory. The major developments of graph theory occurred by the ever growing importance of Computer Science and its connection with graph theory.

Now the question arises "what is a graph"? Consider the example. Suppose there are four sales persons Brown, Abraham, Peterson, Clark and five territories T_1, T_2, T_3, T_4, T_5. Brown is interested to work in the territories T_1, T_2, T_3. Abraham is interested to work in the territories T_2, T_3. Peterson is interested to work in the territories T_1, T_4, T_5 whereas Clark is interested for the territories T_3, T_4, T_5. This is explained in the following figure. This is nothing but a graph, a concept which we are about to study extensively.

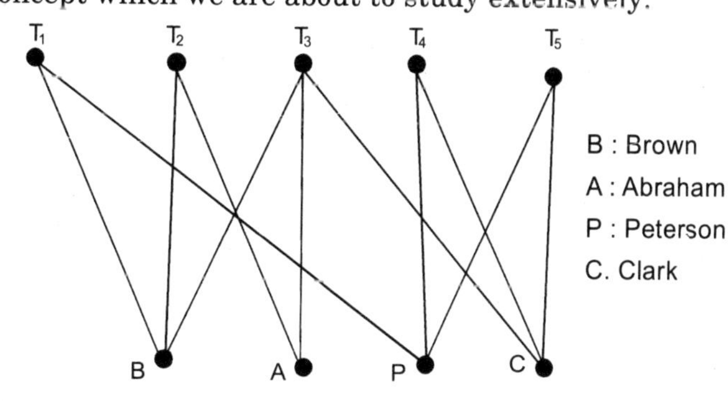

In this chapter, we will study the basic components of graph theory.

12.1 GRAPH

A graph G consists of a finite set of vertices V and a finite set of edges E. Mathematically,

$$G = (V, E)$$

Where, $E = \{(v_i, v_j) \mid v_i, v_j \in V\}$

Let us consider $V = \{1, 2, 3, 4, 5\}$

and $E = \{(1, 2), (1, 3), (2, 3), (2, 4), (3, 4)\}$.

Hence the graph G = (V, E) becomes

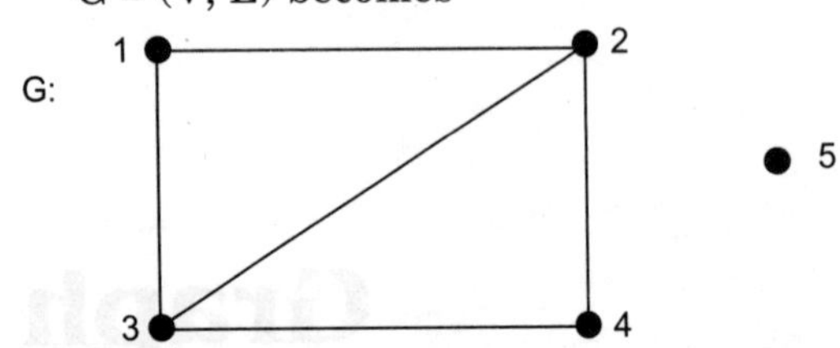

12.1.1 Order and Size

The number of vertices in a graph G(V, E) is called its order, and the number of edges is its size. That is the order of G is $|V|$ and its size $|E|$

Consider the following graph G

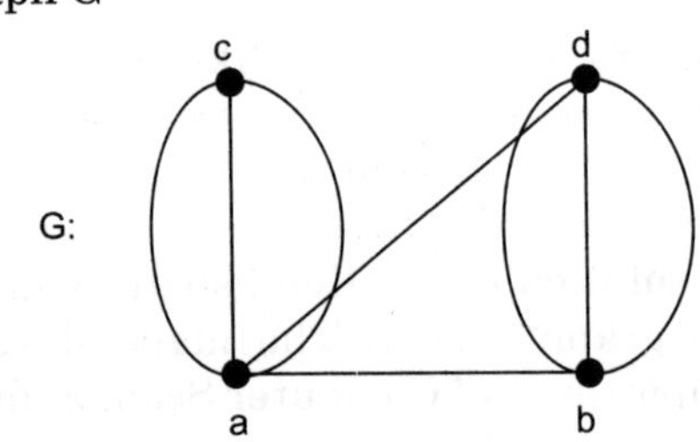

The order of G *i.e.*, $|V| = 4$

The size of G *i.e.*, $|E| = 8$

12.1.2 Adjacent Vertices

Two vertices v_i and v_j are said to be adjacent if there exists an edge (v_i, v_j) in the graph G(V, E).

Consider the graph G as

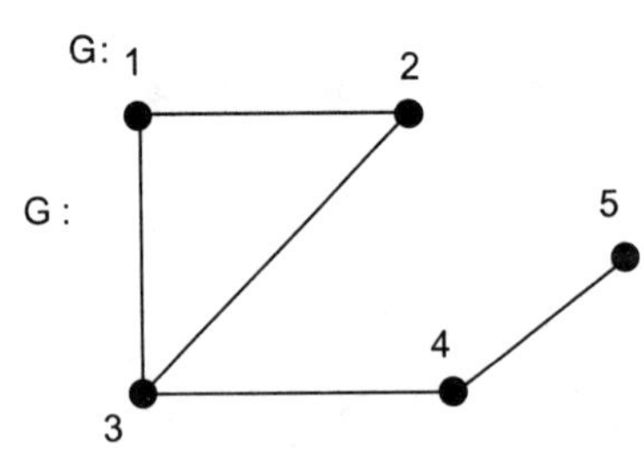

Here the vertices 1 and 2 are adjacent. Similarly, the vertices 1 and 3 are also adjacent.

12.1.3 Parallel Edges

If there is more than one edge between the same pair of vertices, then the edges are termed as parallel edges. Consider the graph G as

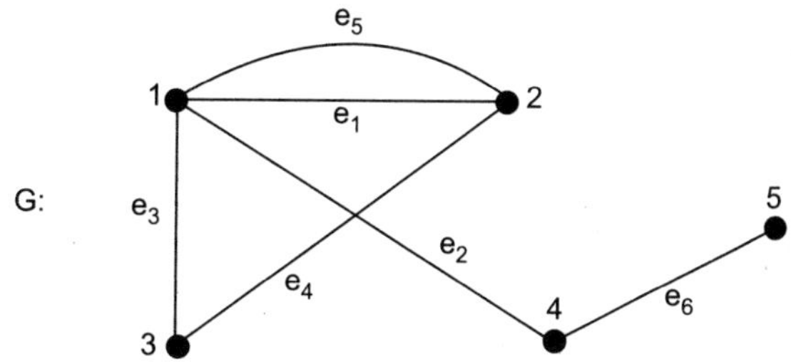

Here the edges e_1 and e_5 are parallel edges.

12.1.4 Loop

An edge whose starting and ending vertex are same is known as a loop. Mathematically $e = (v_i, v_i)$. Consider the graph G as

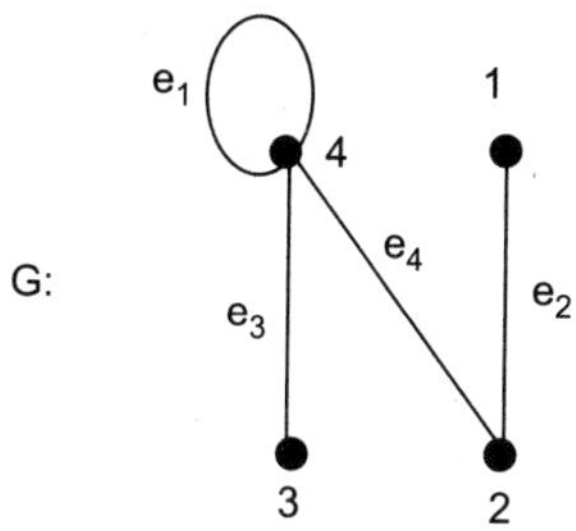

From the graph, it is clear that the edge e_1 is a loop.

12.2 KINDS OF GRAPH

In this section, we will discuss different kinds of graph.

12.2.1 Simple Graph

A graph G(V, E) that has no self-loop or parallel edges is called a simple graph. Consider the graphs G_1 and G_2 as

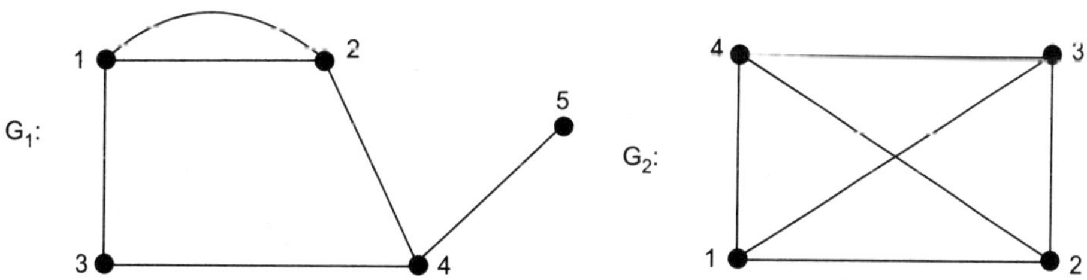

The graph G_1 is not a simple graph because there exists parallel edges between the vertices 1 and 2 whereas the graph G_2 is a simple graph.

12.2.2 Multi Graph

A graph G(V, E) is known as a multi graph if it contains parallel edges, *i.e.*, two or more edges between a pair of vertices. It is to be noted that every simple graph is a multi graph but the converse is not true. Consider the graph G as

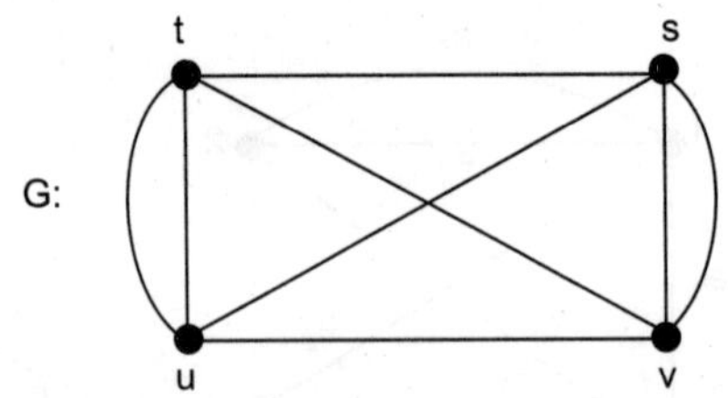

The above graph is a multi graph because there are parallel edges between the vertices u, t and v, s.

12.2.3 Pseudo Graph

A graph G(V,E) is known as a pseudo graph if we allow both parallel edges and loops. It is to be noted that every simple graph and multi graph are pseudo graph but the converse is not true.

Consider the graph G as

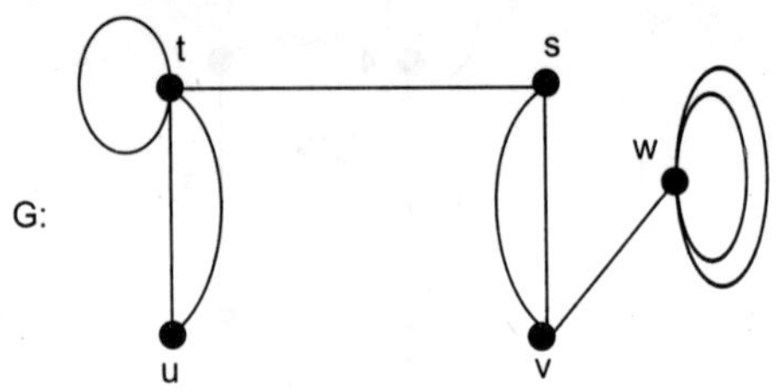

■ 12.3 DIGRAPH

A graph G(V, E) where V is the set of nodes or vertices and E is the set of edges having direction. If (v_i, v_j) is an edge, then there is an edge from the vertex v_i to the vertex v_j. A digraph is also called a directed graph. Let us consider

$$V = \{1, 2, 3, 4, 5\} \text{ and } E = \{(1, 2), (1, 3), (2, 3), (2, 4), (3, 4), (4, 5)\}$$

Hence, the digraph G becomes

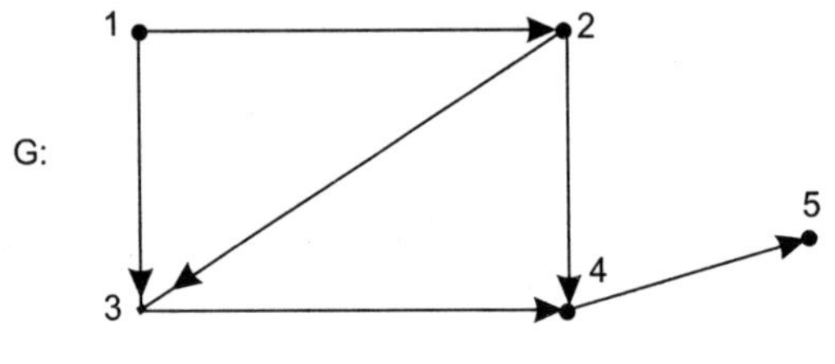

■ 12.4 WEIGHTED GRAPH

A graph (or digraph) is known as a weighted graph (or digraph) if each edge of the graph has some weights. Let us consider

$$V = \{1, 2, 3, 4, 5\} \text{ and } E = \{e_1, e_2, e_3, e_4, e_5\}$$

Where $e_1 = (1, 2), e_2 = (1, 3), e_3 = (2, 4), e_4 = (3, 4), e_5 = (4, 5)$

and $w(e_1) = 5, w(e_2) = 6, w(e_3) = 1, w(e_4) = 6, w(e_5) = 3$

Hence, the weighted graph G becomes

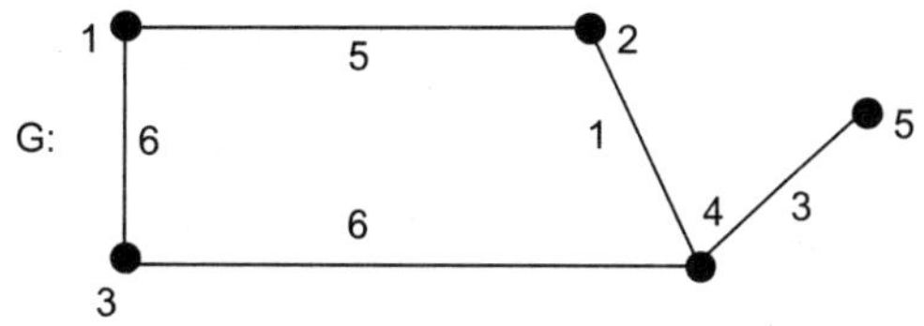

■ 12.5 DEGREE OF A VERTEX

The number of edges connected to the vertex 'v' is known as degree of vertex 'v', generally denoted by degree (v). In case of a digraph, there are two degrees *i.e.*, indegree and outdegree. The number of edges coming to the vertex 'v' is known as indegree of 'v' whereas the number of edges emanating from the vertex 'v' is known as outdegree of 'v'. Generally, the indegree is denoted by indegree (v) and the outdegree is denoted by outdegree (v).

Note: In case of a loop, it contributes 2 to the degree of a vertex.

12.5.1 Isolated Vertex

A vertex is said to be an isolated vertex if there is no edge connected from any other vertex to the vertex. In other words a vertex is said to be an isolated vertex if the degree of that vertex is zero.

i.e., If degree (v) = 0, then v is isolated.

Consider the graph G as

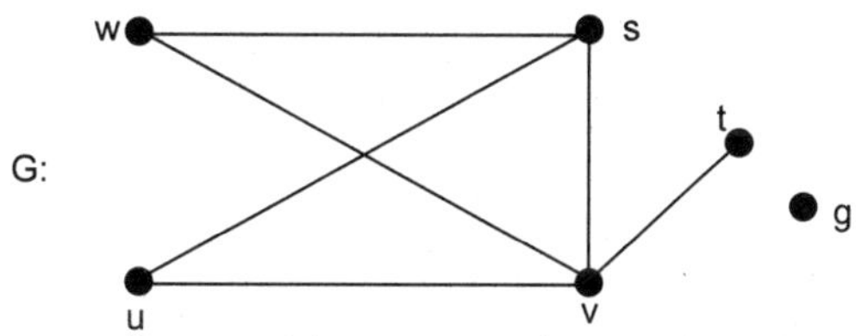

Now, degree (u) = 2; degree (v) = 4; degree (t) = 1
degree (g) = 0; degree (s) = 3; degree (w) = 2

Therefore, it is clear that 'g' is an isolated vertex.

■ 12.6 PATH

A path in a graph is a sequence $v_1, v_2, \ldots, v_k$ of vertices each adjacent to the next, and a choice of an edge between each 'v_i' to 'v_{i+1}' so that no edge is chosen more than once.

Consider the graph G as

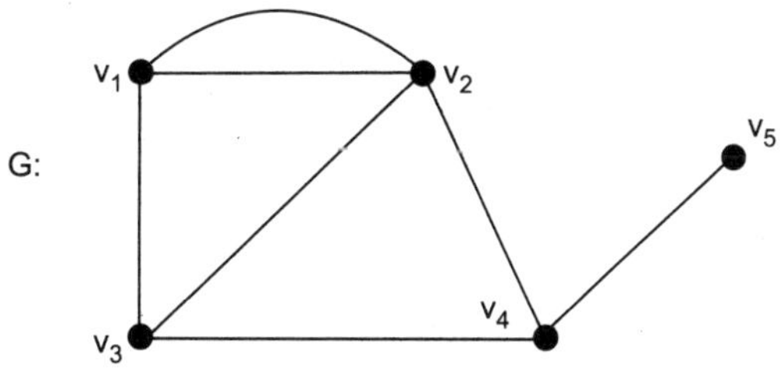

Here one path is $v_1\, v_2\, v_1\, v_3\, v_4\, v_5$.

■ 12.7 COMPLETE GRAPH

A graph (digraph) G is said to be complete if each vertex 'u' is adjacent to every other vertex 'v' in G. In other words, there are edges from any vertex to all other vertices. Consider the graph G as

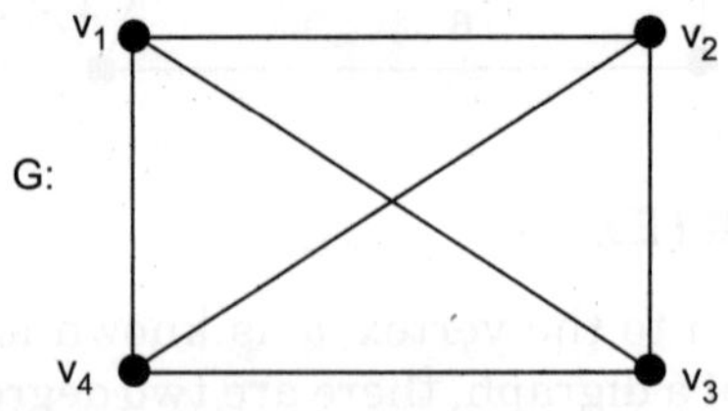

The above graph G is a complete graph.

■ 12.8 REGULAR GRAPH

A graph G (V, E) is said to be regular if the degree of every vertex are equal. Mathematically, G is denoted as regular if

$$\text{degree}(v_i) = \text{degree}(v_j) \ \forall\ i, j.$$

Where, $v_i, v_j \in G(V, E)$.

Consider the graph G as

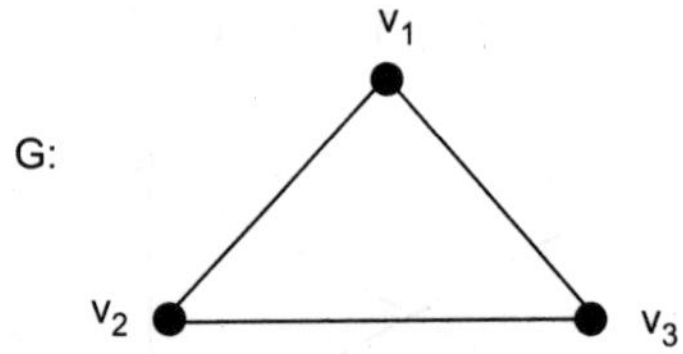

In the above graph, degree (v_1) = degree (v_2) = degree (v_3) = 2. Therefore, the graph G is regular (2 regular). The above graph is also complete.

Consider another example G_1 as

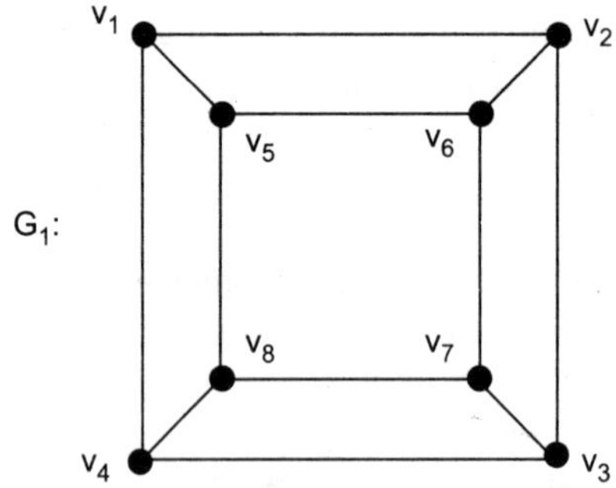

Here the degree of every vertex is 3. So, the above graph is 3-regular but not complete.

■ 12.9 CYCLE

If there is a path containing one or more edges which starts from a vertex 'v' and terminates into the same vertex, then the path is known as a cycle. Consider the graph G as

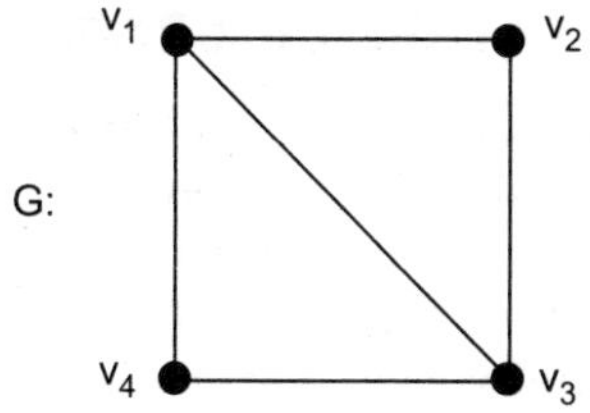

In the above graph G, one cycle is $v_1 v_2 v_3 v_1$. Similarly, another cycle is $v_1 v_2 v_3 v_4 v_1$.

■ 12.10 PENDANT VERTEX

A vertex 'v' in a graph G is said to be a pendant vertex if the degree (v) =1. In case of a digraph, a vertex 'v' is said to be a pendant vertex if the indegree (v) = 1 and outdegree (v) = 0. In the graph 'G(figure 1)' given below, indegree of the vertices v_4 , v_5 , v_6 and v_7 is equal to 1 and the outdegree is equal to 0. Therefore, these vertices are pendant vertices. Similarly, in the graph 'G(figure 2)' given below the vertices v_1, v_5 and v_6 are pendent vertices.

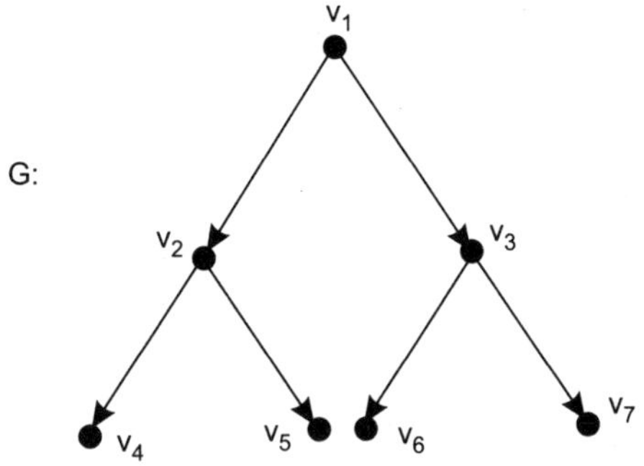

Figure 1

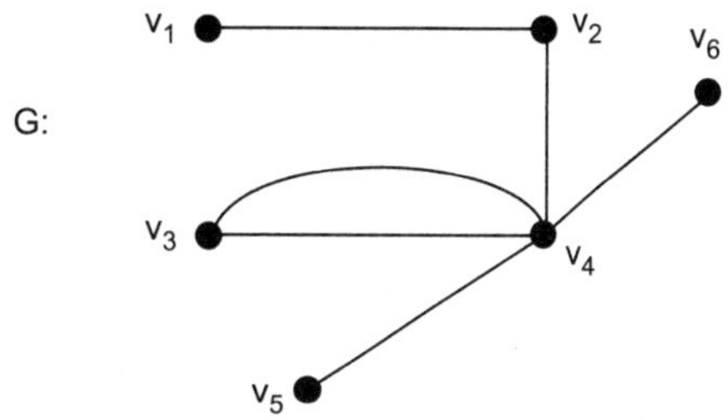

Figure 2

■ 12.11 ACYCLIC GRAPH

A graph (digraph) which does not have any cycle is known as an acyclic graph (digraph). Consider the graph G as

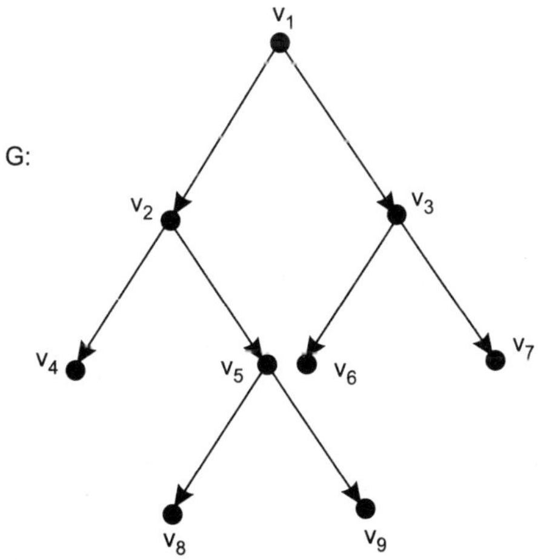

Here, G is an acyclic graph.

12.12 MATRIX REPRESENTATION OF GRAPHS

There are many ways to represent a graph in computer. Generally, graphs are represented diagrammatically, but this is possible only when the number of vertices and edges are reasonably small. So, the concept of matrix representation of graphs is developed. The major advantage of this representation is that the calculation of paths and cycles in graph theoretical problems such as communication networks, power distribution, transportation etc. However, the disadvantage is that this representation takes away from the visual aspect of graphs.

12.12.1 Adjacency Matrix

The most useful way of representing any graph is the matrix representation. It is a square matrix of order $(n \times n)$ where n is the number of vertices in the graph G. Generally denoted by A $[a_{ij}]$ where a_{ij} is the ith row and jth column element. The general form of adjacency matrix is given as below:

$$A = \begin{bmatrix} a_{11} & a_{12} & a_{13} & \cdots & a_{1n} \\ a_{21} & a_{22} & a_{23} & \cdots & a_{2n} \\ a_{31} & a_{32} & a_{33} & \cdots & a_{3n} \\ \cdot & \cdot & \cdot & \cdots & \cdot \\ \cdot & \cdot & \cdot & \cdots & \cdot \\ \cdot & \cdot & \cdot & \cdots & \cdot \\ a_{n1} & a_{n2} & a_{n3} & \cdots & a_{nn} \end{bmatrix}$$

where,
$$a_{ij} = \begin{cases} 1; & \text{if there is an edge from '}v_i\text{' to '}v_j\text{'} \\ 0; & \text{Otherwise} \end{cases}$$

This matrix is termed as adjacency matrix, because an entry stores the information whether two vertices are adjacent or not. This is also known as bit matrix or Boolean matrix as each entry is either 1 or 0.

Notes: 1. In the adjacency matrix if the main diagonal elements are zero, then the graph is said to be a simple graph.

2. In case of a multi graph the adjacency matrix can be found out with the relation.

$$a_{ij} = \begin{cases} n; & n \text{ be the number of edges from '}v_i\text{' to '}v_j\text{'} \\ 0; & \text{Otherwise} \end{cases}$$

3. In case of a weighted graph the adjacency matrix can be found out with the relation

$$a_{ij} = \begin{cases} w; & w \text{ is the weight of the edges from '}v_i\text{' to '}v_j\text{'} \\ 0; & \text{Otherwise} \end{cases}$$

Consider the graph G as

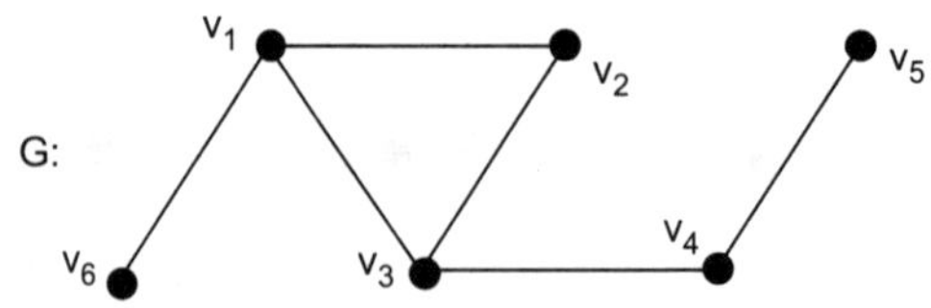

Hence, the adjacency matrix is given as

$$A = \begin{array}{c} \\ v_1 \\ v_2 \\ v_3 \\ v_4 \\ v_5 \\ v_6 \end{array} \begin{array}{c} \begin{matrix} v_1 & v_2 & v_3 & v_4 & v_5 & v_6 \end{matrix} \\ \begin{bmatrix} 0 & 1 & 1 & 0 & 0 & 1 \\ 1 & 0 & 1 & 0 & 0 & 0 \\ 1 & 1 & 0 & 1 & 0 & 0 \\ 0 & 0 & 1 & 0 & 1 & 0 \\ 0 & 0 & 0 & 1 & 0 & 0 \\ 1 & 0 & 0 & 0 & 0 & 0 \end{bmatrix} \end{array}$$

Consider the graph G as

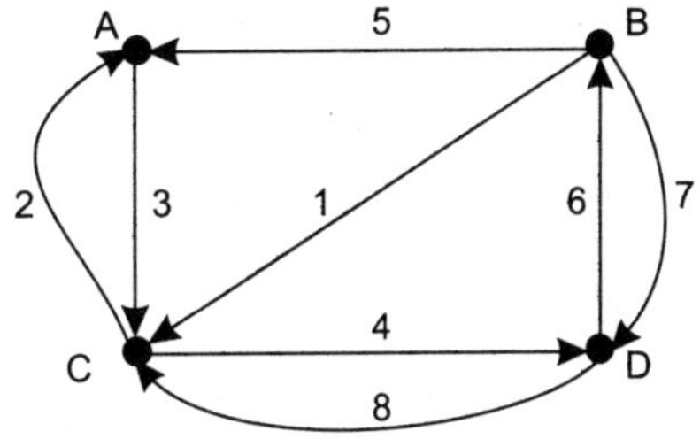

The adjacency matrix of the above graph with respect to the ordering A, B, C and D is given below:

$$A = \begin{bmatrix} 0 & 0 & 3 & 0 \\ 5 & 0 & 1 & 7 \\ 2 & 0 & 0 & 4 \\ 0 & 6 & 8 & 0 \end{bmatrix}.$$

Consider the graph G as

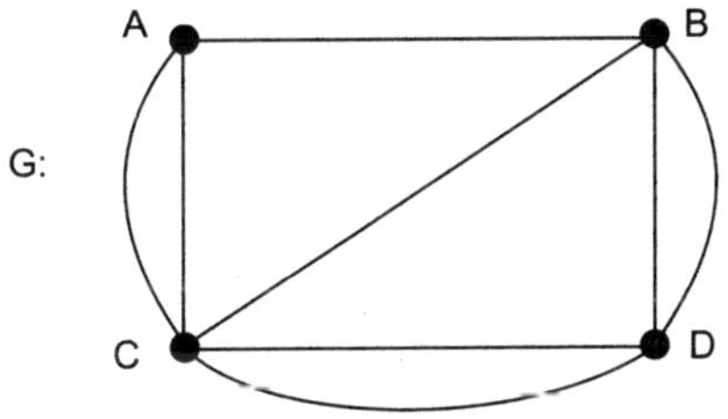

The adjacency matrix of the above graph with respect to the ordering A, B, C and D is given below:

$$A = \begin{bmatrix} 0 & 1 & 2 & 0 \\ 1 & 0 & 1 & 2 \\ 2 & 1 & 0 & 2 \\ 0 & 2 & 2 & 0 \end{bmatrix}$$

12.12.2 Incidence Matrix

Suppose that G be a simple undirected graph with m vertices and n edges, then the incidence matrix I $[a_{ij}]$ is a matrix of order $(m \times n)$ where the element a_{ij} is defined as

$$a_{ij} = \begin{cases} 1; & \text{If vertex } i \text{ belongs to edges } j. \\ 0; & \text{Otherwise} \end{cases}$$

Consider the graph G as

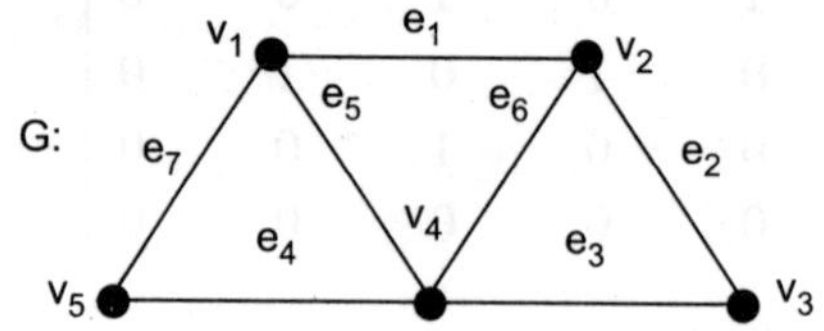

Hence, the incidence matrix of the graph G is of order (5×7). The incidence matrix relative to the ordering v_1, v_2, v_3, v_4, v_5 and $e_1, e_2, e_3, e_4, e_5, e_6, e_7$ is given below:

$$I = \begin{bmatrix} 1 & 0 & 0 & 0 & 1 & 0 & 1 \\ 1 & 1 & 0 & 0 & 0 & 1 & 0 \\ 0 & 1 & 1 & 0 & 0 & 0 & 0 \\ 0 & 0 & 1 & 1 & 1 & 1 & 0 \\ 0 & 0 & 0 & 1 & 0 & 0 & 1 \end{bmatrix}$$

12.12.3 Path Matrix

Suppose that G be simple graph with n-vertices. Then the $(n \times n)$ matrix $P = [p_{ij}]_{(n \times n)}$ defined by

$$p_{ij} = \begin{cases} 1; & \text{if there is a path from } v_i \text{ to } v_j \\ 0; & \text{Otherwise} \end{cases}$$

is known as the path matrix or reachability matrix of the graph G.

Consider the graph G as

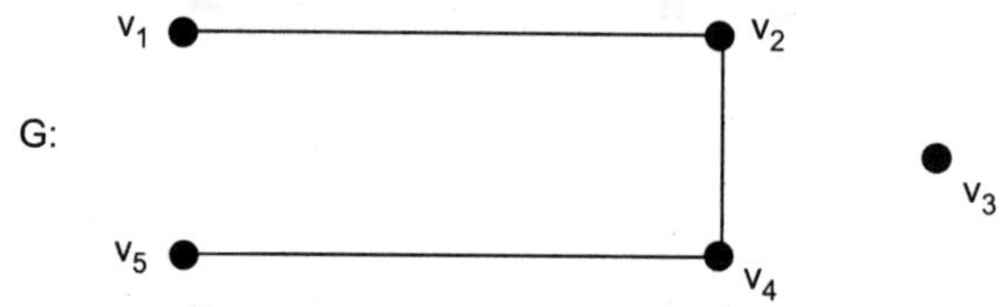

Therefore, the path matrix of the above graph relative to the ordering v_1, v_2, v_3, v_4, v_5 is given as

$$P = \begin{bmatrix} 0 & 1 & 0 & 1 & 1 \\ 1 & 0 & 0 & 1 & 1 \\ 0 & 0 & 0 & 0 & 0 \\ 1 & 1 & 0 & 0 & 1 \\ 1 & 1 & 0 & 1 & 0 \end{bmatrix}$$

■ 12.13 CONNECTED GRAPH

A graph (not digraph) G (V, E) is said to be connected if for every pair of distinct vertices 'u' and 'v' in G, there is a path. A directed graph is said to be strongly connected if for every pair of distinct vertices 'u' and 'v' in G, there is a directed path from 'u' to 'v' and also from 'v' to 'u'. A directed graph is said to be weakly connected if for every pair of distinct vertices, there is a path without taking the direction.

Consider the following graphs

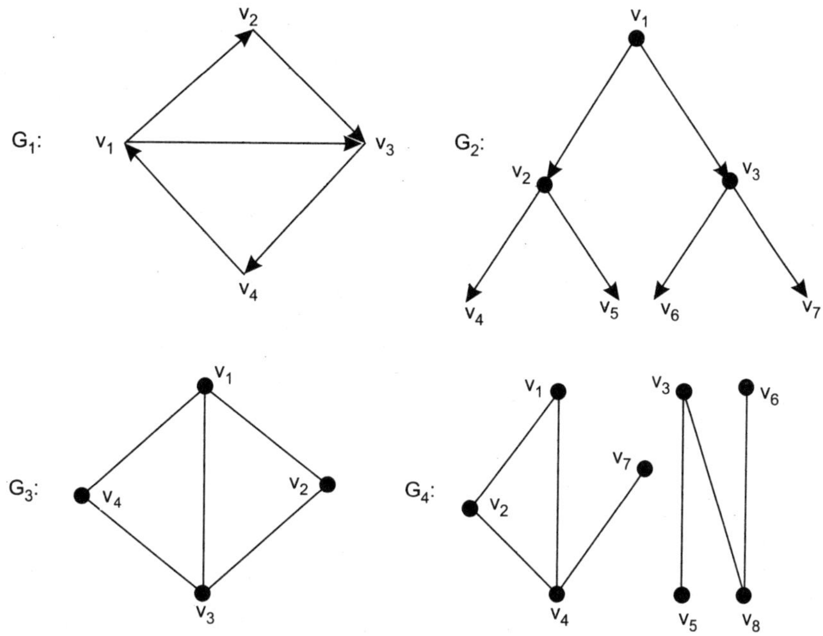

From the above graphs, it is clear that

G_1 : Strongly Connected; G_2: Weakly Connected

G_3 : Connected; G_4: Disconnected

12.13.1 Theorem

Suppose that G be a graph with n-vertices $v_1, v_2, \ldots, v_n$ and let A be the adjacency matrix of G. Let us define $B = [b_{ij}]$ such that

$$B = A + A^2 + A^3 + \ldots + A^{n-1}.$$

If for every pair of distinct indices i and j, $b_{ij} \neq 0$, then the graph is said to be connected.

The proof of the above theorem is beyond the scope of this book.

Consider the graph G as

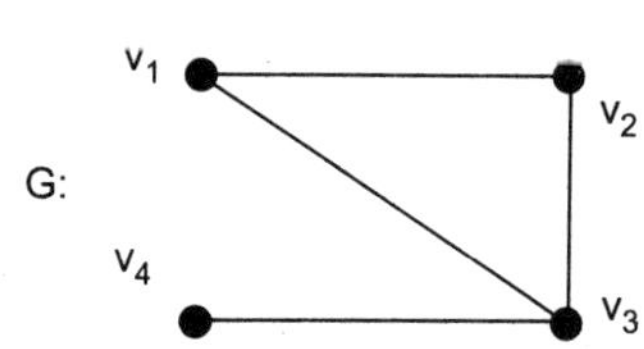

Hence, the adjacency matrix A is given as

$$A = \begin{pmatrix} 0 & 1 & 1 & 0 \\ 1 & 0 & 1 & 0 \\ 1 & 1 & 0 & 1 \\ 0 & 0 & 1 & 0 \end{pmatrix}$$

Here, number of vertices $(n) = 4$. Therefore, $B = A + A^2 + A^3$

Now,
$$A^2 = \begin{pmatrix} 2 & 1 & 1 & 1 \\ 1 & 2 & 1 & 1 \\ 1 & 1 & 3 & 0 \\ 1 & 1 & 0 & 1 \end{pmatrix}; A^3 = A^2 A = \begin{pmatrix} 2 & 3 & 4 & 1 \\ 3 & 2 & 4 & 1 \\ 4 & 4 & 2 & 3 \\ 1 & 1 & 3 & 0 \end{pmatrix}$$

Therefore,
$$B = A + A^2 + A^3 = \begin{pmatrix} 4 & 5 & 6 & 2 \\ 5 & 4 & 6 & 2 \\ 6 & 6 & 5 & 4 \\ 2 & 2 & 4 & 1 \end{pmatrix}$$

Since, all $b_{ij} \neq 0$ for $i \neq j$; the graph G is connected. All elements except the diagonal elements must not be zero for connected graph.

■ 12.14 GRAPH ISOMORPHISM

Suppose $G_1 : (V_1, E_1)$ and $G_2 : (V_2, E_2)$ be two graphs. Then the two graphs G_1 and G_2 are said to be isomorphic if there is one to one correspondence between the edges E_1 of G_1 and E_2 of G_2 which indicates that if $(u_1, v_1) \in G_1$, then $(u_1, v_1) \in G_2$.

Such a pair of correspondence is known as graph isomorphism. The different way of representing the same graph is known as graph isomorphism. Consider graph G_1 and G_2 as

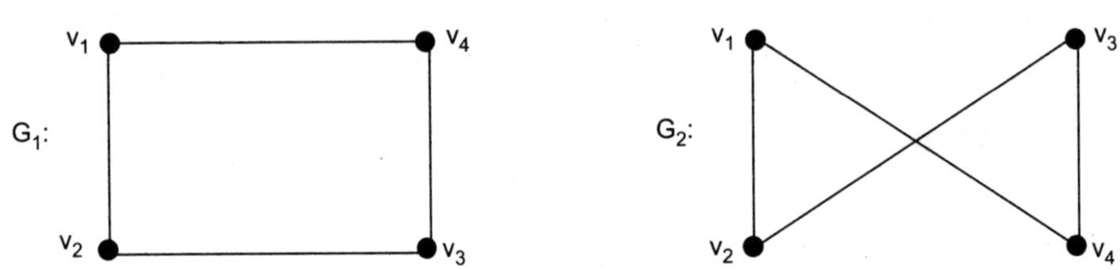

Therefore, the graphs G_1 and G_2 are isomorphic to each other.

■ 12.15 BIPARTITE GRAPH

Suppose that G: (V, E) be the graph. If the vertex set V can be partitioned into two non-empty disjoint sets V_1 and V_2 such that each edge of the graph G has one end in V_1 and other end in V_2, then the graph is said to be bipartite graph.

Consider the graph G as

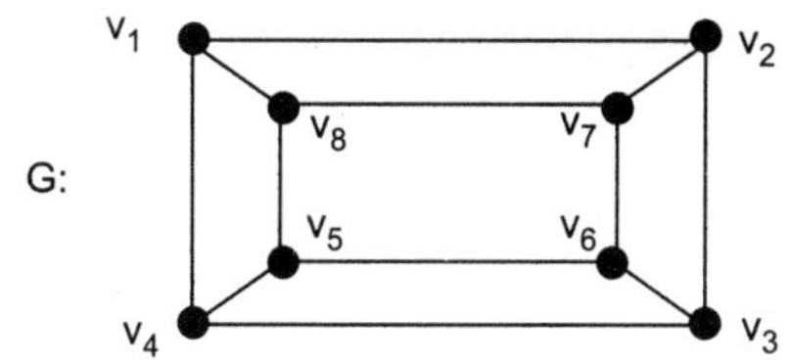

Let $V_1 = \{v_1, v_3, v_5, v_7\}$ and $V_2 = \{v_4, v_2, v_6, v_8\}$

Now, $(V_1 \cap V_2) = \phi$ and each edge of G has one vertex in V_1 and other vertex at V_2. So, G is said to be a bipartite graph.

12.15.1 Complete Bipartite Graph

Suppose that G: (V, E) be the graph. If the vertex set $V = (V_1 \cup V_2)$ and $V_1, V_2 \neq \phi$, $(V_1 \cap V_2) = \phi$, such that each edge of the graph G has one end in V_1 and other end in V_2, then the graph G is termed as bipartite.

If every vertex of V_1 is joined to every vertex of V_2, then the graph G is termed as complete bipartite graph. Consider the graph G as

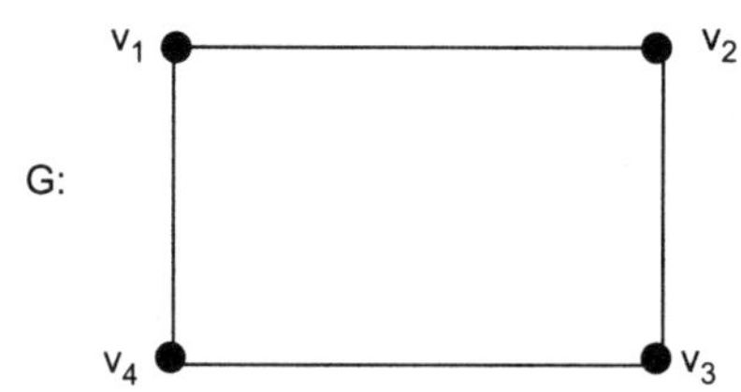

Let $V_1 = \{v_1, v_3\}$ and $V_2 = \{v_2, v_4\}$.

Therefore, $V = (V_1 \cup V_2)$; $V_1, V_2 \neq \phi$, and $(V_1 \cap V_2) = \phi$.

Also, every vertex of V_1 is joined to every vertex of V_2. So, G is a complete bipartite graph.

■ 12.16 SUBGRAPH

Suppose that G and H be two graphs with vertex sets V(G) and V(H). Let the edge sets be E(G) and E(H). Now H is said to be subgraph of G if

$$V(H) \subseteq V(G) \text{ and } E(H) \subseteq E(G)$$

Consider two graphs G and H as

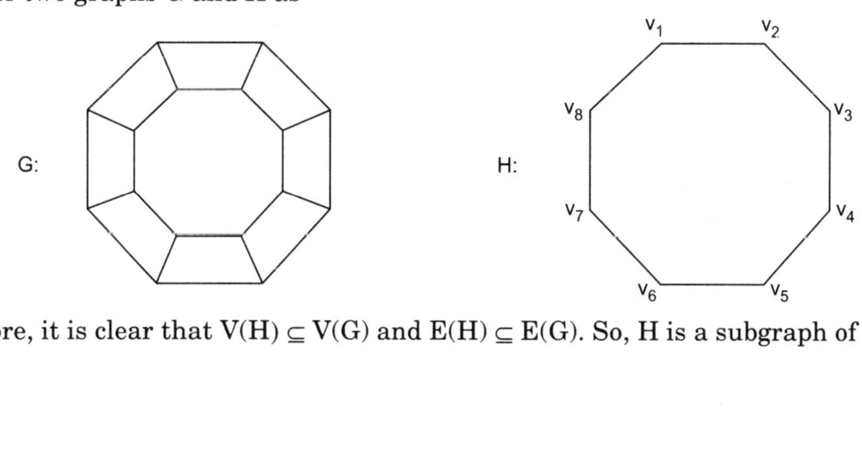

Therefore, it is clear that $V(H) \subseteq V(G)$ and $E(H) \subseteq E(G)$. So, H is a subgraph of G.

12.16.1 Vertex Deleted Subgraph

Suppose that G(V, E) be a graph. If we delete a subset U of the set V and all the edges, which have a vertex in U as an end, then the resultant graph is termed as vertex deleted subgraph of G. Consider the graph G as

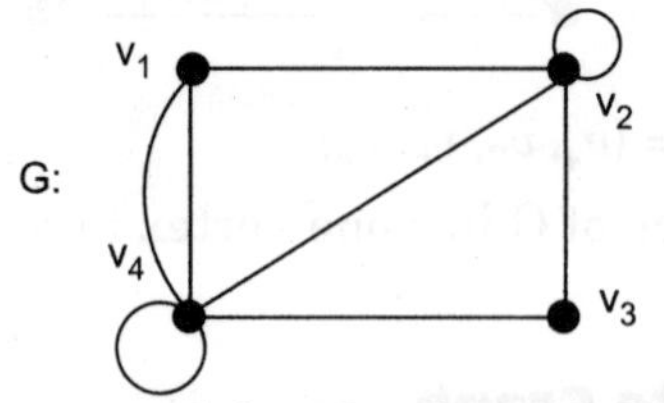

On deleting the vertex v_1, the vertex deleted subgraph H is given as

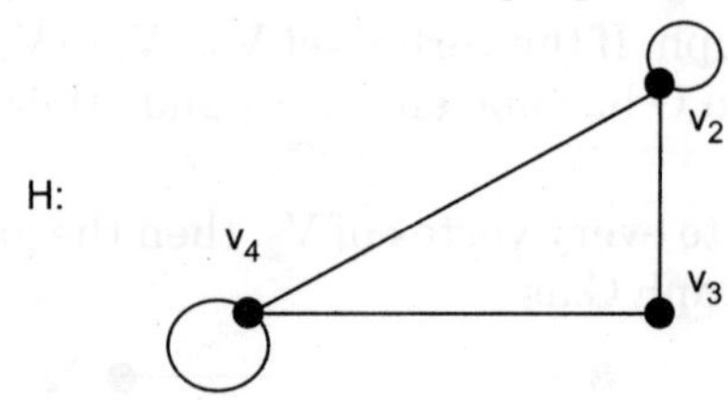

12.16.2 Edge Deleted Subgraph

Suppose that G: (V, E) be a graph. If a subset F from the set of edges E is deleted from the graph G, then the resultant graph is edge deleted subgraph of G. Consider the graph G as

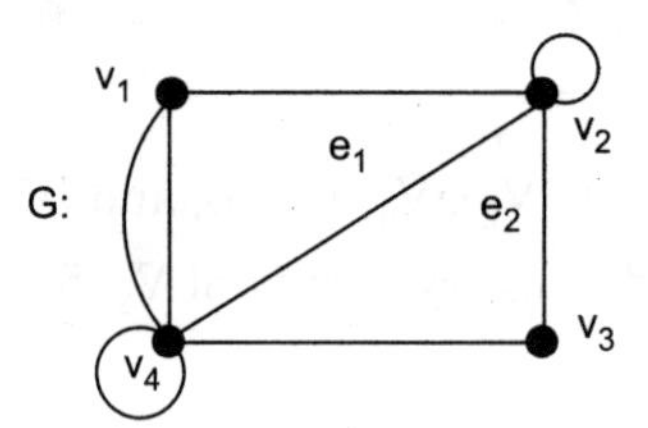

On deleting the edges e_1 and e_2, the edge deleted subgraph is given as

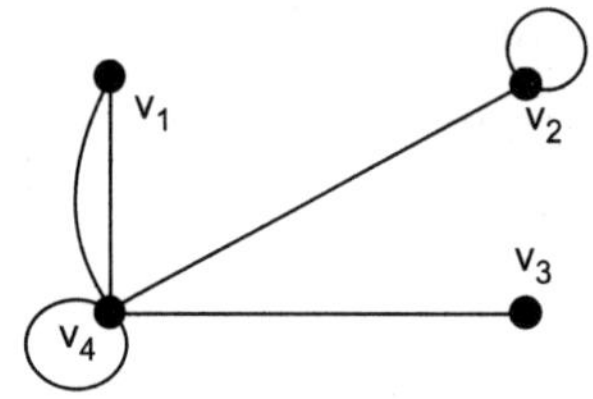

12.17 WALKS

Let G be a graph, then a walk W in a graph G is a finite sequence W = $v_0\, e_1\, v_1 e_2\, v_2\, e_3$ .... $v_{i-1}\, e_i\, v_i \ldots v_{k-1}\, e_k\, v_k$. Whose terms are alternately vertices and edges such that for $1 \le i \le k$, the edge e_i has ends v_{i-1} and v_1 . The starting vertex v_0 is the origin and the end vertex v_k is the terminus of the walk W. The vertices $v_1, v_2, \ldots . v_{k-1}$ are known as internal vertices. The walk is termed as $v_0 - v_k$ walk.

The number of edges present in the walk W is known as the length of walk W. Note that in a walk W there may be repetition of vertices and edges. In a simple graph, a walk $W = v_0e_1\, v_1e_2 \ldots e_kv_k$ is determined by a sequence of vertices $v_0\, v_1\, v_2 \ldots v_{k-1}v_k$ because each pair of vertices $v_{i-1}\, v_i$ has one edge only. Even if a graph is not simple, a walk is often simply denoted by a sequence of vertices $v_0\ v_1\ v_2\ \ldots\ v_{k-1}v_k$ where the consecutive vertices are adjacent.

Notes : 1. A walk containing no edges is known as a trivial walk.

2. A walk containing no repeated edges is termed as a trail.
3. A walk containing no repeated vertices is termed as a path. Which indicates that if the sequence of vertices $v_0\, v_1\, v_2 \ldots\ldots v_{k-1}v_k$ of the walk $W = v_0\, e_1 v_1\, e_2\, v_2 \ldots\ldots\ldots v_{k-1}\, e_k\, v_k$ are distinct, then the walk is a path.
4. Every path is a trail but the converse is not true always.

Consider the graph G as

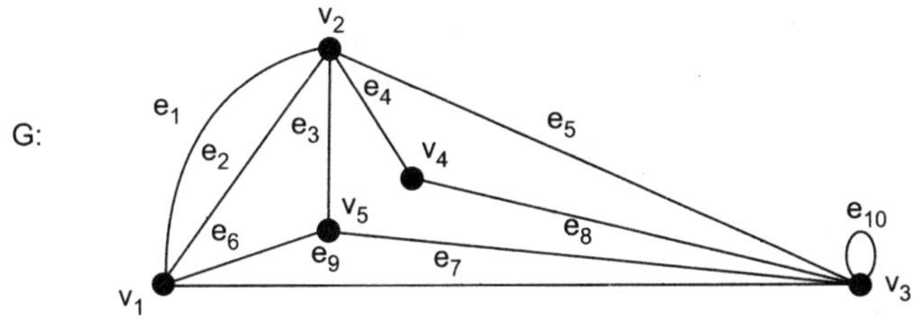

Consider the following walks

$$W_1 = v_1e_1\, v_2e_2\, v_1e_6\, v_5e_7\, v_3e_{10}\, v_3e_8\, v_4$$

$$W_2 = v_1e_1\, v_2e_1\, v_1e_1\, v_2e_2\, v_1e_1\, v_2$$

$$W_3 = v_3e_{10}\, v_3e_9\, v_1e_1\, v_2e_2\, v_1$$

$$W_4 = v_1e_2\, v_2e_5\, v_3e_7\, v_5$$

The length of W_1 is 6. Similarly, the length of other walks can be found out. Here W_1 and W_2 are walks; W_3 is trail and W_4 is a path.

12.17.1 Open and Closed Walk

Suppose that u and v be two vertices of a graph. An $u - v$ walk is said to be open or closed according to $u \neq v$ or $u = v$ respectively. In other words a walk is closed if the starting vertex (u) and the terminus (v) are same otherwise it is open.

■ 12.18 OPERATIONS ON GRAPHS

There are many operations that gives new graphs from old ones. They are mainly separated into three categories such as elementary operation, unary operation and binary operation. In elementary operation, a new graph may be produced from the original graph by a simple local change such as addition or deletion of a vertex or an edge. In unary operation we create a significantly new graph from the old one whereas in binary operation we create a new graph from two initial graphs.

12.18.1 Union

If G_1 and G_2 be two graphs, then their union $(G_1 \cup G_2)$ is a graph with $V(G_1 \cup G_2) = V(G_1) \cup V(G_2)$ and $E(G_1 \cup G_2) = E(G_1) \cup E(G_2)$.

12.18.2 Intersection

If G_1 and G_2 be two graphs with at least one vertex in common, then their intersection $(G_1 \cap G_2)$ is a graph with

$$V(G_1 \cap G_2) = V(G_1) \cap V(G_2)$$

and
$$E(G_1 \cap G_2) = E(G_1) \cap E(G_2)$$

12.18.3 Complement

Suppose that G be a simple graph with n-vertices. Then the complement of G is given by $\overline{G}$ and is defined to be the simple graph with the same vertices of G and where two vertices (u, v) are adjacent in $\overline{G}$, if u and v are not adjacent in G. In other words the complement of G can be obtained from the complete graph K_n by deleting all the edges of G. Consider the graph G as

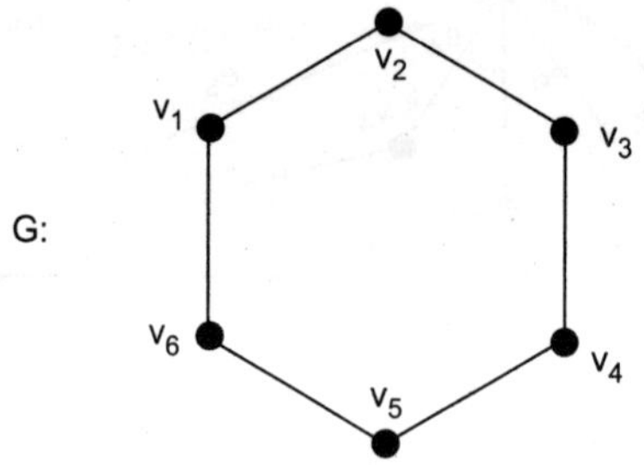

To obtain the complement of G construct the complete graph with the same vertices and then delete the edges of the graph G. The complement graph of G *i.e.*, $\overline{G}$ is given below.

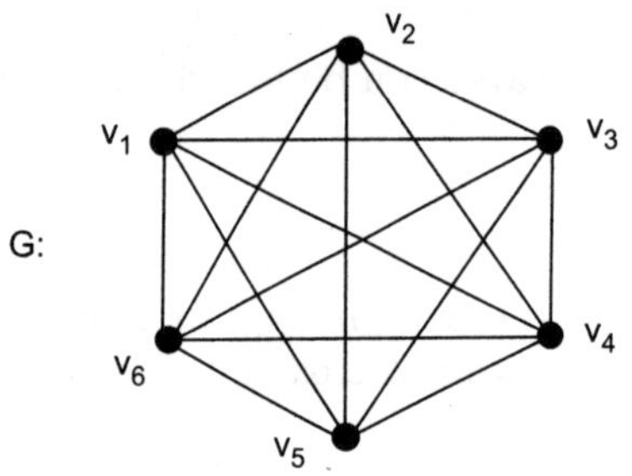

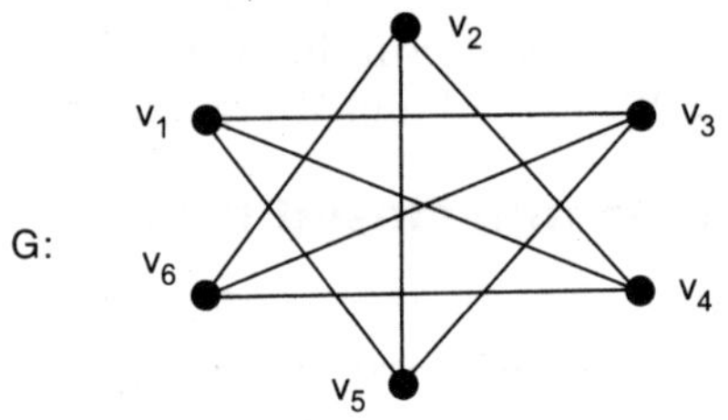

12.18.4 Product of Graphs

Suppose that $G_1: (V_1, E_1)$ and $G_2: (V_2, E_2)$ be two graphs. Then the product of graphs G_1 and G_2 is given as $(G_1 \times G_2)$ and is defined as $(G_1 \times G_2): (V, E)$. Where $V = (V_1 \times V_2)$ and the edge set E can be found out from the following relation.

If (u_1, u_2) and (v_1, v_2) be two vertices of $(G_1 \times G_2)$. Then there is an edge between them if

(*i*) $(u_1 = v_1$ and u_2 is adjacent to $v_2)$ or

(*ii*) $(u_1$ is adjacent to v_1 and $u_2 = v_2)$.

Consider the graphs G_1 and G_2 as

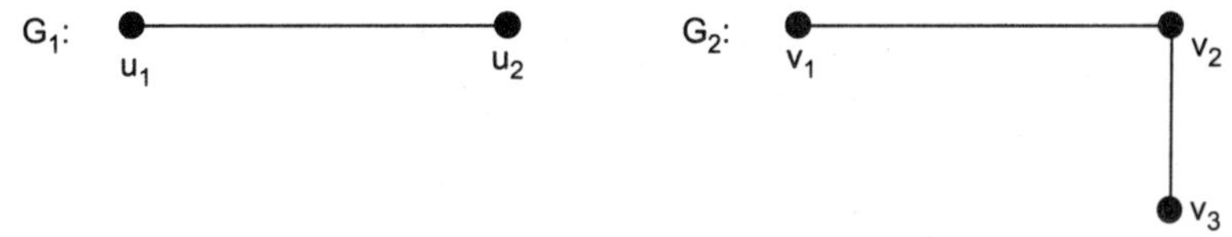

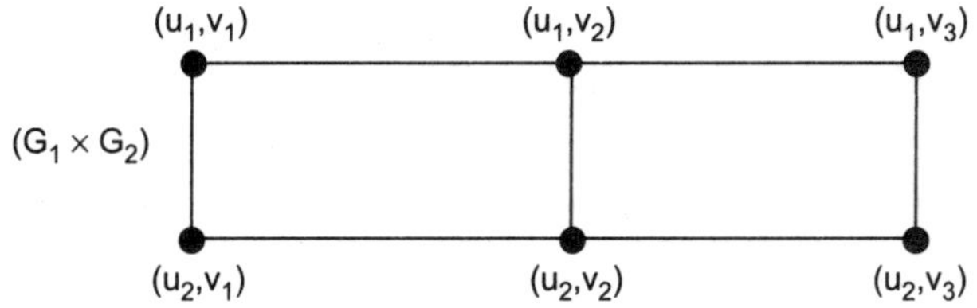

12.18.5 Composition

Suppose that G_1: (V_1, E_1) and G_2: (V_2, E_2) be two graphs. Then the composition of $G_1[G_2]$ and is defined as

$$G_1[G_2]: (V, E)$$

Where, $V = (V_1 \times V_2)$ and the edge set E can be found out from the following relation. If (u_1, u_2) and (v_1, v_2) be two vertices of $G_1[G_2]$, then there is an edge between them if

(i) u_1 is adjacent to v_1 or

(ii) ($u_1 = v_1$ and u_2 is adjacent to v_2)

Consider the graphs G_1 and G_2 as

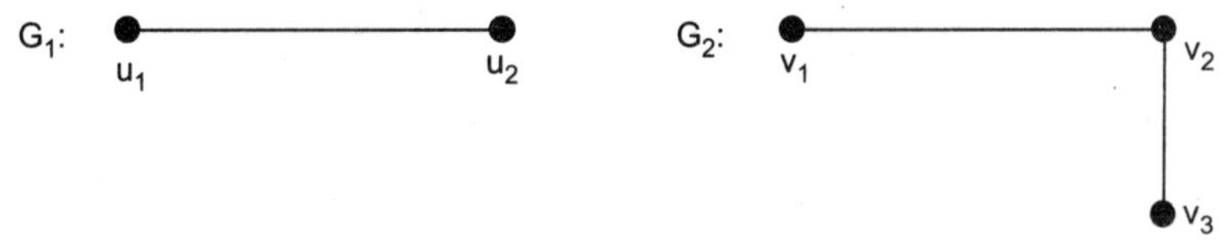

The composition $G_1[G_2]$ is defined as

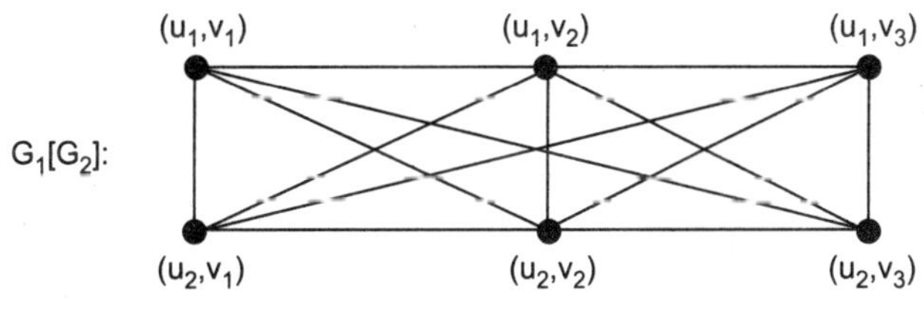

■ 12.19 FUSION OF GRAPHS

Let u and v be distinct vertices of a graph G, we can construct a new graph G_1 by fusing the two vertices. This means by replacing them by a single new vertex 'x' such that every edge that was incident with either 'u' or 'v' is now incident with x.

Consider the graph G as

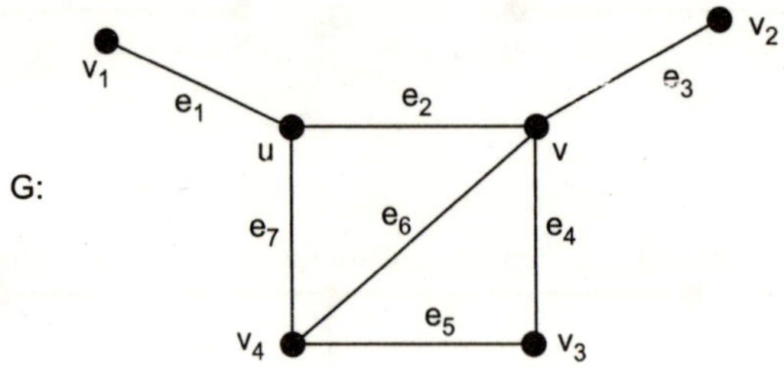

On fusing the vertices 'u' and 'v' the graph becomes

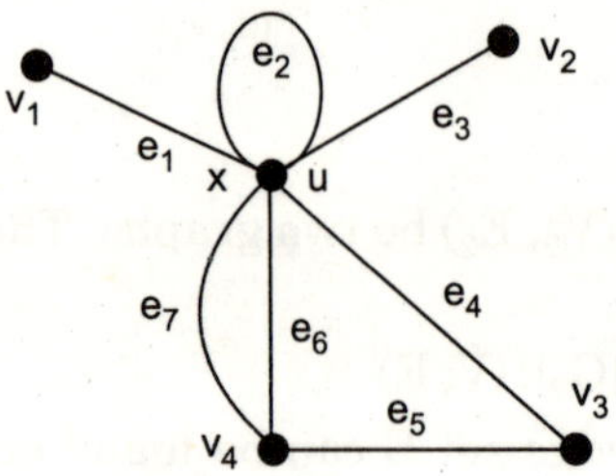

12.19.1 Adjacency Matrix (After fusion of two adjacent vertices)

The following steps are used to find the new adjacency matrix after fusion of two adjacent vertices 'u' and 'v':

Step 1. Replace the uth row by the sum of uth row and vth row. Similarly, replace the uth column by the sum of uth column and vth column.

Step 2. Delete the row and column corresponding to the vertex v. The resulting matrix is the new adjacency matrix.

Consider the graph G as

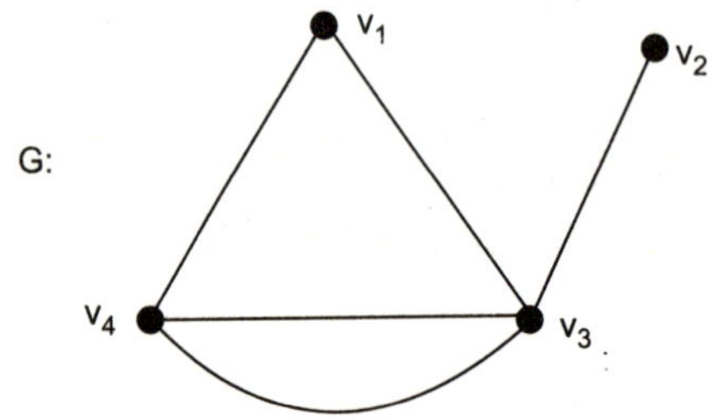

After fusing v_1 and v_4 we have the new graph G_1 as

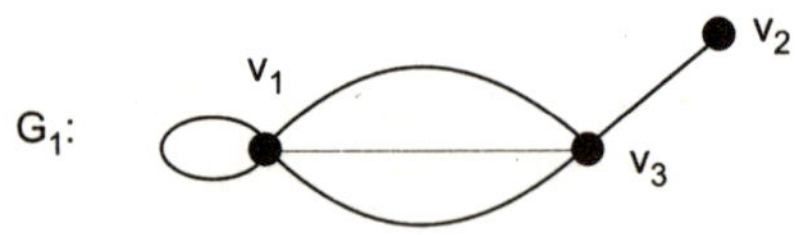

Relative to the ordering v_1, v_2, v_3 and v_4 we have $A(G) = \begin{pmatrix} 0 & 0 & 1 & 1 \\ 0 & 0 & 1 & 0 \\ 1 & 1 & 0 & 2 \\ 1 & 0 & 2 & 0 \end{pmatrix}$

Now on replacing Row $(v_1) \leftarrow$ Row (v_1) + Row (v_4) and Col $(v_1) \leftarrow$ Col (v_1) + Col (v_4), we get

$$A(G) = \begin{pmatrix} 1 & 0 & 3 & 1 \\ 0 & 0 & 1 & 0 \\ 3 & 1 & 0 & 2 \\ 1 & 0 & 2 & 0 \end{pmatrix}$$

On deleting the row and column corresponding to v_4 the adjacency matrix of G_1 is given as

$$A(G_1) = \begin{pmatrix} 1 & 0 & 3 \\ 0 & 0 & 1 \\ 3 & 1 & 0 \end{pmatrix}$$

12.19.2 Fusion Algorithm (Connectedness)

The following steps are used to check the connectedness of a graph G:

Step 1. Replace the graph G by its underlying simple graph. The adjacency matrix can be obtained by replacing all non-zero entries off the diagonal by 1 and all entries on the diagonal by 0.

Step 2. Fuse vertex v_1 to the first of the vertices $v_2, v_3 \ldots, v_n$ with which it is adjacent to give a new graph. Denote it by G in which the new vertex is also denoted by v_1.

Step 3. Carry out step1 on the new graph G.

Step 4. Carry out step 2 to step 3 repeatedly with v_1 until v_1 is not adjacent to any of the other vertices.

Step 5. Carry out steps 2 to 4 on the vertex v_2 of the latest graph and than on all the remaining vertices of the resulting graphs in turn. The final graph is empty and the number of its isolated vertices is the number of connected components of the initial graph G.

Consider the following graph G.

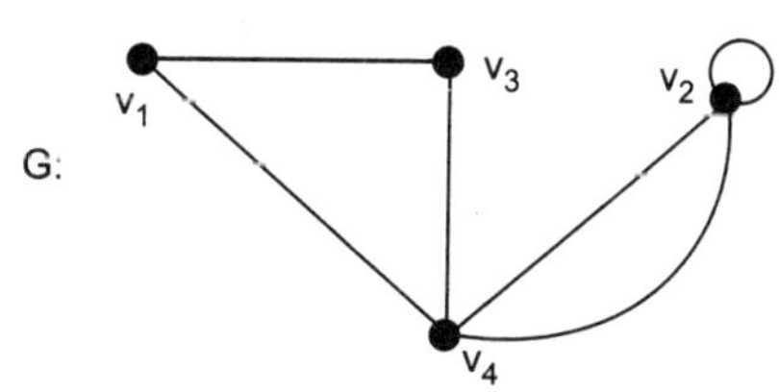

The adjacency matrix A (G) relative to the ordering v_1, v_2, v_3 and v_4 becomes

$$A(G) = \begin{pmatrix} 0 & 0 & 1 & 1 \\ 0 & 1 & 0 & 2 \\ 1 & 0 & 0 & 1 \\ 1 & 2 & 1 & 0 \end{pmatrix}$$

The underlying simple graph of G is given as

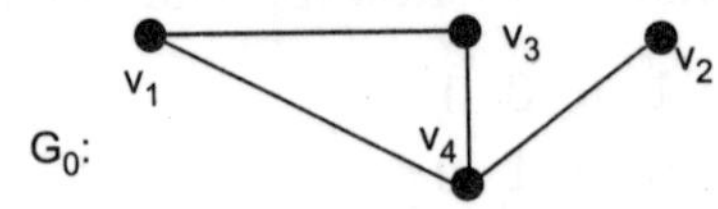

The adjacency matrix becomes

$$A(G_0) = \begin{pmatrix} 0 & 0 & 1 & 1 \\ 0 & 0 & 0 & 1 \\ 1 & 0 & 0 & 1 \\ 1 & 1 & 1 & 0 \end{pmatrix}$$

On fusing v_1 with v_3 we have the graph G as

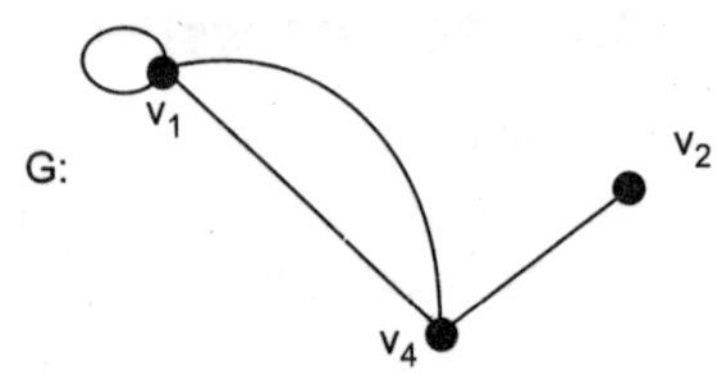

Therefore, on replacing Row $(v_1) \leftarrow$ Row (v_1) + Row (v_3); Col $(v_1) \leftarrow$ Col. (v_1) + Col. (v_3) and on removing the row and column corresponding to v_3 the adjacency matrix relative to the ordering v_1, v_2 and v_4 becomes

$$A(G) = \begin{pmatrix} 1 & 0 & 2 \\ 0 & 0 & 1 \\ 2 & 1 & 0 \end{pmatrix}$$

The underlying simple graph of G is given as

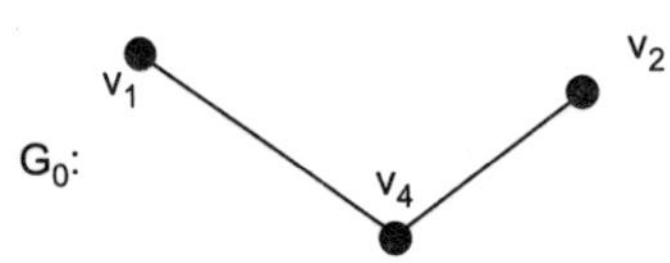

The adjacency matrix becomes

$$A(G_0) = \begin{pmatrix} 0 & 0 & 1 \\ 0 & 0 & 1 \\ 1 & 1 & 0 \end{pmatrix}$$

On fusing v_1 with v_4 we have the graph G as

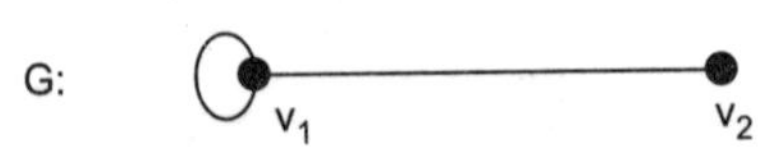

Therefore, on replacing Row $(v_1) \leftarrow$ Row (v_1) + Row (v_4); Col $(v_1) \leftarrow$ Col (v_1) + Col. (v_4) and on removing the row and column corresponding to v_4, the adjacency matrix relative to the ordering v_1 and v_2 becomes

$$A(G) = \begin{pmatrix} 1 & 1 \\ 1 & 0 \end{pmatrix}$$

The underlying spanning graph of G is given as

G_0: 

The adjacency matrix becomes

$$A(G_0) = \begin{pmatrix} 0 & 1 \\ 1 & 0 \end{pmatrix}$$

On fusing v_1 with v_2 we have the graph G as

G: 

Therefore, on replacing Row $(v_1) \leftarrow$ Row (v_1) + Row (v_2); Col $(v_1) \leftarrow$ Col (v_1) + Col (v_2) and on removing the row and column corresponding to v_2, the adjacency matrix relative to v_1 becomes

$$A(G) = (1)$$

The underlying spanning graph of G is given as

G_0: 

The adjacency matrix becomes

$$A(G_0) = (0)$$

As the final graph is empty, the process terminates. Here the number of isolated point is one. So, the graph is said to be connected.

SOLVED EXAMPLES

Example 1 *Draw the graph having the following matrix as its adjacency matrix*

$$\begin{pmatrix} 0 & 1 & 2 & 3 \\ 1 & 0 & 3 & 2 \\ 2 & 3 & 0 & 1 \\ 3 & 2 & 1 & 0 \end{pmatrix}$$

Solution: Given that the adjacency matrix is

$$\begin{pmatrix} 0 & 1 & 2 & 3 \\ 1 & 0 & 3 & 2 \\ 2 & 3 & 0 & 1 \\ 3 & 2 & 1 & 0 \end{pmatrix}$$

The order of the adjacency matrix is (4×4). So, the graph G has four vertices, say v_1, v_2, v_3 and v_4. Relative to the ordering v_1, v_2, v_3 and v_4 the graph G is given below.

G:

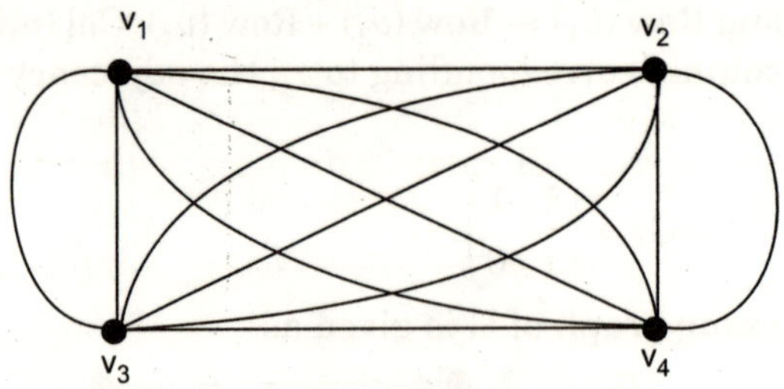

Example 2 *Write down the path matrix of the following graph.*

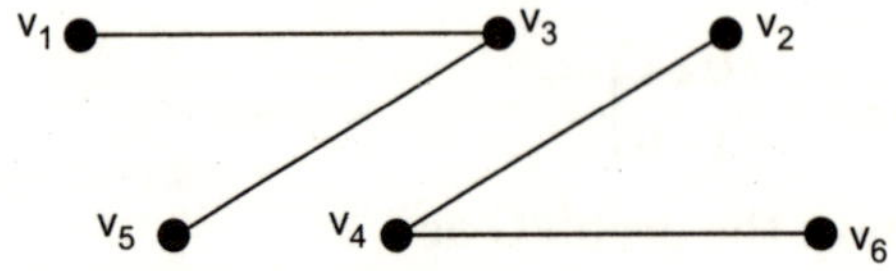

Solution: Given that the graph is

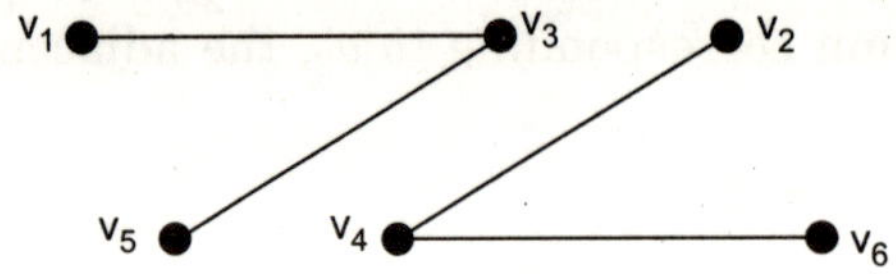

The path matrix relative to the ordering v_1, v_2, v_3, v_4, v_5 and v_6 is given as

$$P(G) = \begin{pmatrix} 0 & 0 & 1 & 0 & 1 & 0 \\ 0 & 0 & 0 & 1 & 0 & 1 \\ 1 & 0 & 0 & 0 & 1 & 0 \\ 0 & 1 & 0 & 0 & 0 & 1 \\ 1 & 0 & 1 & 0 & 0 & 0 \\ 0 & 1 & 0 & 1 & 0 & 0 \end{pmatrix}$$

Example 3 *Write the adjacency matrix of the following graphs*

(a)

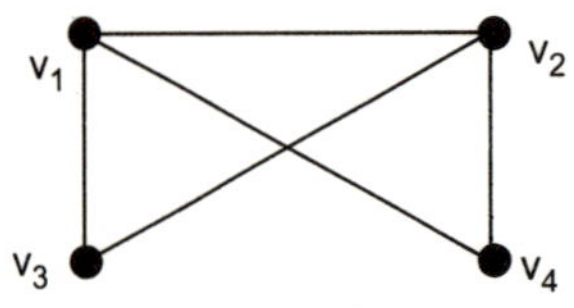

(b)

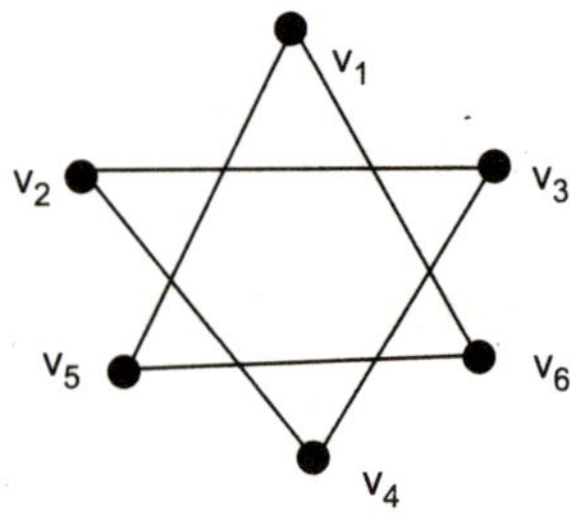

Solution: (*a*) Given graph is

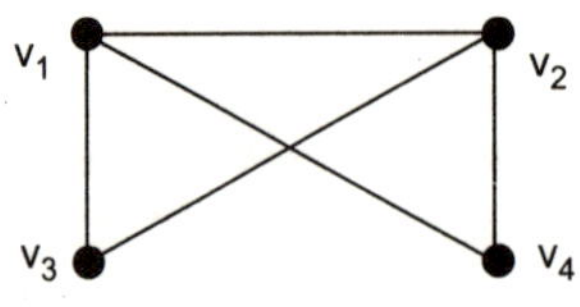

The adjacency matrix relative to the ordering v_1, v_2, v_3 and v_4 is given as

$$A(G) = \begin{pmatrix} 0 & 1 & 1 & 1 \\ 1 & 0 & 1 & 1 \\ 1 & 1 & 0 & 0 \\ 1 & 1 & 0 & 0 \end{pmatrix}$$

(*b*) Given graph is

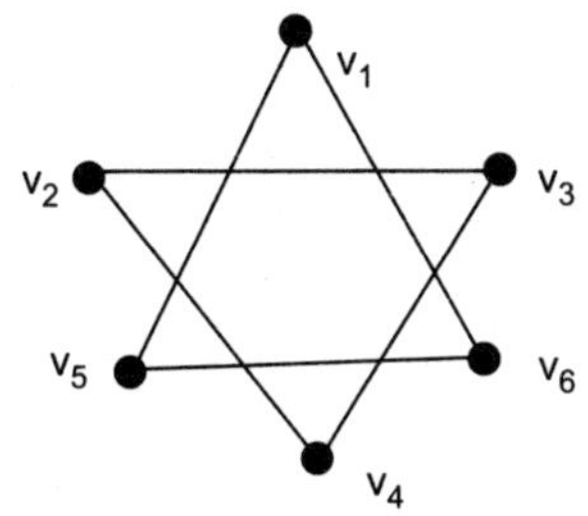

The adjacency matrix relative to the ordering v_1, v_2, v_3, v_4, v_5 and v_6 is given as

$$A(G) = \begin{pmatrix} 0 & 0 & 0 & 0 & 1 & 1 \\ 0 & 0 & 1 & 1 & 0 & 0 \\ 0 & 1 & 0 & 1 & 0 & 0 \\ 0 & 1 & 1 & 0 & 0 & 0 \\ 1 & 0 & 0 & 0 & 0 & 1 \\ 1 & 0 & 0 & 0 & 1 & 0 \end{pmatrix}$$

Example 4 *Write the adjacency matrix of the following weighted graph.*

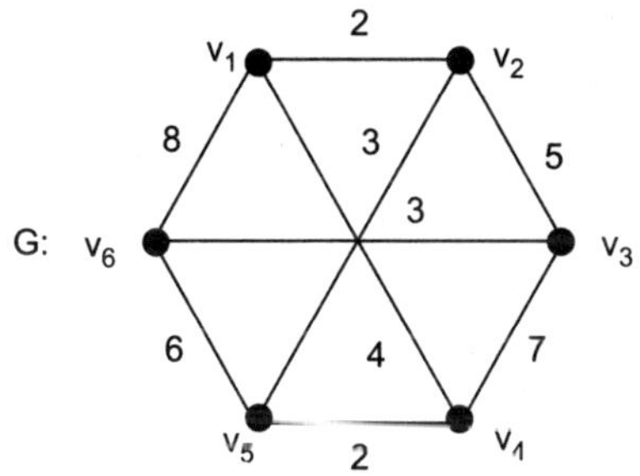

Solution: The weighted graph is

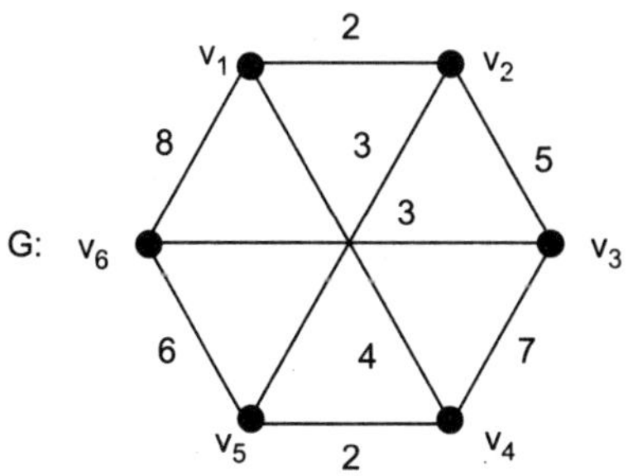

The adjacency matrix relative to the ordering v_1, v_2, v_3, v_4, v_5 and v_6 is given as

$$A(G) = \begin{pmatrix} 0 & 2 & 0 & 4 & 0 & 8 \\ 2 & 0 & 5 & 0 & 3 & 0 \\ 0 & 5 & 0 & 7 & 0 & 3 \\ 4 & 0 & 7 & 0 & 2 & 0 \\ 0 & 3 & 0 & 2 & 0 & 6 \\ 8 & 0 & 3 & 0 & 6 & 0 \end{pmatrix}$$

Example 5 *Write down the incidence matrix of the following graph G.*

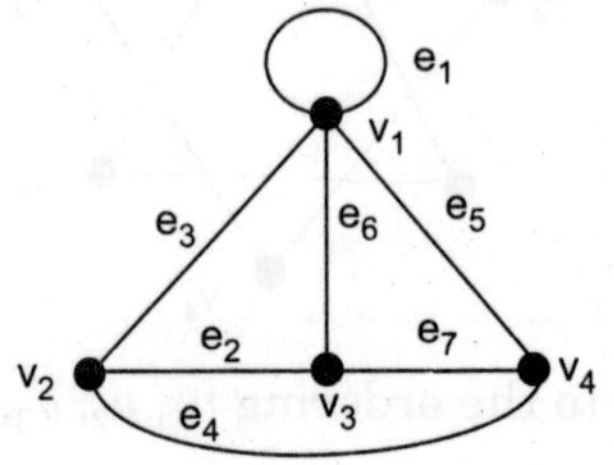

Solution: In the above graph G, V = $\{v_1, v_2, v_3, v_4\}$ and E = $\{e_1, e_2, e_3, e_4, e_5, e_6, e_7\}$. Therefore the order of incidence matrix is (4×7). Relative to the ordering of V and E, the incidence matrix is given as

$$I(G) = \begin{pmatrix} 1 & 0 & 1 & 0 & 1 & 1 & 0 \\ 0 & 1 & 1 & 1 & 0 & 0 & 0 \\ 0 & 1 & 0 & 0 & 0 & 1 & 1 \\ 0 & 0 & 0 & 1 & 1 & 0 & 1 \end{pmatrix}$$

Example 6 *Find the union of the following graphs.*

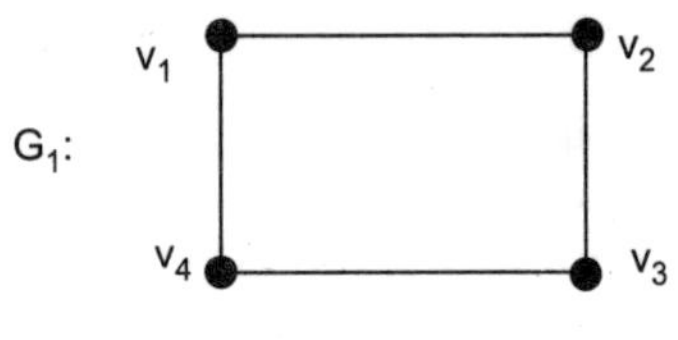

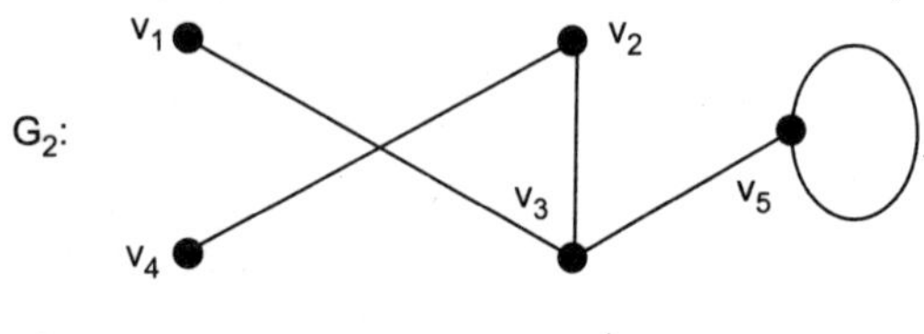

Solution: Here, $V(G_1) = \{v_1, v_2, v_3, v_4\}$

and $V(G_2) = \{v_1, v_2, v_3, v_4, v_5\}$.

Therefore, $V(G_1 \cup G_2) = \{v_1, v_2, v_3, v_4, v_5\}$.

Similarly, $E(G_1 \cup G_2) = \{(v_1, v_2), (v_2, v_3), (v_3, v_4), (v_4, v_1), (v_1, v_3), (v_2, v_4), (v_3, v_5), (v_5, v_5)\}$.

Therefore, the graph $(G_1 \cup G_2)$ becomes

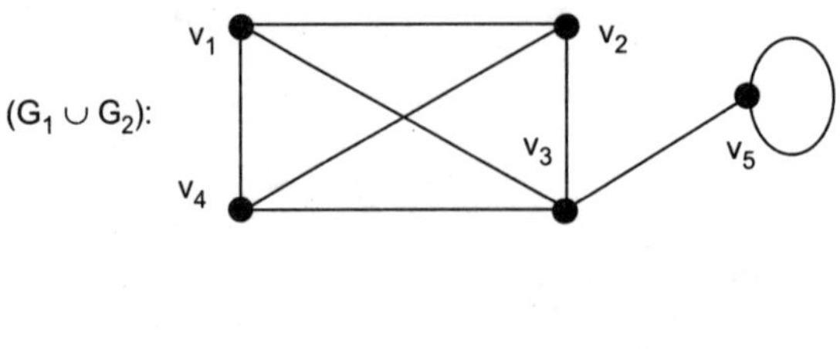

Example 7 *Write the adjacency matrix of the following directed weighted graph*

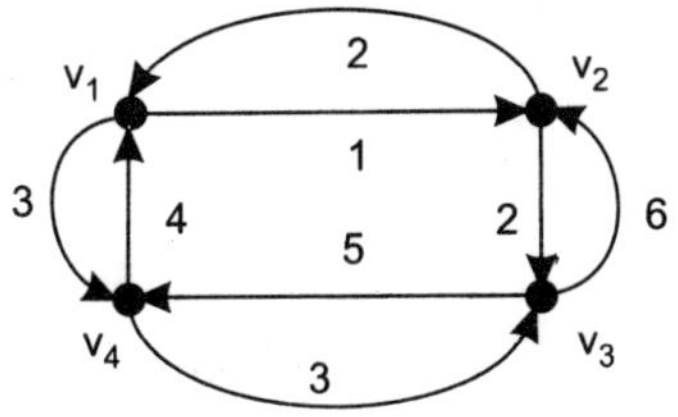

Solution: In the above directed weighted graph the total number of vertices are 4. So, the adjacency matrix is of order (4 × 4). The adjacency matrix relative to the ordering v_1, v_2, v_3 and v_4 is given as below.

$$A(G) = \begin{pmatrix} 0 & 1 & 0 & 3 \\ 2 & 0 & 2 & 0 \\ 0 & 6 & 0 & 5 \\ 4 & 0 & 3 & 0 \end{pmatrix}$$

Example 8 *Find the intersection of the following graphs.*

G_1:

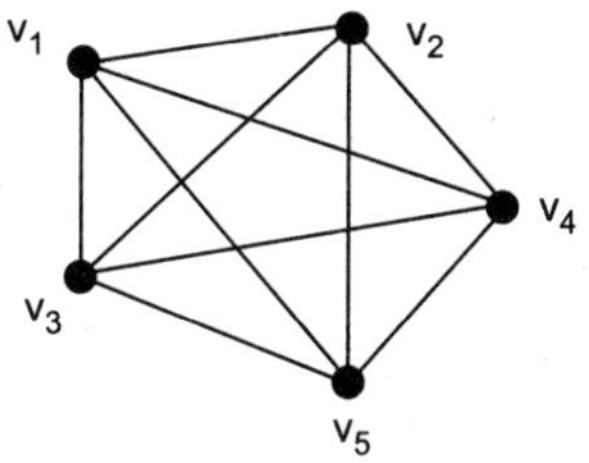

G_2:

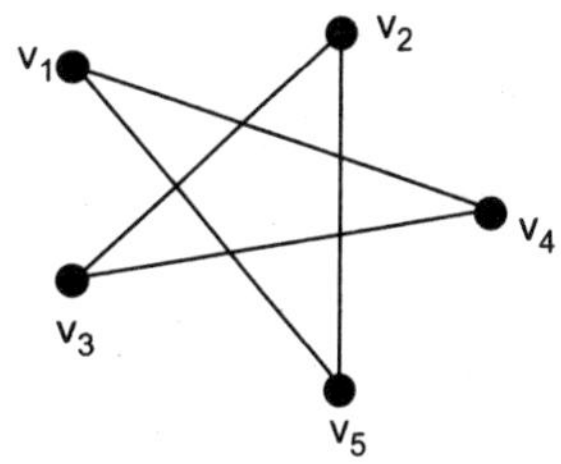

Solution: Here, $V(G_1) = \{v_1, v_2, v_3, v_4, v_5\}$

and $V(G_2) = \{v_1, v_2, v_3, v_4, v_5\}$.

Therefore, $V(G_1 \cap G_2) = \{v_1, v_2, v_3, v_4, v_5\}$.

Similarly, $E(G_1 \cap G_2) = \{(v_1, v_4), (v_4, v_3), (v_3, v_2), (v_2, v_5), (v_5, v_1)\}$.

Therefore, the graph $(G_1 \cap G_2)$ becomes

$(G_1 \cap G_2)$:

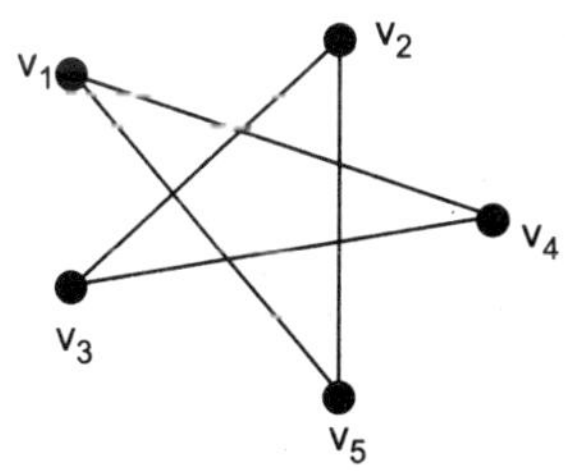

Example 9 *Find the complement of the following graphs.*

(a)

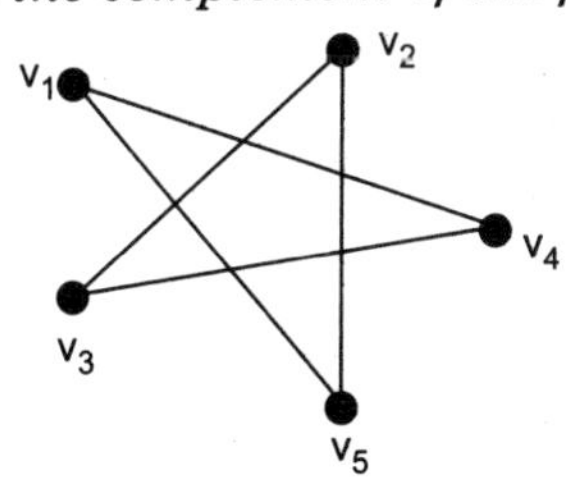

(b)

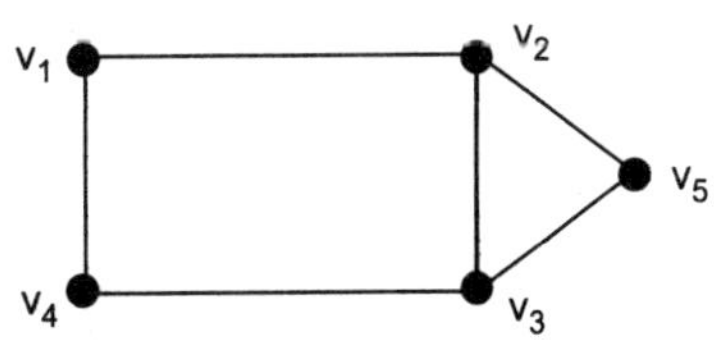

Solution: (*a*) To obtain the complement of G, find the complete graph with the same vertices. This is given below.

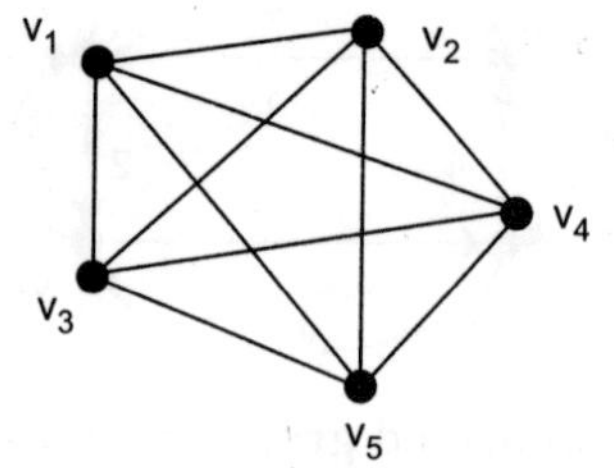

On deleting the edges of G, the complement $\overline{G}$ of G is given below.

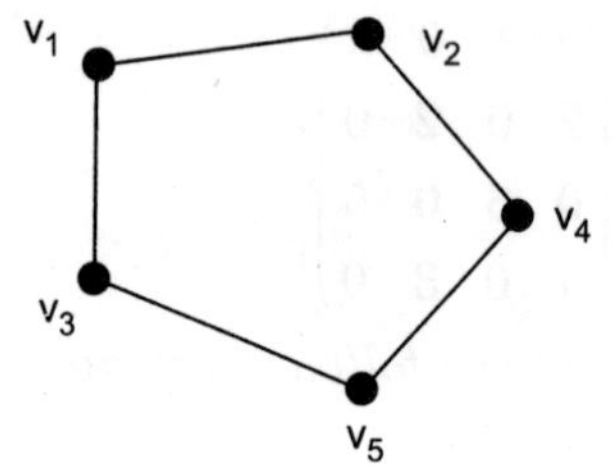

(*b*) To obtain the complement of G, find the complete graph with the same vertices. This is given below.

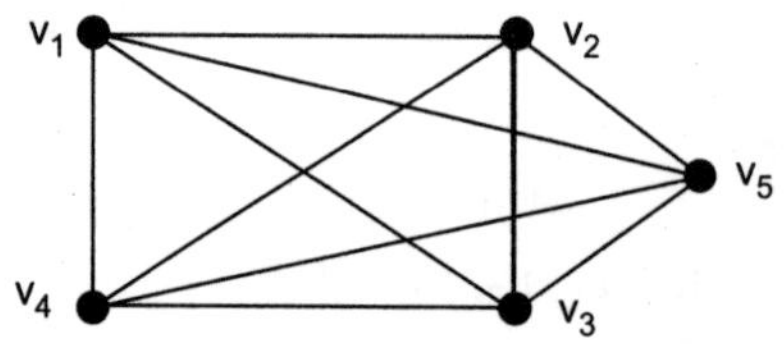

On deleting the edges of G, the complement $\overline{G}$ of G is given below.

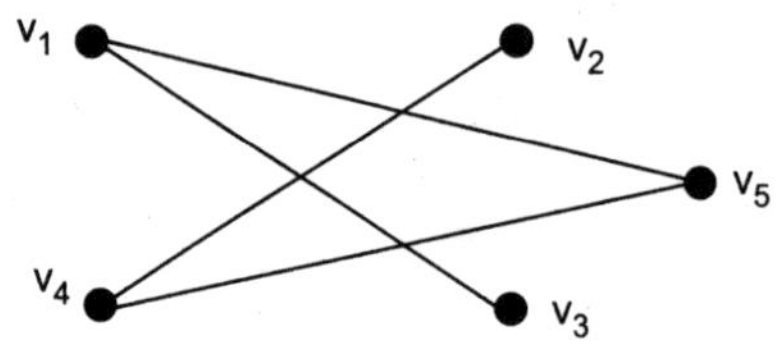

Example 10 *If G_1 and G_2 be two graphs given below, then find the product of graphs $(G_1 \times G_2)$. Where*

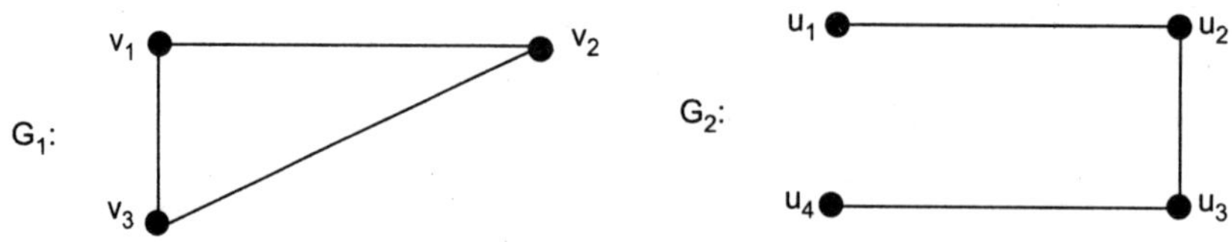

Solution: Here $V(G_1) = \{v_1, v_2, v_3\}$

and $V(G_2) = \{u_1, u_2, u_3, u_4\}$.

Therefore, $V(G_1 \times G_2) = \{(v_1, u_1), (v_1, u_2), (v_1, u_3), (v_1, u_4), (v_2, u_1), (v_2, u_2), (v_2, u_3), (v_2, u_4), (v_3, u_1), (v_3, u_2), (v_3, u_3), (v_3, u_4)\}$

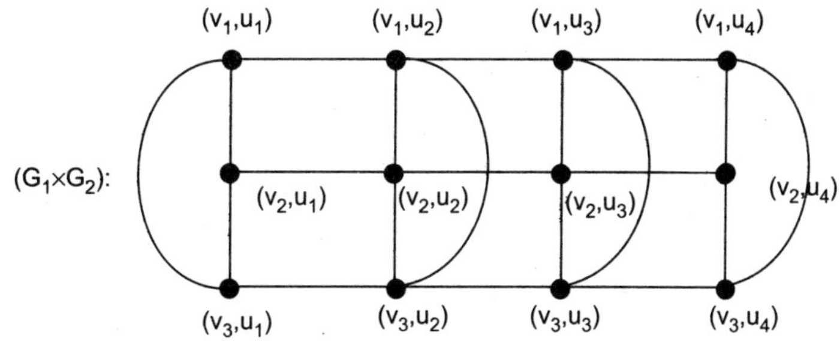

Example 11 *Given G_1 and G_2 be two graphs. Find the composition $G_1[G_2]$ where*

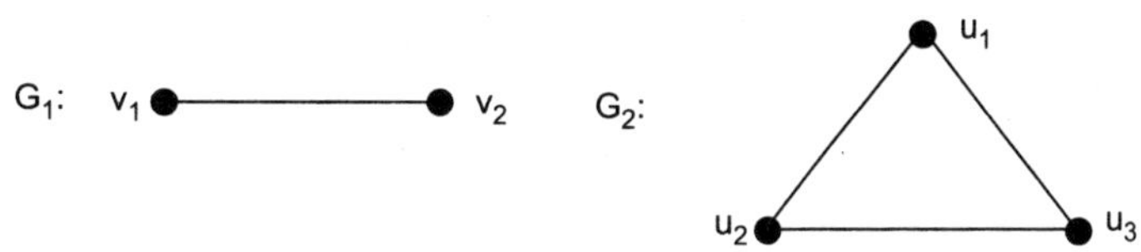

Solution: In the above graph

$$V(G_1) = \{v_1, v_2\}$$

and $$V(G_2) = \{u_1, u_2, u_3\}.$$

Therefore, the vertex set of $G_1[G_2]$ is $\{(v_1, u_1), (v_1, u_2), (v_1, u_3), (v_2, u_1), (v_2, u_2), (v_2, u_3)\}$. Thus the composition graph $G_1[G_2]$ is given below.

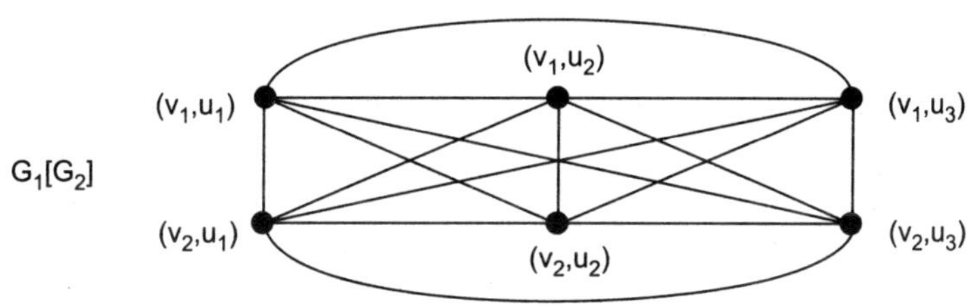

Example 12 *Let G be the graph given below.*

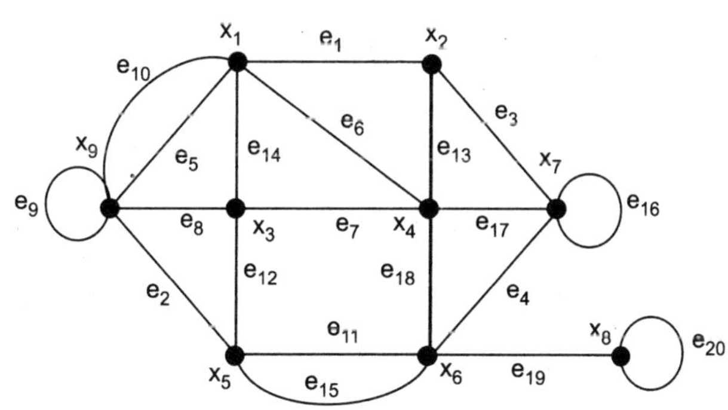

(a) *Find G – U; where* $U = \{x_1, x_3, x_5, x_7\}$

(b) Find G(U); where $U = \{x_1, x_3, x_4, x_9\}$

(c) Find G – V; where $V = \{e_2, e_5, e_8, e_{12}, e_{14}, e_1, e_6, e_{18}, e_4, e_{19}, e_{20}\}$

(d) Find G(V); where $V = \{e_1, e_6, e_7, e_{11}, e_{15}\}$

Solution: (*a*) Given $U = \{x_1, x_3, x_5, x_7\}$.

Therefore, G – U becomes

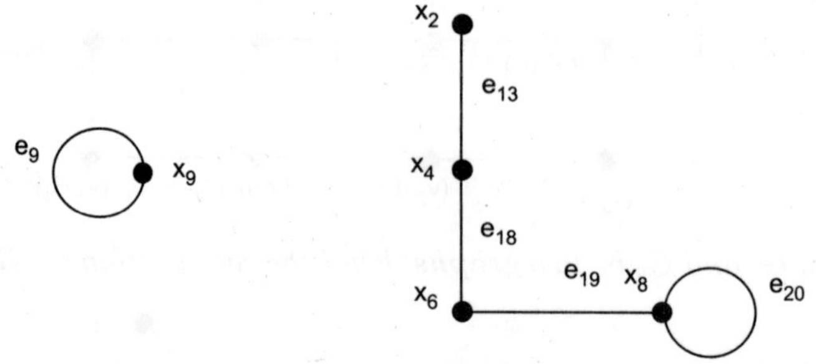

(*b*) Given: $U = \{x_1, x_3, x_4, x_9\}$. Therefore, G(U) becomes

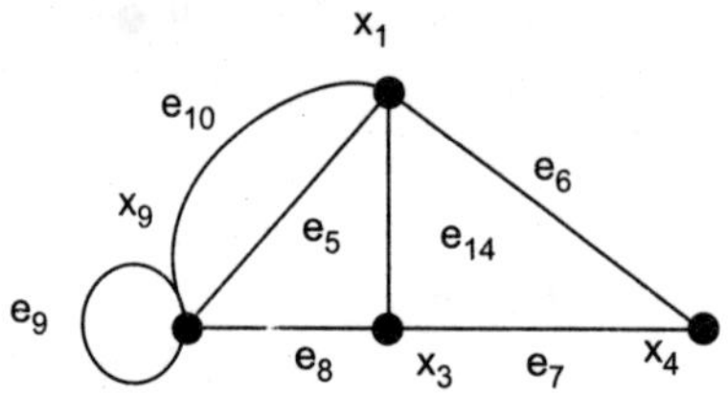

(*c*) Given: $V = \{e_2, e_5, e_8, e_{12}, e_{14}, e_1, e_6, e_{18}, e_4, e_{19}, e_{20}\}$. Therefore, G – V becomes

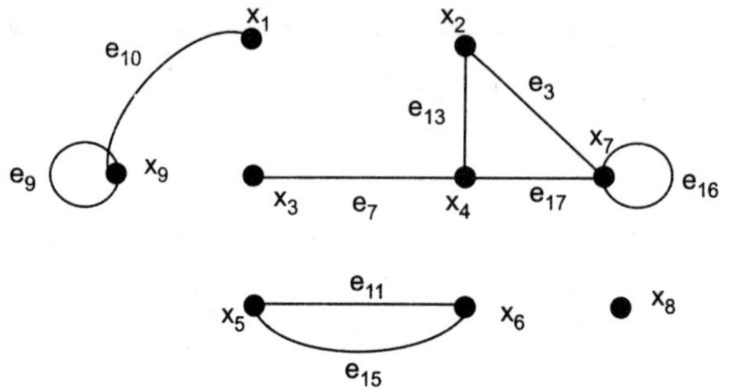

(*d*) Given: $V = \{e_1, e_6, e_7, e_{11}, e_{15}\}$. Therefore, G(V) becomes

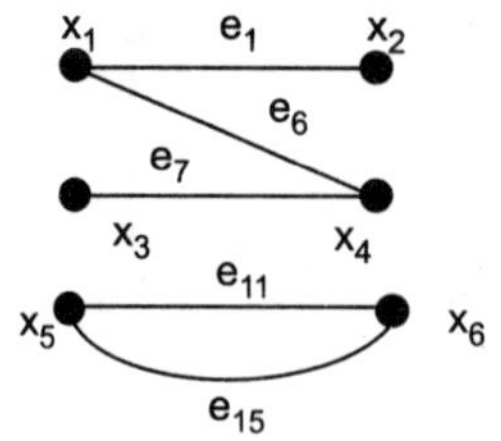

Example 13 *Determine whether the graph given below by its adjacency matrix is connected or not.*

$$\begin{pmatrix} 0 & 1 & 1 & 1 \\ 1 & 0 & 1 & 0 \\ 1 & 1 & 0 & 1 \\ 1 & 0 & 1 & 0 \end{pmatrix}$$

Solution: The adjacency matrix A is given as

$$A = \begin{pmatrix} 0 & 1 & 1 & 1 \\ 1 & 0 & 1 & 0 \\ 1 & 1 & 0 & 1 \\ 1 & 0 & 1 & 0 \end{pmatrix}$$

Here, the number of vertices $(n) = 4$. Let $B = A + A^2 + A^3$

Now,
$$A^2 = \begin{pmatrix} 0 & 1 & 1 & 1 \\ 1 & 0 & 1 & 0 \\ 1 & 1 & 0 & 1 \\ 1 & 0 & 1 & 0 \end{pmatrix}\begin{pmatrix} 0 & 1 & 1 & 1 \\ 1 & 0 & 1 & 0 \\ 1 & 1 & 0 & 1 \\ 1 & 0 & 1 & 0 \end{pmatrix} = \begin{pmatrix} 3 & 1 & 2 & 1 \\ 1 & 2 & 1 & 2 \\ 2 & 1 & 3 & 1 \\ 1 & 2 & 1 & 2 \end{pmatrix}$$

Again,
$$A^3 = A^2A = \begin{pmatrix} 3 & 1 & 2 & 1 \\ 1 & 2 & 1 & 2 \\ 2 & 1 & 3 & 1 \\ 1 & 2 & 1 & 2 \end{pmatrix}\begin{pmatrix} 0 & 1 & 1 & 1 \\ 1 & 0 & 1 & 0 \\ 1 & 1 & 0 & 1 \\ 1 & 0 & 1 & 0 \end{pmatrix} = \begin{pmatrix} 4 & 5 & 5 & 5 \\ 5 & 2 & 5 & 2 \\ 5 & 5 & 4 & 5 \\ 5 & 2 & 5 & 2 \end{pmatrix}$$

Therefore,
$$B = A + A^2 + A^3 = \begin{pmatrix} 7 & 7 & 8 & 7 \\ 7 & 4 & 7 & 4 \\ 8 & 7 & 7 & 7 \\ 7 & 4 & 7 & 4 \end{pmatrix}$$

As all $b_{ij} \neq 0$ for $i \neq j$, the graph G is connected.

Example 14 *Draw a complete bipartite graph on two and four vertices.*

Solution: Complete bipartite graph on m and n is the simple graph whose vertex set is partitioned into sets V_1 and V_2 with m and n vertices respectively. Generally, denoted by K_{mn}.

The complete bipartite graph on two and four vertices is shown in the following figure.

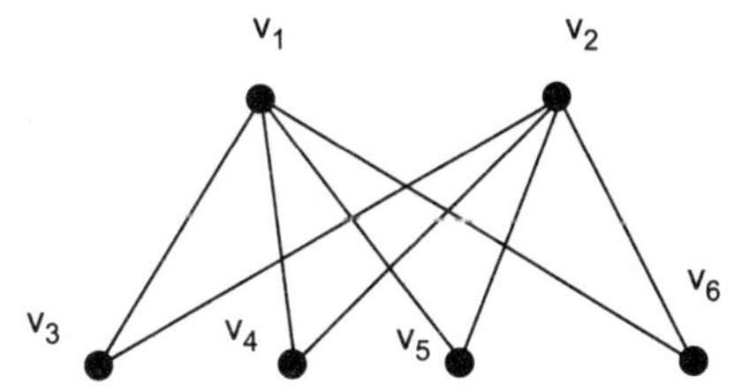

Example 15 *Use the fusion algorithm to determine whether the graph given below by its adjacency matrix is connected or not.*

$$\begin{pmatrix} 0 & 2 & 1 & 0 & 0 & 0 \\ 2 & 0 & 0 & 1 & 0 & 0 \\ 1 & 0 & 0 & 2 & 0 & 0 \\ 0 & 1 & 2 & 0 & 0 & 0 \\ 0 & 0 & 0 & 0 & 0 & 3 \\ 0 & 0 & 0 & 0 & 3 & 0 \end{pmatrix}$$

Solution: The adjacency matrix of the graph G is given as

$$A(G) = \begin{pmatrix} 0 & 2 & 1 & 0 & 0 & 0 \\ 2 & 0 & 0 & 1 & 0 & 0 \\ 1 & 0 & 0 & 2 & 0 & 0 \\ 0 & 1 & 2 & 0 & 0 & 0 \\ 0 & 0 & 0 & 0 & 0 & 3 \\ 0 & 0 & 0 & 0 & 3 & 0 \end{pmatrix}$$

Therefore, the graph G becomes

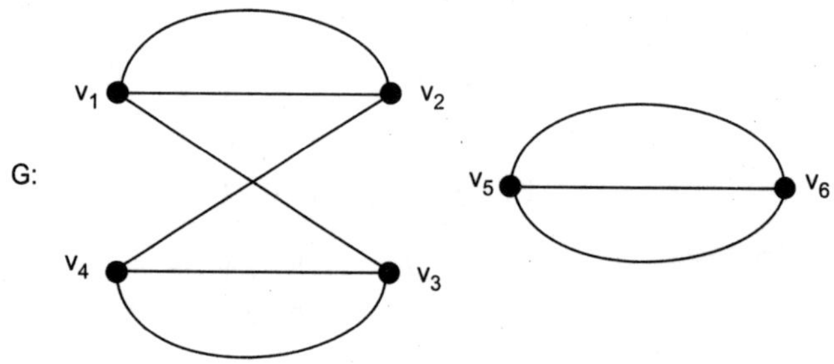

The underlying simple graph of G is given as

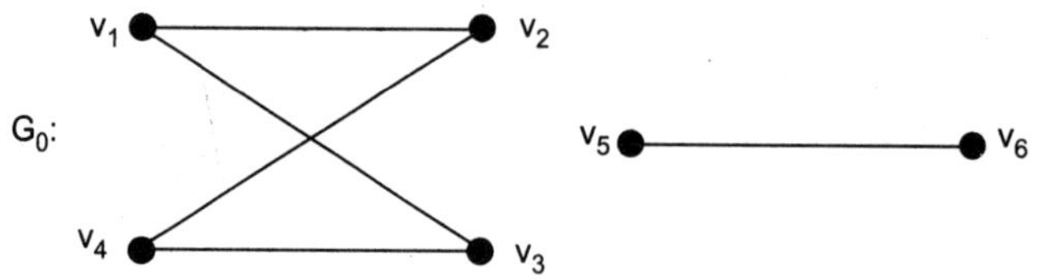

The adjacency matrix is given as

$$A(G_0) = \begin{pmatrix} 0 & 1 & 1 & 0 & 0 & 0 \\ 1 & 0 & 0 & 1 & 0 & 0 \\ 1 & 0 & 0 & 1 & 0 & 0 \\ 0 & 1 & 1 & 0 & 0 & 0 \\ 0 & 0 & 0 & 0 & 0 & 1 \\ 0 & 0 & 0 & 0 & 1 & 0 \end{pmatrix}$$

On fusing vertex v_1 with v_2 we have the graph G as

G:

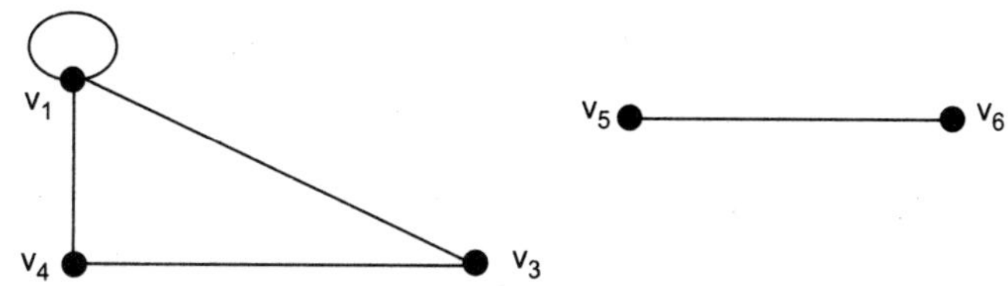

So, on replacing Row $(v_1) \leftarrow$ Row (v_1) + Row (v_2); Col $(v_1) \leftarrow$ Col (v_1) + Col (v_2) and on removing the row and column corresponding to v_2 the adjacency matrix becomes

$$A(G) = \begin{pmatrix} 1 & 1 & 1 & 0 & 0 \\ 1 & 0 & 1 & 0 & 0 \\ 1 & 1 & 0 & 0 & 0 \\ 0 & 0 & 0 & 0 & 1 \\ 0 & 0 & 0 & 1 & 0 \end{pmatrix}$$

The underlying simple graph of G is given as

G_0:

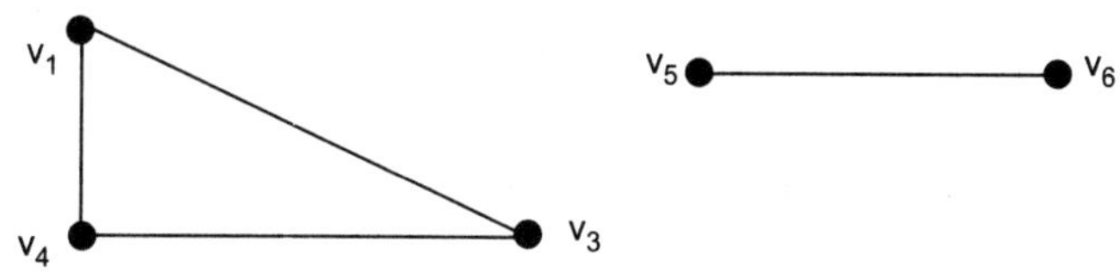

The adjacency matrix becomes

$$A(G_0) = \begin{pmatrix} 0 & 1 & 1 & 0 & 0 \\ 1 & 0 & 1 & 0 & 0 \\ 1 & 1 & 0 & 0 & 0 \\ 0 & 0 & 0 & 0 & 1 \\ 0 & 0 & 0 & 1 & 0 \end{pmatrix}$$

On fusing v_1 with v_3 we have the graph G as

G:

So, on replacing Row $(v_1) \leftarrow$ Row (v_1) + Row (v_3); Col $(v_1) \leftarrow$ Col (v_1) + Col (v_3) and on removing the row and column corresponding to v_3 the adjacency matrix becomes

$$A(G) = \begin{pmatrix} 1 & 2 & 0 & 0 \\ 2 & 0 & 0 & 0 \\ 0 & 0 & 0 & 1 \\ 0 & 0 & 1 & 0 \end{pmatrix}$$

The underlying simple graph of G is given as

G_0:

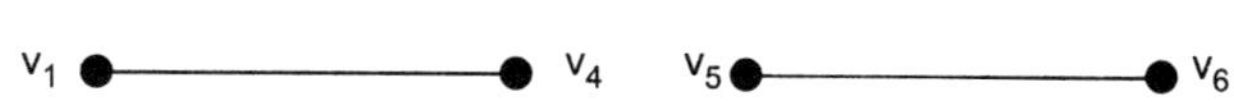

The adjacency matrix becomes

$$A(G_0) = \begin{pmatrix} 0 & 1 & 0 & 0 \\ 1 & 0 & 0 & 0 \\ 0 & 0 & 0 & 1 \\ 0 & 0 & 1 & 0 \end{pmatrix}$$

On fusing v_1 with v_4 we have the graph G as

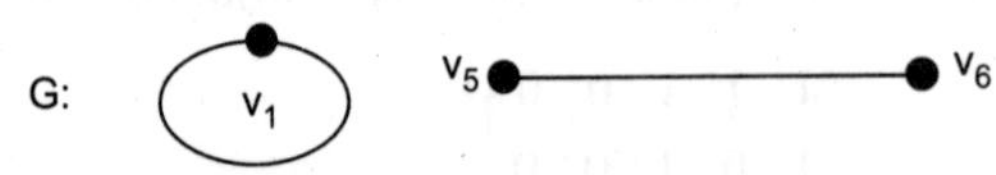

So, on replacing Row $(v_1) \leftarrow$ Row (v_1) + Row (v_4); Col $(v_1) \leftarrow$ Col (v_1) + Col (v_4) and on removing the row and column corresponding to v_4 the adjacency matrix becomes

$$A(G) = \begin{pmatrix} 1 & 0 & 0 \\ 0 & 0 & 1 \\ 0 & 1 & 0 \end{pmatrix}$$

The underlying simple graph of G is given as

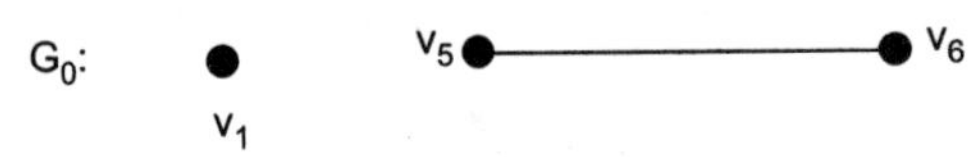

The adjacency matrix becomes

$$A(G_0) = \begin{pmatrix} 0 & 0 & 0 \\ 0 & 0 & 1 \\ 0 & 1 & 0 \end{pmatrix}$$

On fusing v_5 with v_6 we have the graph G as

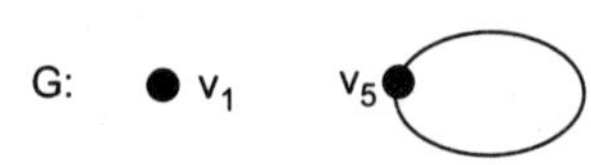

So, on replacing Row $(v_5) \leftarrow$ Row (v_5) + Row (v_6); Col $(v_5) \leftarrow$ Col (v_5) + Col (v_6) and on removing the row and column corresponding to v_6 the adjacency matrix becomes

$$A(G) = \begin{pmatrix} 0 & 0 \\ 0 & 1 \end{pmatrix}$$

The underlying simple graph of G is given as

The adjacency matrix becomes

$$A(G_0) = \begin{pmatrix} 0 & 0 \\ 0 & 0 \end{pmatrix}$$

As the final graph is empty, the process terminates. The number of isolated points is the order of the matrix *i.e.*, two. So, the graph is not connected.

Example 16 *Draw the following graphs.*

(*i*) *3 regular but not complete*
(*ii*) *3 regular and complete*
(*iii*) *4 regular but not complete*
(*iv*) *2 regular but not complete.*

Solution:

(*i*) In the graph given below, the degree of every vertex is 3 but for vertices v_1 and v_6 there is no edge. Hence, the graph is 3 regular but not complete.

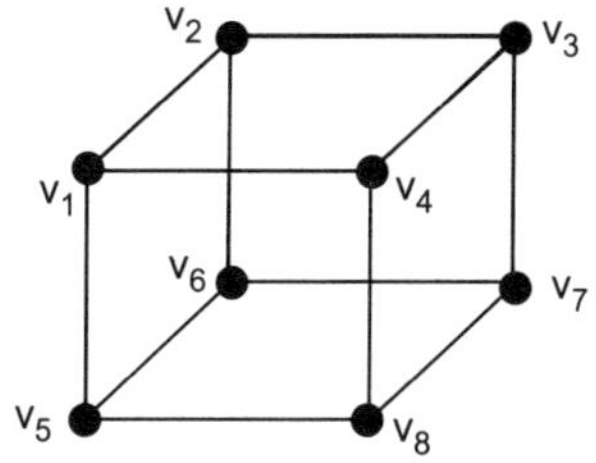

(*ii*) In the graph given below, the degree of every vertex is 3 and for any two vertices v_i and v_j there is an edge. Hence the graph is 3 regular and complete.

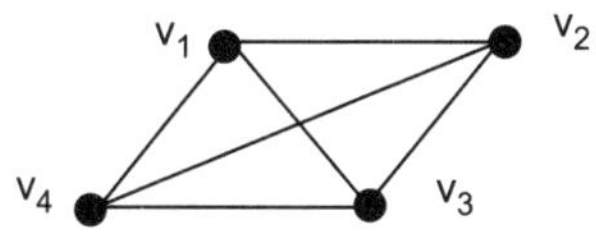

(*iii*) In the graph given below, the degree of every vertex is 4 and for the vertices v_2 and v_5 there exists no edge. Hence, the graph is 4 regular but not complete.

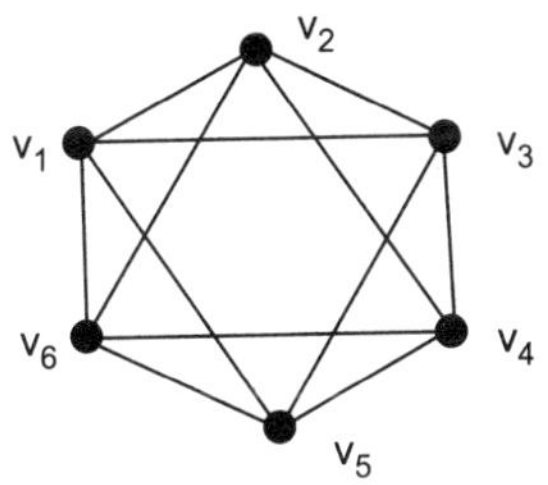

(*iv*) In the graph given below, the degree of every vertex is 2 and for the vertices v_2 and v_5 there exists no edge. Hence, the graph is 2 regular but not complete.

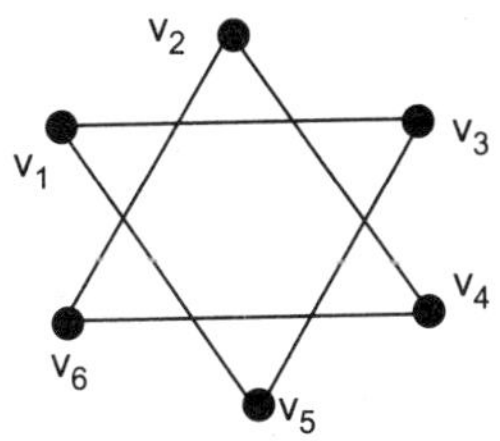

Example 17 *Find whether the graph given below is connected or not.*

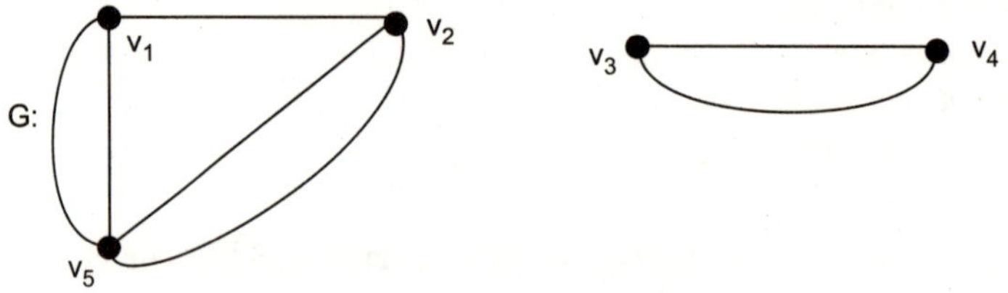

Solution: The adjacency matrix A(G) of the above graph relative to the ordering v_1, v_2, v_3, v_4 and v_5 is given as

$$A(G) = \begin{pmatrix} 0 & 1 & 0 & 0 & 2 \\ 1 & 0 & 0 & 0 & 2 \\ 0 & 0 & 0 & 2 & 0 \\ 0 & 0 & 2 & 1 & 0 \\ 2 & 2 & 0 & 0 & 0 \end{pmatrix}$$

Here, the number of vertices $(n) = 5$. Let $B = A + A^2 + A^3 + A^4$. Therefore, we get

$$A^2 = AA = \begin{pmatrix} 5 & 4 & 0 & 0 & 2 \\ 4 & 5 & 0 & 0 & 2 \\ 0 & 0 & 4 & 2 & 0 \\ 0 & 0 & 2 & 5 & 0 \\ 2 & 2 & 0 & 0 & 8 \end{pmatrix}; A^3 = A^2 A = \begin{pmatrix} 8 & 9 & 0 & 0 & 18 \\ 9 & 8 & 0 & 0 & 18 \\ 0 & 0 & 4 & 10 & 0 \\ 0 & 0 & 10 & 9 & 0 \\ 18 & 18 & 0 & 0 & 8 \end{pmatrix}$$

$$A^4 = A^3 A = \begin{pmatrix} 45 & 44 & 0 & 0 & 34 \\ 44 & 45 & 0 & 0 & 34 \\ 0 & 0 & 20 & 18 & 0 \\ 0 & 0 & 18 & 29 & 0 \\ 34 & 34 & 0 & 0 & 72 \end{pmatrix}$$

Therefore, $$B = A + A^2 + A^3 + A^4$$

$$= \begin{pmatrix} 58 & 58 & 0 & 0 & 56 \\ 58 & 58 & 0 & 0 & 56 \\ 0 & 0 & 28 & 32 & 0 \\ 0 & 0 & 32 & 44 & 0 \\ 56 & 56 & 0 & 0 & 88 \end{pmatrix}$$

As some $b_{ij} = 0$ for $i \neq j$, the graph G is not connected.

Example 18 *Find the complement graph of the following graph G, where*

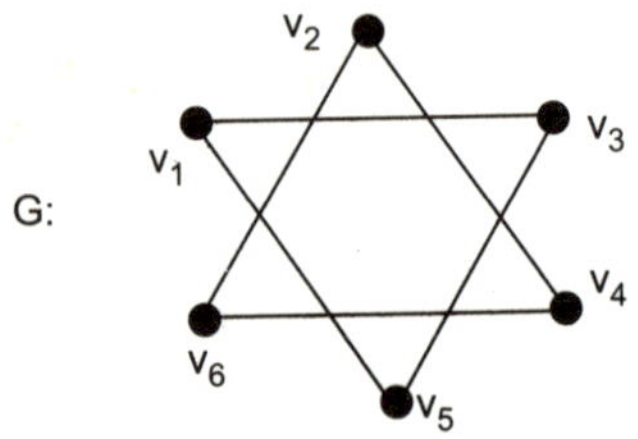

Solution: On considering the above graph G, construct the complete graph with the same vertices as G. The graph is given below.

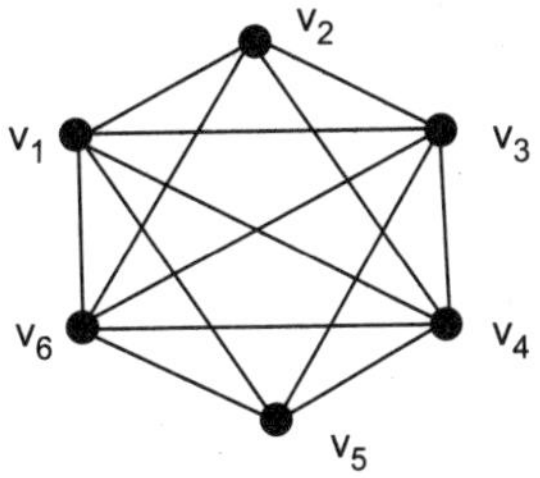

On deleting the edges of the graph G, the complement $\overline{G}$ of G is given below.

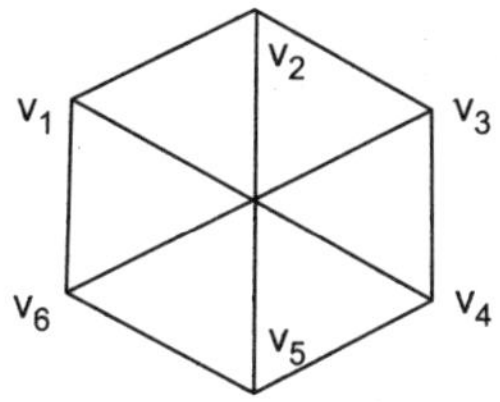

EXERCISES

1. Using fusion algorithm determine whether the graph given below is connected or not.

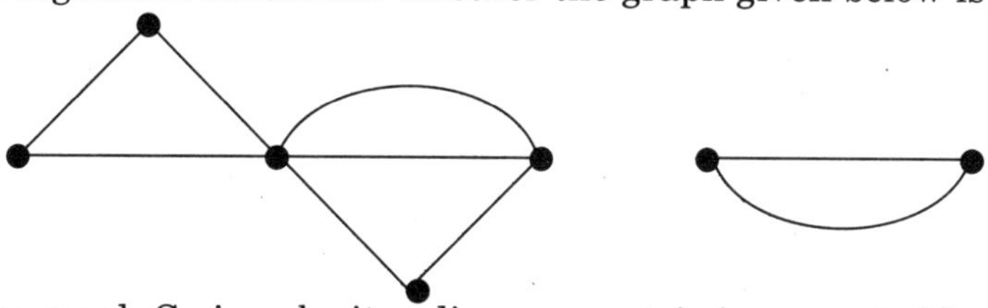

2. Show that the graph G given by its adjacency matrix is connected by using fusion algorithm.

$$A(G) = \begin{pmatrix} 0 & 1 & 0 & 1 \\ 1 & 0 & 0 & 2 \\ 0 & 0 & 1 & 1 \\ 1 & 2 & 1 & 0 \end{pmatrix}$$

3. Find the complement of the following graphs.

(a) 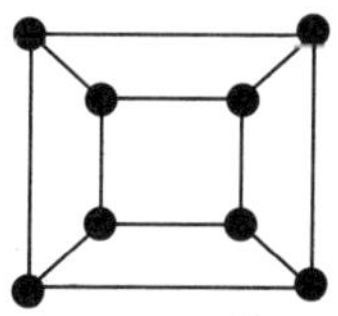

(b) 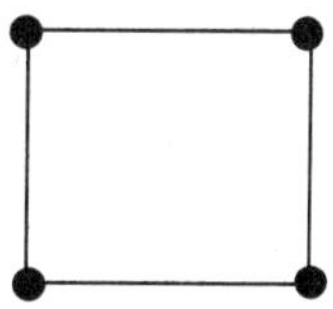

(c) 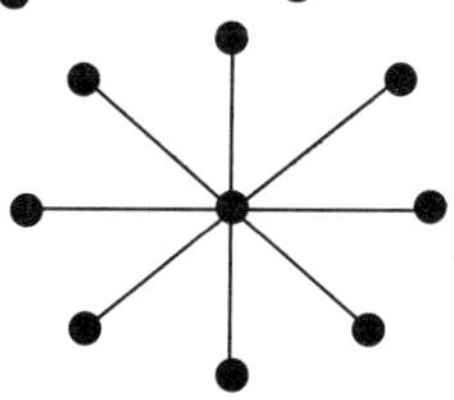

(d) 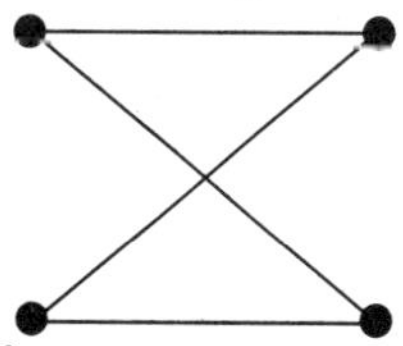

4. Find the product graph where G_1 and G_2 are given below.

(a) G_1:

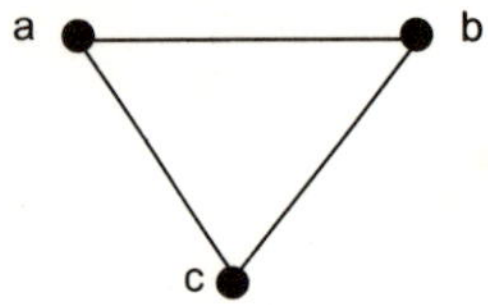

G_2:

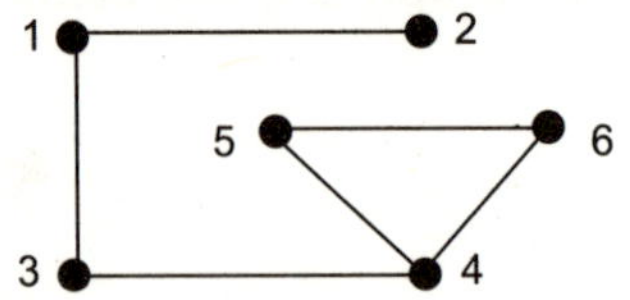

(b) G_1:

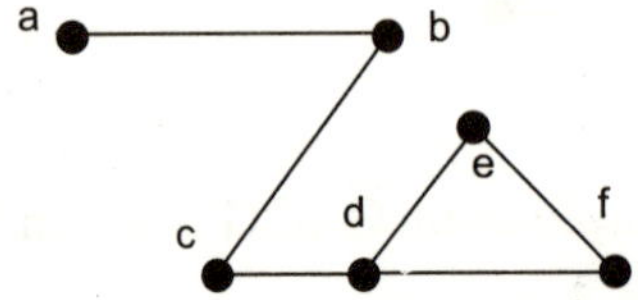

G_2:

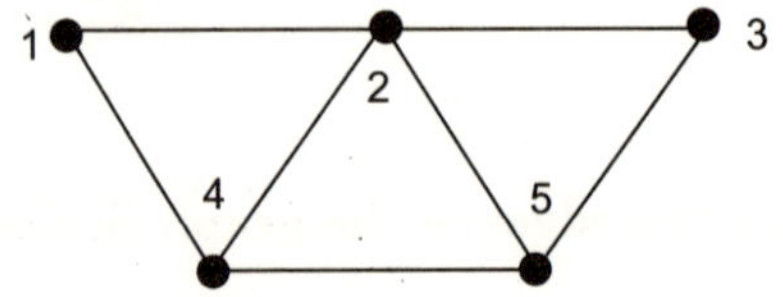

5. Construct a graph of order 5, whose vertices have degrees 1, 2, 2, 3 and 4. What is the size of this graph?

6. Construct a 3-regular graph G. Determine the complement of G. Show that $\overline{G}$ is also regular.

7. Write the graph which is the composition of the following graphs G_1 and G_2.

(a) G_1:

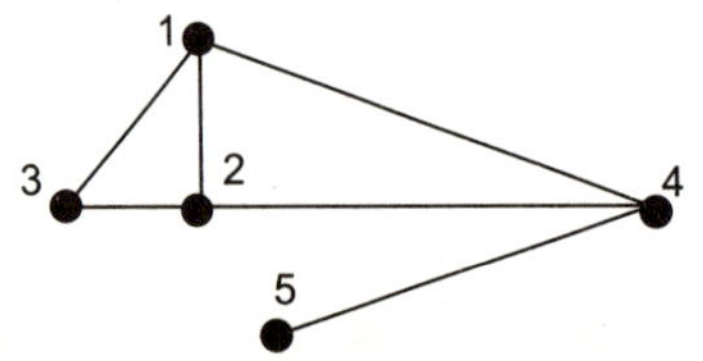

G_2:

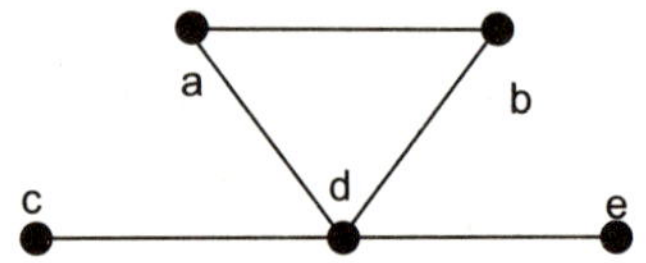

(b) G_1:

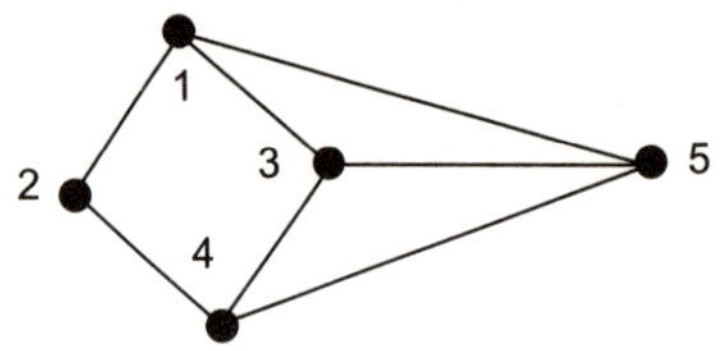

G_2:

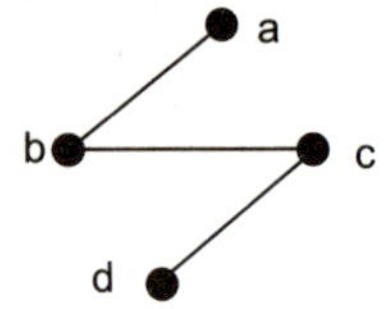

8. For the following graphs, find the adjacency matrix.

(a)

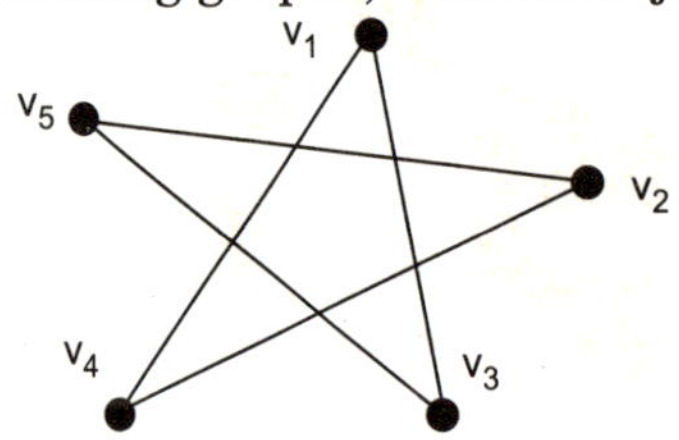

(b)

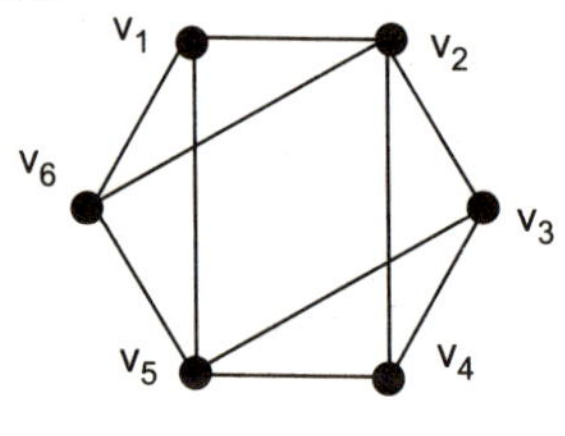

(c)

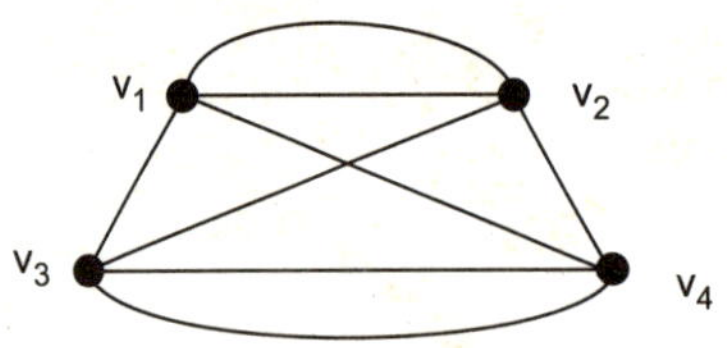

(d)

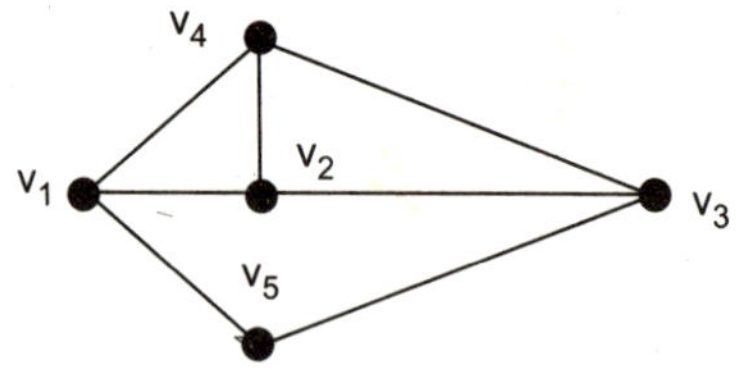

(e)

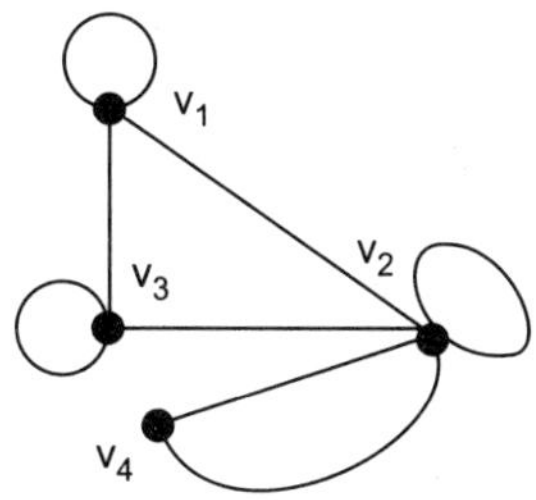

(f)

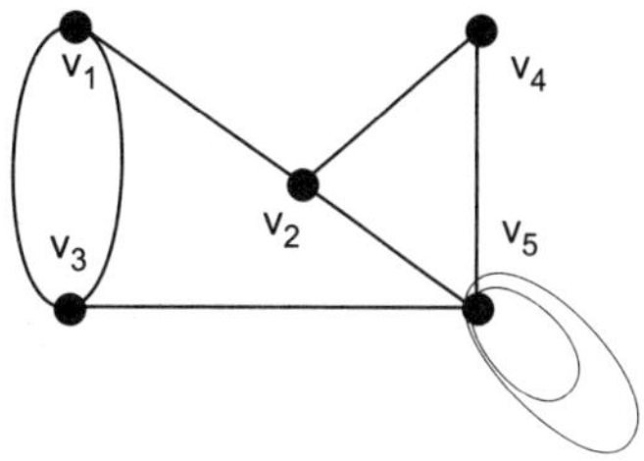

9. Find the path matrix of the following graphs.

(a)

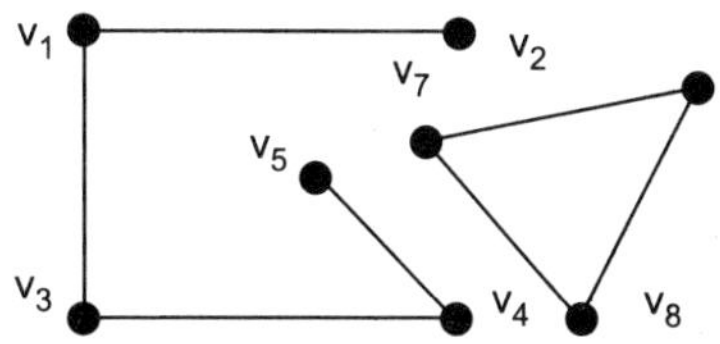

(b)

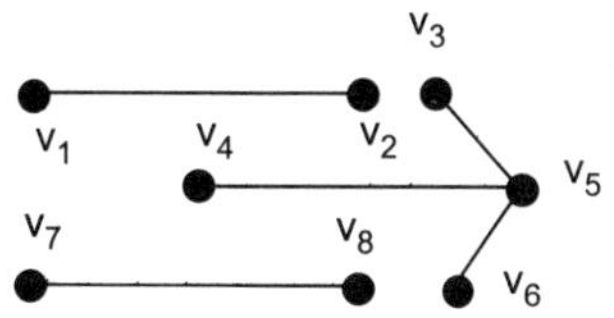

(c)

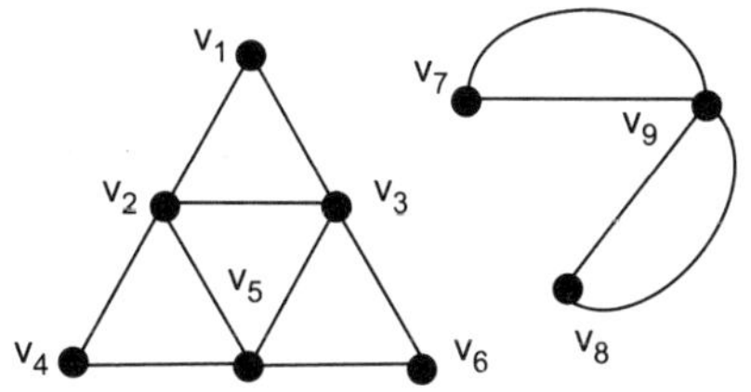

(d)

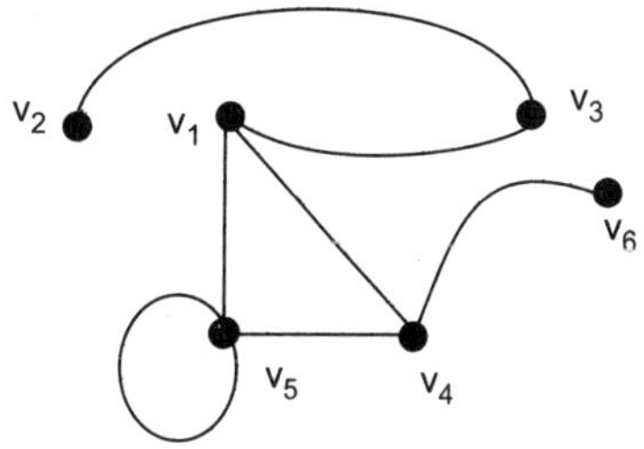

10. Write the incidence matrix of the following graphs.

(a)

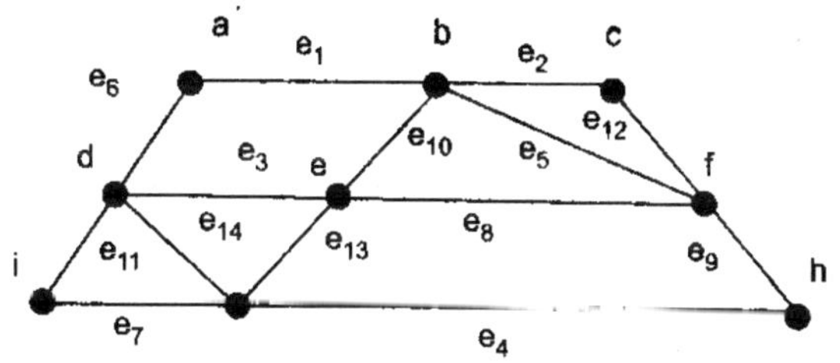

(b)

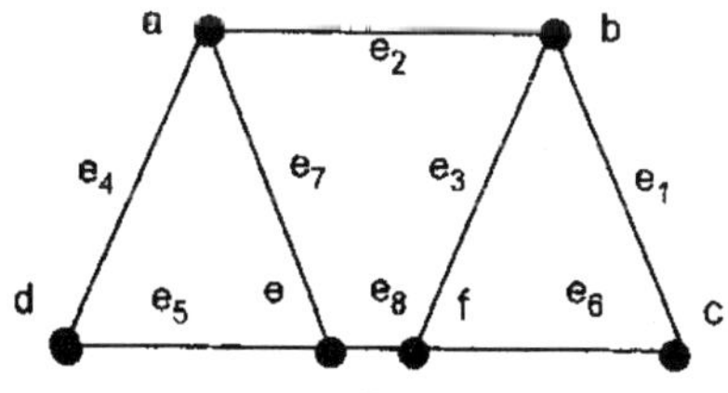

(c)

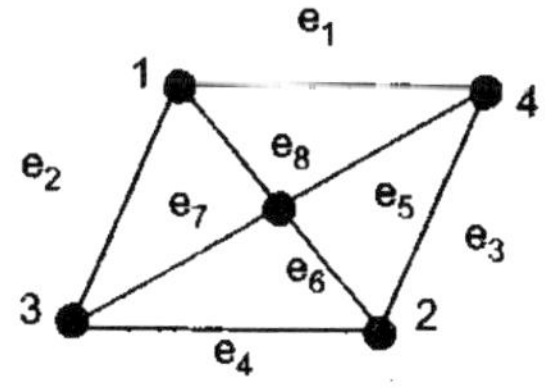

(d)

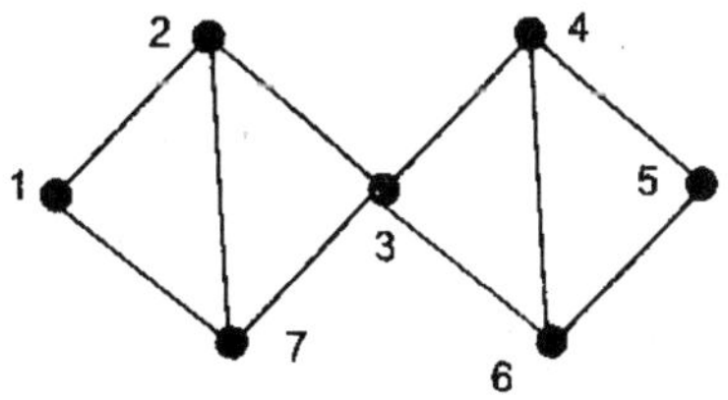

11. Let the graph G is given below. Find the followings.

(*a*) $G - V_1$; where $V_1 = \{1, 3, 5, 6, 7, 8\}$
(*b*) $G - E_1$; where $E_1 = \{a, c, d, f, g, i, j, m, n, q, r, t\}$
(*c*) $G - V_2$; where $V_2 = \{1, 3, 5, 7, 9, 11, 13\}$
(*d*) $G - E_2$; where $E_2 = \{m, l, n, k, o, j, f, e, d\}$
(*e*) G(U); where $U = \{1, 2, 3, 4\}$
(*f*) G(V); where $V = \{a, b, c, d, e, f\}$

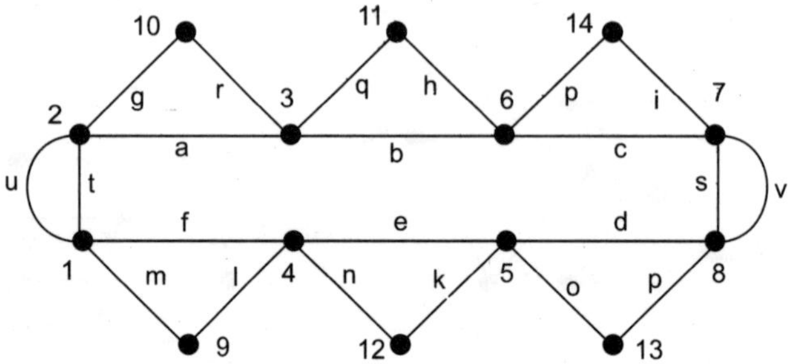

12. Write the union and intersection of the following graphs.

(a) G_1:

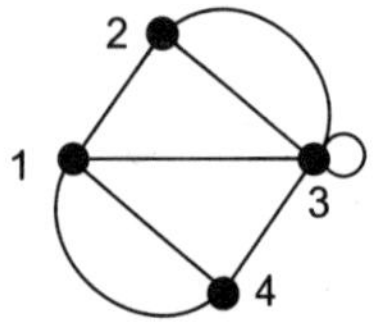

G_2:

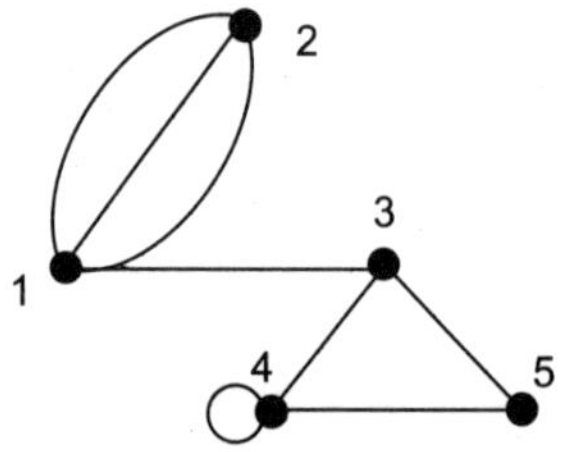

(b) G_1:

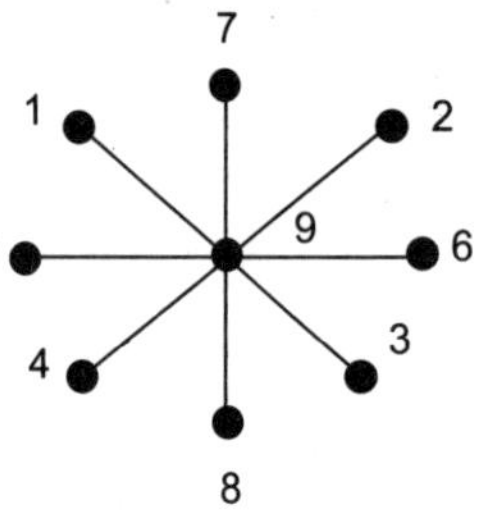

G_2:

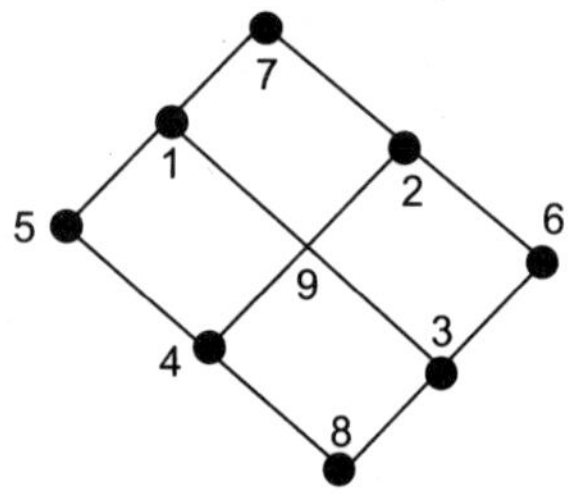

13. Let G be the set of all graphs. Show that the relation "is isomorphic" is an equivalence relation on the set G.

14. Find the degree of every vertex for the following graphs.

(a)

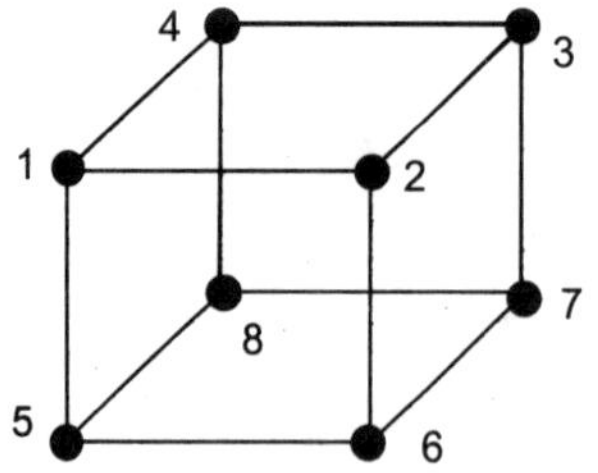

(b)

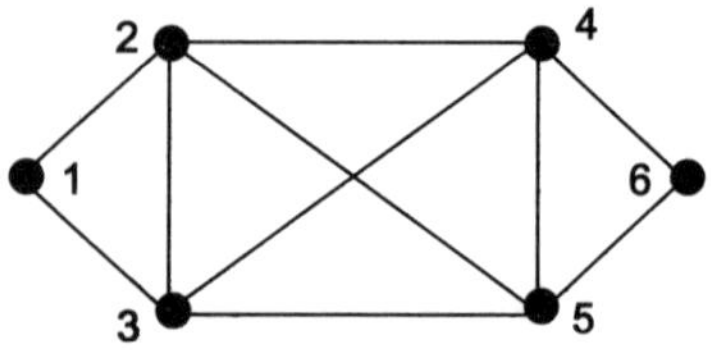

(c) 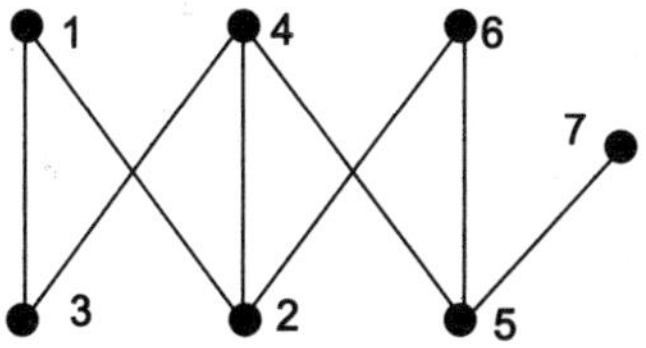

(d) 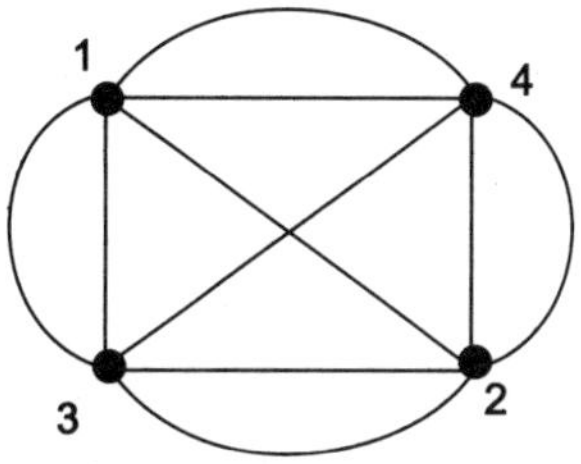

15. Determine the degrees of the vertices v_i ; I = 1, 2, 3, 4, 5 and 6 of the graph G shown below. Compute $\sum_i deg\,(v_i)$. Use this to determine the size of G.

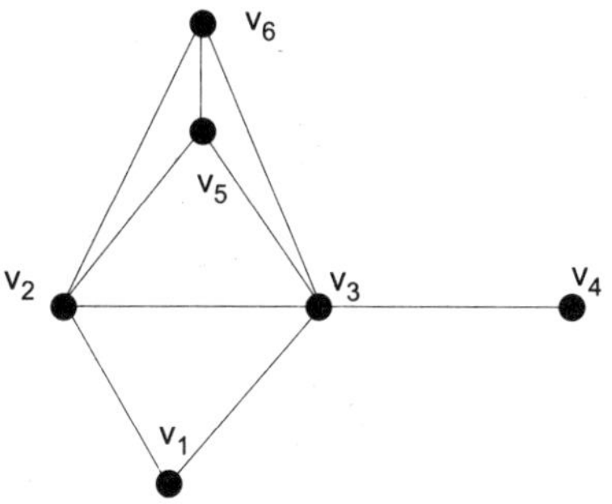

16. Determine which pairs of the graphs G_1, G_2 and G_3 are isomorphic.

G_1: 

G_2: 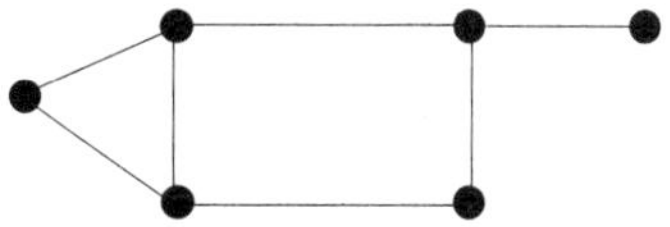

G_3: 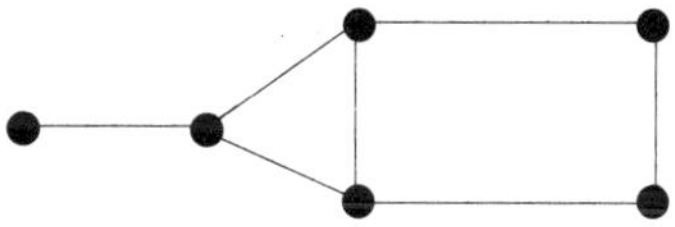

17. From the graphs given below identify

(*i*) Regular graphs

(*ii*) Complete graphs and

(*iii*) Neither regular nor complete graphs.

(a) 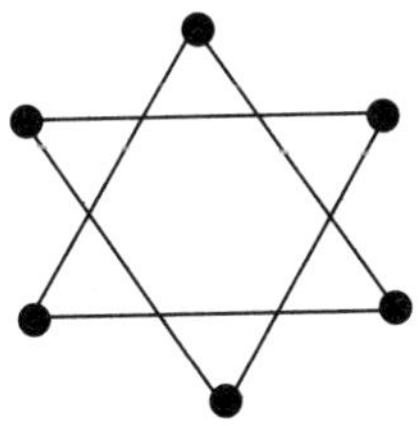

(b) 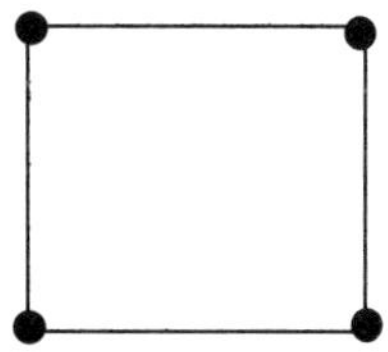

(c) 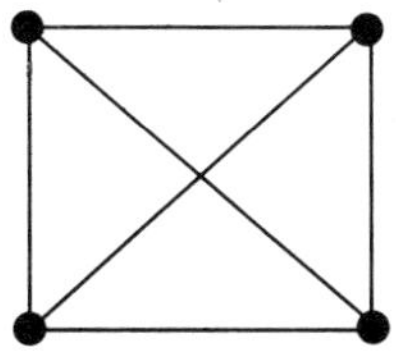

(d) 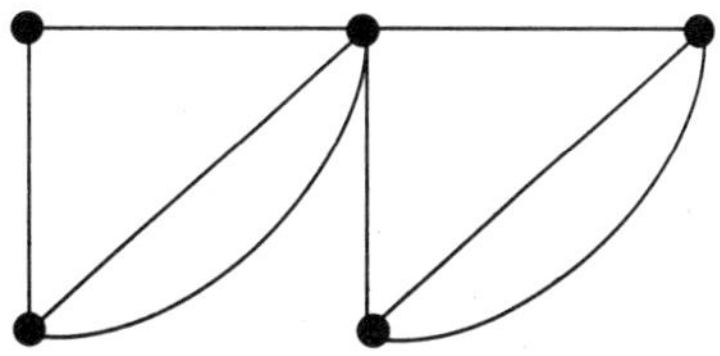

(c)

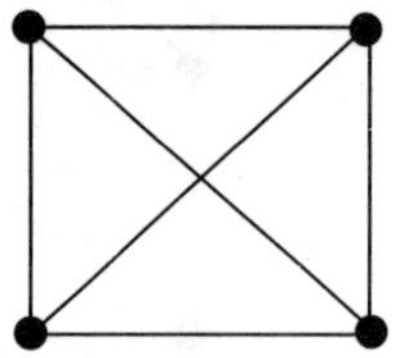

(d)

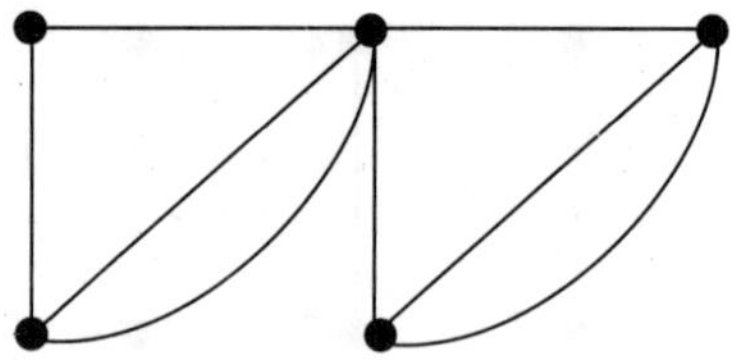

(e)

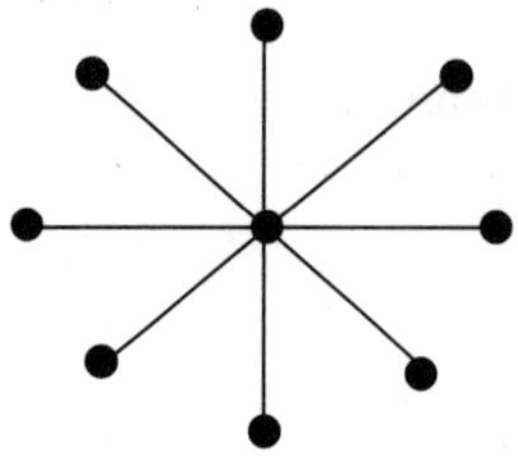

(f)

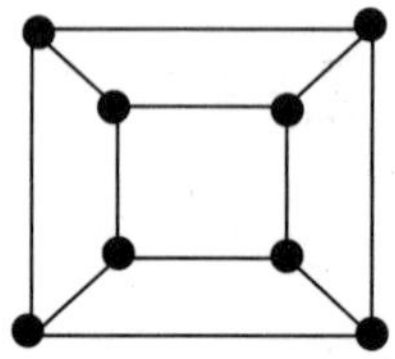

18. Determine whether the graphs G_1 and G_2 shown below are isomorphic.

G_1:

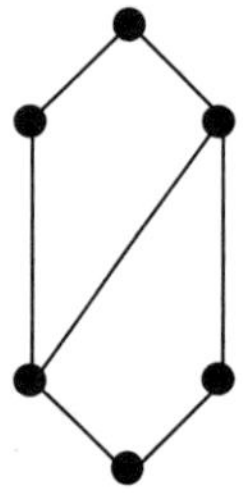

G_2:

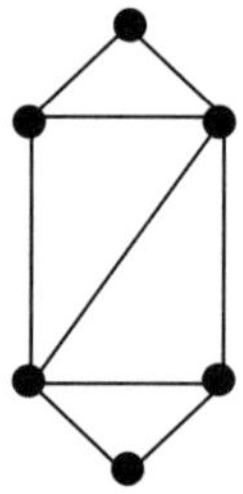

19. Determine whether the graph G shown below is strongly connected or weakly connected.

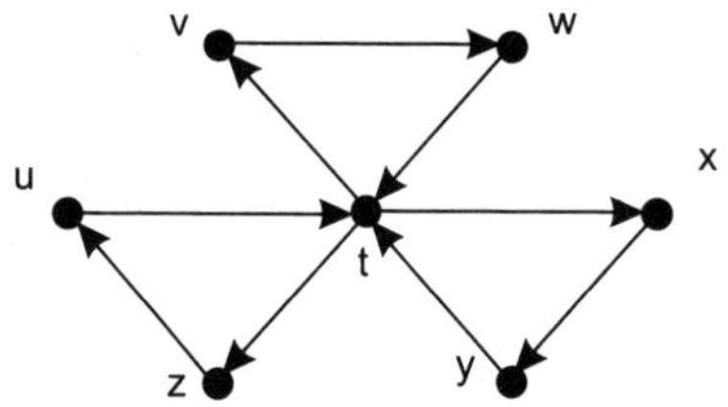

20. In the digraph G shown below, find the indegree and outdegree of every vertex.

G:

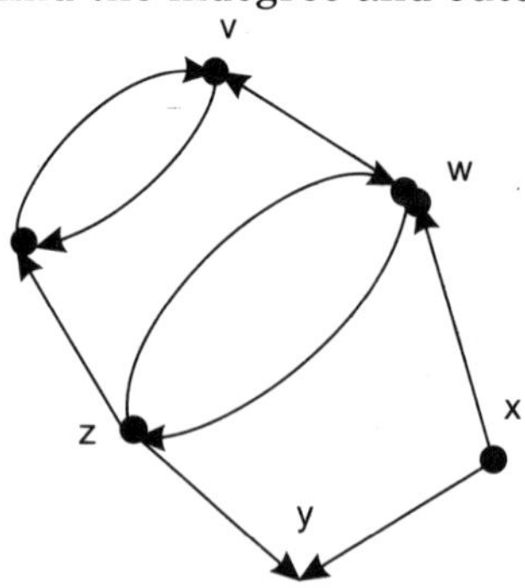

13

Tree

■ 13.0 INTRODUCTION

Another very simple and important graph is tree. Computer science makes extensive use of trees. Trees are useful in organizing and relating data in a database. These are interesting not only for their applications to computer science but also for their theoretical properties. Let us consider a single elimination tournament, which means when a player loses, he/she is out of the tournament. Winners continue to play until only one person, the champion, remains. The following figure shows that Mary defeated Finzi and Smith defeated Blake. The winners Mary and Smith, then played, and Smith defeated Mary. *i.e.,* Smith became champion. This is nothing but a tree.

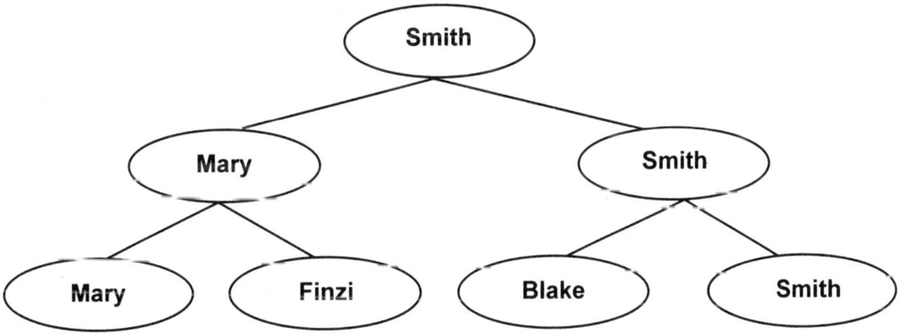

■ 13.1 TREE

A connected acyclic graph G is called a tree. A tree T is a finite set of one or more nodes such that

(*i*) There is a specially designated node called the root.

(*ii*) Remaining nodes are partitioned into k disjoint sets $T_1, T_2, ..., T_k$; $k > 0$. Where each T_i for $i = 1, 2, ... , k$ is a tree. $T_1, T_2, ... , T_k$ are called subtrees of the root.

In the example given below, the tree T has 14 number of nodes. Which are partitioned into three sets T_1, T_2 and T_3 called the subtrees of T.

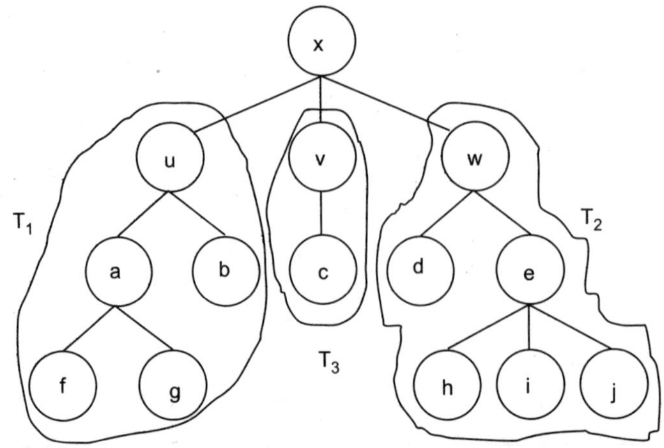

13.2 FUNDAMENTAL TERMINOLOGIES

A tree has the following fundamental terminologies:

13.2.1 Node

The main component of a tree is the node. This stores the actual data and links to the other node.

13.2.2 Child

Child of a node is the immediate successor of a node. Child, which is at the left side, is called left child and the child, which is at the right side, is called right child.

13.2.3 Parent

Parent of a node is the immediate predecessor of a node.

In the figure given below '*x*' is the parent of '*a*' and '*b*'. Where, '*a*' is the left child and '*b*' is the right child of '*x*'.

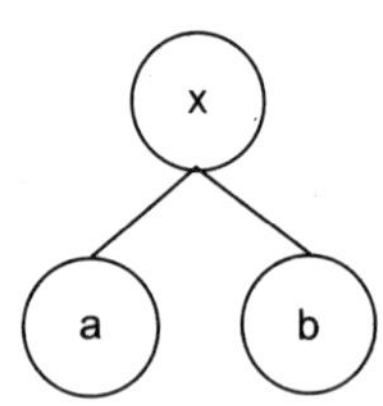

13.2.4 Root

A node that has no parent is termed as the root of the tree. In the above figure '*x*' is termed as the root.

13.2.5 Leaf

The node which is at the end and which does not have any child is called leaf node. Leaf node is also termed as terminal node and external node.

13.2.6 Level

The rank of the hierarchy is known as level. The root node is termed as level 0. If a node is at level n, then its parent is at level $(n - 1)$ and child is at level $(n + 1)$.

13.2.7 Height

The height h of a tree T is defined as maximum number of nodes that is present in a path starting from root node to a leaf node. The height of a tree is also termed as depth of tree. Mathematically,

$$h = \text{Maximum level} + 1$$

Consider the example of a tree

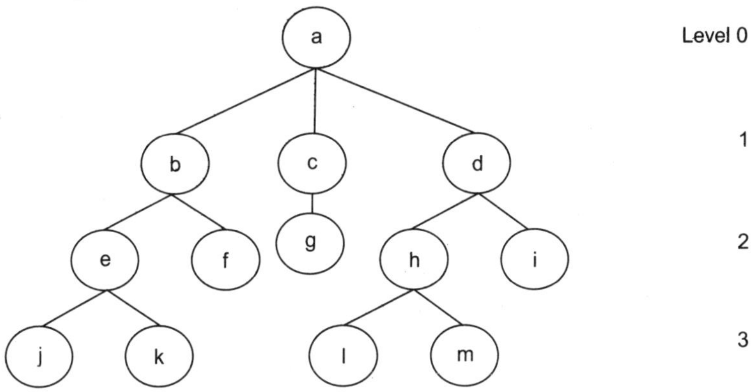

Height of the tree = Maximum level + 1 = 3 + 1 = 4.

13.2.8 Sibling

The nodes, which have the same parent, are termed as siblings. In the above figure h and i are siblings. Similarly l and m are siblings.

■ 13.3 BINARY TREE

A binary tree T is a finite set of nodes such that

(*i*) T is empty or

(*ii*) T contains a specially designed node called the root of T and the remaining nodes of T form two disjoint binary trees T_1 and T_2. This implies that in case of a binary tree a node may have at most two children.

Consider the following simple binary tree T as

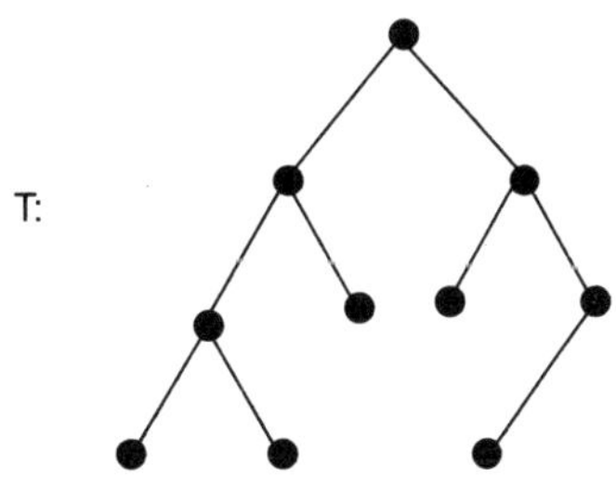

13.3.1 Full Binary Tree

A binary tree T is said to be a full binary tree if it contains maximum possible number of nodes in all level. This indicates that, for the level 'n' of the tree it must contain 2^n number of nodes.

13.3.2 Complete Binary Tree

A binary tree T is said to be a complete binary tree if it contains maximum possible number of nodes in all levels except the last level.

Consider the following examples. In the figure given below T_1 is a full binary tree whereas T_2 is a complete binary tree.

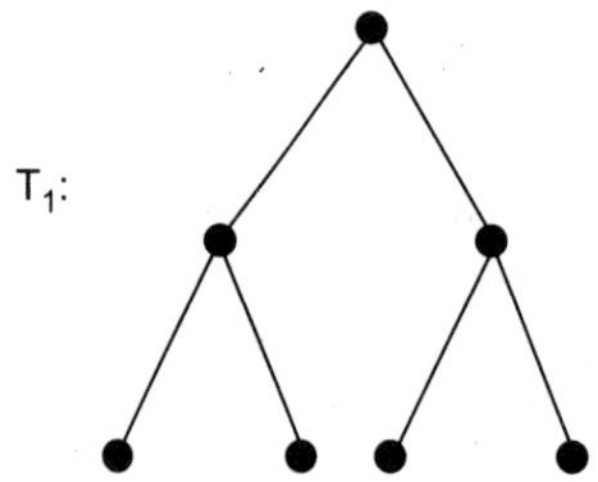

(Full Binary Tree)

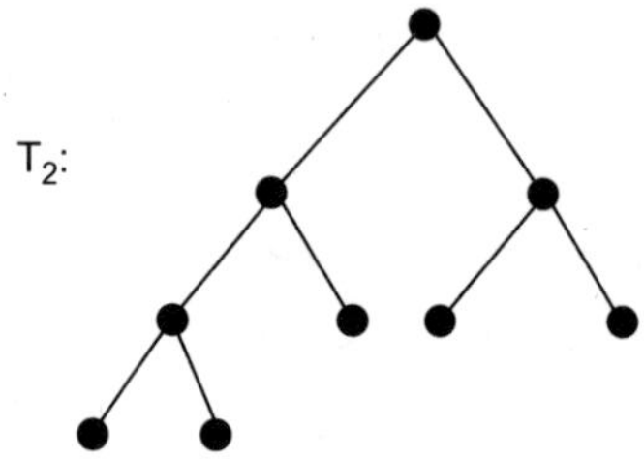

(Complete Binary Tree)

■ 13.4 BRIDGE

An edge of a graph G (V, E) is said to be a bridge if we remove the edge from the graph G, then the graph G has more connected components. Consider the graph G as

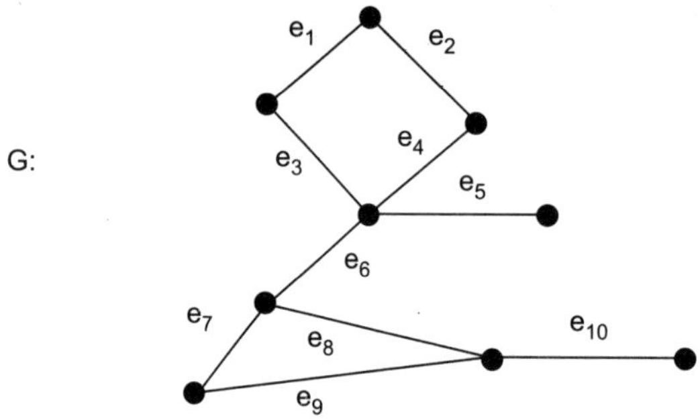

On removing the edge e_6 from the above graph G, the graph has two connected components such as

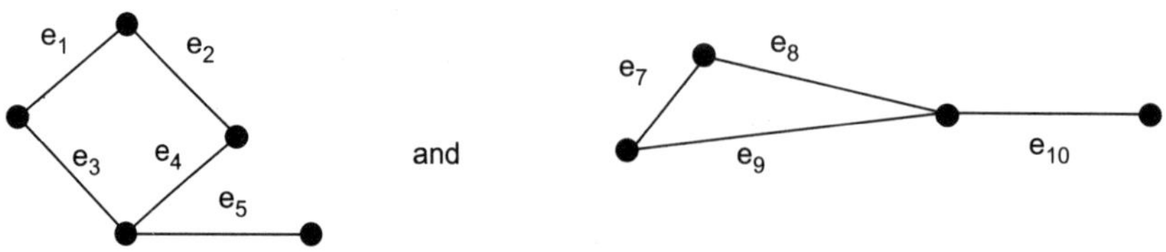

Hence, the edge e_6 is called as a bridge. In the above figure e_5 and e_{10} are also bridges. The bridge is also known as cut edge.

13.4.1 Theorem

A tree of order n has size $(n - 1)$.

Proof: We prove this by the method of induction.

For $n = 1$ we have a single vertex and hence size is 0.

For $n = 2$, the tree T contains two vertices, so size is 1.

Hence the result follows for $n = 1$ and 2. Assume that the result is true for all trees of order less than k. Let T be a tree of order $n = k$ and size q, and let e be an edge of T.

We have already observed that e is a bridge of T, so that $(T - e)$ is a forest. Let the two components of $(T - e)$ are T_1 and T_2, where T_i is a tree of order n_i and size q_i for $i = 1$ and 2.

As, $n_i < k$ for $i = 1$ and 2 so we have $q_1 = (n_1 - 1)$ and $q_2 = (n_2 - 1)$. Since, $n = (n_1 + n_2)$ and $q = (q_1 + q_2 + 1)$ we get

$$q = (n_1 - 1) + (n_2 - 1) + 1 = (n_1 + n_2) - 1 = (n - 1)$$

Therefore, by induction the size of a tree is $(n - 1)$, *i.e.*, one less than its order.

13.4.2 Theorem

Every non-trivial tree contains at least two end vertices.

Proof: Suppose that T be a tree of order n and size q. Let $d_1, d_2, \ldots, d_n$ denote the degrees of its vertices, ordered so that $d_1 \le d_2 \le d_3 \le \ldots \le d_n$. Since T is connected and non-trivial, $d_i \ge 1$ for each i; $1 \le i \le n$.

Assume that T does not contain two end-vertices. Hence $d_1 \ge 1$ and $d_i \ge 2$ for $2 \le i \le n$. Thus,

$$\sum_{i=1}^{n} d_i = d_1 + d_2 + d_3 + \ldots + d_n \ge 1 + 2(n - 1) = 2n - 1 \qquad \ldots (i)$$

But we know $$\sum_{i=1}^{n} d_i = 2q = 2(n - 1) = 2n - 2$$

This contradicts inequality (i). So our assumption is wrong. Hence, T contains at least two end-vertices.

13.5 DISTANCE AND ECCENTRICITY

In general, a graph's edges are stretchable. As a result, we cannot measure distances in a graph with linear measures. When we look at a graph, however, we can still sense that some parts of the graph are further apart than others. So, in order to quantity a graph the concept of distance and eccentricity is developed.

Let u and v be two vertices of the graph G. The distance between u and v is denoted by $d(u, v)$ and is defined as the length of a shortest $u - v$ path. If there is no path between u and v, then $d(u, v) = \infty$.

Let V be the vertex set of G. Let $v \in V$. The eccentricity of v is denoted as $e(v)$ and is defined as

$$e(v) = \text{Max}\,\{d(u, v) : u \in V \text{ and } u \ne v\}$$

13.5.1 Radius and Diameter

Let G be the graph, then the radius of G is denoted as rad (G) and is defined as

$$\text{rad}(G) = \text{Min}\,\{e(v): v \in V\}.$$

Let V be the vertex set of the graph G. The diameter of G is denoted as diam (G) and is defined as

$$\text{diam}(G) = \text{Max}\,\{e(v): v \in V\}.$$

■ 13.6 CENTRAL POINT AND CENTRE

Let V be the vertex set of the graph G. Then $v \in V$ is said to be a central point if $e(v) = \text{rad}(G)$. The set of all central points of G is known as centre of G.

Consider the following graph G where each edge is of length 1.

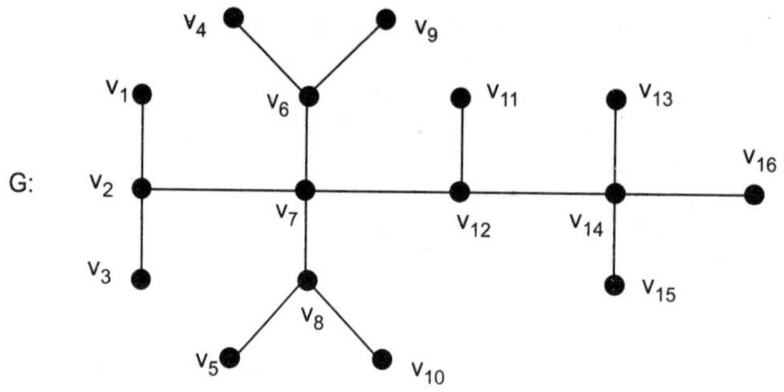

Here $V = \{v_1, v_2, v_3, \ldots, v_{16}\}$. Now, we get

$d(v_1, v_2) = 1$; $d(v_1, v_3) = 2$; $d(v_1, v_4) = 4$; $d(v_1, v_5) = 4$; $d(v_1, v_6) = 3$;
$d(v_1, v_7) = 2$; $d(v_1, v_8) = 3$; $d(v_1, v_9) = 4$; $d(v_1, v_{10}) = 4$; $d(v_1, v_{11}) = 4$;
$d(v_1, v_{12}) = 3$; $d(v_1, v_{13}) = 5$; $d(v_1, v_{14}) = 4$; $d(v_1, v_{15}) = 5$; $d(v_1, v_{16}) = 3$.

Therefore, $e(v_1) = \text{Max}\,\{1, 2, 3, 4, 5\} = 5$.

$d(v_2, v_1) = 1$; $d(v_2, v_3) = 1$; $d(v_2, v_4) = 3$; $d(v_2, v_5) = 3$; $d(v_2, v_6) = 2$; $d(v_2, v_7) = 1$;
$d(v_2, v_8) = 2$; $d(v_2, v_9) = 3$; $d(v_2, v_{10}) = 3$; $d(v_2, v_{11}) = 3$; $d(v_2, v_{12}) = 2$; $d(v_2, v_{13}) = 4$;
$d(v_2, v_{14}) = 3$; $d(v_2, v_{15}) = 4$; $d(v_2, v_{16}) = 4$.

Therefore, $e(v_2) = \text{Max}\,\{1, 2, 3, 4\} = 4$. Proceeding in this manner, we will get

$e(v_3) = 5$; $e(v_4) = 5$; $e(v_5) = 5$; $e(v_6) = 4$; $e(v_7) = 3$;
$e(v_8) = 4$; $e(v_9) = 5$; $e(v_{10}) = 5$; $e(v_{11}) = 4$; $e(v_{12}) = 3$;
$e(v_{13}) = 5$; $e(v_{14}) = 4$; $e(v_{15}) = 5$; $e(v_{16}) = 5$.

Now, radius = rad (G) = Min $\{e(v), v \in V\}$ = Min $\{5, 4, 3\}$
= 3 and

diameter = diam (G) = Max $\{e(v), v \in V\}$
= Max $\{5, 4, 3\} = 5$.

Therefore, the central points are v_7 and v_{12} and center = $\{v_7, v_{12}\}$.

■ 13.7 SPANNING TREE

Suppose G = (V, E) be a graph. A sub graph H of G is said to be a spanning sub graph of G if both H and G has same vertex set. A spanning tree of a graph G is a tree which is a spanning sub graph of G. In this section we will discuss the algorithms for finding minimum spanning tree.

13.7.1 Prim's Algorithm

The following steps are used in Prim's algorithm for finding a minimum spanning tree of a graph G. Assume that the graph G has n vertices.

1. Choose any vertex v_1 of G
2. Choose an edge $e_1 = v_1 v_2$ of G such that $v_1 \neq v_2$ and e_1 has smallest weight among the edges of G incident with v_1.
3. If edges $e_1, e_2, \ldots\ldots, e_i$ have been chosen involving vertices $v_1, v_2, \ldots, v_{i+1}$, then choose an edge $e_{i+1} = u\,v$ with $u \in \{v_1, v_2, \ldots, v_{i+1}\}$ and $v \notin \{v_1, v_2, \ldots, v_{i+1}\}$ such that e_{i+1} has smallest weight among the edges of G.
4. The step 3 is to be repeated until we are getting the total $(n - 1)$ edges.

Consider the following connected weighted graph G. Here the number of vertices $n = 7$

G:

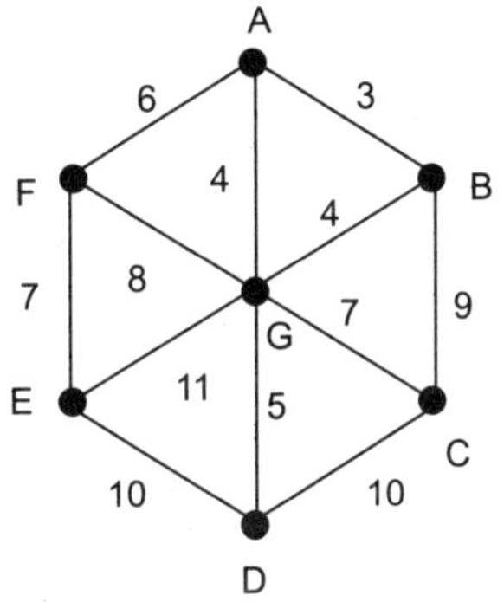

(V_1= A ; One can choose any other vertex)

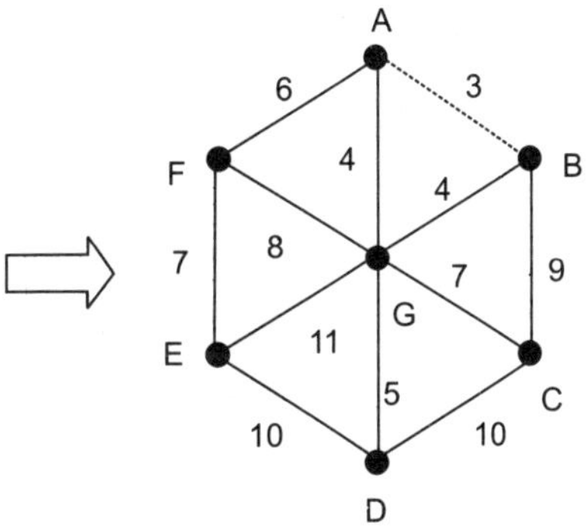

(e_1 = AB, So that v_2 = B)

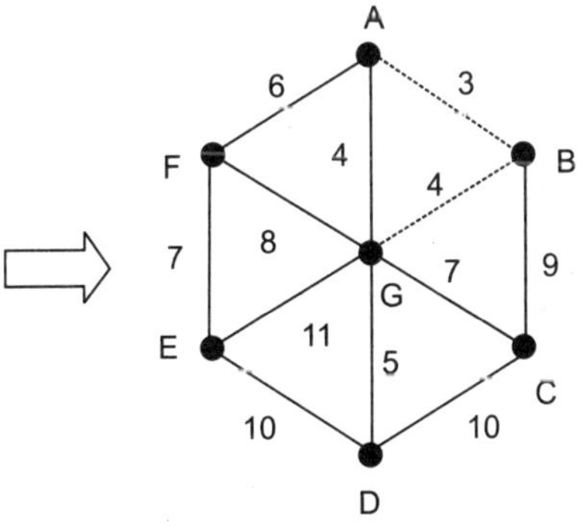

(e_2 = BG ; So that v_3 = G ; An alternative choice is AG)

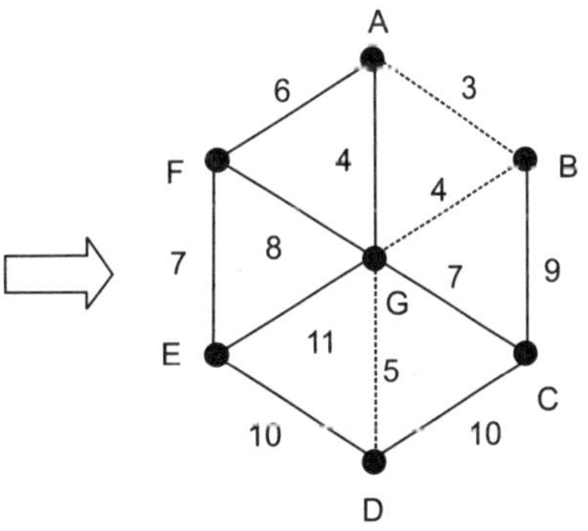

(e_3 = GD ; So that v_4 = D; No alternative choice)

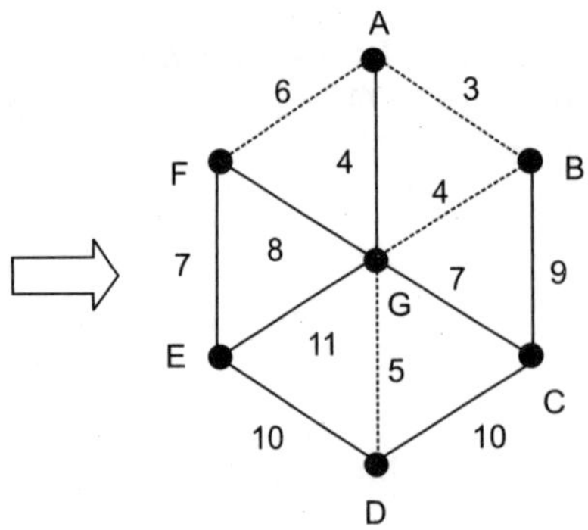

(e_4 = AF ; So that v_4 = F ; No alternative choice)

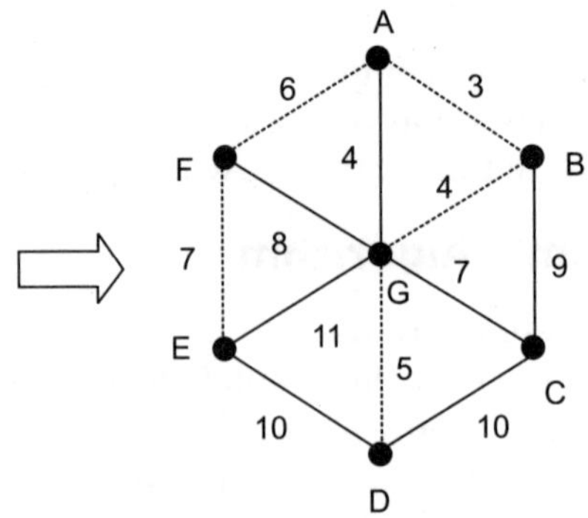

(e_5 = FE ; So that v_5 = E ; An alternative choice is GC)

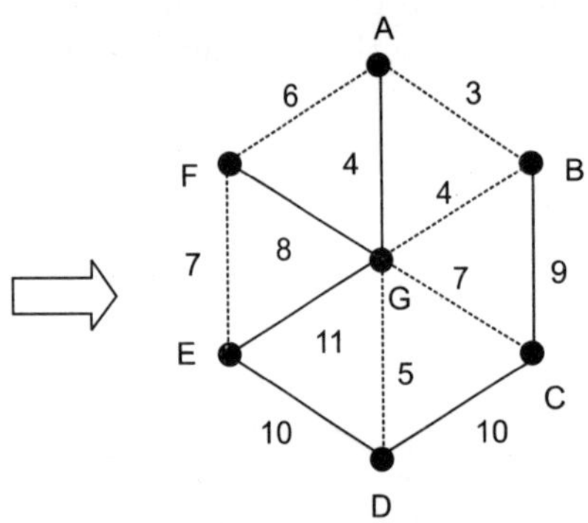

(e_6 = GC ; So that v_6 = C; No alternative choice)

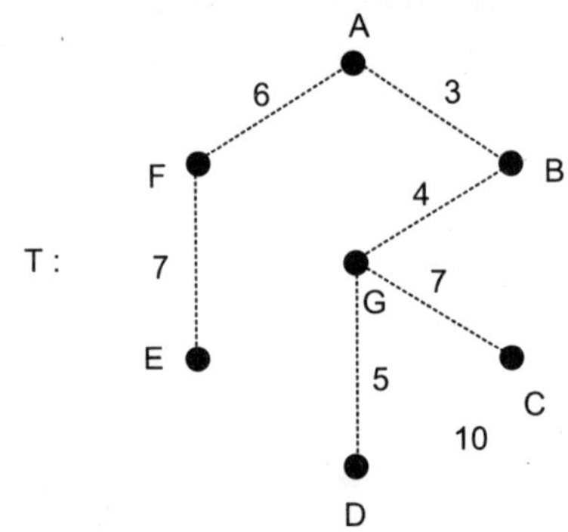

w(T) = 32

Since the total edges are 6 = (7 – 1), the process terminates. Hence, the minimum spanning tree T is given as shown in the above figure.

13.7.2 Kruskal's Algorithm

The following steps are used in Kruskal's algorithm for finding a minimum spanning tree of a graph G. Assume that the graph G has n vertices.

1. Choose an edge e_1 of G, which is as small as possible and e_1 must not be a loop.
2. Suppose the edges e_1, e_2, ..., e_k have been chosen. Then the edge e_{k+1} (not already chosen) can be chosen such that
   (*i*) The induced sub graph G[{e_1, e_2 ,.......... , e_{k+1}}] is acyclic and
   (*ii*) Weight of e_{k+1} is as small as possible.
3. The step 2 is to be repeated until we are getting the total (n – 1) edges.

Consider the following connected weighted graph G. Here the number of vertices n = 5. On applying Kruskal's algorithm we have the following stages.

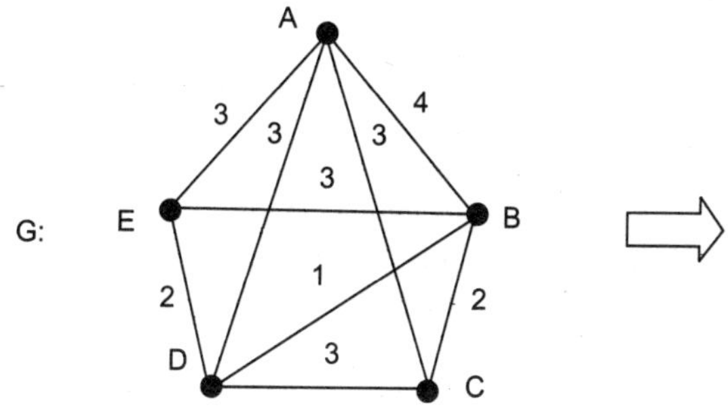

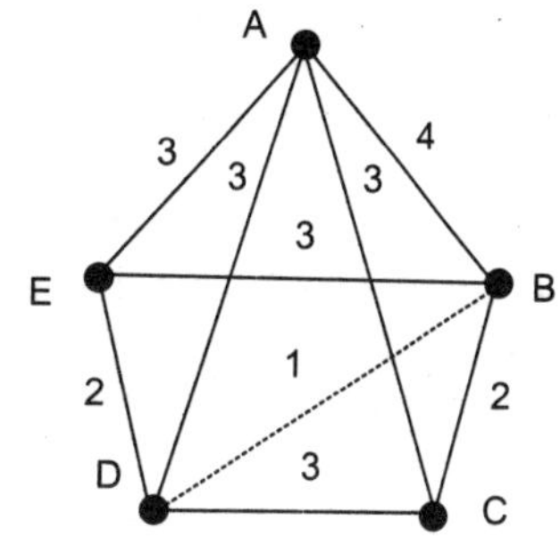

(e_1 = BD ; w (e_1) = 1 ; No alternative choice)

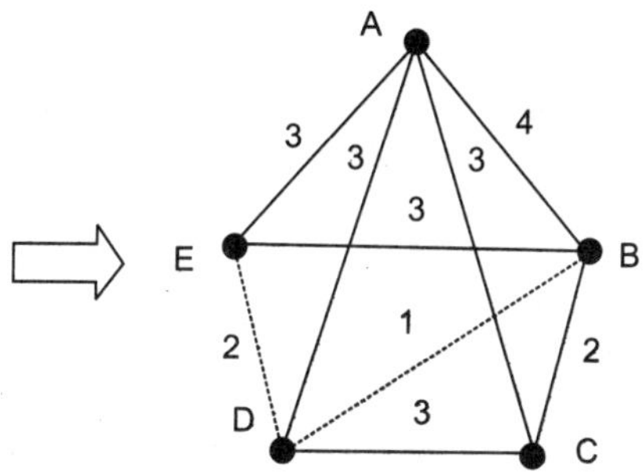

(e_2 = DE ; w (e_2) = 2; An alternative choice is BC)

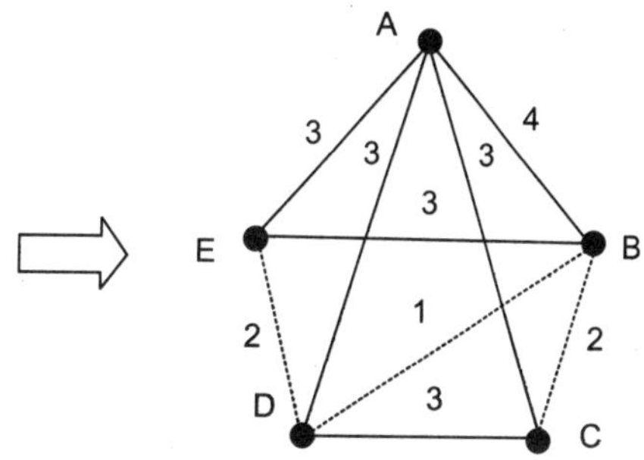

(e_3 = BC; w (e_3) = 2 ; No alternative choice)

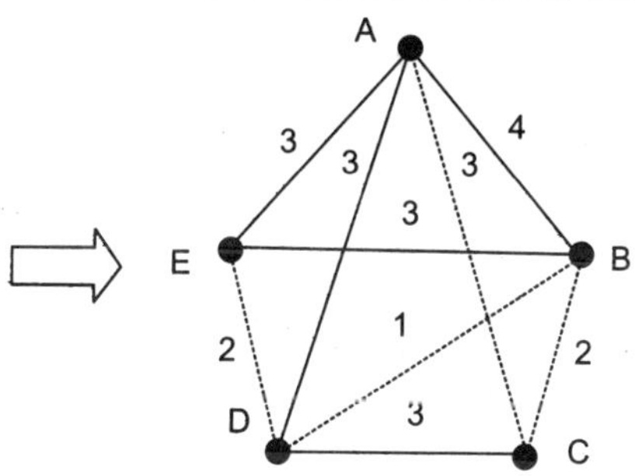

(e_4 = AC; w (e_4) = 3 ; Alternative choices are AE and AD)

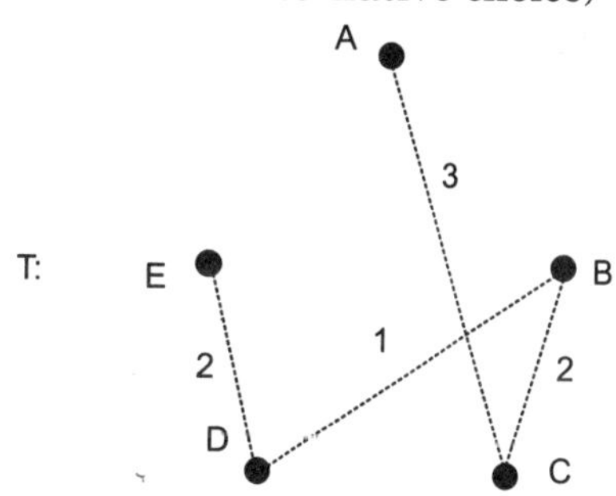

w (T) = 1 + 2 + 2 + 3 = 8

Since the total edges are 4 = (5 – 1), the process terminates. Hence, the minimum spanning tree T is given as shown in the above figure.

■ 13.8 SEARCHING ALGORITHMS

This section presents methods for searching a graph. This means systematically following the edges of the graph so as to visit the vertices of the graph. The graph searching algorithms can discover much about the structure of a graph. Here we present two algorithms, depth first search and breadth first search. In addition, we will discuss to create a breadth first and depth first tree.

13.8.1 Breadth First Search

Breadth first search is one of the simplest algorithms for searching a graph. Given a graph G(V, E) and a distinguished source vertex s, breadth first search systematically explores the edges of G to discover every vertex that is reachable from s. It computes the distance (fewest number of edges) from s to all such reachable vertices. Breadth first search is so named because it expands the frontier between discovered and undiscovered vertices uniformly across the breadth of the frontier. It constructs a breadth first tree, initially containing only its root, that is the source vertex s.

Suppose that v_i, v_j, be two specified vertices of G. We will now describe a method of finding a path from v_i to v_j which uses the least number of edges. Such a path is known as shortest path, if it exists. The method assigns labels 0, 1, 2, ... to the vertices of G and is called the Breadth First Search (BFS) technique. The BFS algorithm consists of the following steps:

1. Label the vertex v_i with 0. Set $i = 0$
2. Find all unlabelled vertices in G, which are adjacent to vertices, labeled i. If there are no such vertices, then v_i is not connected to v_j else label them by $(i + 1)$.
3. If v_j is labeled go to step 4, else replace i by $(i + 1)$ and go to step 2.
4. The length of shortest path from v_i to v_j is $(i + 1)$ then stop.

Consider the following graph G. Now we have to find out the shortest path from the source vertex a to the vertex z. On using the BFS technique, we get the following stages.

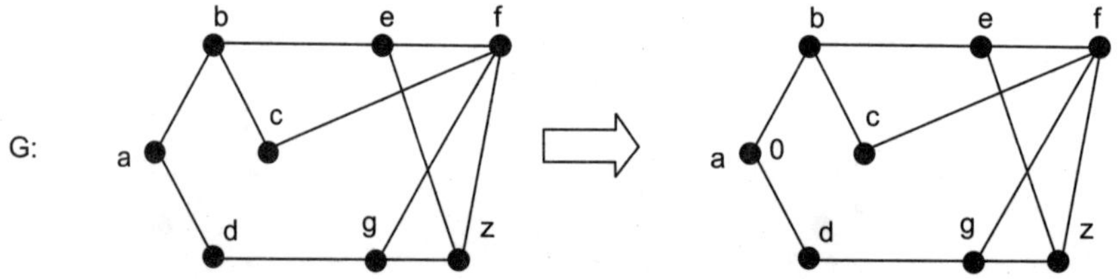

(Label (a) = 0 and set $i = 0$)

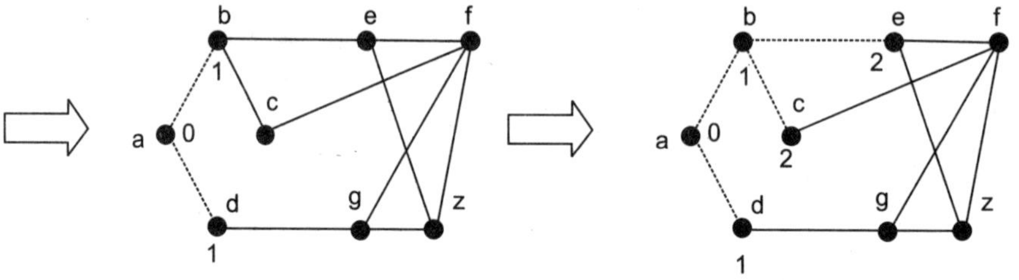

In the above figure the adjacent vertices of a are b and d. Therefore we get label $(b) = i + 1 = 0 + 1 = 1$ and label $(d) = i + 1 = 0 + 1 = 1$. Similarly, adjacent vertices of b are c and e. Therefore, label $(c) = i + 1 = 1 + 1 = 2$ and label $(e) = i + 1 = 1 + 1 = 2$.

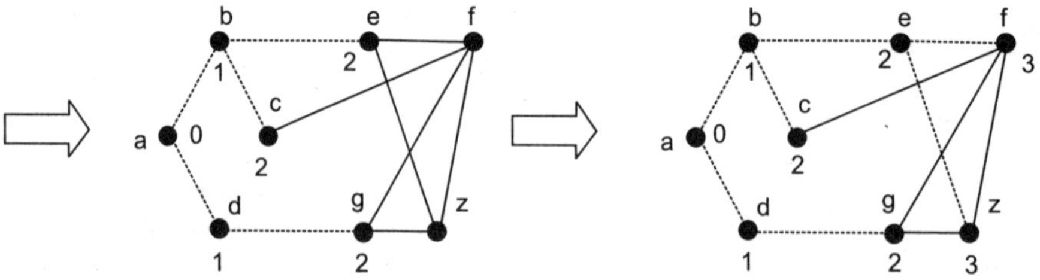

In the above figure the adjacent vertices of d is g. Therefore we get label (g) = $i + 1 = 1 + 1 = 2$. Similarly, adjacent vertices of e are f and z. Therefore, label (f) = $i + 1 = 2 + 1 = 3$ and label (z) = $i + 1 = 2 + 1 = 3$. Hence, the breadth first tree T becomes

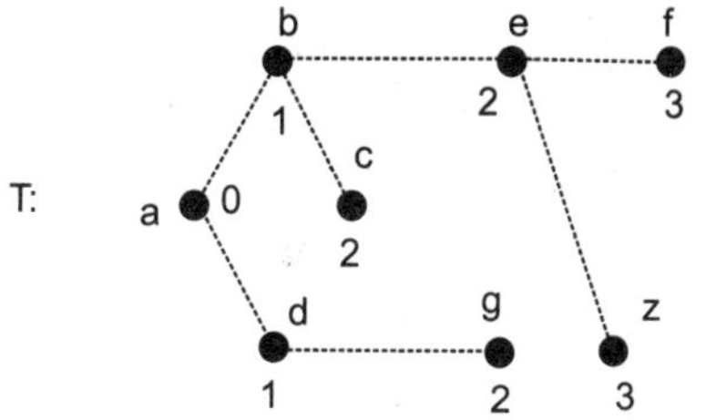

13.8.2 Back-Tracking Algorithm

The following steps are used in back-tracking algorithm:

1. Set $\lambda(t) = i$ and assign $v_i = t$, where 't' is the terminating node.
2. Find a vertex 'u' which is adjacent to v_i and with $\lambda(u) = (i - 1)$. Set $v_{i-1} = u$.
3. If $i = 1$, then stop else replace i by $(i - 1)$ and go to step 2.

Consider the following graph G. Now we have to find out the shortest path from the source vertex 'a' to the vertex 'z'. On using the BFS technique, we get.

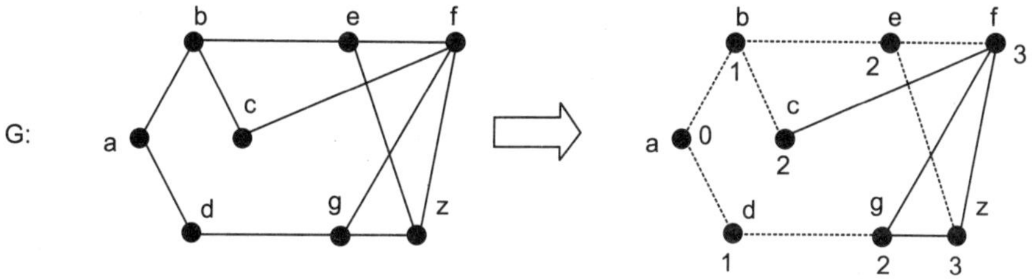

On using back-tracking algorithm we have

1. Set $i = \lambda(z) = 3$ and $v_i = v_3 = z$
2. The adjacent to $v_3 = z$ is e and $\lambda(e) = (i - 1) = 2$. Set $v_2 = e$.
3. As $i = 3 \neq 1$, so $i = (i - 1) = 2$, Go to step 2.
   2. The adjacent to $v_2 = e$ is b and $\lambda(b) = (i - 1) = 1$. Set $v_1 = b$.
   3. As $i = 2 \neq 1$, so $i = (i - 1) = 1$, Go to step 2.
      2. The adjacent to $v_1 = b$ is a and $\lambda(a) = (i - 1) = 0$. Set $v_0 = a$.
      3. As $i = 1$, so the process terminates.

Therefore, the shortest path from 'a' to z is given as '$a\ b\ e\ z$'. Besides that, there could be several paths from 'a' to 'z'.

13.8.3 Depth First Search

Basic philosophy in depth first search is that first all vertices reachable from the vertex 'v' are searched before proceeding to see the siblings. In depth first search, edges are explored out of the most recently discovered vertex 'v' that still has unexplored edges leaving it. When all of v's edges have been explored, then the search "backtracks" to explore edges leaving the vertex from which 'v' was discovered. The process is being continued until we have discovered all the vertices that are reachable from the original source vertex. If any undiscovered vertices remain, then one of them is selected as a new source vertex and the search is repeated. This process is repeated until all vertices are discovered.

Consider the graph G as below. Let us consider the source vertex as '*a*'. On using the DFS technique, the order in which the vertices are being visited is described below by the sequence of graphs.

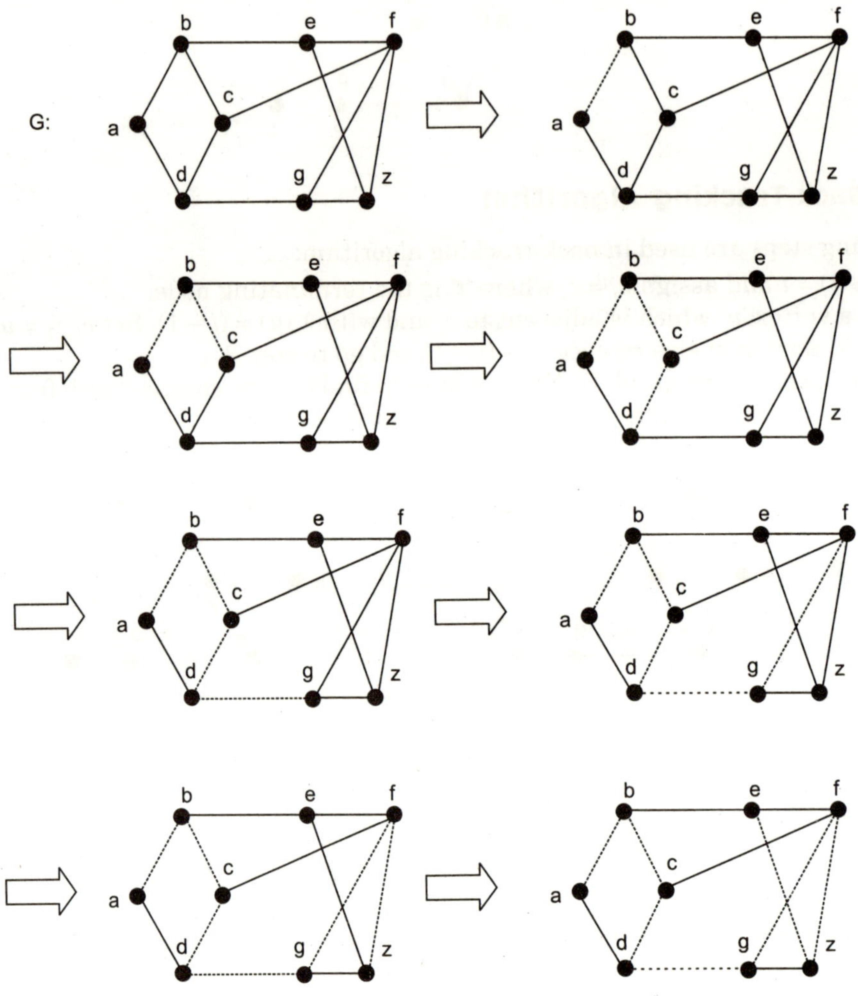

Therefore, the depth first tree T is given below. Besides that, there could be several depth first trees from the same vertex '*a*'. This indicates that the depth first tree is not unique.

T:

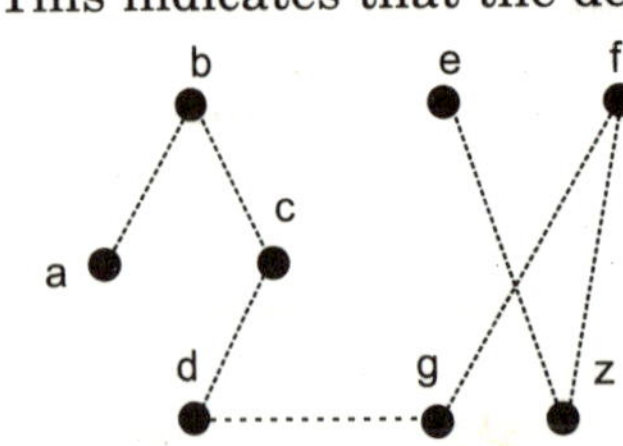

13.9 SHORTEST PATH ALGORITHMS

This section presents methods for finding shortest path from a source vertex to a terminating vertex in a graph G. This problem is a real life problem, where cities are connected through roads, rails and air routes and we want to find out the shortest path between the vertices. Here we present two algorithms Dijkastra's Algorithm and Floyd-Warshall Algorithm.

13.9.1 Dijkstra's Algorithm

In a given graph G(V, E), we want to find a shortest path from a given source vertex '$s \in V$' to every vertex '$v \in V$'. This is otherwise known as single source shortest path problem. Dijkastra's Algorithm solves the single source shortest path problems in weighted graphs that to non-negative weights. Therefore, we assume that $w(uv) \geq 0$ for each edge $(uv) \in E$.

1. Set $\lambda(v_s) = 0$ and for all vertices $v_i \neq v_s$ $\lambda(v_i) = \infty$. Set T = V; where V is the set of vertices of G and T is the set of uncoloured vertices.
2. Let u be the vertex in T for which $\lambda(u)$ is minimum.
3. If $u = v_t$ (Terminating node), then stop. Else, go to step 4.
4. For every edge $e = uv$, incident with u, if $v \in T$, then replace $\lambda(v)$ with Min $\{\lambda(v), \lambda(u) + w(uv)\}$.

i.e, $$\lambda(v) = \text{Min}\,\{\lambda(v), \lambda(u) + w(uv)\}$$

5. Change T by T – $\{u\}$ and go to step 2.

Consider the following graph G. Let us consider the source vertex as e and the terminating vertex as f. We have to find out the shortest distance between the vertices e and f.

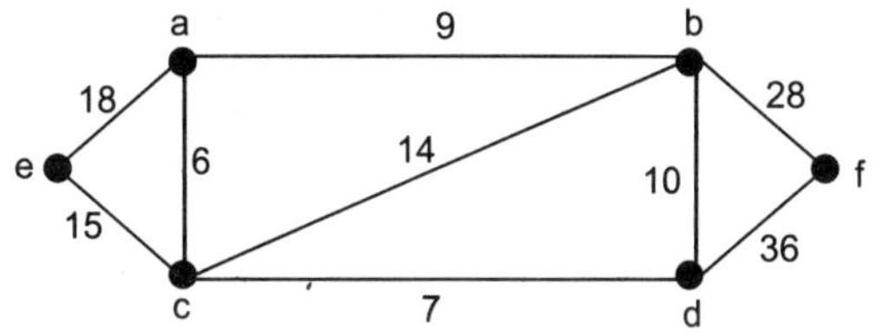

In the above graph G, the source vertex is $v_s = e$ and $v_t = f$. Set $\lambda(e) = 0$ and $\lambda(a) = \lambda(b) = \lambda(c) = \lambda(d) = \lambda(f) = \infty$. T = V = $\{e, a, b, c, d, f\}$. Hence, we have the following table

Vertex	e	a	b	c	d	f
$\lambda(v)$	0	∞	∞	∞	∞	∞
T	e	a	b	c	d	f

Now, $u = e$ as $\lambda(u) = \lambda(e) = 0$ which is minimum. The edges incident on $u = e$ are ea and ec. Therefore,

$$\lambda(a) = \text{Min}\,[\lambda(a), \lambda(e) + w(ea)]$$
$$= \text{Min}\,[\infty, 18] = 18.$$
$$\lambda(c) = \text{Min}\,[\lambda(c), \lambda(e) + w(ec)]$$
$$= \text{Min}\,[\infty, 15] = 15.$$

Again, T = T – $\{u = e\}$ = $\{a, b, c, d, f\}$. Thus, we have the following table

Vertex	e	a	b	c	d	f
$\lambda(v)$	0	18	∞	15	∞	∞
T		a	b	c	d	f

Now, $u = c$ as $\lambda(u) = \lambda(c) = 15$ which is minimum. The edges incident with $u = c$ are ca, cb and cd. Therefore,

$$\lambda(a) = \text{Min } [\lambda(a), \lambda(c) + w(ca)]$$
$$= \text{Min } [18, 21] = 18.$$
$$\lambda(b) = \text{Min } [\lambda(b), \lambda(c) + w(cb)]$$
$$= \text{Min } [\infty, 29] = 29.$$
$$\lambda(d) = \text{Min } [\lambda(d), \lambda(c) + w(cd)]$$
$$= \text{Min } [\infty, 22] = 22.$$

Again, $T = T - \{u = c\} = \{a, b, d, f\}$. Thus, we have the following table

Vertex	e	a	b	c	d	f
$\lambda(v)$	0	18	29	15	22	∞
T		a	b		d	f

Now, $u = a$ as $\lambda(u) = \lambda(a) = 18$ which is minimum. The edges incident with $u = a$ is ab. Therefore,

$$\lambda(b) = \text{Min } [\lambda(b), \lambda(a) + w(ab)]$$
$$= \text{Min } [29, 27] = 27.$$

Again, $T = T - \{u = a\} = \{b, d, f\}$. Thus, we have the following table

Vertex	e	a	b	c	d	f
$\lambda(v)$	0	18	27	15	22	∞
T			b		d	f

Now, $u = d$ as $\lambda(u) = \lambda(d) = 22$ which is minimum. The edges incident with $u = d$ are db and df. Therefore,

$$\lambda(b) = \text{Min } [\lambda(b), \lambda(d) + w(db)]$$
$$= \text{Min } [27, 32] = 27.$$
$$\lambda(f) = \text{Min } [\lambda(f), \lambda(d) + w(df)]$$
$$= \text{Min } [\infty, 58] = 58.$$

Again, $T = T - \{u = d\} = \{b, f\}$. Thus, we have the following table.

Vertex	e	a	b	c	d	f
$\lambda(v)$	0	18	27	15	22	58
T			b			f

Now, $u = b$ as $\lambda(u) = \lambda(b) = 27$ which is minimum. The edges incident with $u = b$ is bf. Therefore,

$$\lambda(f) = \text{Min } [\lambda(f), \lambda(b) + w(bf)]$$
$$= \text{Min } [58, 55] = 55.$$

Again, $T = T - \{u = b\} = \{f\}$. Thus, we have the following table

Vertex	e	a	b	c	d	f
$\lambda(v)$	0	18	27	15	22	55
T						f

Now, $u = f$ and f is the terminating node, so the process terminates. Hence, the shortest distances from e to a, b, c, d and f are 18, 27, 15, 22, 55 respectively. The shortest distance between e and f is given in the following figure.

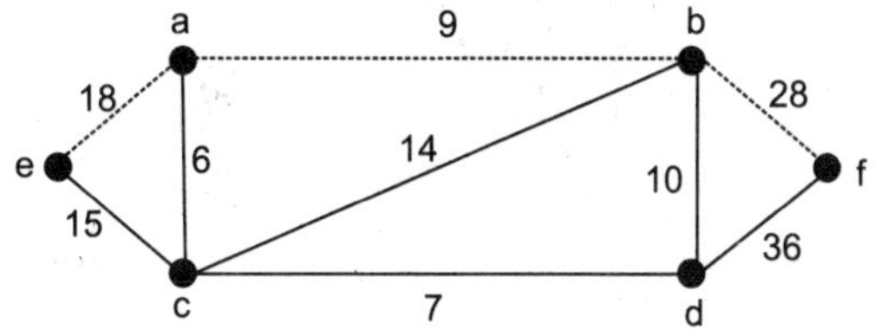

13.9.2 Floyd-Warshall Algorithm

Floyd-Warshall algorithm solves all-pairs shortest paths problem on a directed weighted graph G = (V, E). The weighted graph may contain negative weight edges, but we shall assume that there are no negative weight cycles. In this algorithm, we use the adjacency matrix of the graph to find out the shortest distance between any pair of vertices.

Suppose that G(V, E) be a graph. Let W be the adjacency matrix of the weighted directed graph G. The algorithm has the following steps:

1. n = Rows [W]
2. $D^{(0)} = W$
3. For $k = 1$ to n
4. Do for $i = 1$ to n
5. Do for $j = 1$ to n
6. $d_{ij}^{(k)} = \text{Min}\left(d_{ij}^{(k-1)}, d_{ik}^{(k-1)} + d_{kj}^{(k-1)}\right)$
7. Write $D^{(n)}$

where, $D^{(n)} = \left(d_{ij}^{n}\right)$.

■ 13.10 CUT VERTICES

Suppose that G(V, E) be the graph. A vertex 'v' of a graph G is called a cut vertex of G if the number of component of (G – v) is greater than the number of components of G.

i.e. w (G – v) > $w(g)$, where w(G) represents number of component of G.

Consider the graph G as below. In the graph G, 'v_4' is a cut vertex as $w(G - v_4) = 3 > w(G) = 1$.

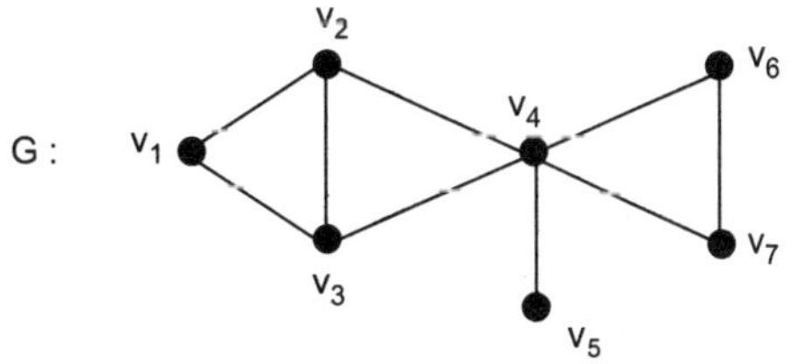

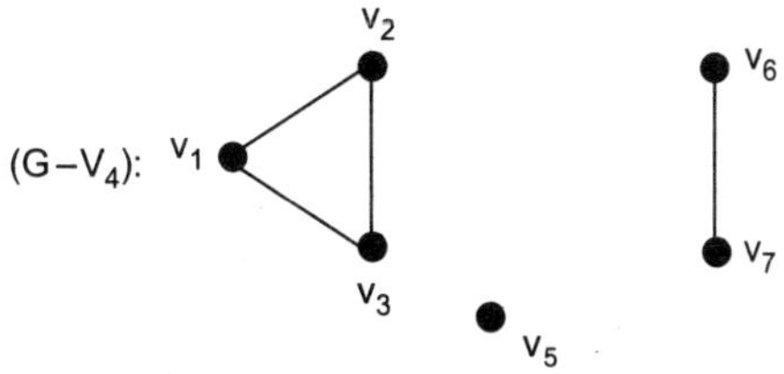

■ 13.11 EULER GRAPH

A tour is a closed walk of G, which include every edge of G at least once. An Euler tour is a closed walk of G, which include every edge of G exactly once. A graph G is said to be an Euler or Eulerian if the graph G has an Euler tour.

In this section, we will discuss two algorithms *i.e.*, Fleury's algorithm and Hierholzer's algorithm to construct Euler tour in a Euler graph.

13.11.1 Fleury's Algorithm

This algorithm is generally developed to construct an Euler tour in a Euler graph. The following steps are used in this algorithm:

1. Choose any vertex v_0 in the Euler graph G and set $W_0 = v_0$.
2. If the trail $W_i = v_0\, e_1\, v_1\, e_2\, v_2 \ldots e_i\, v_i$ has been chosen, then choose an edge e_{i+1} different from $e_1, e_2, \ldots, e_i$ such that
   (*i*) e_{i+1} is incident with v_i and
   (*ii*) unless there is no alternative, e_{i+1} is not a bridge of the edge deleted subgraph $G - \{e_1, e_2, \ldots, e_i\}$
3. Stop if w_i contains every edge of G; otherwise go to step 2.

Consider the following Euler graph G. We have to find out the Euler tour using Fleury's algorithm for the Euler graph G.

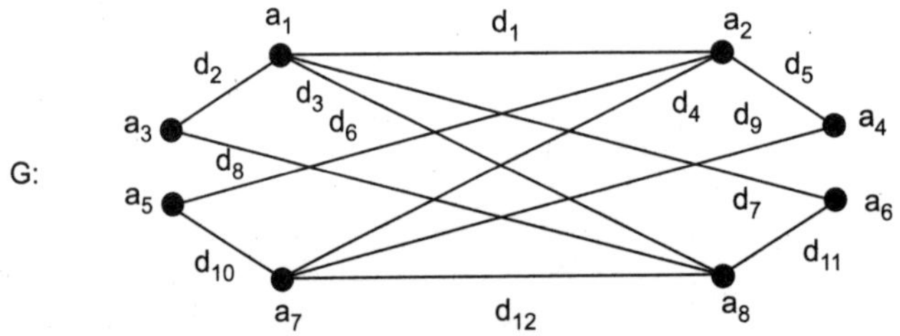

1. Let us choose $v_0 = a_1$ and set $w_0 = a_1$
2. Choose edge $e_1 = d_1$ such that $W_1 = v_0\, e_1 = a_1\, d_1\, a_2$
3. As W_1 contains only one edge, so go to step 2.
   2. Choose edge $e_2 = d_6$ such that $W_2 = a_1\, d_1\, a_2\, d_6\, a_5$
   3. As W_2 contains 2 edges, so go to step 2.
      2. Choose edge $e_3 = d_{10}$ such that $W_3 = a_1\, d_1\, a_2\, d_6\, a_5\, d_{10}\, a_7$
      3. As W_3 contains 3 edges, so go to step 2.
         2. Choose edge $e_4 = d_9$ such that $W_4 = a_1\, d_1\, a_2\, d_6\, a_5\, d_{10}\, a_7\, d_9\, a_4$
         3. As W_4 contains 4 edges, so go to step 2.
            2. Choose edge $e_5 = d_5$ such that $W_5 = a_1\, d_1\, a_2\, d_6\, a_5\, d_{10}\, a_7\, d_9\, a_4\, d_5\, a_2$
            3. As W_5 contains 5 edges, so go to step 2.

Proceeding in this manner, we will get

$$W_{12} = a_1\, d_1\, a_2\, d_6\, a_5\, d_{10}\, a_7\, d_9\, a_4\, d_5\, a_2\, d_4\, a_7\, d_{12}\, a_8\, d_8\, a_3\, d_2\, a_1\, d_3\, a_8\, d_{11}\, a_6\, d_7\, a_1$$

As W_{12} contains all the 12 edges once, so the process terminates. Thus the Euler tour produced is given as

$$a_1\, d_1\, a_2\, d_6\, a_5\, d_{10}\, a_7\, d_9\, a_4\, d_5\, a_2\, d_4\, a_7\, d_{12}\, a_8\, d_8\, a_3\, d_2\, a_1\, d_3\, a_8\, d_{11}\, a_6\, d_7\, a_1$$

13.11.2 Hierholzer's Algorithm

Like Fleury's algorithm, this algorithm is also developed to construct an Euler tour in a Euler graph. The following steps are used in this algorithm:

1. Choose any vertex v in G and choose any closed trail W_0 in G. Set $i = 0$.
2. If $E(W_i) = E(G)$, then stop and W_i is an Euler tour of G; else chose a vertex v_i on W_i which is incident with an edge in G but not in W_i. Choose a closed trail W_i*in the subgraph G - $E(W_i)$, starting at the vertex v_i. Where W_i* is the detour trail.
3. Let W_{i+1} be the closed trail consisting of the edges of both W_i and W_i* obtained by starting at the vertex v, traversing the trail W_i until v_i is reached, then traversing the closed trail W_i* and returning to v_i , completing the rest of the trail W_i. Replace i by $(i + 1)$ and go to step 2.

Consider the graph G. We have to find out the Euler tour using Hierholzer's algorithm for the Euler graph G.

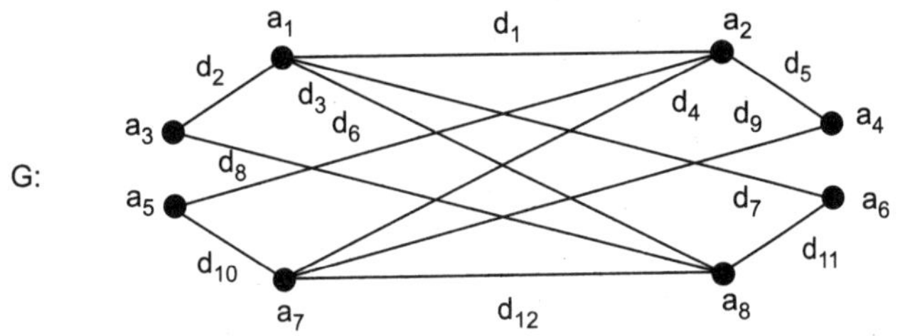

1. Let $v = a_1$ choose the closed trail W_0 as $W_0 = a_1\, d_1\, a_2\, d_5\, a_4\, d_9\, a_7\, d_{12}\, a_8\, d_3\, a_1$. Set $i = 0$.
2. As $E(W_0) \neq E(G)$, choose a_2 on W_0 incident with d_6 not in W_0. Choose $W_0{}^* = a_2\, d_6\, a_5\, d_{10}\, a_7\, d_4\, a_2$; where all $d_i \in G - E(W_0)$; $i = 6, 4, 10$.
3. Now, we have $W_1 = W_{0+1} = a_1\, d_1\, a_2\, d_6\, a_5\, d_{10}\, a_7\, d_4\, a_2\, d_5\, a_4\, d_9\, a_7\, d_{12}\, a_8\, d_3\, a_1$ and $i = (i + 1) = 0 + 1 = 1$. Go to step 2.
   2. As $E(W_1) \neq E(G)$, choose a_1 on W_1 incident with d_2 not in W_1. Choose $W_1{}^* = a_1\, d_2\, a_3\, d_8\, a_8\, d_{11}\, a_6\, d_7\, a_1$; where all $d_i \in G - E(W_1)$; $i = 2, 8, 11, 7$.
   3. Now, we get $W_2 = W_{1+1} = a_1\, d_2\, a_3\, d_8\, a_8\, d_{11}\, a_6\, d_7\, a_1\, d_1\, a_2\, d_6\, a_5\, d_{10}\, a_7\, d_4\, a_2\, d_5\, a_4\, d_9\, a_7\, d_{12}\, a_8\, d_3 a_1$ and $i = (i + 1) = 2$. Go to step 2. Since, $E(W_2) = E(G)$; the process terminates. Therefore, the Euler tour is given as

$$a_1\, d_2\, a_3\, d_8\, a_8\, d_{11}\, a_6\, d_7\, a_1\, d_1\, a_2\, d_6\, a_5\, d_{10}\, a_7\, d_4\, a_2\, d_5\, a_4\, d_9\, a_7\, d_{12}\, a_8\, d_3\, a_1.$$

13.11.3 Euler Trail

Suppose that G be the graph. A trail in G is said to be an Euler trail if it contains every edge of G exactly once. So every Euler tour is a closed Euler Trail.

Consider the graph G as below. One Euler trail in the graph G is given as

$$v_1\, e_1\, v_2\, e_2\, v_5\, e_3\, v_4\, e_4\, v_2\, e_5\, v_3\, e_6\, v_4\, e_7\, v_6\, e_8\, v_5\, e_9\, v_1\, e_{10}\, v_6.$$

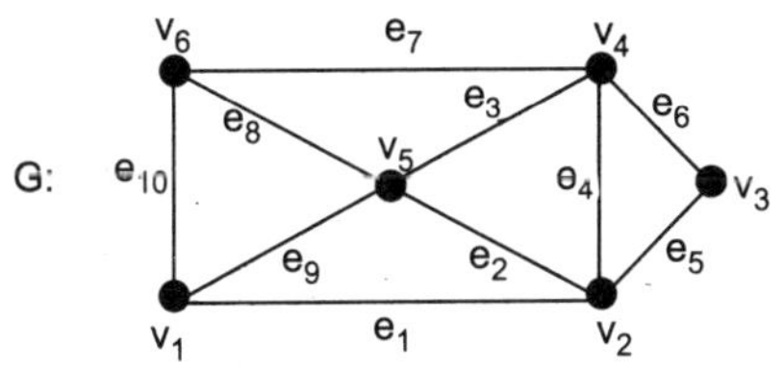

Consider another graph G as below. In the graph G the closed Euler trail is given as

$$v_1\, e_1\, v_2\, e_2\, v_5\, e_3\, v_4\, e_6 v_3\, e_5\, v_2\, e_4\, v_4\, e_7\, v_6\, e_8\, v_5\, e_9\, v_1\, e_{10}\, v_6\, e_{11}\, v_7\, e_{12}\, v_1.$$

This is known as an Euler Tour.

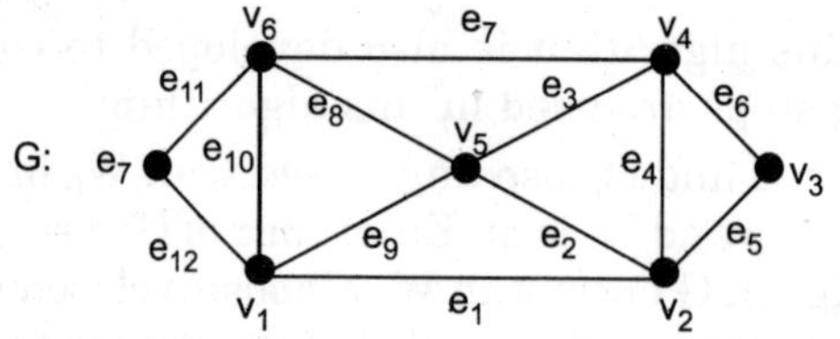

13.12 HAMILTONIAN PATH

A path of a graph G(V, E) which contains every vertex of G exactly once is known as Hamiltonian path. Consider the following graphs:

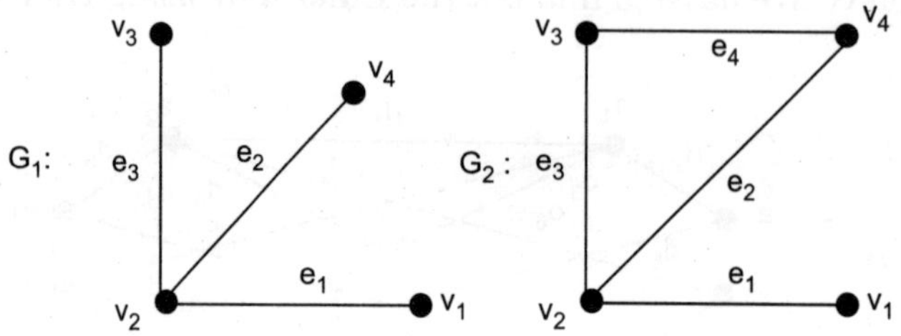

The graph G_1 has no Hamiltonian path where as G_2 has a Hamiltonian path, *i.e.*,

$$v_1\, e_1\, v_2\, e_3\, v_3\, e_4\, v_4\,.$$

13.12.1 Hamiltonian Graph

A cycle in a graph G, which contains every vertex of G only once, is known as a Hamiltonian cycle. It is to be noted that no vertex of a cycle is repeated apart from the final vertex, which is same as the starting vertex. A graph G is said to be Hamiltonian if it has a Hamiltonian cycle. Consider the graph G as

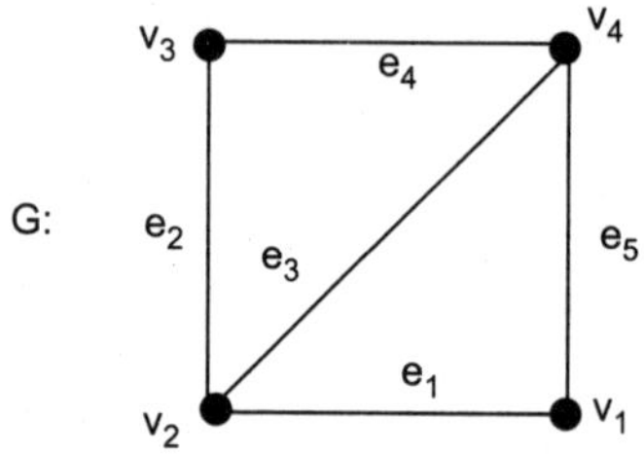

The Hamiltonian cycle is $v_1\, e_1\, v_2\, e_2\, v_3\, e_4\, v_4\, e_5\, v_1$. Therefore, the graph G is a Hamiltonian graph.

13.13 CLOSURE OF A GRAPH

Let G be a simple graph. If there are two non-adjacent vertices u_1 and v_1 in G such that $d(u_1) + d(v_1) \geq n$ (number of vertices in G) then join u_1 and v_1 by an edge to get the super graph G_1 of G. Continue this process recursively joining pairs of non-adjacent vertices whose degree sum is at least n until no such pair remains. The final super graph thus obtained is called the closure of G denoted by C(G).

Consider the graph G as

G:

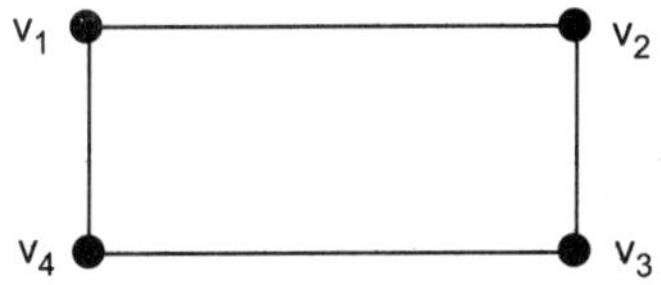

Here, V = { v_1, v_2, v_3, v_4 } and $n = 4$ (number of vertices). Now for the non-adjacent vertices v_1 and v_3 we get

$$d(v_1) + d(v_3) = 2 + 2 = 4 \geq n = 4.$$

Therefore, there exists an edge between v_1 and v_3. Similarly, for the non-adjacent vertices v_2 and v_4 we get $d(v_2) + d(v_4) = 2 + 2 = 4 \geq 4 = n$. So, there exists an edge between v_2 and v_4. Thus, the final super graph is given as below. This is nothing but the closure of G, *i.e.*, C(G).

C(G):

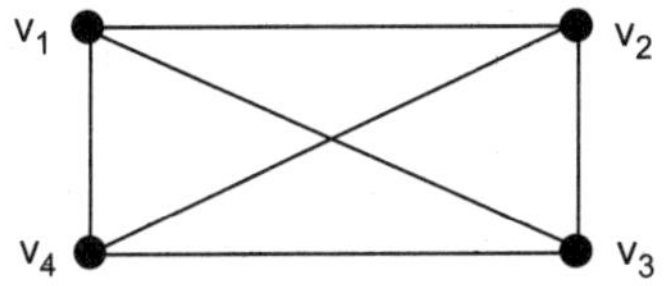

■ 13.14 TRAVELLING SALESMAN PROBLEM

The job of a travelling salesman is to visit all the towns linked with roads in a particular territory. He has to visit all the towns exactly once in such a manner that the total distance travelled by himself will be minimum.

In graph theory we denote nodes as towns joined by a weighted edge if and only if road connects them which does not pass through any of the other towns. In travelling salesman problem, we have to construct a minimum Hamiltonian cycle. The following algorithms provide minimum Hamiltonian cycle in case of a complete weighted graph:

(*i*) Two optimal algorithm and
(*ii*) Closest insertion algorithm

13.14.1 Two-Optimal Algorithm

Suppose that G(V, E) be a complete weighted graph. Where V{$v_1, v_2, \ldots, v_n$}. Here we choose a Hamiltonian cycle C and perform a sequence of modifications to C to find a smaller weight. The following steps are used in two-optimal algorithm:

1. Let $C = v_1 v_2 \ldots\ldots v_n v_1$ be a Hamiltonian cycle of the complete weighted graph G. Calculate the weight w of C by the relation
$$w = w(v_1, v_2) + w(v_2, v_3) + \ldots + w(v_n, v_1).$$
Where, $w(v_i, v_j)$ denote the weight of the edge joining v_i and v_j.
2. Set $i = 1$
3. Set $j = i + 2$
4. Let C_{ij} denote the Hamiltonian cycle as
$$C_{ij} = v_1 v_2 v_3 \ldots v_i v_j v_{j-1} v_{j-2} \ldots v_{i+1} v_{j+1} \ldots v_n v_1.$$

Calculate w_{ij} of C_{ij}, where $w_{ij} = w - w(v_i v_{i+1}) - w(v_j v_{j+1}) + w(v_i v_j) + w(v_{i+1} v_{j+1})$.

5. If $w_{ij} < w$, then replace C by C_{ij} and w by w_{ij} . Also relabel the vertices of C_{ij} in the order $v_1 v_2 v_3 \ldots\ldots v_n v_1$; else go to step 6.
6. Set $j = (j + 1)$. If $j \leq n$, go to step 4 else set $i = (i + 1)$.
7. If $i \leq (n - 2)$, go to step 3 else stop.

13.14.2 The Closest Insertion Algorithm

In this algorithm we gradually build up a sequence of cycles in the graph which involve more and more vertices until all the vertices are chosen up. In this case one more vertex is inserted into the cycle each time in cheapest possible way. The description uses the idea of the distance of a vertex v from a walk W. The following steps are used in this algorithm:

1. Choose any vertex v_1 as a starting vertex.
2. Choose the 2nd vertex v_2 which is closest to v_1 from the $(n - 1)$ vertices not chosen so far. Let $w_2 = v_1v_2v_1$ denote the walk.
3. Choose the 3rd vertex v_3 which is closest to the walk $w_2 = v_1v_2v_1$ from the $(n - 2)$ vertices not chosen so far. Let $w_3 = v_1v_2v_3v_1$ denote the walk.
4. Choose the 4th vertex v_4 which is closest to the walk $w_3 = v_1v_2v_3v_1$ from the $(n - 3)$ vertices not chosen so far. Find the shortest walk from the walks $v_1\, v_2\, v_3\, v_4\, v_1$; $v_1\, v_2\, v_4\, v_3\, v_1$; $v_1\, v_4\, v_2\, v_3\, v_1$. Let w_4 denote the shortest walk. Relabel the vertices as $v_1\, v_2\, v_3\, v_4\, v_1$ if necessary.
5. Choose the 5th vertex v_5 which is closest to the walk w_4 from the $(n - 4)$ vertices not chosen so far. Find the shortest walk from the walks $v_1\, v_2\, v_3\, v_4\, v_5\, v_1$; $v_1\, v_2\, v_3\, v_5\, v_4\, v_1$; $v_1\, v_2\, v_5\, v_3\, v_4\, v_1$; $v_1\, v_5\, v_2\, v_3\, v_4\, v_1$. Let w_5 denote the shortest walk. Relabel the vertices as $v_1\, v_2\, v_3\, v_4\, v_5\, v_1$ if necessary.
6. The process is being repeated until all the vertices are included in the cycle. Therefore the walk w_n is the Hamiltonian cycle of the graph G.

Note: Both the algorithms *i.e.,* Two optimal algorithm and Closest insertion algorithm provide reasonably good solutions. Therefore, both are approximately optimal.

SOLVED EXAMPLES

Example 1 *If u and v are distinct vertices of a tree T, then T contains exactly one u – v path.*

Solution: Suppose, to the contrary, the tree T contains two $u - v$ paths. Let us assume that the two $u - v$ paths are Q and S.

Since Q and S are different $u - v$ paths, there must exists a vertex x belonging to both Q and S such that the vertex immediately following x on Q is different from the vertex immediately following x on S. This can be easily understandable from the figure shown below.

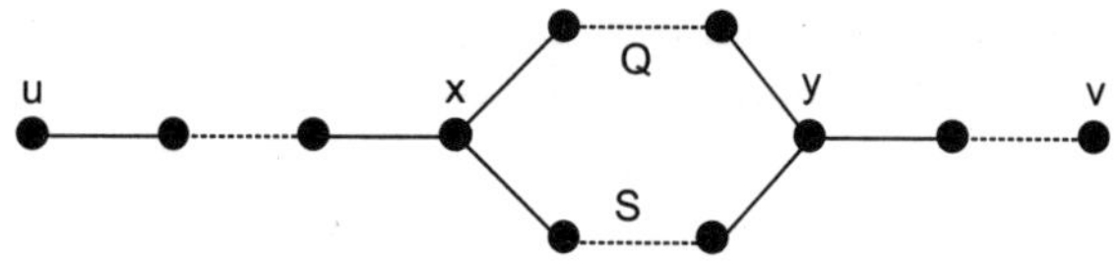

Let us assume that y be the first vertex of Q following x, which also belongs to S. This implies that there exists two $x - y$ paths that have only x and y in common. It is clear that these two paths produce a cycle in T. This is a contradiction. This contradict to the fact that T is a tree.

Therefore, our supposition is wrong. Hence, T has only one $u - v$ path.

Example 2 *For the following weighted graph G apply Floyd-Warshall algorithm to find the shortest path between any pair of vertices a, b, c, d and e. Show at least one iteration in details.*

G:

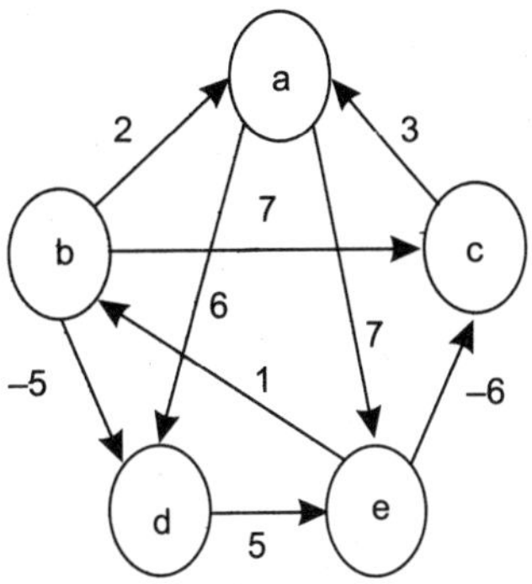

Solution: The adjacency matrix W with respect to the nodes *b*, *a*, *c*, *e*

and *d* is given as $\begin{pmatrix} 0 & 2 & 7 & \infty & -5 \\ \infty & 0 & \infty & 7 & 6 \\ \infty & 3 & 0 & \infty & \infty \\ 1 & \infty & -6 & 0 & \infty \\ \infty & \infty & \infty & 5 & 0 \end{pmatrix}$

Hence, $n = \text{Row}[W] = 5$

$$D^{(0)} = \left(d_{ij}^{(0)}\right) = \begin{pmatrix} 0 & 2 & 7 & \infty & -5 \\ \infty & 0 & \infty & 7 & 6 \\ \infty & 3 & 0 & \infty & \infty \\ 1 & \infty & -6 & 0 & \infty \\ \infty & \infty & \infty & 5 & 0 \end{pmatrix}$$

For $k = 1$, $i = 1$ and $j = 1$ to 5 we get

$$d_{11}^1 = \text{Min}(d_{11}^0, d_{11}^0 + d_{11}^0) = \text{Min}(0, 0+0) = 0$$
$$d_{12}^1 = \text{Min}(d_{12}^0, d_{11}^0 + d_{12}^0) = \text{Min}(2, 0+2) = 2$$
$$d_{13}^1 = \text{Min}(d_{13}^0, d_{11}^0 + d_{13}^0) = \text{Min}(7, 0+7) = 7$$
$$d_{14}^1 = \text{Min}(d_{14}^0, d_{11}^0 + d_{14}^0) = \text{Min}(\infty, 0+\infty) = \infty$$
$$d_{15}^1 = \text{Min}(d_{15}^0, d_{11}^0 + d_{15}^0) = \text{Min}(-5, 0-5) = -5$$

For $k = 1$, $i = 2$ and $j = 1$ to 5 we get

$$d_{21}^1 = \text{Min}(d_{21}^0, d_{21}^0 + d_{11}^0) = \text{Min}(\infty, \infty+0) = \infty$$
$$d_{22}^1 = \text{Min}(d_{22}^0, d_{21}^0 + d_{12}^0) = \text{Min}(0, \infty+2) = 0$$
$$d_{23}^1 = \text{Min}(d_{23}^0, d_{21}^0 + d_{13}^0) = \text{Min}(\infty, \infty+7) = \infty$$
$$d_{24}^1 = \text{Min}(d_{24}^0, d_{21}^0 + d_{14}^0) = \text{Min}(7, \infty+\infty) = 7$$
$$d_{25}^1 = \text{Min}(d_{25}^0, d_{21}^0 + d_{15}^0) = \text{Min}(6, \infty-5) = 6$$

For $k = 1$, $i = 3$ and $j = 1$ to 5 we get

$$d_{31}^1 = \text{Min}(d_{31}^0, d_{31}^0 + d_{11}^0) = \text{Min}(\infty, \infty + 0) = \infty$$
$$d_{32}^1 = \text{Min}(d_{32}^0, d_{31}^0 + d_{12}^0) = \text{Min}(3, \infty + 2) = 3$$
$$d_{33}^1 = \text{Min}(d_{33}^0, d_{31}^0 + d_{13}^0) = \text{Min}(0, \infty + 7) = 0$$
$$d_{34}^1 = \text{Min}(d_{34}^0, d_{31}^0 + d_{14}^0) = \text{Min}(\infty, \infty + \infty) = \infty$$
$$d_{35}^1 = \text{Min}(d_{35}^0, d_{31}^0 + d_{15}^0) = \text{Min}(\infty, \infty - 5) = \infty$$

For $k = 1$, $i = 4$ and $j = 1$ to 5 we get

$$d_{41}^1 = \text{Min}(d_{41}^0, d_{41}^0 + d_{11}^0) = \text{Min}(1, 1 + 0) = 1$$
$$d_{42}^1 = \text{Min}(d_{42}^0, d_{41}^0 + d_{12}^0) = \text{Min}(\infty, 1 + 2) = 3$$
$$d_{43}^1 = \text{Min}(d_{43}^0, d_{41}^0 + d_{13}^0) = \text{Min}(-6, 1 + 7) = -6$$
$$d_{44}^1 = \text{Min}(d_{44}^0, d_{41}^0 + d_{14}^0) = \text{Min}(0, 1 + \infty) = 0$$
$$d_{45}^1 = \text{Min}(d_{45}^0, d_{41}^0 + d_{15}^0) = \text{Min}(\infty, 1 - 5) = -4$$

For $k = 1$, $i = 5$ and $j = 1$ to 5 we get

$$d_{51}^1 = \text{Min}(d_{51}^0, d_{51}^0 + d_{11}^0) = \text{Min}(\infty, \infty + 0) = \infty$$
$$d_{52}^1 = \text{Min}(d_{52}^0, d_{51}^0 + d_{12}^0) = \text{Min}(\infty, \infty + 2) = \infty$$
$$d_{53}^1 = \text{Min}(d_{53}^0, d_{51}^0 + d_{13}^0) = \text{Min}(\infty, \infty + 7) = \infty$$
$$d_{54}^1 = \text{Min}(d_{54}^0, d_{51}^0 + d_{14}^0) = \text{Min}(5, \infty + \infty) = 5$$
$$d_{55}^1 = \text{Min}(d_{55}^0, d_{51}^0 + d_{15}^0) = \text{Min}(0, \infty - 5) = 0$$

Therefore, we have

$$D^{(1)} = \begin{pmatrix} 0 & 2 & 7 & \infty & -5 \\ \infty & 0 & \infty & 7 & 6 \\ \infty & 3 & 0 & \infty & \infty \\ 1 & 3 & -6 & 0 & -4 \\ \infty & \infty & \infty & 5 & 0 \end{pmatrix}$$

Similarly for $k = 2$, $i = 1$ to 5 and $j = 1$ to 5, we get

$$D^{(2)} = \begin{pmatrix} 0 & 2 & 7 & 9 & -5 \\ \infty & 0 & \infty & 7 & 6 \\ \infty & 3 & 0 & 10 & 9 \\ 1 & 3 & -6 & 0 & -4 \\ \infty & \infty & \infty & 5 & 0 \end{pmatrix}$$

Similarly for $k = 3$, $i = 1$ to 5 and $j = 1$ to 5, we get

$$D^{(3)} = \begin{pmatrix} 0 & 2 & 7 & 9 & -5 \\ \infty & 0 & \infty & 7 & 6 \\ \infty & 3 & 0 & 10 & 9 \\ 1 & -3 & -6 & 0 & -4 \\ \infty & \infty & \infty & 5 & 0 \end{pmatrix}$$

Similarly for $k = 4, i = 1$ to 5 and $j = 1$ to 5, we get

$$D^{(4)} = \begin{pmatrix} 0 & 2 & 3 & 9 & -5 \\ 8 & 0 & 1 & 7 & 3 \\ 11 & 3 & 0 & 10 & 6 \\ 1 & -3 & -6 & 0 & -4 \\ 6 & 2 & -1 & 5 & 0 \end{pmatrix}$$

Similarly for $k = 5, i = 1$ to 5 and $j = 1$ to 5, we get

$$D^{(5)} = \begin{pmatrix} 0 & -3 & -6 & 0 & -5 \\ 8 & 0 & 1 & 7 & 3 \\ 11 & 3 & 0 & 10 & 6 \\ 1 & -3 & -6 & 0 & -4 \\ 6 & 2 & -1 & 5 & 0 \end{pmatrix}$$

From the above matrix, the shortest distance for any pair of vertices can be found out.

Example 3 *Find the closure of the graph G where*

G:

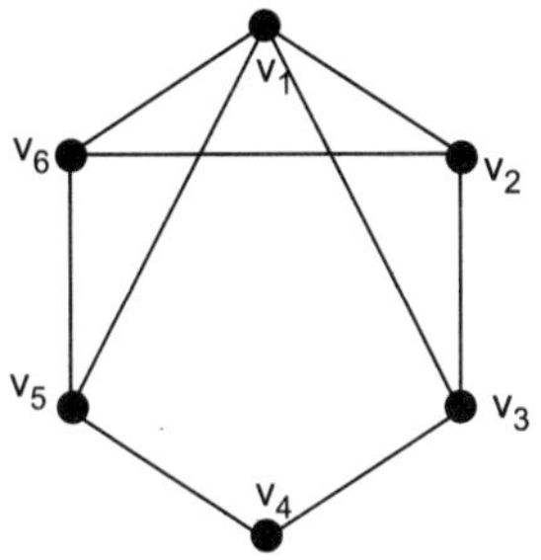

Solution: In the above graph G we have V = $\{v_1, v_2, v_3, v_4, v_5, v_6\}$ and number of vertices $(n) = 6$. Now for the non adjacent vertices v_1 and v_4 we have

$$d(v_1) + d(v_4) = 4 + 2 = 6 \geq n = 6.$$

Therefore, there exists an edge between v_1 and v_4. Let the super graph G_1 be

G_1:

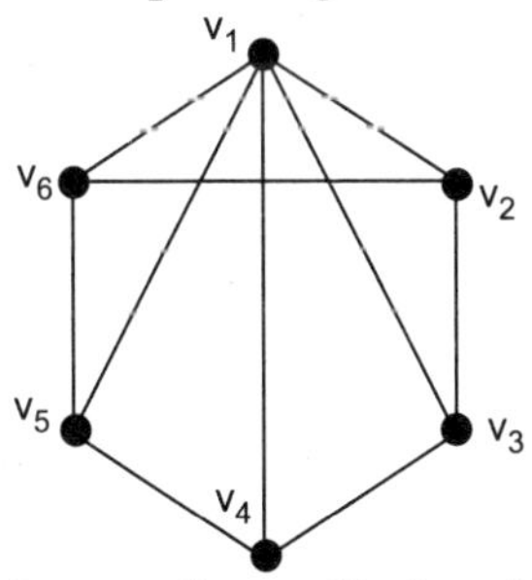

For the non-adjacent vertices v_2 and v_4 we have $d(v_2) + d(v_4) = 3 + 3 = 6 \geq n$. Therefore, there exists an edge between v_2 and v_4 . Let the super graph G_2 be

G_2:

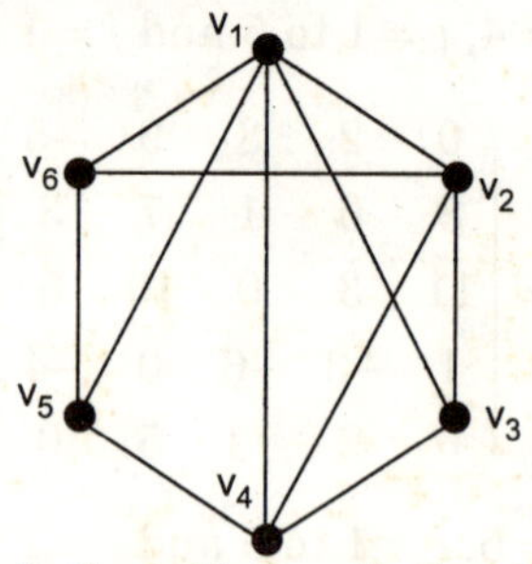

Again, v_2 and v_5 are non-adjacent such that $d(v_2) + d(v_5) = 4 + 3 = 7 \geq n = 6$. Therefore, there exists an edge between v_2 and v_5. Let the super graph G_3 be

G_3:

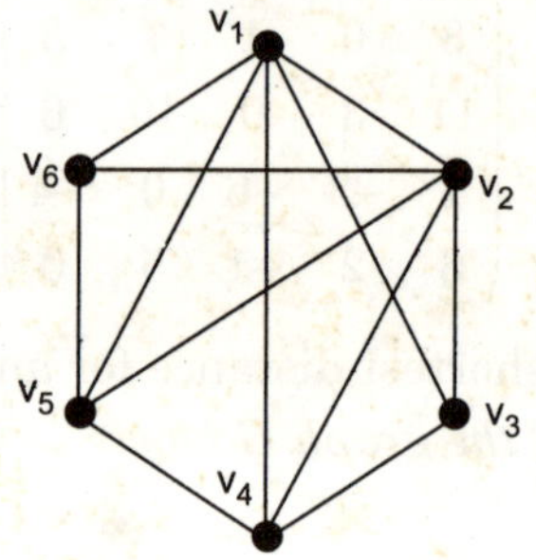

For the non-adjacent vertices v_3 and v_5 we have $d(v_3) + d(v_5) = 3 + 4 = 7 \geq n$. Therefore, there exists an edge between v_3 and v_5. Let the super graph G_4 be

G_4:

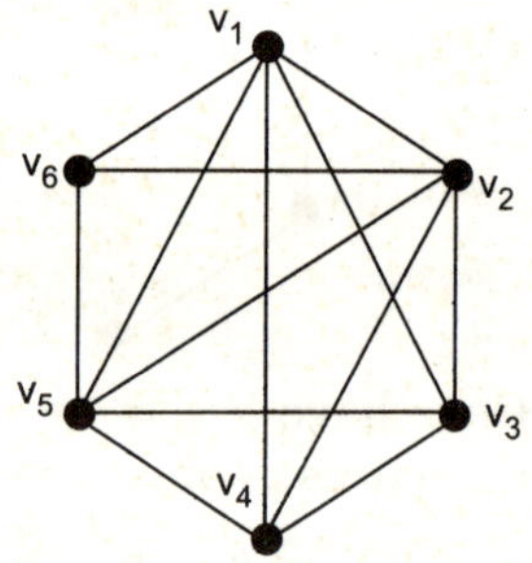

For the non-adjacent vertices v_3 and v_6 we have $d(v_3) + d(v_6) = 4 + 3 = 7 \geq n$. Therefore, there exists an edge between v_3 and v_6 . Let the super graph G_5 be.

G_5:

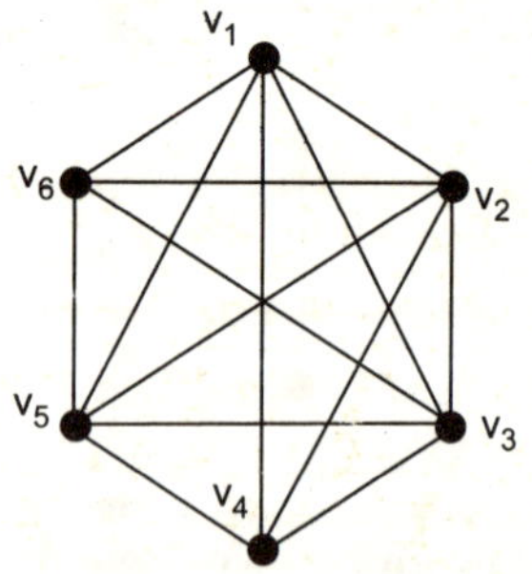

For the non-adjacent vertices v_4 and v_6 we have $d(v_4) + d(v_6) = 4 + 4 = 8 \geq n$. Therefore, there exists an edge between v_4 and v_6 . Let the super graph be G_6 . In the above graph, there is no two non-adjacent vertices. Thus, G_6 is the final super graph. Therefore, the closure of the graph G is given as

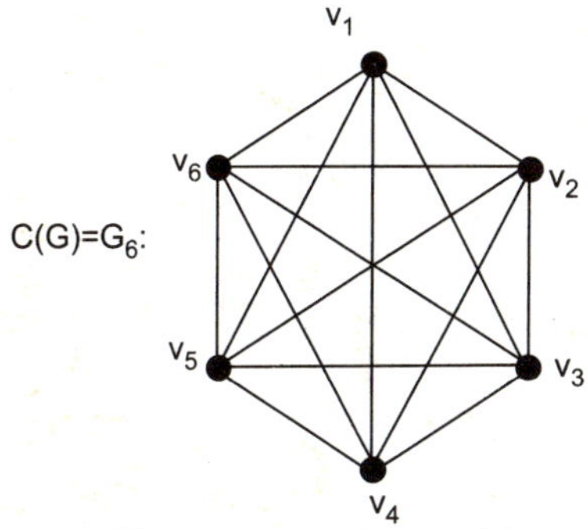

Example 4 *For the following travelling salesman problem, carry out the closest insertion algorithm.*

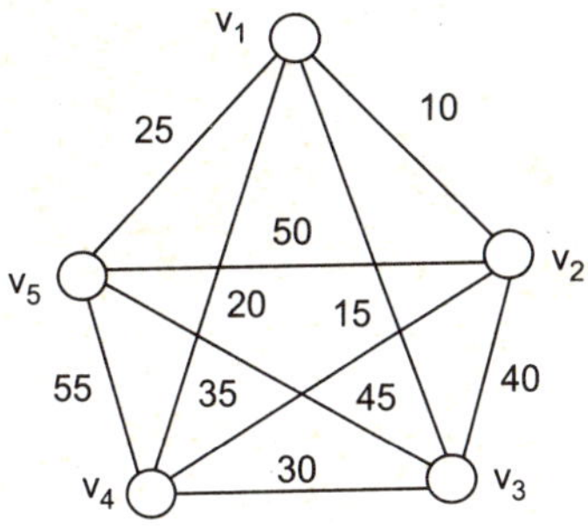

Solution: Given that the complete weighted graph G as

1. Choose the vertex v_1
2. Choose the vertex v_2, which is closest to v_1. So, $w_2 = v_1\, v_2\, v_1$
3. Choose the vertex v_3, which is close to w_2. So, $w_3 = v_1\, v_2\, v_3\, v_1$
4. Choose the vertex v_4, which is close to w_3. Hence, we have the following cases.

$$w_4 = v_1\, v_2\, v_3\, v_4\, v_1 \text{ or}$$
$$= v_1\, v_2\, v_4\, v_3\, v_1 \text{ or}$$
$$= v_1\, v_4\, v_2\, v_3\, v_1$$

Now length of $v_1\, v_2\, v_3\, v_4\, v_1 = 10 + 40 + 30 + 20 = 100$

Length of $v_1\, v_2\, v_4\, v_3\, v_1 = 10 + 45 + 30 + 15 = 100$

Length of $v_1\, v_4\, v_2\, v_3\, v_1 = 20 + 45 + 40 + 15 = 120$

Therefore, $w_4 = v_1\, v_2\, v_3\, v_4\, v_1$ is minimum.

5. Choose the vertex v_5, which is close to w_4. Hence, we have the following cases. The length of following cycles is given as below.

$$v_1\, v_2\, v_3\, v_4\, v_5\, v_1 = 10 + 40 + 30 + 55 + 25 = 160$$
$$v_1\, v_2\, v_3\, v_5\, v_4\, v_1 = 10 + 40 + 35 + 55 + 20 = 160$$
$$v_1\, v_2\, v_5\, v_3\, v_4\, v_1 = 10 + 50 + 35 + 30 + 20 = 145$$
$$v_1\, v_5\, v_2\, v_3\, v_4\, v_1 = 25 + 50 + 40 + 30 + 20 = 165$$

As all the vertices are included in the cycle, so the process terminates. Hence , the shortest Hamiltonian cycle is given as $v_1\, v_2\, v_5\, v_3\, v_4\, v_1$.

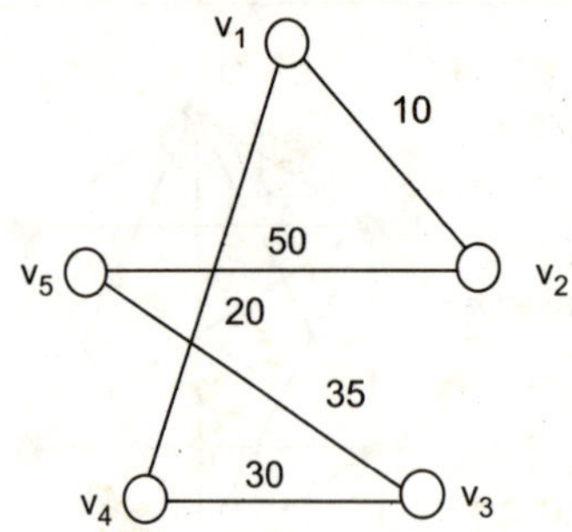

Example 5 *For the travelling salesman problem given in example 4, carry out the two optimal algorithm.*

Solution: For the complete weighted graph G given above, the number of vertices $(n) = 5$. According to the two optimal algorithm we have the following steps:

1. Let $C = v_1, v_2, v_3, v_4, v_5, v_1$ be a Hamiltonian cycle.
   Therefore, we get
   $$w = w(v_1v_2) + w(v_2v_3) + w(v_3v_4) + w(v_4v_5) + w(v_5v_1)$$
   $$= 10 + 40 + 30 + 55 + 25 = 160$$
2. Set $i = 1$
3. Set $j = i + 2 = 3$
4. Set $C_{ij} = C_{13} = v_1\, v_3\, v_2\, v_4\, v_5\, v_1$
   $$w_{13} = w - w(v_1v_2) - w(v_3v_4) + w(v_1v_3) + w(v_2v_4)$$
   $$= 160 - 10 - 30 + 15 + 45 = 180$$
5. As $w_{13} \not< w$; Go to step 6.
6. Set $j = (j + 1) = 4$ and $4 \leq n = 5$. Go to step 4.
4. Set $C_{ij} = C_{14} = v_1\, v_4\, v_3\, v_2\, v_5\, v_1$
   $$w_{14} = w - w(v_1v_2) - w(v_4v_5) + w(v_1v_4) + w(v_2v_5) = 165$$
5. As $w_{14} = 165 \not< 160 = w$; Go to step 6.
6. Set $j = (j + 1) = 5$ and $5 \leq n = 5$. Go to step 4.
4. Set $C_{ij} = C_{15} = v_1\, v_5\, v_4\, v_3\, v_2\, v_1$
   $$w_{15} = w - w(v_1v_2) - w(v_5v_1) + w(v_1v_5) + w(v_2v_1) = 160$$
5. As $w_{15} = 160 \not< 160 = w$; Go to step 6
6. Set $j = (j + 1) = 6$ and $6 \not< n = 5$. Go to step 7 with $i = (i + 1) = 2$.
7. As $i = 2 \leq (n - 2) = 3$, Go to step 3.
3. Set $j = (i + 2) = 2 + 2 = 4$
4. Set $C_{ij} = C_{24} = v_1\, v_2\, v_4\, v_3\, v_5\, v_1$
   $$W_{24} = w - w(v_2\, v_3) - w(v_4\, v_5) + w(v_2\, v_4) + w(v_3\, v_5) = 145$$
5. As $w_{24} = 145 < 160 = w$; go to step 1
   1. $C = C_{24} = v_1\, v_2\, v_4\, v_3\, v_5\, v_1$ with $w = w_{24} = 145$.
   After re-labeling the vertices we have
   $$C = C_{24} = v_1\, v_2\, v_3\, v_4\, v_5\, v_1.$$

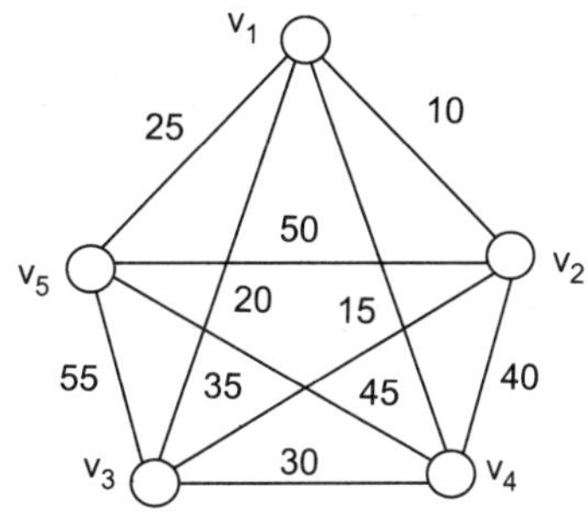

2. Set $i = 1$
3. Set $j = (i + 2) = 3$
4. Set $C_{ij} = C_{13} = v_1\, v_3\, v_2\, v_4\, v_5\, v_1$

$$w_{13} = w - w(v_1\, v_2) - w(v_3\, v_4) + w(v_1\, v_3) + w(v_2\, v_4) = 165$$

5. As $w_{13} = 165 \not< 145 = w$; go to step 6
6. Set $j = (j + 1) = 4$ and $4 \le n = 5$. Go to step 4
4. Set $C_{ij} = C_{14} = v_1\, v_4\, v_3\, v_2\, v_5\, v_1$

$$w_{14} = w - w(v_1\, v_2) - w(v_4\, v_5) + w(v_1\, v_4) + w(v_2\, v_5) = 165$$

5. As $w_{14} = 165 \not< 145 = w$; go to step 6
6. Set $j = (j + 1) = 5$ and $5 \le n = 5$. Go to step 4
4. Set $C_{ij} = C_{15} = v_1\, v_5\, v_4\, v_3\, v_2\, v_1$

$$w_{15} = w - w(v_1\, v_2) - w(v_5\, v_1) + w(v_1\, v_5) + w(v_2\, v_1) = 145$$

5. As $w_{15} = 145 \not< 145 = w$; go to step 6
6. Set $j = (j + 1) = 6 \not< n = 5$ with $i = (i + 1) = 2$.
7. As $i = 2 \le (n - 2) = 3$, Go to step 3
3. Set $j = (i + 2) = 2 + 2 = 4$
4. Set $C_{ij} = C_{24} = v_1\, v_2\, v_4\, v_3\, v_5\, v_1$

$$w_{24} = w - w(v_2\, v_3) - w(v_4\, v_5) + w(v_2\, v_4) + w(v_3\, v_5) = 160$$

5. As $w_{24} = 160 \not< 145 = w$; go to step 6
6. Set $j = (j + 1) = 5 \le 5 = n$, go to step 4
4. Set $C_{ij} = C_{25} = v_1\, v_2\, v_5\, v_4\, v_3\, v_1$

$$w_{25} = w - w(v_2\, v_3) - w(v_5\, v_1) + w(v_2\, v_5) + w(v_3\, v_1) = 145$$

5. As $w_{25} - 145 \not< 145 = w$; go to step 6
6. Set $j = (j + 1) = 6 \not< n = 5$ with $i = (i + 1) = 3$. go to step 7
7. As $i = 3 \le (n - 2) = 3$, go to step 3
3. Set $j = (i + 2) = 5$
4. Set $C_{ij} = C_{35} = v_1\, v_2\, v_3\, v_5\, v_4\, v_1$

$$w_{35} = w - w(v_3\, v_4) - w(v_5\, v_1) + w(v_3\, v_5) + w(v_4\, v_1) = 160$$

5. As $w_{35} = 160 \not< 145 = w$; go to step 6
6. Set $j = (j + 1) = 6 \not< n = 5$ with $i = (i + 1) = 4$, go to step 7
7. As $i = 4 \not< (n - 2) = 3$, therefore the process terminates. Hence, the minimum Hamiltonian path is given as $v_1\, v_2\, v_3\, v_4\, v_5\, v_1$. The path for travelling salesman is given below.

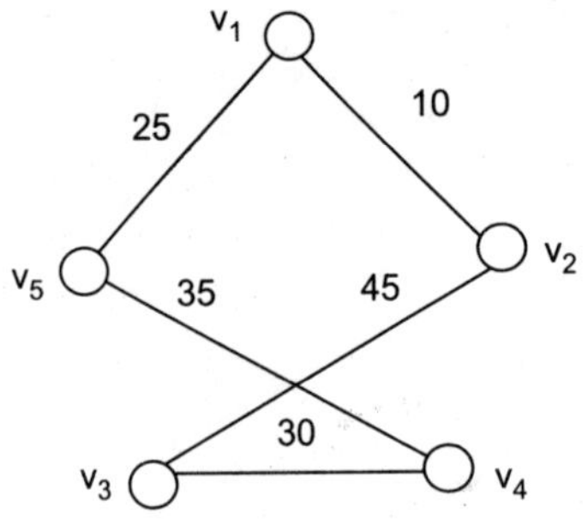

Example 6 *Find the eccentricity of all vertices, radius, diameter and centre of the graph given below. It is given that the distance between any two adjacent vertices is 1.*

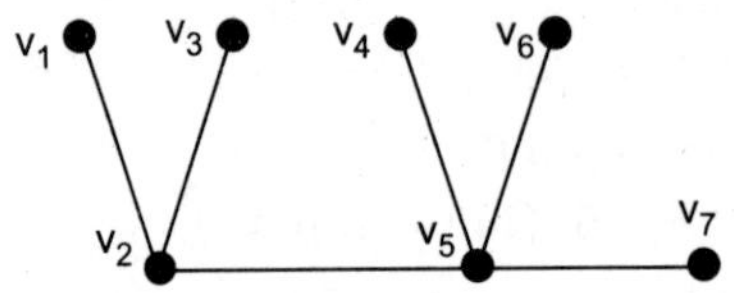

Solution: In the graph given above $V = \{v_1, v_2, v_3, v_4, v_5, v_6, v_7\}$. It is also given that length of each edge is 1. Now,

$d(v_1, v_2) = 1$; $d(v_1, v_3) = 2$; $d(v_1, v_4) = 3$;
$d(v_1, v_5) = 2$; $d(v_1, v_6) = 3$; $d(v_1, v_7) = 3$;

Therefore, $e(v_1) = \text{Max}\{1, 2, 3\} = 3.$

Similarly, we get

$d(v_2, v_1) = 1$; $d(v_2, v_3) = 1$; $d(v_2, v_4) = 2$;
$d(v_2, v_5) = 1$; $d(v_2, v_6) = 2$; $d(v_2, v_7) = 2$;

Therefore, $e(v_2) = \text{Max}\{1, 2\} = 2.$

Proceeding in this way we get

$$e(v_3) = 3; e(v_4) = 3; \; e(v_5) = 2; \; e(v_6) = 3; \; e(v_7) = 3.$$

Now, radius = rad (G) = Min $\{e(v): v \in V\}$ = Min (2, 3) = 2.

Diameter = diam (G) = Max $\{e(v): v \in V\}$ = Max (2, 3) = 3.

So, the central points are v_2, v_5 and centre $\{v_2, v_5\}$.

Example 7 *Let T be a tree of order p and size q having p_i vertices of degree i (i = 1, 2, 3, ...).*

$$\text{Let } \sum_i p_i = p \text{ and } \sum_i ip_i = 2q = 2(p-1).$$

Show that $p_1 = p_3 + 2p_4 + 3p_5 + 4p_6 + \ldots + 2.$

Solution: Given that T is a tree of order p and size q. It is also given that

$$\sum_i p_i = p \text{ and } \sum_i ip_i = 2(p-1)$$

i.e., $$p_1 + 2p_2 + 3p_3 + 4p_4 + \ldots = 2p - 2$$

i.e., $$p_1 + 2p_2 + 3p_3 + 4p_4 + \ldots = 2\sum_i p_i - 2$$

i.e., $\quad p_1 + 2p_2 + 3p_3 + 4p_4 + \ldots = 2(p_1 + p_2 + p_3 + \ldots) - 2$

i.e., $$p_1 = p_3 + 2p_4 + 3p_5 + 4p_6 + \ldots + 2.$$

Example 8 *If T is a binary tree of height h and order p, then*

$$(h + 1) \le p \le 2^{(h+1)} - 1$$

Solution: Let p_k denotes the number of vertices of T at level k for $0 \le k \le h$.

Therefore, we get

$$\sum_{k=0}^{h} p_k = p$$

Since $p_k \ge 1$ for each k, and $p_k \le 2p_{(k-1)}$ for $1 \le k \le h$, it follows, inductively, that $p_k \le 2^k$. Again,

$$\sum_{k=0}^{h} 2^k = 1 + 2 + 2^2 + 2^3 + \ldots + 2^h = 2^{h+1} - 1$$

Again, $$\sum_{k=0}^{h} p_k \le \sum_{k=0}^{h} 2^k = 2^{h+1} - 1$$

i.e., $$p \le 2^{h+1} - 1 \qquad \ldots (i)$$

Also, $$\sum_{k=0}^{h} 1 \le \sum_{k=0}^{h} p_k = p$$

i.e., $$(h + 1) \le p \qquad \ldots (ii)$$

On combining equations (*i*) and (*ii*), we get

$$(h + 1) \le p \le 2^{h+1} - 1.$$

Example 9 *Construct the binary tree for the arithmetic expression*

(A(B – C)) / ((D – E)(F + G – H)).

Solution: Given arithmetic expression is

(A(B – C))/((D – E)(F + G – H)).

The binary tree corresponding to the above expression is given below.

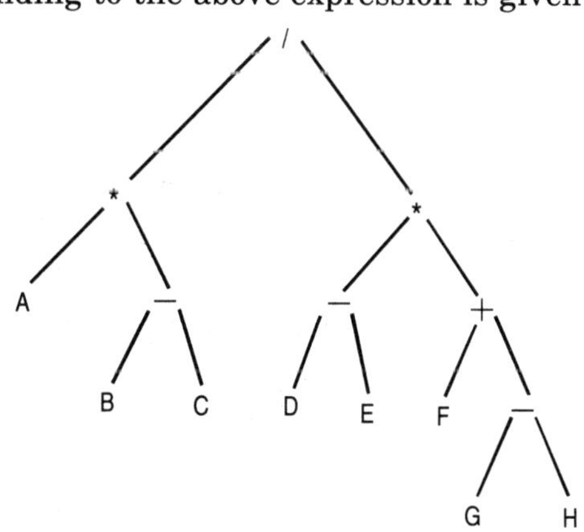

Example 10 *For the graph G shown below, use Dijkstra's algorithm to compute the shortest path between a and f.*

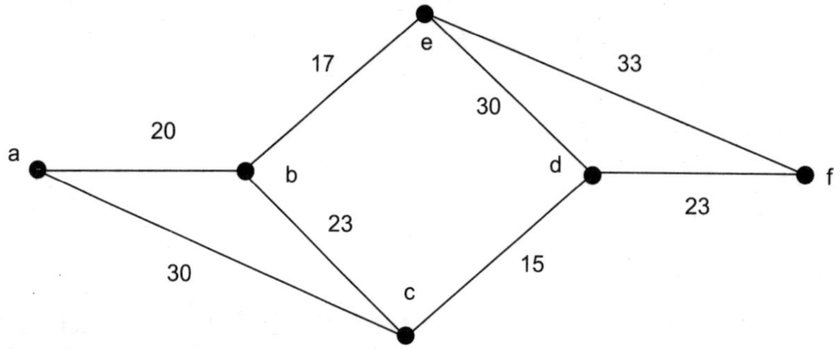

Solution: In the above graph G, the source vertex is

$v_s = a$ and $v_t = f$. Set $\lambda(a) = 0$

and $\lambda(b) = \lambda(c) = \lambda(d) = \lambda(e) = \lambda(f) = \infty$. T = V = $\{a, b, c, d, e, f\}$.

Hence, we have the following table

Vertex	a	b	c	d	e	f
$\lambda(v)$	0	∞	∞	∞	∞	∞
T	a	b	c	d	e	f

Now, $u = a$ as $\lambda(u) = \lambda(a) = 0$

which is minimum.

The edges incident on $u = a$ are ab and ac.

Therefore,

$$\lambda(b) = \text{Min } [\lambda(b), \lambda(a) + w(ab)]$$
$$= \text{Min } [\infty, 20] = 20.$$
$$\lambda(c) = \text{Min } [\lambda(c), \lambda(a) + w(ac)]$$
$$= \text{Min } [\infty, 30] = 30.$$

Again, $T = T - \{u = a\} = \{b, c, d, e, f\}$.

Therefore, we have the following table

Vertex	a	b	c	d	e	f
$\lambda(v)$	0	20	30	∞	∞	∞
T		b	c	d	e	f

Now, $u = b$ as $\lambda(u) = \lambda(b) = 20$

which is minimum.

The edges incident with $u = b$ are bc and be. Therefore,

$$\lambda(c) = \text{Min } [\lambda(c), \lambda(b) + w(bc)]$$
$$= \text{Min } [30, 43] = 30.$$
$$\lambda(e) = \text{Min } [\lambda(e), \lambda(b) + w(be)]$$
$$= \text{Min } [\infty, 37] = 37.$$

Again, $T = T - \{u = b\} = \{c, d, e, f\}$.

Thus, we have the following table

Vertex	a	b	c	d	e	f
$\lambda(v)$	0	20	30	∞	37	∞
T			c	d	e	f

Now, $u = c$ as $\lambda(u) = \lambda(c) = 30$ which is minimum.

The edges incident with $u = c$ is cd.

Therefore, $\lambda(d) = \text{Min }[\lambda(d), \lambda(c) + w(cd)]$

$= \text{Min }[\infty, 45] = 45.$

Again, $\text{T} = \text{T} - \{u = c\} = \{d, e, f\}$.

Thus, we have the following table

Vertex	a	b	c	d	e	f
$\lambda(v)$	0	20	30	45	37	∞
T				d	e	f

Now, $u = e$ as $\lambda(u) = \lambda(e) = 37$ which is minimum.

The edges incident with $u = e$ are ed and ef.

Therefore, $\lambda(d) = \text{Min }[\lambda(d), \lambda(e) + w(ed)]$

$= \text{Min }[45, 67] = 45.$

$\lambda(f) = \text{Min }[\lambda(f), \lambda(e) + w(ef)]$

$= \text{Min }[\infty, 70] = 70.$

Again, $\text{T} = \text{T} - \{u = e\} = \{d, f\}$.

Thus, we have the following table

Vertex	a	b	c	d	e	f
$\lambda(v)$	0	20	30	45	37	70
T				d		f

Now, $u = d$ as $\lambda(u) = \lambda(d) = 45$ which is minimum.

The edges incident with $u = d$ is df. Therefore,

$\lambda(f) = \text{Min }[\lambda(f), \lambda(d) + w(df)]$

$= \text{Min }[70, 68] = 68.$

Again, $\text{T} = \text{T} - \{u = d\} = \{f\}$.

Thus, we have the following table

Vertex	a	b	c	d	e	f
$\lambda(v)$	0	20	30	45	37	68
T						f

Now, $u = f$ and f is the terminating node, so the process terminates.

Hence, the shortest distances from a to b, c, d, e and f are 20, 30, 45, 37, 68 respectively. The shortest distance between a and f is given in the figure further.

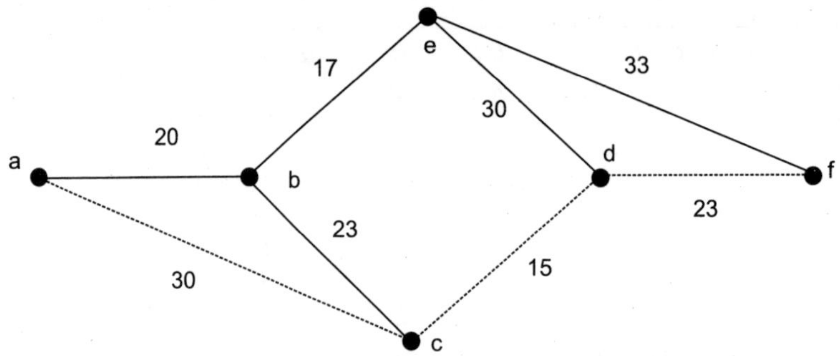

Example 11 *Construct the following graphs.*

(a) Eulerian but not Hamiltonian *(b) Hamiltonian but not Eulerian*
(c) Neither Eulerian nor Hamiltonian *(d) Eulerian and Hamiltonian*

Solution: The different graphs are given below.

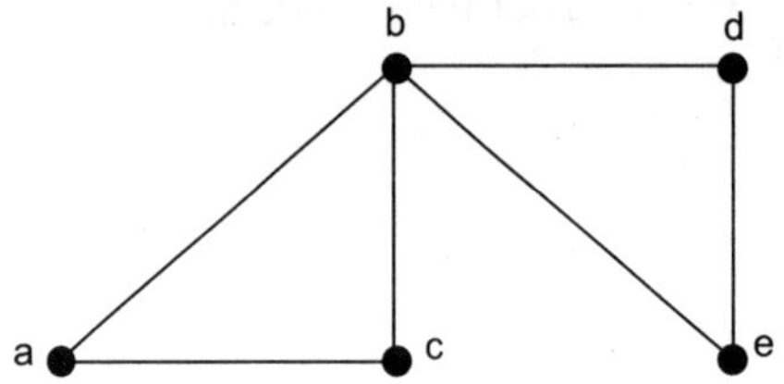

Eulerian but not Hamiltonian

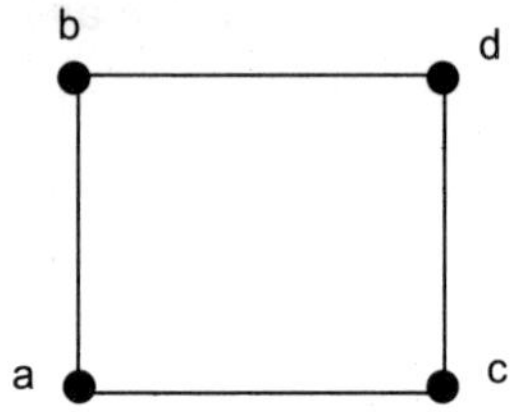

Hamiltonian but not Eulerian

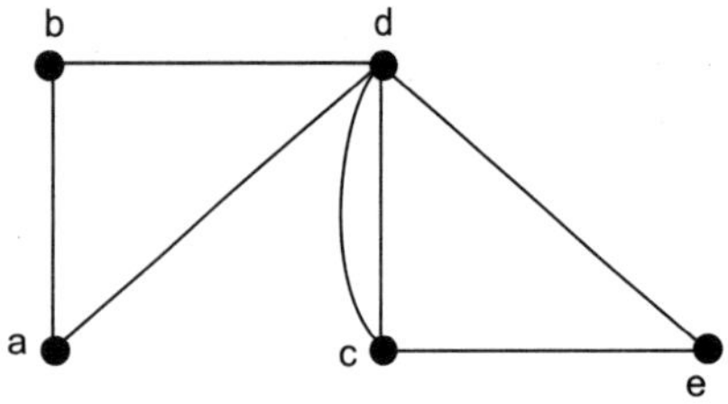

Neither Eulerian nor Hamiltonian

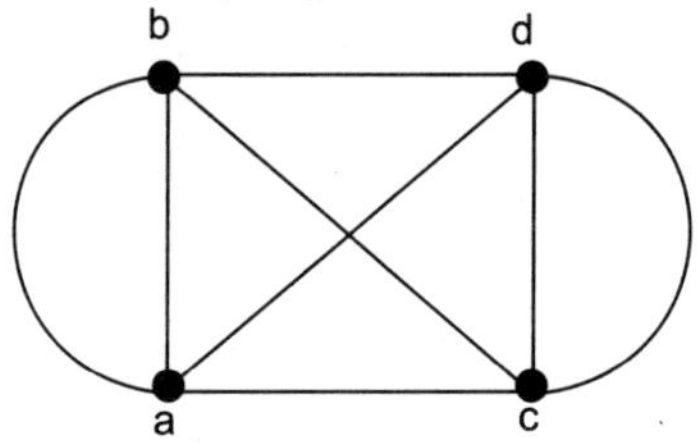

Hamiltonian and Eulerian

Example 12 *For the graph G shown below, find the depth first search forest.*

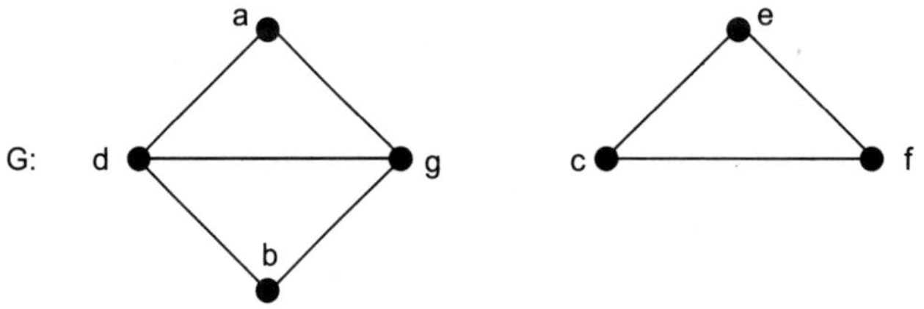

Solution: Let us consider the source vertex as '*a*' in the above graph G. On using the DFS technique, the order in which the vertices are being visited is described below by the sequence of graphs.

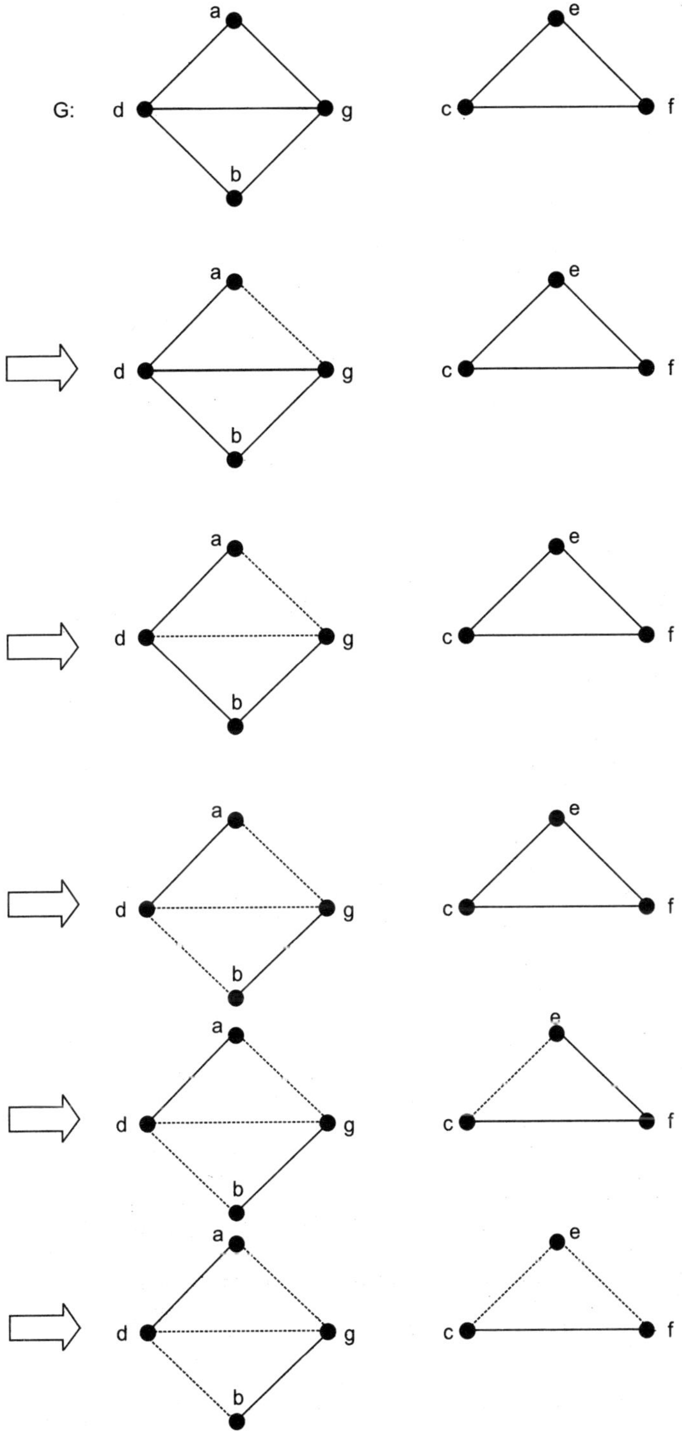

Therefore, the dotted graph shown above is the depth first forest T of the graph G. Besides that there could be several depth first forest from the same vertex '*a*'. This indicates that the depth first forest is not unique.

Example 13 *For the graph G shown below, find the breadth first search tree.*

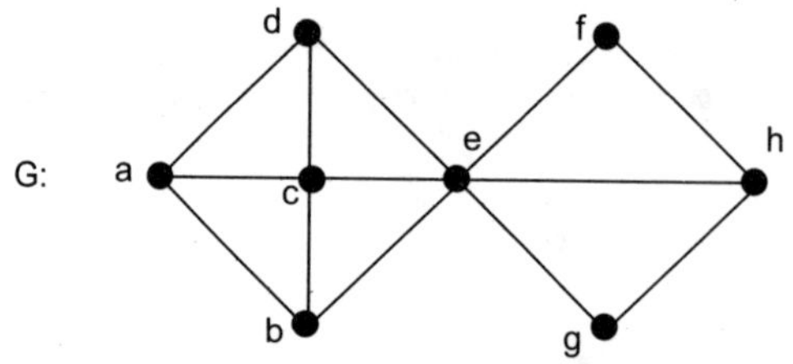

Solution: Consider the graph G given above. Now we have to find out the shortest path from the source vertex a to the vertex h. On using the BFS technique, we get the following stages.

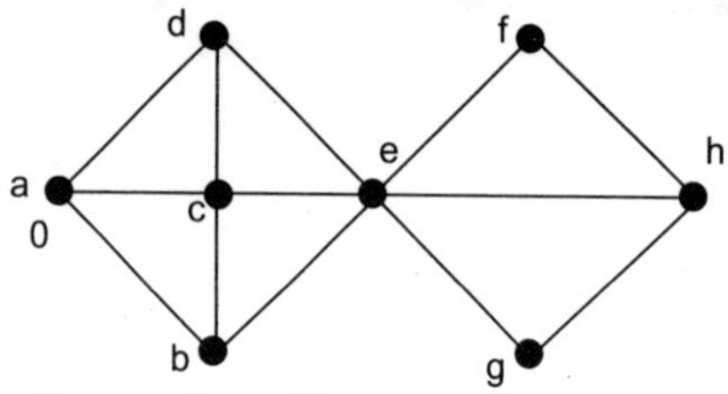

(Label $(a) = 0$ and set $i = 0$)

In the above figure the adjacent vertices of a are b, c and d. Therefore we get label

label $(b) = i + 1 = 0 + 1 = 1$;

label $(c) = i + 1 = 0 + 1 = 1$ and label $(d) = i + 1 = 0 + 1 = 1$.

Similarly, the adjacent vertex of d is e.

Therefore we get label $(e) = i + 1 = 1 + 1 = 2$.

Therefore, we have

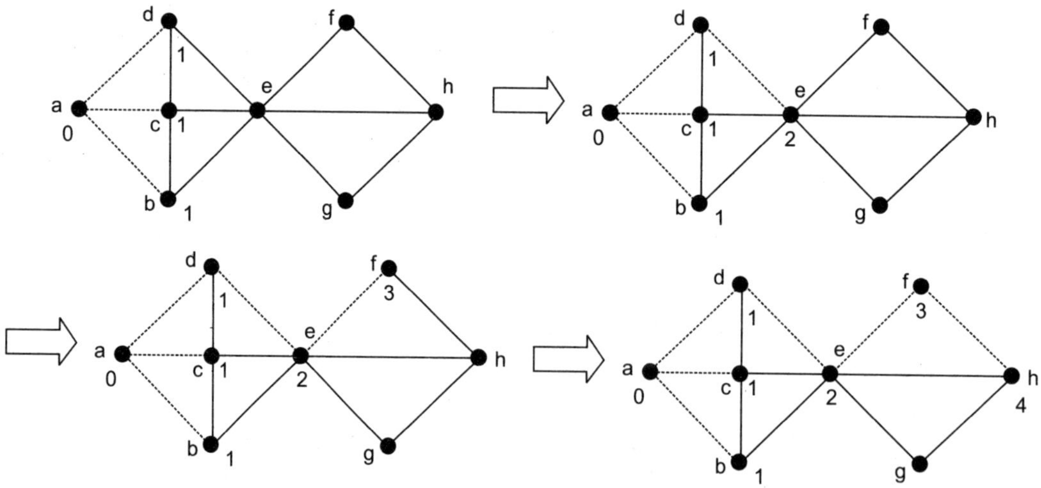

In the above figure, the adjacent vertex of e are f, g and h. Therefore we get,

label $(f) = i + 1 = 2 + 1 = 3$.

label $(g) = i + 1 = 2 + 1 = 3$.

label $(h) = i + 1 = 2 + 1 = 3$.

Therefore, the breadth first search tree is given as below.

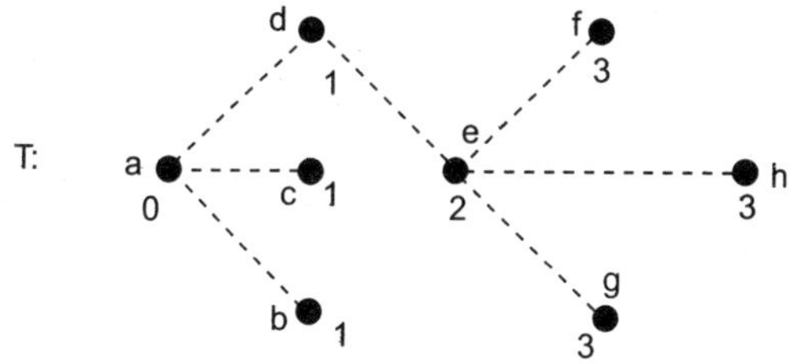

EXERCISES

1. With reference to the given tree T find the followings.
   (*a*) Height of the tree
   (*b*) Degree of the tree
   (*c*) Longest path of the tree
   (*d*) Level (L); Level (H); Level (N)
   (*e*) Parent (M); Sibling (B); Child (D)

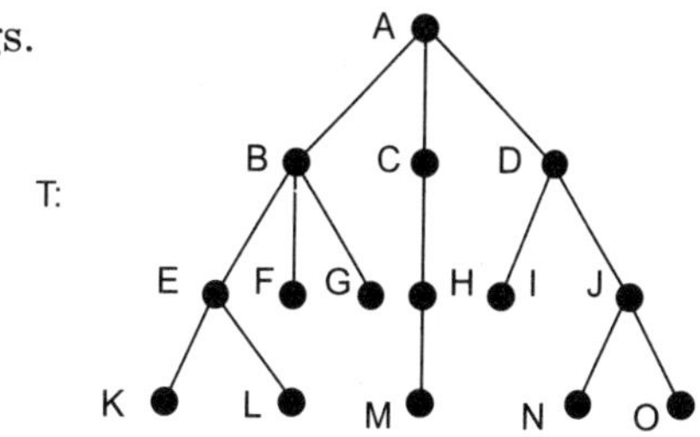

2. (*a*) Draw all trees of order 5
   (*b*) Draw all trees of order 7 and $\Delta(T) \geq 4$, where $\Delta(T)$ represents maximum degree of tree T.
3. In a binary tree of height h, there are at most 2^{h-1} leaf nodes.
4. If T is a binary tree of height h and order p, then $h \geq \lceil g((p+1)/2) \rceil$. The equality holds if T is a balanced complete binary tree.
5. Find the eccentricity of all vertices, radius, diameter and centre of the graph G given below. It is given that the distance between any two adjacent vertices is 1.

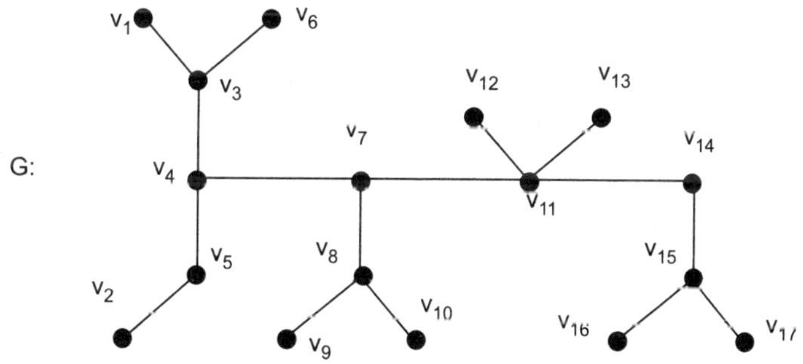

6. Construct the following graphs.
   (*a*) Eulerian but not Hamiltonian
   (*b*) Hamiltonian but not Eulerian
   (*c*) Neither Eulerian nor Hamiltonian
   (*d*) Eulerian and Hamiltonian.
7. For the graph G shown below, find the depth-first search tree.

(a)

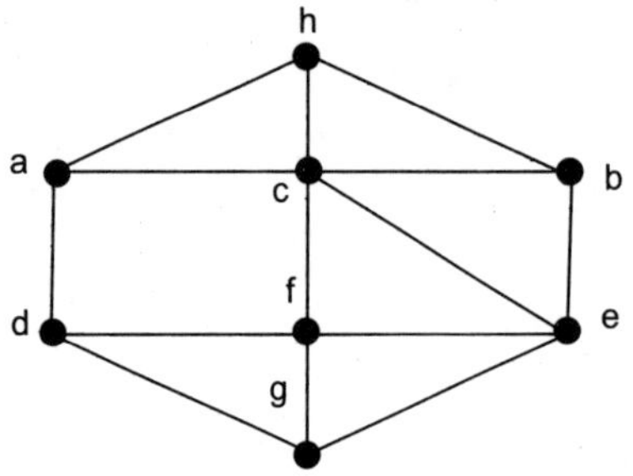

(b)

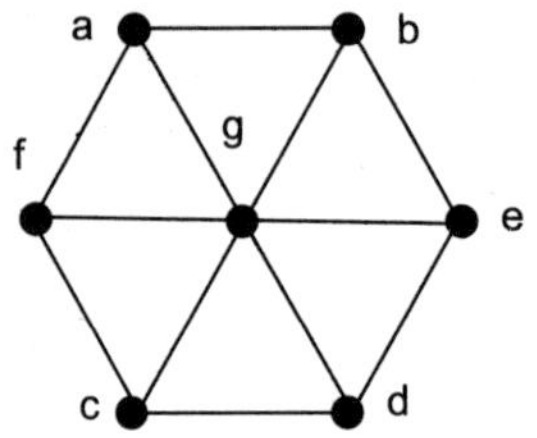

(c)

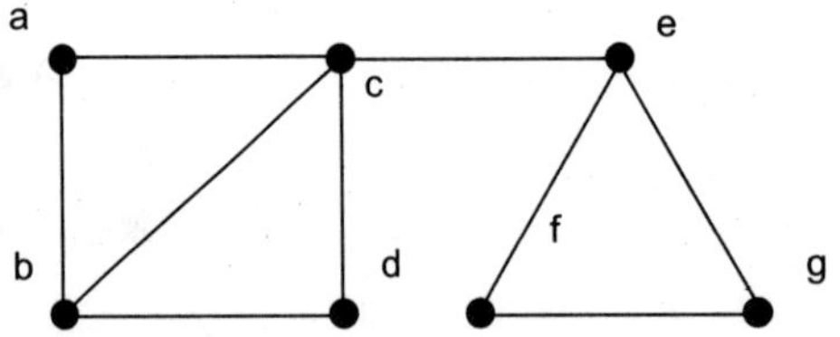

(d)

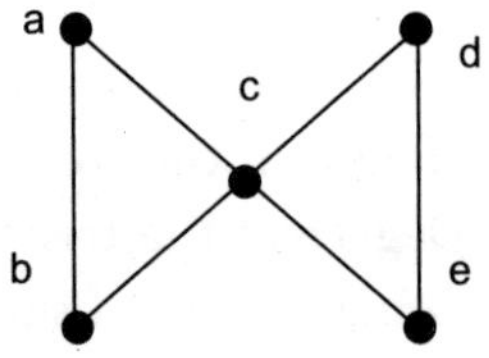

8. For the graphs given on No. 7, find the breadth first search tree.

9. Solve the travelling salesman problem for the complete weighted graph G given below by using
   (*a*) Closest insertion algorithm and
   (*b*) Two optimal algorithm.

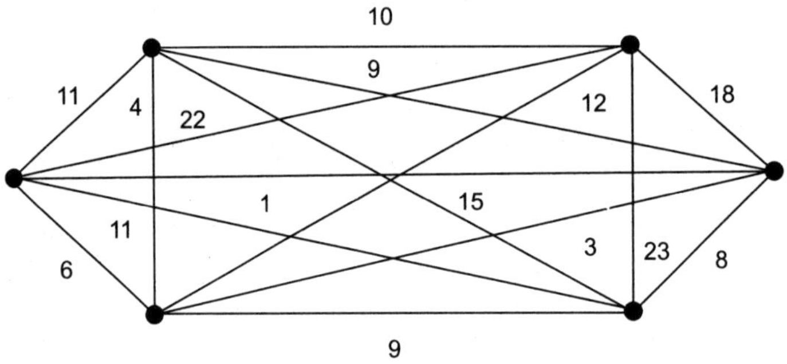

10. Let G be the weighted graph shown below. Use Dijkstra's algorithm to compute the shortest distance between u and v.

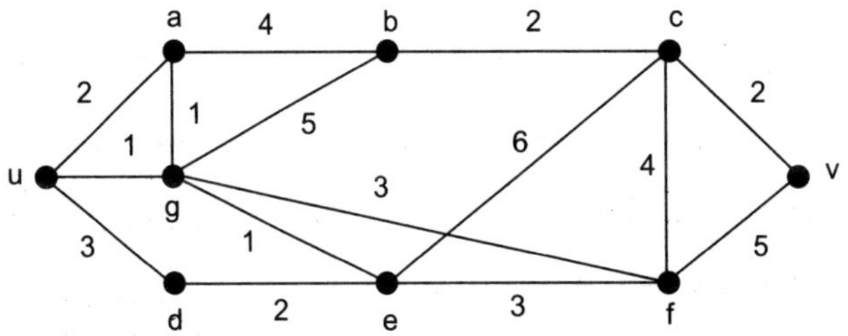

11. Determine which of the graphs given below are Euler graph by using the following algorithms.
   (*a*) Fleury's algorithm and
   (*b*) Hierholzer's algorithm

(a)

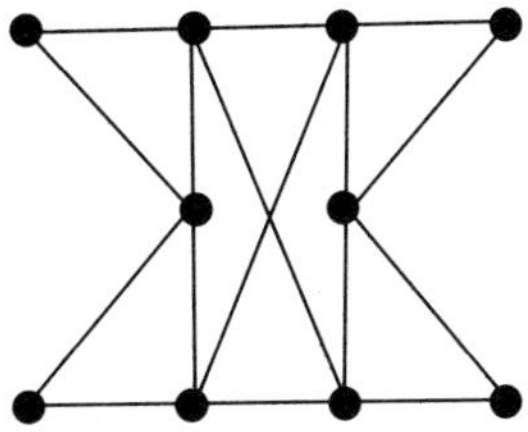

(b)

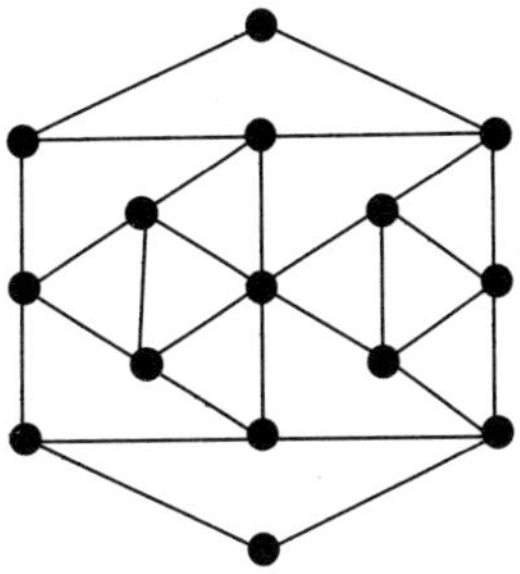

12. Find the closure graph C(G) for the graphs shown below.

(a)

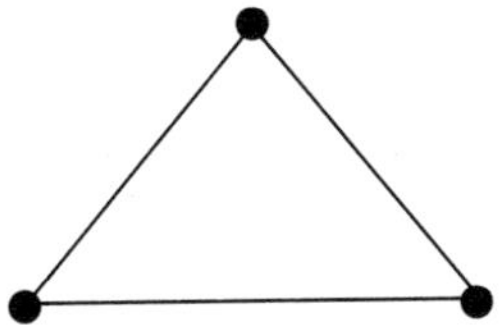

(b)

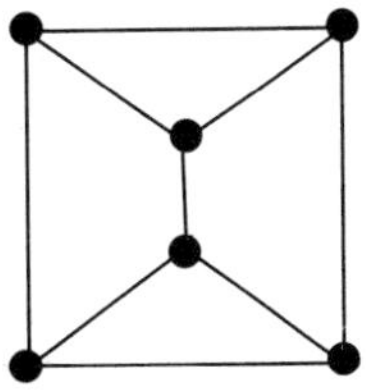

(c)

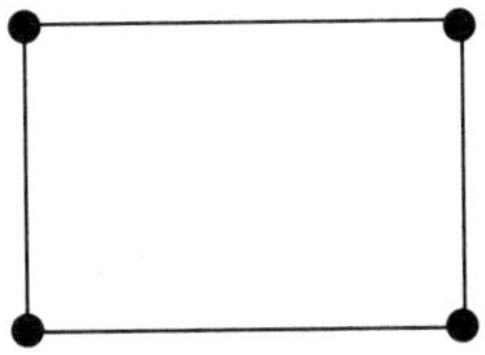

(d)

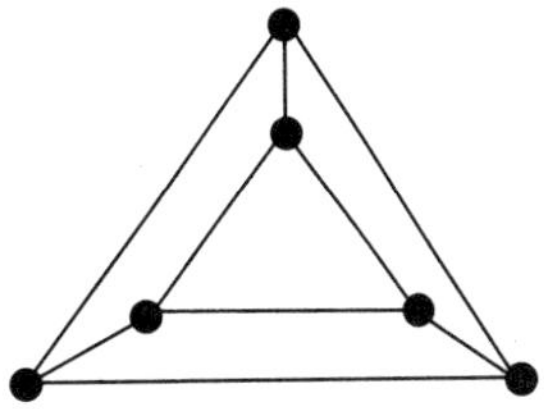

(e)

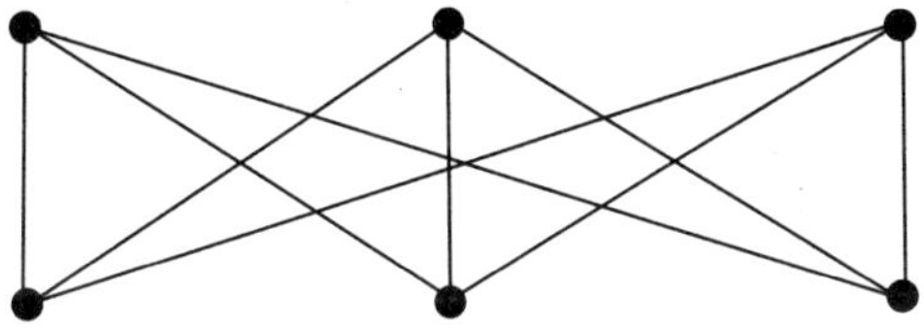

13. Find the binary tree representation of the followings.

(*a*) $(4x + 2)(2x + xy)$ (*b*) $(x + 3y) - ((5x + y)/4)$

14. Let G be a connected weighted graph. Use Dijkstra's algorithms to find the length of shortest paths from the vertex a to each of the other vertices.

(a)

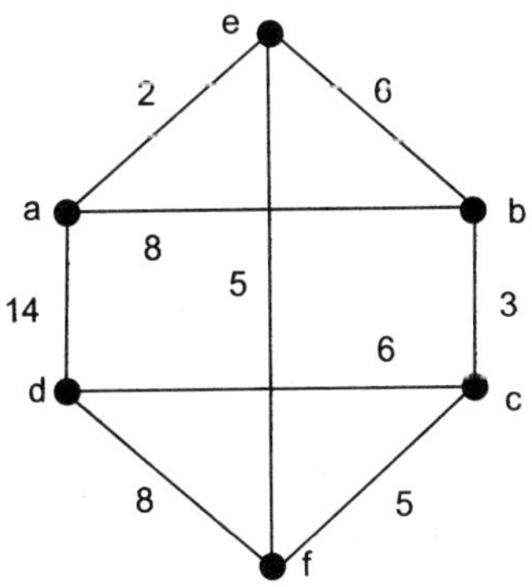

(b)

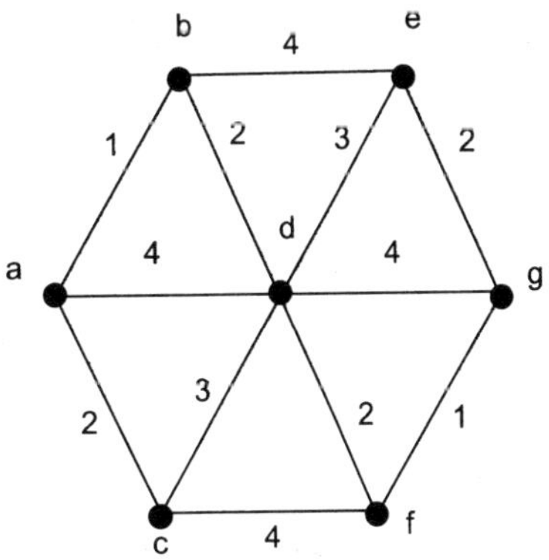

15. Apply Dijkstra's algorithm to the weighted graph G below to find the shortest distance for each vertex from the source vertex a.

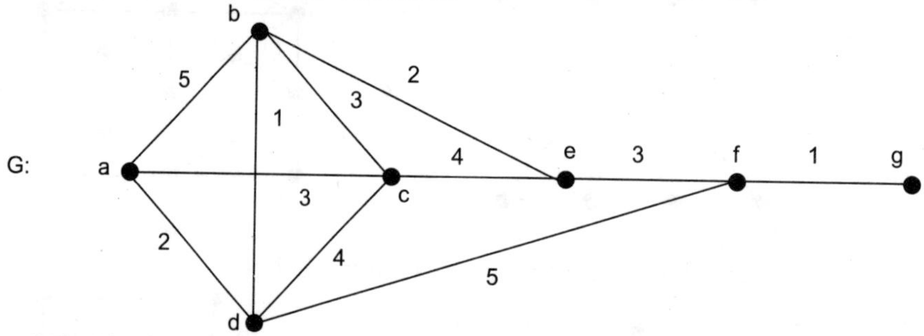

16. Use Floyd-Warshall algorithm on the weighted, directed graph G shown below to find out shortest path between any pair of vertices. Show the matrix D(k) that results for each iteration.

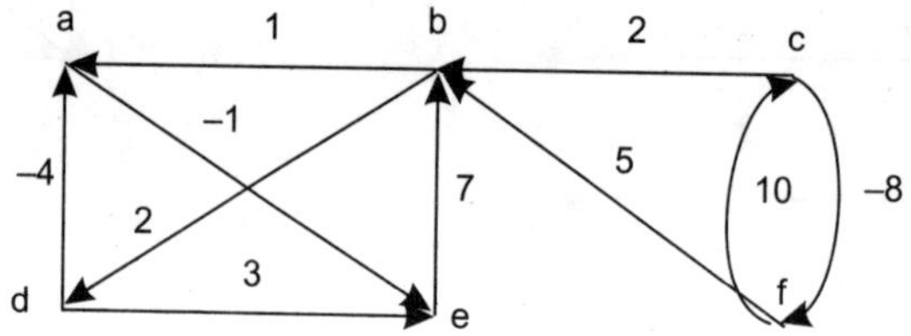

17. Find the minimum spanning tree of the graphs shown below by using
(a) Prim's algorithm and (b) Kruskal's algorithm.

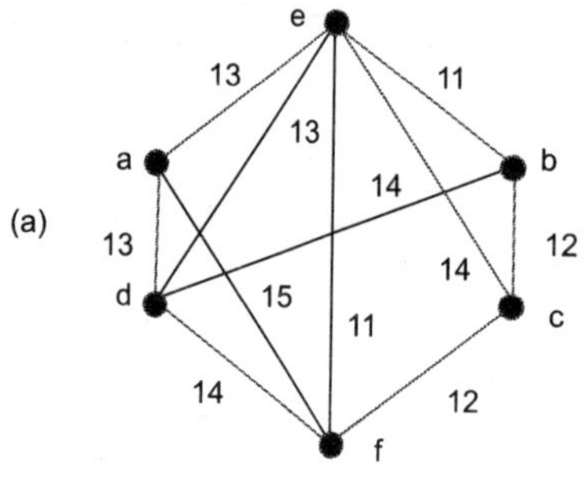

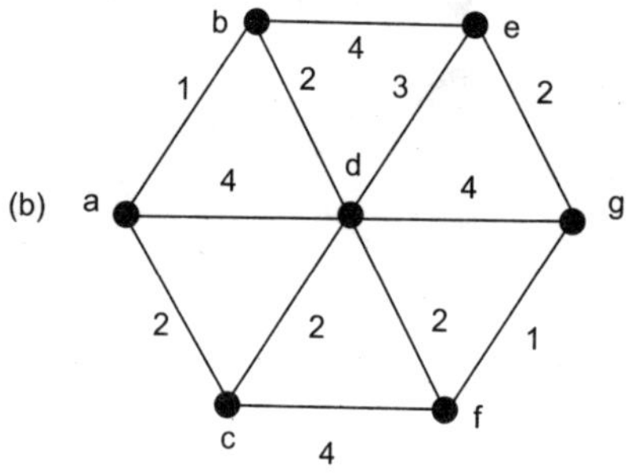

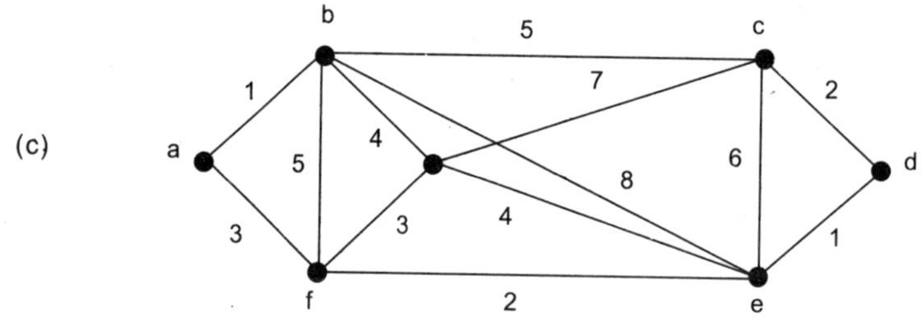

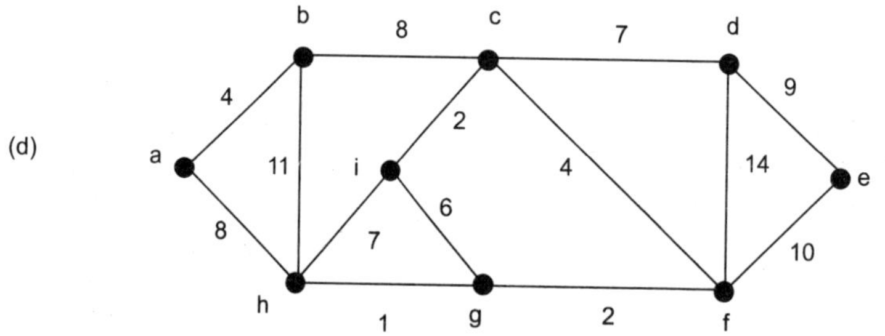

18. Find a maximal spanning tree for each of the graphs of No. 17. using either Prim's algorithm or Kruskal's algorithm. [**Hint:** To get the maximal spanning tree replace the weight of each edge of the graph by M – $w(e)$, where M is any number greater than the weight $w(e)$ of every edge e of the graph. Then apply Prim's algorithm or Kruskal's algorithm. The corresponding spanning tree in the original weighted graph is a maximal spanning tree.]

19. Find the closure graph C(G) of the following graphs.

(a)

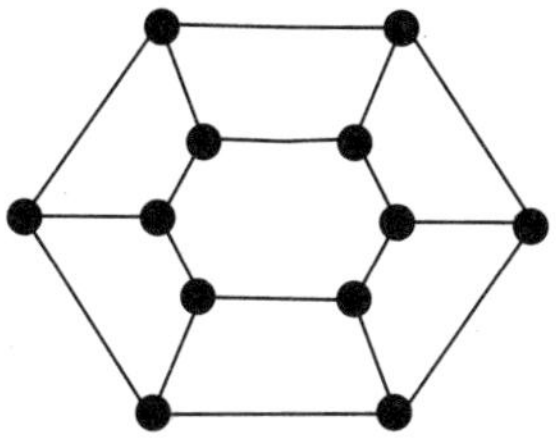

(b)

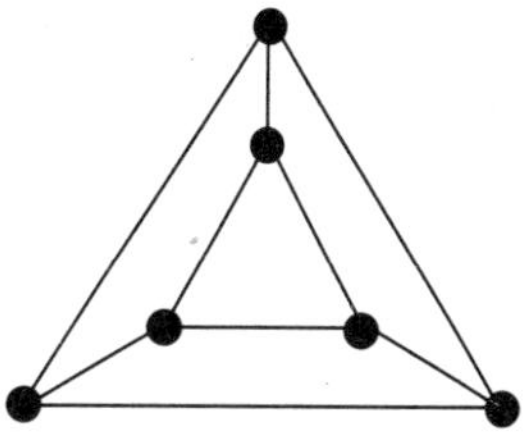

20. Let G be a connected weighted graph. Use Floyd-Warshall algorithm to find the length of shortest path between any pair of vertices.

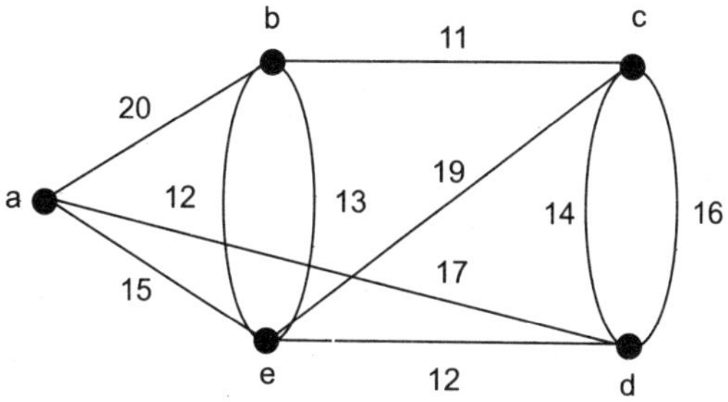

14

Fuzzy Set Theory

14.0 INTRODUCTION

Problems in the real world quite often turn out to be complex owing to an element of uncertainty. This uncertainty is due to parameters which define the problem or the situations in which the problem occurs. Although probability theory is an effective tool to handle uncertainty but it can be applied only to situations whose characteristics are based on random processes. However, in real life situations a large class of problems arise whose uncertainty is characterized by a nonrandom processes. Generally, this nonrandom process arises due to lack of information about the problem. Consider an example: "Is Adams Honest?" The answer to this question need not be a definite yes or no. Human brain always processes such uncertainty. Such type of data processing is always carried out by the world's biggest computer "human brain". In order to make this processing faster than human brain, we need computer. It can be fed with such data which it can process. Therefore, it can be made more intelligent artificially. For this reason, a study of vague concepts or vague knowledge or uncertainty is necessary. An excellent mathematical tool to handle the uncertainty due to vague concepts is Fuzzy set theory. In this chapter our objective is to define the basic concepts in a clear manner so that at a latter stage application of fuzzy sets in artificial intelligence, expert system etc., can be carried out more efficiently. This fuzzy set theory was propounded by Lotfi A. Zadeh. It is however, the Japanese who exploited the potential by commercializing the technology.

14.1 FUZZY VERSUS CRISP

Consider a statement, "Sun rises in the east." Definitely, the truth value of this statement is either true or false. If true is accorded as 1 and false is accorded as 0, then this statement results in a 0 / 1 type of situation. Such type of logic whose value is either 1 or 0 is termed as crisp in the domain of fuzzy set theory.

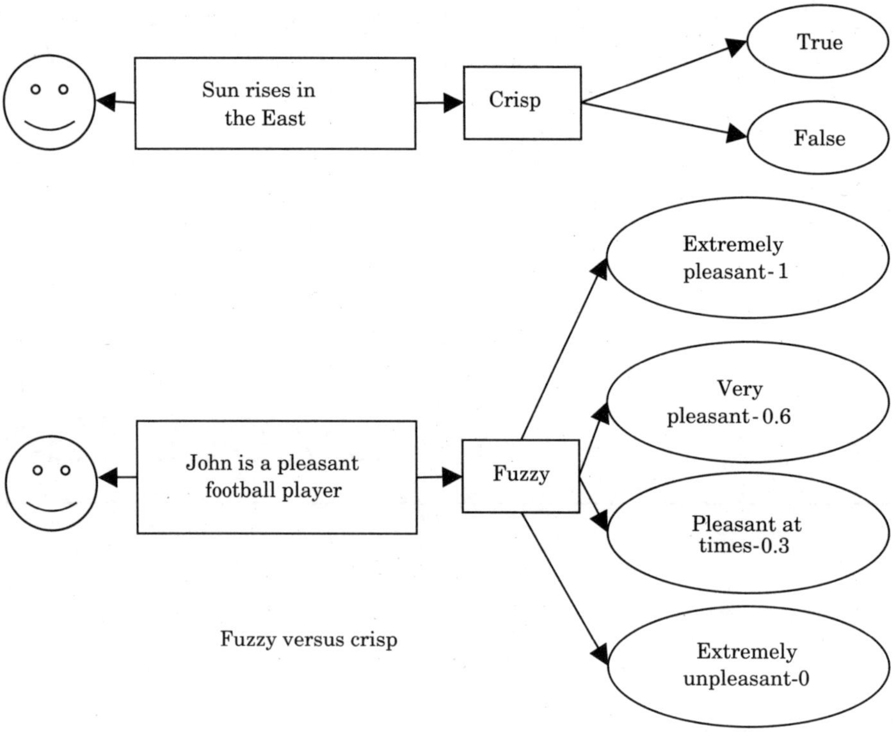

Fuzzy versus crisp

On the other hand, consider the statement "John is a pleasant football player." The truth values of this statement need not be a definite true or false. It varies from person to person on considering the degree to which one knows John. In such situation, a variety of answers spanning a range, such as "extremely pleasant", "extremely unpleasant", "very pleasant", "pleasant at times" could be generated. If "extremely pleasant" is accorded a value of 1 at the high end, then "extremely unpleasant" has a value 0 at the low end. It indicates that "very pleasant" and "pleasant at times" will be accorded a value in between 0 and 1. Such a situation is termed as fuzzy. A diagrammatic approach to illustrate fuzzy and crisp situation is given above. Crisp set theory or classical set theory is fundamental to the study of fuzzy set theory which we have already discussed in chapter 2.

■ 14.2 FUZZY SETS

In Cantor's definition, a set is defined as a collection of well defined objects. Thus, collection of all students in a particular class is a set. But the collection of all tall students in that particular class is not a set. This is because the qualification for an object to become a member is "tall" which is not well defined or ill-defined. Therefore, it is difficult to identify which are members and which are not. In fuzzy set theory, we assume that all are members and they belong to the set up to certain extent. Some elements belong to 90%, some elements belong to 75%, some elements belong to 60%, etc. This gives the degree of belongingness or measure of belongingness. Thus, we define fuzzy set formally as below.

If X is a universe of discourse and x is a particular element of X, then a fuzzy set $\tilde{A}$ of the universe of discourse X is defined as a collection of ordered pairs $(x, \mu_{\tilde{A}}(x)$, where $\mu_{\tilde{A}}(x): X \to [0,1]$ is a mapping known as the membership function of $\tilde{A}$.

Mathematically, we write

$$\tilde{A} = \{(x,\ \mu_{\tilde{A}}(x)) : x \in X\}$$

i.e., If $X = \{x_1,\ x_2,\ x_3, \cdots, x_n\}$, then a fuzzy set $\tilde{A}$ of X could be written as

$$\tilde{A} = \{(x_1, \mu_{\tilde{A}}(x_1)), (x_2, \mu_{\tilde{A}}(x_2)), (x_3, \mu_{\tilde{A}}(x_3)), \cdots\cdots, (x_n, \mu_{\tilde{A}}(x_n))\}$$

It is clear that each pair $(x,\ \mu_{\tilde{A}}(x))$ is called a singleton. An alternative way in which we represent a fuzzy set symbolically is given below.

$$\tilde{A} = \left\{\frac{\mu_{\tilde{A}}(x_1)}{x_1}, \frac{\mu_{\tilde{A}}(x_2)}{x_2}, \frac{\mu_{\tilde{A}}(x_3)}{x_3}, \ \cdots\cdots\ , \frac{\mu_{\tilde{A}}(x_n)}{x_n}\right\}.$$

14.2.1 Membership Function

In crisp set theory, characteristic function assigns a value of either 1 or 0 to each element in the universal set X. This function can be generalized such that the values assigned to the elements of the universal set fall within a specified range and point to the degree of belongingness of these elements in the set in question. Larger values indicate the higher degree precision of set membership whereas smaller values indicate the lower degree precision of set membership. Such a function is called a membership function, and the set defined by it is called a fuzzy set.

As defined in the previous section, a fuzzy set $\tilde{A}$ of the universe of discourse X is defined as a collection of ordered pairs $x, \mu_{\tilde{A}}(x))$ where $x \in X$, $\mu_{\tilde{A}}(x): X \to [0,1]$ is a mapping called the membership function of $\tilde{A}$ and $\mu_{\tilde{A}}(x)$ is the grade of membership or degree of belonging- ness or degree of membership of $x \in X$ in $\tilde{A}$.

The membership function values always need not be described by discrete values. Frequently, these turn out to be as described by a continuous function. The fuzzy membership function for the fuzzy set of a class of real numbers that are close to 3 may turn out to be as illustrated below:

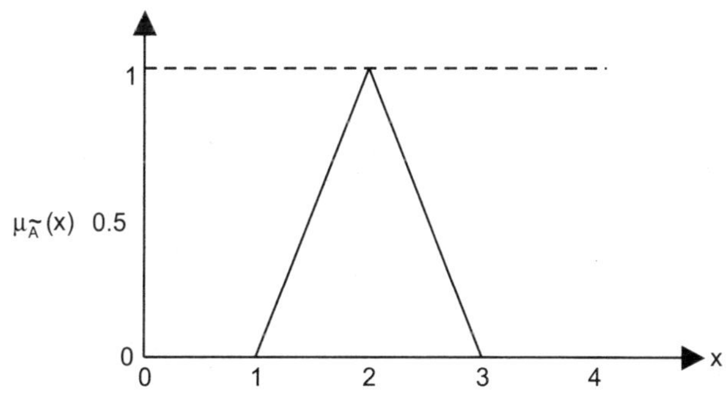

A membership function can also be defined mathematically as further:

$$\mu_{\tilde{A}}(x) = \frac{1}{(1+x)^2};\ x \in \Re\ \textit{(Set of real numbers)}$$

The different shapes of membership functions are given below. The shapes could be trapezoidal, triangular, curved or their variations as shown in the following figure.

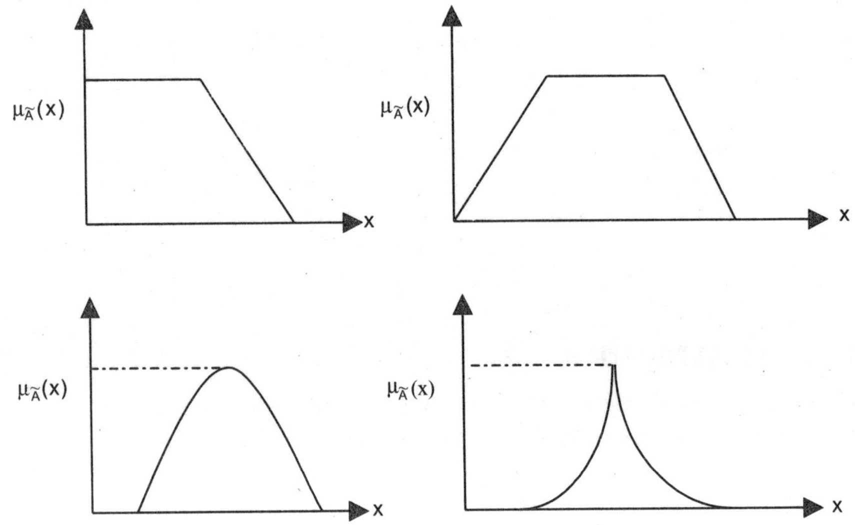

14.3 BASIC DEFINITIONS

In this section we will discuss the fundamental definitions that we use frequently in fuzzy sets.

14.3.1 Equality

If $\tilde{A}$ and $\tilde{B}$ be two fuzzy sets with $\mu_{\tilde{A}}(x)$ and $\mu_{\tilde{B}}(x)$ as their respective membership functions defined over the universe of discourse X, then the two fuzzy sets $\tilde{A}$ and $\tilde{B}$ are said to be equal *i.e.*, $\tilde{A} = \tilde{B}$, if and only if

$$\mu_{\tilde{A}}(x) = \mu_{\tilde{B}}(x)\ \ \forall\ \ x \in X$$

For example, if $\tilde{A} = \{(x_1,\ 0.2),\ (x_2,\ 0.6)\}$, $\tilde{B} = \{(x_1,\ 0.2),\ (x_2,\ 0.6)\}$ and $\tilde{C} = \{(x_1,\ 0.5),\ (x_2,\ 0.3)\}$ be three fuzzy sets defined over the universe of discourse $X = \{x_1,\ x_2\}$, then $\tilde{A} = \tilde{B}$ since,

$$\mu_{\tilde{A}}(x_1) = \mu_{\tilde{B}}(x_1) = 0.2 \text{ and } \mu_{\tilde{A}}(x_2) = \mu_{\tilde{B}}(x_2) = 0.6$$

Again from the above example it is clear that $\tilde{A} \neq \tilde{C}$ since, $\mu_{\tilde{A}}(x_1) \neq \mu_{\tilde{C}}(x_1)$ and $\mu_{\tilde{A}}(x_2) \neq \mu_{\tilde{C}}(x_2)$. Similarly, it can be shown that $\tilde{B} \neq \tilde{C}$.

14.3.2 Containment

If $\tilde{A}$ and $\tilde{B}$ be two fuzzy sets with $\mu_{\tilde{A}}(x)$ and $\mu_{\tilde{B}}(x)$ as their respective membership functions defined over the universe of discourse X, then we say that the fuzzy set $\tilde{A}$ is contained in the fuzzy set $\tilde{B}$ *i.e.*, $\tilde{A} \subseteq \tilde{B}$, if and only if

$$\mu_{\tilde{A}}(x) \leq \mu_{\tilde{B}}(x)\ \ \forall\ \ x \in X$$

For example, consider two fuzzy sets $\tilde{A} = \{(x_1, 0.3), (x_2, 0.4), (x_3, 0.6)\}$ and $\tilde{B} = \{(x_1, 0.5), (x_2, 0.6), (x_3, 0.8)\}$ defined over the universe of discourse $X = \{x_1, x_2, x_3\}$, then it is clear that,

$$\mu_{\tilde{A}}(x_1) = 0.3 \le \mu_{\tilde{B}}(x_1) = 0.5;$$
$$\mu_{\tilde{A}}(x_2) = 0.4 \le \mu_{\tilde{B}}(x_2) = 0.6 \text{ and}$$
$$\mu_{\tilde{A}}(x_3) = 0.6 \le \mu_{\tilde{B}}(x_3) = 0.8$$

Therefore, the fuzzy set $\tilde{A}$ is contained in the fuzzy set $\tilde{B}$ *i.e,* $\tilde{A} \subseteq \tilde{B}$.

14.3.3 Normal Fuzzy Set

If $\tilde{A}$ be a fuzzy set defined over the universe of discourse X with the membership function $\mu_{\tilde{A}}(x)$, then we say that the fuzzy set $\tilde{A}$ is a normal fuzzy set, if and only if

$$\max_{x \in X} \mu_{\tilde{A}}(x) = 1$$

i.e., $\mu_{\tilde{A}}(x) = 1$ for at least one $x \in X$.

For example, the fuzzy set $\tilde{A} = \{(x_1, 0.3), (x_2, 1), (x_3, 0.6)\}$ is a normal fuzzy set since $\mu_{\tilde{A}}(x_2) = 1$.

14.3.4 Support of a Fuzzy Set

If $\tilde{A}$ be a fuzzy set defined over the universe of discourse X with the membership function $\mu_{\tilde{A}}(x)$, then the support of $\tilde{A}$ is the crisp or classical set defined as

$$\text{Support}(\tilde{A}) = \{x \in X : \mu_{\tilde{A}}(x) > 0\}$$

Thus, it is clear that support of $\tilde{A}$ is a subset of X. *i.e.*, Support $(\tilde{A}) \subseteq X$.

For example, if the fuzzy set $\tilde{A} = \{(x_1, 0.5), (x_2, 0), (x_3, 0.8)\}$ is defined over the universe of discourse, $X = \{x_1, x_2, x_3\}$, then the support of the fuzzy set $\tilde{A}$ is given as

$$\text{Support}(\tilde{A}) = \{x_1, x_3\}$$

Similarly, if $\tilde{A} = \{(x_1, 0.6), (x_2, 0.1), (x_3, 0.9), (x_4, 0.3\}$ is defined over the universe of discourse, $X = \{x_1, x_2, x_3, x_4\}$, then the support of $\tilde{A}$ is same as the universe of discourse X.

14.3.5 α-Level Cut

If $\tilde{A}$ be a fuzzy set defined over the universe of discourse X with the membership function $\mu_{\tilde{A}}(x)$, then the α-level cut or α-cut of $\tilde{A}$ is the classical set A_α and is defined as

$$A_\alpha = \{x \in X : \mu_{\tilde{A}}(x) > \alpha\}$$

For example, if $\tilde{A} = \{(x_1, 0.2), (x_2, 0.1), (x_3, 0.4), (x_4, 0.6), (x_5, 0.7)\}$ be the fuzzy set defined over the universe $X = \{x_1, x_2, x_3, x_4, x_5\}$, then

$$A_{0.2} = \{x_1, x_3, x_4, x_5\} \qquad A_{0.4} = \{x_3, x_4, x_5\}$$
$$A_{0.1} = \{x_1, x_2, x_3, x_4, x_5\} \qquad A_{0.7} = \{x_5\}$$

14.3.6 Product of a Fuzzy Set by a Crisp Number

If $\tilde{A}$ be a fuzzy set defined over the universe of discourse X with the membership function $\mu_{\tilde{A}}(x)$ and k be crisp number, then the product of $\tilde{A}$ by a crisp number k results in a new fuzzy set product $k.\ \tilde{A}$ with the membership function $\mu_{k.\tilde{A}}(x)$ defined as

$$\mu_{k.\tilde{A}}(x) = k.\mu_{\tilde{A}}(x)\ \forall\ x \in X$$

For example, if $\tilde{A} = \{(x_1, 0.3), (x_2, 0.2), (x_3, 0.5), (x_4, 0.1), (x_5, 0.8)\}$ be a fuzzy set defined over the universe of discourse $X = \{x_1, x_2, x_3, x_4, x_5\}$ and $k = 0.3$, then $k.\ \tilde{A}$ is defined as

$$k.\tilde{A} = \{(x_1, 0.09), (x_2, 0.06), (x_3, 0.15), (x_4, 0.03), (x_5, 0.24)\} \text{ since,}$$

$$\mu_{k.\tilde{A}}(x_1) = k.\mu_{\tilde{A}}(x_1) = (0.3)(0.3) = 0.09$$
$$\mu_{k.\tilde{A}}(x_2) = k.\mu_{\tilde{A}}(x_2) = (0.3)(0.2) = 0.06$$
$$\mu_{k.\tilde{A}}(x_3) = k.\mu_{\tilde{A}}(x_3) = (0.3)(0.5) = 0.15$$
$$\mu_{k.\tilde{A}}(x_4) = k.\mu_{\tilde{A}}(x_4) = (0.3)(0.1) = 0.03$$
$$\mu_{k.\tilde{A}}(x_5) = k.\mu_{\tilde{A}}(x_5) = (0.3)(0.8) = 0.24$$

14.3.7 Power of a Fuzzy Set

If $\tilde{A}$ be a fuzzy set defined over the universe of discourse X with the membership function $\mu_{\tilde{A}}(x)$, then the m power of a fuzzy set $\tilde{A}$ is a new fuzzy set $\tilde{A}^m$ whose membership function is given as

$$\mu_{\tilde{A}^m}(x) = (\mu_{\tilde{A}}(x))^m \quad \forall\ x \in X$$

Raising a fuzzy set to its second power *i.e.*, $m = 2$ is called concentration (CON) whereas taking the square root, *i.e.*, $m = \frac{1}{2}$, is called dilation (DIL).

Let us consider a fuzzy set $\tilde{A} = \{(x_1, 0.2), (x_2, 0.1), (x_3, 0.4), (x_4, 0.5)\}$ defined over the universe of discourse $X = \{x_1, x_2, x_3, x_4\}$ and $m = 3$. Then $\tilde{A}^3$ is defined as

$$\tilde{A}^3 = \{(x_1, 0.008), (x_2, 0.001), (x_3, 0.064), (x_4, 0.125)\} \text{ since,}$$

$$\mu_{\tilde{A}^3}(x_1) = (\mu_{\tilde{A}}(x_1))^3 = (0.2)^3 = 0.008$$
$$\mu_{\tilde{A}^3}(x_2) = (\mu_{\tilde{A}}(x_2))^3 = (0.1)^3 = 0.001$$
$$\mu_{\tilde{A}^3}(x_3) = (\mu_{\tilde{A}}(x_3))^3 = (0.4)^3 = 0.064$$
$$\mu_{\tilde{A}^3}(x_4) = (\mu_{\tilde{A}}(x_4))^3 = (0.5)^3 = 0.125$$

14.4 BASIC OPERATIONS ON FUZZY SETS

Like crisp set theory, when we have two or more fuzzy sets describing a given problem, analytical solutions often require operations among fuzzy sets. Therefore, the concept of basic operations on fuzzy sets is highly essential to study the real life problems. In this section we will discuss the basic fuzzy set operations such as union, intersection, complement and product of two fuzzy sets. These fuzzy set operations are widely studied in the literature.

14.4.1 Union

If $\tilde{A}$ and $\tilde{B}$ be two fuzzy sets with $\mu_{\tilde{A}}(x)$ and $\mu_{\tilde{B}}(x)$ as their respective membership functions defined over the universe of discourse X, then their union $\tilde{A} \cup \tilde{B}$ is a new fuzzy set defined also on X with the membership function $\mu_{\tilde{A} \cup \tilde{B}}(x)$ defined as

$$\mu_{\tilde{A} \cup \tilde{B}}(x) = \max(\mu_{\tilde{A}}(x), \mu_{\tilde{B}}(x))$$

For example, if $\tilde{A} = \{(x_1, 0.4), (x_2, 0.3), (x_3, 0.7), (x_4, 0.1)\}$ and $\tilde{B} = \{(x_1, 0.9), (x_2, 0.1), (x_3, 0.6), (x_4, 0.3)\}$ are two fuzzy sets defined over the universe, $X = \{x_1, x_2, x_3, x_4\}$, then the union $\tilde{A} \cup \tilde{B}$ is defined as

$$\tilde{A} \cup B = \{(x_1, 0.9), (x_2, 0.3), (x_3, 0.7), (x_4, 0.3)\} \text{ since,}$$
$$\mu_{\tilde{A} \cup \tilde{B}}(x_1) = \max(\mu_{\tilde{A}}(x_1), \mu_{\tilde{B}}(x_1)) = \max(0.4, 0.9) = 0.9$$
$$\mu_{\tilde{A} \cup \tilde{B}}(x_2) = \max(\mu_{\tilde{A}}(x_2), \mu_{\tilde{B}}(x_2)) = \max(0.3, 0.1) = 0.3$$
$$\mu_{\tilde{A} \cup \tilde{B}}(x_3) = \max(\mu_{\tilde{A}}(x_3), \mu_{\tilde{B}}(x_3)) = \max(0.7, 0.6) = 0.7$$
$$\mu_{\tilde{A} \cup \tilde{B}}(x_4) = \max(\mu_{\tilde{A}}(x_4), \mu_{\tilde{B}}(x_4)) = \max(0.1, 0.3) = 0.3$$

14.4.2 Intersection

If $\tilde{A}$ and $\tilde{B}$ be two fuzzy sets with $\mu_{\tilde{A}}(x)$ and $\mu_{\tilde{B}}(x)$ as their respective membership functions defined over the universe of discourse X, then the intersection of $\tilde{A}$ and $\tilde{B}$ and is a new fuzzy set $\tilde{A} \cap \tilde{B}$ whose membership function $\mu_{\tilde{A} \cap \tilde{B}}(x)$ is defined as

$$\mu_{\tilde{A} \cap \tilde{B}}(x) = \min(\mu_{\tilde{A}}(x), \mu_{\tilde{B}}(x))$$

For example, if $\tilde{A} = \{(x_1, 0.4), (x_2, 0.7), (x_3, 0.2), (x_4, 1)\}$ and $\tilde{B} - \{(x_1, 0.9), (x_2, 0.1), (x_3, 0.6), (x_4, 0.3)\}$ are two fuzzy sets defined over the universe of discourse $X = \{x_1, x_2, x_3, x_4\}$, then the intersection $\tilde{A} \cap \tilde{B}$ is defined as

$$\tilde{A} \cap \tilde{B} = \{(x_1, 0.4), (x_2, 0.1), (x_3, 0.2), (x_4, 0.3)\} \text{ since,}$$

$$\mu_{\tilde{A} \cap \tilde{B}}(x_1) = \min\ (\mu_{\tilde{A}}(x_1), \mu_{\tilde{B}}(x_1)) = \min\ (0.4, 0.9) = 0.4$$

$$\mu_{\tilde{A} \cap \tilde{B}}(x_2) = \min\ (\mu_{\tilde{A}}(x_2), \mu_{\tilde{B}}(x_2)) = \min\ (0.7, 0.1) = 0.1$$

$$\mu_{\tilde{A} \cap \tilde{B}}(x_3) = \min\ (\mu_{\tilde{A}}(x_3), \mu_{\tilde{B}}(x_3)) = \min\ (0.2, 0.6) = 0.2$$

$$\mu_{\tilde{A} \cap \tilde{B}}(x_4) = \min\ (\mu_{\tilde{A}}(x_4), \mu_{\tilde{B}}(x_4)) = \min\ (1, 0.3) = 0.3$$

14.4.3 Complement

If $\tilde{A}$ be a fuzzy set with $\mu_{\tilde{A}}(x)$ as membership function defined over the universe of discourse X, then the complement of $\tilde{A}$ is a new fuzzy set $\tilde{A}^c$ whose membership function $\mu_{\tilde{A}^c}(x)$ is defined as

$$\mu_{\tilde{A}^c}(x) = 1 - \mu_{\tilde{A}}(x)$$

For example, if $\tilde{A} = \{(x_1, 0.4), (x_2, 0.7), (x_3, 0.2)\}$ be a fuzzy set defined over the universe of discourse, $X = \{x_1, x_2, x_3\}$, then we have

$$\tilde{A}^c = \{(x_1, 0.6), (x_2, 0.3), (x_3, 0.8)\} \text{ since,}$$

$$\mu_{\tilde{A}^c}(x_1) = 1 - \mu_{\tilde{A}}(x_1) = 1 - 0.4 = 0.6$$

$$\mu_{\tilde{A}^c}(x_2) = 1 - \mu_{\tilde{A}}(x_2) = 1 - 0.7 = 0.3$$

$$\mu_{\tilde{A}^c}(x_3) = 1 - \mu_{\tilde{A}}(x_3) = 1 - 0.2 = 0.8$$

14.4.4 Product

If $\tilde{A}$ and $\tilde{B}$ be two fuzzy sets with $\mu_{\tilde{A}}(x)$ and $\mu_{\tilde{B}}(x)$ as their respective membership functions defined over the universe of discourse X, then the product of two fuzzy sets $\tilde{A}$ and $\tilde{B}$ is a new fuzzy set $\tilde{A} . \tilde{B}$ whose membership function $\mu_{\tilde{A} . \tilde{B}}(x)$ is defined as

$$\mu_{\tilde{A} . \tilde{B}}(x) = \mu_{\tilde{A}}(x)\ \mu_{\tilde{B}}(x)$$

For example, let us consider $\tilde{A} = \{(x_1, 0.3), (x_2, 0.6), (x_3, 0.2)\}$ and $\tilde{B} = \{(x_1, 0.5), (x_2, 0.1), (x_3, 0.4)\}$ are two fuzzy sets defined over the universe of discourse, $X = \{x_1, x_2, x_3\}$, then the product $\tilde{A} . \tilde{B}$ is given as

$$\tilde{A} . \tilde{B} = \{(x_1, 0.15), (x_2, 0.06), (x_3, 0.08)\} \text{ since,}$$

$$\mu_{\tilde{A} . \tilde{B}}(x_1) = \mu_{\tilde{A}}(x_1) \cdot \mu_{\tilde{B}}(x_1) = (0.3).(0.5) = 0.15$$

$$\mu_{\tilde{A} . \tilde{B}}(x_2) = \mu_{\tilde{A}}(x_2) \cdot \mu_{\tilde{B}}(x_2) = (0.6).(0.1) = 0.06$$

$$\mu_{\tilde{A} . \tilde{B}}(x_3) = \mu_{\tilde{A}}(x_3) \cdot \mu_{\tilde{B}}(x_3) = (0.2).(0.4) = 0.08$$

14.4.5 Difference

If $\tilde{A}$ and $\tilde{B}$ be two fuzzy sets with $\mu_{\tilde{A}}(x)$ and $\mu_{\tilde{B}}(x)$ as their respective membership functions defined over the universe of discourse X, then the difference of two fuzzy sets $\tilde{A}$ and $\tilde{B}$ is a new fuzzy set $\tilde{A} - \tilde{B}$ defined as

$$\tilde{A} - \tilde{B} = (\tilde{A} \cap \tilde{B}^c)$$

For example, let us consider $\tilde{A} = \{(x_1, 0.5), (x_2, 0.1), (x_3, 0.4), (x_4, 0.8)\}$ and $\tilde{B} = \{(x_1, 0.7), (x_2, 0.2), (x_3, 0.5), (x_4, 0.3)\}$ are two fuzzy sets defined over the universe of discourse, $X = \{x_1, x_2, x_3, x_4\}$, then we get $\tilde{B}^c = \{(x_1, 0.3), (x_2, 0.8), (x_3, 0.5), (x_4, 0.7)\}$. Therefore, we have

$$\begin{aligned}\tilde{A} - \tilde{B} &= (\tilde{A} \cap \tilde{B}^c)\\ &= \{(x_1, 0.3), (x_2, 0.1), (x_3, 0.4), (x_4, 0.7)\}\end{aligned}$$

14.4.6 Disjunctive Sum

The disjunctive sum of two fuzzy sets $\tilde{A}$ and $\tilde{B}$ defined over the universe of discourse X, is a new fuzzy set $\tilde{A} \oplus \tilde{B}$ defined as

$$\tilde{A} \oplus \tilde{B} = (\tilde{A} \cap \tilde{B}^c) \cup (\tilde{A}^c \cap \tilde{B})$$

For example, let us consider $\tilde{A} = \{(x_1, 0.5), (x_2, 0.2), (x_3, 0.3), (x_4, 0.7)\}$ and $\tilde{B} = \{(x_1, 0.1), (x_2, 0.2), (x_3, 0.4), (x_4, 0.3)\}$ are two fuzzy sets defined over the universe of discourse, $X = \{x_1, x_2, x_3, x_4\}$, then we have

$$\tilde{A}^c = \{(x_1, 0.5), (x_2, 0.8), (x_3, 0.7), (x_4, 0.3)\}$$
$$\tilde{B}^c = \{(x_1, 0.9), (x_2, 0.8), (x_3, 0.6), (x_4, 0.7)\}$$
$$(\tilde{A} \cap \tilde{B}^c) = \{(x_1, 0.5), (x_2, 0.2), (x_3, 0.3), (x_4, 0.7)\}$$
$$(\tilde{A}^c \cap \tilde{B}) = \{(x_1, 0.1), (x_2, 0.2), (x_3, 0.4), (x_4, 0.3)\}$$

Therefore, we get

$$\begin{aligned}\tilde{A} \oplus \tilde{B} &= (\tilde{A} \cap \tilde{B}^c) \cup (\tilde{A}^c \cap \tilde{B})\\ &= \{(x_1, 0.5), (x_2, 0.2), (x_3, 0.4), (x_4, 0.7)\}\end{aligned}$$

14.5 PROPERTIES OF FUZZY SETS

Fuzzy set satisfy some of the properties of crisp set or classical set. Here, we have listed the different properties satisfied by fuzzy sets. We have also listed the laws of excluded middle that do not hold for fuzzy sets. These are listed below:

Commutative Laws: $\tilde{A} \cup \tilde{B} = \tilde{B} \cup \tilde{A}$ and $\tilde{A} \cap \tilde{B} = \tilde{B} \cap \tilde{A}$

Associative Laws: $\tilde{A} \cup (\tilde{B} \cup \tilde{C}) = (\tilde{A} \cup \tilde{B}) \cup \tilde{C}$ and

$\tilde{A} \cap (\tilde{B} \cap \tilde{C}) = (\tilde{A} \cap \tilde{B}) \cap \tilde{C}$

Distributive Laws: $\tilde{A} \cup (\tilde{B} \cap \tilde{C}) = (\tilde{A} \cup \tilde{B}) \cap (\tilde{A} \cup \tilde{C})$ and
$\tilde{A} \cap (\tilde{B} \cup \tilde{C}) = (\tilde{A} \cap \tilde{B}) \cup (\tilde{A} \cap \tilde{C})$

Idempotent Laws: $\tilde{A} \cup \tilde{A} = \tilde{A}$ and $\tilde{A} \cap \tilde{A} = \tilde{A}$

Identity Laws: $\tilde{A} \cup \phi = \tilde{A}$; $\tilde{A} \cap X = \tilde{A}$
$\tilde{A} \cap \phi = \phi$ and $\tilde{A} \cup X = X$

Transitivity Laws: If $\tilde{A} \subset \tilde{B}$ and $\tilde{B} \subset \tilde{C}$, then $\tilde{A} \subset \tilde{C}$

Involution Laws: $(\tilde{A}^c)^c = \tilde{A}$

De Morgan's Laws: $(\tilde{A} \cup \tilde{B})^c = \tilde{A}^c \cap \tilde{B}^c$ and
$(\tilde{A} \cap \tilde{B})^c = \tilde{A}^c \cup \tilde{B}^c$

The laws of excluded middle do not hold well because fuzzy sets can overlap each other. Therefore, we have $\tilde{A} \cup \tilde{A}^c \neq X$ and $\tilde{A} \cap \tilde{A}^c \neq \phi$.

■ 14.6 INTERVAL VALUED FUZZY SET

This is an extension of the concept of a fuzzy set by an interval valued fuzzy set. It is developed because the membership value of an element $x \in X$ may not be always assigned a particular real number. In such cases, the membership functions are defined only approximately. For example, we may only be able to identify lower and upper bounds of membership grades for each element $x \in X$. Therefore, either we have to suppress the identification uncertainty by considering the middle values between the lower and upper bounds or we may accept the uncertainty and include it in the definition of the membership function. The latter approach assigns a closed interval of real numbers to each element of the universe of discourse between the identified lower and upper bounds. Fuzzy sets of this type are known as interval valued fuzzy set or *i-v* fuzzy set. This extension was carried out by L. A. Zadeh in 1975. Interval valued fuzzy sets are defined formally by functions of the form

$$\mu_{\tilde{A}}(x): X \to \varepsilon\,([0, 1])$$

where ε ([0, 1]) denotes the family of all closed intervals of real numbers in [0, 1]. For example, the membership function of this type is given in the following figure. In this figure, $\mu_{\tilde{A}}(x)$ for each x is represented by the segment between two curves. It is also clear that $\mu_{\tilde{A}}(x) = [a_1, a_2]$ for $x_1 \in X$.

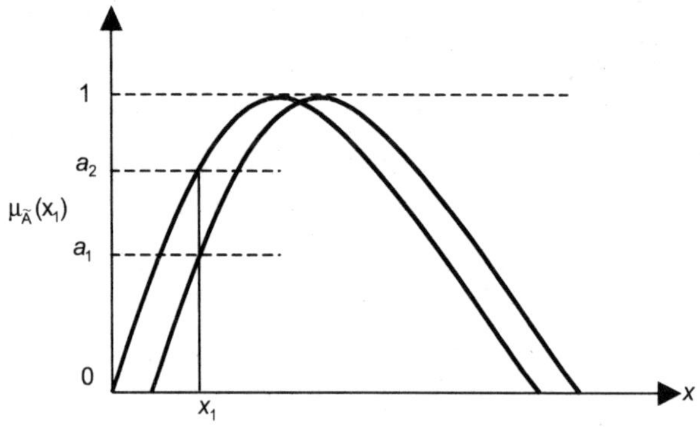

(Interval valued fuzzy set $\mu_{\tilde{A}}(x_1) = [a_1, a_2]$)

Let $\tilde{A}$ denote an *interval-valued fuzzy set*. Then,

$$\tilde{A} = \left(x,\ [L_{\tilde{A}}(x),\ U_{\tilde{A}}(x)]\right),\ x \in X$$

where, $L_{\tilde{A}}$, $U_{\tilde{A}}$ are fuzzy sets of X that are called the lower bound and upper bound of $\tilde{A}$, respectively. It is clear that $\tilde{A}$ becomes an ordinary fuzzy set when $L_{\tilde{A}} = U_{\tilde{A}}$.

Let $\bar{\mu}_{\tilde{A}}(x) = [L_{\tilde{A}}(x),\ U_{\tilde{A}}(x)],\ x \in X$ and let D[0, 1] denotes the family of all closed intervals contained in the interval [0, 1]. Thus, $[x_1,\ x_2] \in D[0,\ 1]$ for all $x_1,\ x_2 \in [0,\ 1];\ x_1 < x_2$. Therefore, the membership function is defined as $\bar{\mu}_{\tilde{A}}(x) : X \to D[0,\ 1]$. Hence, an *interval-valued fuzzy set* $\tilde{A}$ is defined as

$$\tilde{A} = (x,\ \bar{\mu}_{\tilde{A}}(x)),\ x \in X.$$

■ 14.7 OPERATIONS ON *i-v* FUZZY SETS

In this section we discuss the different kind of basic operations that can be performed on *i-v* fuzzy sets. Here we restrict to the definitions of operations that are essential for further considerations.

14.7.1 Union of Two *i-v* Fuzzy Sets

Let $\tilde{A}$ and $\tilde{B}$ are two interval-valued fuzzy sets defined over the universe of discourse X, then their union $(\tilde{A} \cup \tilde{B})$ is also an interval-valued fuzzy set defined as

$$(\tilde{A} \cup \tilde{B}) = \{(x,\ \bar{\mu}_{\tilde{A} \cup \tilde{B}}(x))\} \text{ where,}$$

$$\bar{\mu}_{\tilde{A} \cup \tilde{B}}(x) = [L_{\tilde{A} \cup \tilde{B}}(x),\ U_{\tilde{A} \cup \tilde{B}}(x)];\ x \in X$$

$$L_{\tilde{A} \cup \tilde{B}}(x) = \max\ [L_{\tilde{A}}(x),\ L_{\tilde{B}}(x)] \text{ and}$$

$$U_{\tilde{A} \cup \tilde{B}}(x) = \max\ [U_{\tilde{A}}(x),\ U_{\tilde{B}}(x)] \text{ for all } x \in X.$$

The above definition of union can be generalized to finite number of interval-valued fuzzy sets.

For example, if $\tilde{A} = \{(x_1,\ [0.2,\ 0.7]),\ (x_2,\ [0.4,\ 0.6]),\ (x_3,\ [0.1,\ 0.5])\}$ and $\tilde{B} = \{(x_1,\ [0.4,\ 0.9]),\ (x_2,\ [0.2,\ 0.8]),\ (x_3,\ [0.4,\ 0.3])\}$ be two interval-valued fuzzy sets defined over the universe of discourse, $X = \{x_1,\ x_2,\ x_3\}$, then

$(\tilde{A} \cup \tilde{B}) = \{(x_1,\ [0.4,\ 0.9]),\ (x_2,\ [0.4,\ 0.8]),\ (x_3,\ [0.4,\ 0.5])\}$ since,

$$\bar{\mu}_{\tilde{A} \cup \tilde{B}}(x_1) = [L_{\tilde{A} \cup \tilde{B}}(x_1),\ U_{\tilde{A} \cup \tilde{B}}(x_1)];$$

$$L_{\tilde{A} \cup \tilde{B}}(x_1) = \max\ [0.2,\ 0.4] = 0.4; \text{ and}$$

$$U_{\tilde{A} \cup \tilde{B}}(x_1) = \max\ [0.7,\ 0.9] = 0.9$$

Similarly, one can find the interval of belongingness for the elements x_2 and x_3 as defined for the element $x_1 \in X$.

14.7.2 Intersection of Two *i-v* Fuzzy Sets

Let $\tilde{A}$ and $\tilde{B}$ are two interval-valued fuzzy sets defined over the universe of discourse X, then their intersection $(\tilde{A} \cap \tilde{B})$ is also an interval-valued fuzzy set. It is given as

$$(\tilde{A} \cap \tilde{B}) = \{(x, \bar{\mu}_{\tilde{A} \cap \tilde{B}}(x))\}$$

where,
$$\bar{\mu}_{\tilde{A} \cap \tilde{B}}(x) = [L_{\tilde{A} \cap \tilde{B}}(x), U_{\tilde{A} \cap \tilde{B}}(x)]; \ x \in X$$
$$L_{\tilde{A} \cap \tilde{B}}(x) = \min \ [L_{\tilde{A}}(x), L_{\tilde{B}}(x)]$$
and
$$U_{\tilde{A} \cap \tilde{B}}(x) = \min \ [U_{\tilde{A}}(x), U_{\tilde{B}}(x)] \text{ for all } x \in X.$$

The above definition of intersection can be generalized to finite number of interval-valued fuzzy sets.

For example, if $\tilde{A} = \{(x_1, [0.2, 0.7]), (x_2, [0.4, 0.6]), (x_3, [0.1, 0.5])\}$ and $\tilde{B} = \{(x_1, [0.4, 0.9]), (x_2, [0.2, 0.8]), (x_3, [0.4, 0.3])\}$ be two interval-valued fuzzy sets defined over the universe of discourse, $X = \{x_1, x_2, x_3\}$, then

$$(\tilde{A} \cap \tilde{B}) = \{(x_1, [0.2, 0.7]), \{(x_2, [0.2, 0.6]), \{(x_1, [0.1, 0.3])\} \text{ since,}$$
$$\bar{\mu}_{\tilde{A} \cap \tilde{B}}(x_1) = [L_{\tilde{A} \cap \tilde{B}}(x_1), U_{\tilde{A} \cap \tilde{B}}(x_1)];$$
$$L_{\tilde{A} \cap \tilde{B}}(x_1) = \min \ [0.2, 0.4] = 0.2; \text{ and}$$
$$U_{\tilde{A} \cap \tilde{B}}(x_1) = \min \ [0.7, 0.9] = 0.7$$

Similarly, one can find the interval of belongingness for the elements x_2 and x_3 as defined for the element $x_1 \in X$.

14.8 FUZZY RELATIONS

Fuzzy relation provides a measure of relation between two fuzzy sets defined in the range [0, 1]. Crisp relation that we have studied in Chapter 3 always may not find a relation between any two objects in the real world whereas in fuzzy relation we always find a relation between any two objects in the real world. It has wide applications such as decision making, expert systems, image processing, medical diagnoses, economics etc.

Fuzzy relation is a fuzzy set defined on the Cartesian product of classical sets $X_1, X_2, X_3, X_4, \cdots\cdots, X_n$ where the n–tuples $(x_1, x_2, x_3, x_4, \cdots\cdots, x_n)$ may have varying degrees of belongingness within the relation. The belongingness values indicate the strength of the relation between the tuples. The membership function is defined as

$$\mu_R(x_1, x_2, \cdots, x_n) : (X_1, X_2, \cdots, X_n) \to [0, 1]$$

For example consider the fuzzy relation R between two sets X_1 and X_2, where the set of diseases is X_1 and the set of symptoms is X_2.

Let X_1 = {*typhoid, common cold, malaria, viral fever*}

X_2 = {*shivering, running nose, high temperature*}

Therefore, the fuzzy relation may be defined as

	shivering	*running nose*	*high temperature*
typhoid	0.8	0.1	0.9
common cold	0.6	0.9	0.3
malaria	0.4	0.1	0.7
viral fever	0.8	0.2	0.9

From the above tabular representation, it is clear that typhoid is related to high temperature with strength 0.9, common cold is related to running nose is 0.9, malaria is related to shivering is 0.4 etc.

14.8.1 Fuzzy Cartesian Product

If $\tilde{A}$ be a fuzzy set defined on the universe of discourse X and $\tilde{B}$ be a fuzzy set defined on the universe of discourse Y, then the Cartesian product between the fuzzy sets $\tilde{A}$ and $\tilde{B}$ is given as $\tilde{A} \times \tilde{B}$ and resulting in a fuzzy relation $\tilde{R}$. It is given by

$$\tilde{R} = \tilde{A} \times \tilde{B} \subseteq X \times Y$$

with the membership function

$$\mu_{\tilde{R}}(x, y) = \mu_{\tilde{A} \times \tilde{B}}(x, y) = \min \left(\mu_{\tilde{A}}(x), \mu_{\tilde{B}}(y)\right)$$

For example, let us consider two fuzzy sets $\tilde{A}$ and $\tilde{B}$ such that $\tilde{A} = \{(x_1, 0.3), (x_2, 0.5), (x_3, 0.8)\}$ and $\tilde{B} = \{(y_1, 0.2), (y_2, 0.7), (y_3, 0.4)\}$ defined over the universe $X = \{x_1, x_2, x_3\}$ and $Y = \{y_1, y_2, y_3\}$ respectively. Then, the fuzzy relation $\tilde{R}$ ensuing out of the fuzzy Cartesian product $\tilde{A} \times \tilde{B}$ is given by

$$\tilde{R} = \tilde{A} \times \tilde{B} = \begin{array}{c} \\ x_1 \\ x_2 \\ x_3 \end{array} \begin{array}{c} \begin{array}{ccc} y_1 & y_2 & y_3 \end{array} \\ \begin{bmatrix} 0.2 & 0.3 & 0.3 \\ 0.2 & 0.5 & 0.4 \\ 0.2 & 0.7 & 0.4 \end{bmatrix} \end{array} \text{ since,}$$

$$\tilde{R}(x_1, y_1) = \min\ (\mu_{\tilde{A}}(x_1), \mu_{\tilde{B}}(y_1)) = \min(0.3, 0.2) = 0.2$$

$$\tilde{R}(x_1, y_2) = \min\ (\mu_{\tilde{A}}(x_1), \mu_{\tilde{B}}(y_2)) = \min(0.3, 0.7) = 0.3$$

$$\tilde{R}(x_1, y_3) = \min\ (\mu_{\tilde{A}}(x_1), \mu_{\tilde{B}}(y_3)) = \min(0.3, 0.4) = 0.3$$

$$\tilde{R}(x_2, y_1) = \min\ (\mu_{\tilde{A}}(x_2), \mu_{\tilde{B}}(y_1)) = \min(0.5, 0.2) = 0.2$$

$$\tilde{R}(x_2, y_2) = \min\ (\mu_{\tilde{A}}(x_2), \mu_{\tilde{B}}(y_2)) = \min(0.5, 0.7) = 0.5$$

$$\tilde{R}(x_2, y_3) = \min\ (\mu_{\tilde{A}}(x_2), \mu_{\tilde{B}}(y_3)) = \min(0.5, 0.4) = 0.4$$

$$\tilde{R}(x_3, y_1) = \min\ (\mu_{\tilde{A}}(x_3), \mu_{\tilde{B}}(y_1)) = \min(0.8, 0.2) = 0.2$$

$$\tilde{R}(x_3, y_2) = \min\ (\mu_{\tilde{A}}(x_3), \mu_{\tilde{B}}(y_2)) = \min(0.8, 0.7) = 0.7$$

$$\tilde{R}(x_3, y_3) = \min\ (\mu_{\tilde{A}}(x_3), \mu_{\tilde{B}}(y_3)) = \min(0.8, 0.4) = 0.4$$

14.9 OPERATIONS ON FUZZY RELATIONS

In this section we discuss the different fuzzy set operations such as union, intersection, complement and composition on fuzzy relations. These are highly useful in real world problems.

14.9.1 Union

If $\tilde{R}$ and $\tilde{S}$ be two fuzzy relations defined on (X × Y) with membership functions $\mu_{\tilde{R}}(x, y)$ and $\mu_{\tilde{S}}(x, y)$ respectively, then their union ($\tilde{R} \cup \tilde{S}$) is a new fuzzy relation with the membership function $\mu_{\tilde{R} \cup \tilde{S}}(x,y)$ defined as

$$\mu_{\tilde{R} \cup \tilde{S}}(x, y) = \max\ (\mu_{\tilde{R}}(x, y),\ \mu_{\tilde{S}}(x, y))$$

For example, let us consider two fuzzy relations $\tilde{R}$ and $\tilde{S}$ on $X \times Y$ where $X = \{x_1, x_2\}$ and $Y = \{y_1, y_2, y_3\}$ such that

$$\tilde{R} = \begin{array}{c} \\ x_1 \\ x_2 \end{array}\!\!\begin{array}{c} \begin{array}{ccc} y_1 & y_2 & y_3 \end{array} \\ \begin{bmatrix} 0.2 & 0.5 & 0.2 \\ 0.3 & 0.7 & 0.4 \end{bmatrix} \end{array} \text{ and } \tilde{S} = \begin{array}{c} \\ x_1 \\ x_2 \end{array}\!\!\begin{array}{c} \begin{array}{ccc} y_1 & y_2 & y_3 \end{array} \\ \begin{bmatrix} 0.5 & 0.3 & 0.8 \\ 0.2 & 0.5 & 0.7 \end{bmatrix} \end{array}$$

Therefore, the union $(\tilde{R} \cup \tilde{S})$ is defined as

$$\tilde{R} \cup \tilde{S} = \begin{array}{c} \\ x_1 \\ x_2 \end{array}\!\!\begin{array}{c} \begin{array}{ccc} y_1 & y_2 & y_3 \end{array} \\ \begin{bmatrix} 0.5 & 0.5 & 0.8 \\ 0.3 & 0.7 & 0.7 \end{bmatrix} \end{array} \text{ since,}$$

$$\mu_{\tilde{R} \cup \tilde{S}}(x_1, y_1) = \max\ (\mu_{\tilde{R}}(x_1, y_1),\ \mu_{\tilde{S}}(x_1, y_1)) = \max(0.2,\ 0.5) = 0.5$$
$$\mu_{\tilde{R} \cup \tilde{S}}(x_1, y_2) = \max\ (\mu_{\tilde{R}}(x_1, y_2),\ \mu_{\tilde{S}}(x_1, y_2)) = \max(0.5,\ 0.3) = 0.5$$
$$\mu_{\tilde{R} \cup \tilde{S}}(x_1, y_3) = \max\ (\mu_{\tilde{R}}(x_1, y_3),\ \mu_{\tilde{S}}(x_1, y_3)) = \max(0.2,\ 0.8) = 0.8$$
$$\mu_{\tilde{R} \cup \tilde{S}}(x_2, y_1) = \max\ (\mu_{\tilde{R}}(x_2, y_1),\ \mu_{\tilde{S}}(x_2, y_1)) = \max(0.3,\ 0.2) = 0.3$$
$$\mu_{\tilde{R} \cup \tilde{S}}(x_2, y_2) = \max\ (\mu_{\tilde{R}}(x_2, y_2),\ \mu_{\tilde{S}}(x_2, y_2)) = \max(0.7,\ 0.5) = 0.7$$
$$\mu_{\tilde{R} \cup \tilde{S}}(x_2, y_3) = \max\ (\mu_{\tilde{R}}(x_2, y_3),\ \mu_{\tilde{S}}(x_2, y_3)) = \max(0.4,\ 0.7) = 0.7$$

14.9.2 Intersection

If $\tilde{R}$ and $\tilde{S}$ be two fuzzy relations defined on X × Y with membership functions $\mu_{\tilde{R}}(x, y)$ and $\mu_{\tilde{S}}(x, y)$ respectively, then their intersection $(\tilde{R} \cap \tilde{S})$ is a new fuzzy relation with the membership function $\mu_{\tilde{R} \cap \tilde{S}}(x,y)$ defined as

$$\mu_{\tilde{R} \cap \tilde{S}}(x, y) = \min\ (\mu_{\tilde{R}}(x, y),\ \mu_{\tilde{S}}(x, y))$$

For example, let us consider two fuzzy relations $\tilde{R}$ and $\tilde{S}$ and on X × Y where $X = \{x_1, x_2\}$ and $Y = \{y_1, y_2, y_3\}$ such that

$$\tilde{R} = \begin{matrix} & y_1 & y_2 & y_3 \\ x_1 & 0.2 & 0.5 & 0.2 \\ x_2 & 0.3 & 0.7 & 0.4 \end{matrix} \quad \text{and} \quad \tilde{S} = \begin{matrix} & y_1 & y_2 & y_3 \\ x_1 & 0.5 & 0.3 & 0.8 \\ x_2 & 0.2 & 0.5 & 0.7 \end{matrix}$$

Therefore, the intersection $\tilde{R} \cap \tilde{S}$ is defined as

$$\tilde{R} \cap \tilde{S} = \begin{matrix} & y_1 & y_2 & y_3 \\ x_1 & 0.2 & 0.3 & 0.2 \\ x_2 & 0.2 & 0.5 & 0.4 \end{matrix} \text{ since,}$$

$$\mu_{\tilde{R}\cap\tilde{S}}(x_1, y_1) = \min(\mu_{\tilde{R}}(x_1, y_1), \mu_{\tilde{S}}(x_1, y_1)) = \min(0.2, 0.5) = 0.2$$
$$\mu_{\tilde{R}\cap\tilde{S}}(x_1, y_2) = \min(\mu_{\tilde{R}}(x_1, y_2), \mu_{\tilde{S}}(x_1, y_2)) = \min(0.5, 0.3) = 0.3$$
$$\mu_{\tilde{R}\cap\tilde{S}}(x_1, y_3) = \min(\mu_{\tilde{R}}(x_1, y_3), \mu_{\tilde{S}}(x_1, y_3)) = \min(0.2, 0.8) = 0.2$$
$$\mu_{\tilde{R}\cap\tilde{S}}(x_2, y_1) = \min(\mu_{\tilde{R}}(x_2, y_1), \mu_{\tilde{S}}(x_2, y_1)) = \min(0.3, 0.2) = 0.2$$
$$\mu_{\tilde{R}\cap\tilde{S}}(x_2, y_2) = \min(\mu_{\tilde{R}}(x_2, y_2), \mu_{\tilde{S}}(x_2, y_2)) = \min(0.7, 0.5) = 0.5$$
$$\mu_{\tilde{R}\cap\tilde{S}}(x_2, y_3) = \min(\mu_{\tilde{R}}(x_2, y_3), \mu_{\tilde{S}}(x_2, y_3)) = \min(0.4, 0.7) = 0.4$$

14.9.3 Complement

If $\tilde{R}$ be a fuzzy relations defined on X × Y with membership function $\mu_{\tilde{R}}(x, y)$, then the complement $\tilde{R}^c$ is a new fuzzy relation with the membership function $\mu_{\tilde{R}^c}(x, y)$ defined as

$$\mu_{\tilde{R}^c}(x, y) = 1 - \mu_{\tilde{R}}(x, y)$$

For example, let us consider a fuzzy relation $\tilde{R}$ on X × Y where $X = \{x_1, x_2\}$ and $Y = \{y_1, y_2, y_3\}$ such that

$$\tilde{R} = \begin{matrix} & y_1 & y_2 & y_3 \\ x_1 & 0.3 & 0.8 & 0.1 \\ x_2 & 0.9 & 0.4 & 0.7 \end{matrix}$$

Therefore, the complement $\tilde{R}^c$ is given as

$$\tilde{R}^c = \begin{matrix} & y_1 & y_2 & y_3 \\ x_1 & 0.7 & 0.2 & 0.9 \\ x_2 & 0.1 & 0.6 & 0.3 \end{matrix}$$

14.9.4 Composition of Relations

If $\tilde{R}$ and $\tilde{S}$ be two fuzzy relations defined on X × Y and Y × Z with membership functions $\mu_{\tilde{R}}(x, y)$ and $\mu_{\tilde{S}}(y, z)$ respectively, then their composition $(\tilde{R} \circ \tilde{S})$ is a new fuzzy relation on X × Z with the membership function defined as

$$\mu_{\tilde{R} \circ \tilde{S}}(x, z) = \max_{y \in Y} \left(\min \left(\mu_{\tilde{R}}(x, y), \mu_{\tilde{S}}(y, z)\right)\right)$$

For example, let us consider two fuzzy relations $\tilde{R}$ and $\tilde{S}$ on X × Y and Y × Z where X = $\{x_1, x_2\}$, Y = $\{y_1, y_2, y_3\}$ and Z = $\{z_1, z_2, z_3\}$ such that

$$\tilde{R} = \begin{array}{c} \\ x_1 \\ x_2 \end{array}\begin{array}{c} \begin{array}{ccc} y_1 & y_2 & y_3 \end{array} \\ \begin{bmatrix} 0.1 & 0.3 & 0.6 \\ 0.4 & 0.7 & 0.8 \end{bmatrix} \end{array} \quad \text{and} \quad \tilde{S} = \begin{array}{c} \\ y_1 \\ y_2 \\ y_3 \end{array}\begin{array}{c} \begin{array}{ccc} z_1 & z_2 & z_3 \end{array} \\ \begin{bmatrix} 0.2 & 0.1 & 0.6 \\ 0.4 & 0.5 & 0.3 \\ 0.2 & 0.6 & 0.8 \end{bmatrix} \end{array}$$

Therefore, the composition $(\tilde{R} \circ \tilde{S})$ is given as

$$\tilde{R} \circ \tilde{S} = \begin{array}{c} \\ x_1 \\ x_2 \end{array}\begin{array}{c} \begin{array}{ccc} z_1 & z_2 & z_3 \end{array} \\ \begin{bmatrix} 0.3 & 0.6 & 0.6 \\ 0.4 & 0.6 & 0.8 \end{bmatrix} \end{array} \quad \text{since,}$$

$$\mu_{\tilde{R} \circ \tilde{S}}(x_1, z_1) = \max_{y \in Y} \big(\min (\mu_{\tilde{R}}(x_1, y_1), \mu_{\tilde{S}}(y_1, z_1)), \min (\mu_{\tilde{R}}(x_1, y_2), \mu_{\tilde{S}}(y_2, z_1)), \min (\mu_{\tilde{R}}(x_1, y_3), \mu_{\tilde{S}}(y_3, z_1))\big)$$

i.e.,
$$\mu_{\tilde{R} \circ \tilde{S}}(x_1, z_1) = \max_{y \in Y} (\min (0.1, 0.2), \min (0.3, 0.4), \min (0.6, 0.2))$$
$$= \max(0.1, 0.3, 0.2) = 0.3$$

Similarly,

$$\mu_{\tilde{R} \circ \tilde{S}}(x_1, z_2) = \max_{y \in Y} (\min (0.1, 0.1), \min (0.3, 0.5), \min (0.6, 0.6))$$
$$= \max(0.1, 0.3, 0.6) = 0.6$$

$$\mu_{\tilde{R} \circ \tilde{S}}(x_1, z_3) = \max_{y \in Y} (\min (0.1, 0.6), \min (0.3, 0.3), \min (0.6, 0.8))$$
$$= \max(0.1, 0.3, 0.6) = 0.6$$

$$\mu_{\tilde{R} \circ \tilde{S}}(x_2, z_1) = \max_{y \in Y} (\min (0.4, 0.2), \min (0.7, 0.4), \min (0.8, 0.2))$$
$$= \max(0.2, 0.4, 0.2) = 0.4$$

$$\mu_{\tilde{R} \circ \tilde{S}}(x_2, z_2) = \max_{y \in Y} (\min (0.4, 0.1), \min (0.7, 0.5), \min (0.8, 0.6))$$
$$= \max(0.1, 0.5, 0.6) = 0.6$$

$$\mu_{\tilde{R} \circ \tilde{S}}(x_2, z_3) = \max_{y \in Y} (\min (0.4, 0.6), \min (0.7, 0.3), \min (0.8, 0.8))$$
$$= \max(0.4, 0.3, 0.8) = 0.8$$

14.10 FUZZY LOGIC

The study of the methods and principles of reasoning in all its possible forms is known as logic. Classical or binary logic deals with propositions that are either true or false *i.e.*, the truth values acquired by statements are bi-valued. This may numerically equivalent to (0, 1). But in our daily life we are coming across many statements whose truth values is not restricted to true or false only. Such type of statements is known as fuzzy propositions. Therefore, it is clear that in fuzzy logic, truth values are multivalued such as very true, partly true, absolutely true, absolutely false, and so on and are numerically equivalent to (0 – 1).

14.10.1 Fuzzy Proposition

A fuzzy proposition is a statement that acquires a fuzzy truth value. Therefore, if $\tilde{P}$ be a fuzzy statement, then $T(\tilde{P})$ denotes the truth value that are numerically equivalent to (0–1) attached to $\tilde{P}$. Therefore, it is clear that, fuzzy propositions are associated with fuzzy sets. The membership value associated with the fuzzy set $\tilde{A}$ for the fuzzy proposition $\tilde{P}$ is treated as the fuzzy truth value $T(\tilde{P})$. Therefore, $T(\tilde{P}) = \mu_{\tilde{A}}(x);\ 0 \le \mu_{\tilde{A}}(x) \le 1$.

For example, let us consider a statement

$\tilde{P} \equiv$ Abraham is a nice boy.

Thus, $T(\tilde{P}) = 0.7$ if $\tilde{P}$ is partially true,

$T(\tilde{P}) = 1$ if $\tilde{P}$ is absolutely true

and $T(\tilde{P}) = 0$ if $\tilde{P}$ is absolutely false.

14.10.2 Fuzzy Connectives

The fundamental connectives of crisp logic or bi-valued logic that we have discussed in Chapter 1 are negation, conjunction and disjunction. Fuzzy logic also supports the same connectives. Let $\tilde{P}$ and $\tilde{Q}$ are two fuzzy propositions with truth values $T(\tilde{P})$ and $T(\tilde{Q})$ respectively. The following table illustrates the symbol and definition of the different connectives.

Symbol	*Connective*	*Usage*	*Definition*
$\neg$	Negation	$\neg\tilde{P}$	$1 - T(\tilde{P})$
$\wedge$	Conjunction	$\tilde{P} \wedge \tilde{Q}$	$\min(T(P), T(\tilde{Q}))$
$\vee$	Disjunction	$\tilde{P} \vee \tilde{Q}$	$\max(T(P), T(\tilde{Q}))$
$\Rightarrow$	Im plication	$\tilde{P} \Rightarrow \tilde{Q}$	$\neg\tilde{P} \vee \tilde{Q}$

For example, consider two fuzzy propositions with their truth values.

$\tilde{P} \equiv$ Smith is honest, $T(\tilde{P}) = 0.7$

$\tilde{Q} \equiv$ Blake is honest, $T(\tilde{Q}) = 0.5$

Therefore, $\neg\tilde{P} \equiv$ Smith is not honest, $T(\neg\tilde{P}) = 1 - T(\tilde{P}) = 0.3$

$(\tilde{P} \wedge \tilde{Q}) \equiv$ Smith is honest and so Blake.

$$T(\tilde{P} \wedge \tilde{Q}) = \min\ (T(\tilde{P}),\ T(\tilde{Q})) = \min\ (0.7,\ 0.5) = 0.5$$

$(\tilde{P} \vee \tilde{Q}) \equiv$ either Smith or Blake is honest.

$$T(\tilde{P} \vee \tilde{Q}) = \max\ (T(\tilde{P}),\ T(\tilde{Q})) = \max\ (0.7,\ 0.5) = 0.7$$

$$\tilde{P} \Rightarrow \tilde{Q} \equiv \text{if Smith is honest then so is Blake.}$$

$$T(\tilde{P} \Rightarrow \tilde{Q}) = T(\neg\tilde{P} \vee \tilde{Q}) = \max\ (T(\neg\tilde{P}),\ T(\tilde{Q}))$$
$$= \max\ (0.3,\ 0.5) = 0.5$$

SOLVED EXAMPLES

Example 1 If $\tilde{A} = \{(x_1,\ 0.21),\ (x_2,\ 0.13),\ (x_3,\ 0.5),\ (x_4,\ 0.4)\}$ *and* $\tilde{B} = \{(x_1,\ 0.7),\ (x_2,\ 0.16),\ (x_3,\ 0.23), (x_4,\ 0.35)\}$ *are two fuzzy sets defined over the universe of discourse* $X = \{x_1,\ x_2,\ x_3,\ x_4\}$, *then compute the followings:*

(a) $(\tilde{A} \cup \tilde{B})$ (b) $(\tilde{A} \cap \tilde{B})$ (c) $(\tilde{A} - \tilde{B})$

Solution: Given that $\tilde{A} = \{(x_1,\ 0.21),\ (x_2,\ 0.13),\ (x_3,\ 0.5),\ (x_4,\ 0.4)\}$ and $\tilde{B} = \{(x_1,\ 0.7),\ (x_2,\ 0.16),\ (x_3,\ 0.23),\ (x_4,\ 0.35)\}$. Therefore,

(a) $(\tilde{A} \cup \tilde{B}) = \{(x_1,\ 0.7),\ (x_2,\ 0.16),\ (x_3,\ 0.5),\ (x_4,\ 0.4)\}$ since,

$$\mu_{\tilde{A}\cup\tilde{B}}(x_1) = \max\ (\mu_{\tilde{A}}(x_1),\ \mu_{\tilde{B}}(x_1)) = \max(0.21,\ 0.7) = 0.7$$
$$\mu_{\tilde{A}\cup\tilde{B}}(x_2) = \max\ (\mu_{\tilde{A}}(x_2),\ \mu_{\tilde{B}}(x_2)) = \max(0.13,\ 0.16) = 0.16$$
$$\mu_{\tilde{A}\cup\tilde{B}}(x_3) = \max\ (\mu_{\tilde{A}}(x_3),\ \mu_{\tilde{B}}(x_3)) = \max(0.5,\ 0.23) = 0.5$$
$$\mu_{\tilde{A}\cup\tilde{B}}(x_4) = \max\ (\mu_{\tilde{A}}(x_4),\ \mu_{\tilde{B}}(x_4)) = \max(0.4,\ 0.35) = 0.4$$

(b) $(\tilde{A} \cap \tilde{B}) = \{(x_1,\ 0.21),\ (x_2,\ 0.13),\ (x_3,\ 0.23),\ (x_4,\ 0.35)\}$ since,

$$\mu_{\tilde{A}\cap\tilde{B}}(x_1) = \min\ (\mu_{\tilde{A}}(x_1),\ \mu_{\tilde{B}}(x_1)) = \min\ (0.21,\ 0.7) = 0.21$$
$$\mu_{\tilde{A}\cap\tilde{B}}(x_2) = \min\ (\mu_{\tilde{A}}(x_2),\ \mu_{\tilde{B}}(x_2)) = \min\ (0.13,\ 0.16) = 0.13$$
$$\mu_{\tilde{A}\cap\tilde{B}}(x_3) = \min\ (\mu_{\tilde{A}}(x_3),\ \mu_{\tilde{B}}(x_3)) = \min\ (0.5,\ 0.23) = 0.23$$
$$\mu_{\tilde{A}\cap\tilde{B}}(x_4) = \min\ (\mu_{\tilde{A}}(x_4),\ \mu_{\tilde{B}}(x_4)) = \min\ (0.4,\ 0.35) = 0.35$$

(c) $(\tilde{A} - \tilde{B}) = (\tilde{A} \cap \tilde{B}^c)$

Now, $\tilde{B}^c = \{(x_1,\ 0.3),\ (x_2,\ 0.84),\ (x_3,\ 0.77),\ (x_4,\ 0.65)\}$. Therefore,

$(\tilde{A} - \tilde{B}) = (\tilde{A} \cap \tilde{B}^c) = \{(x_1,\ 0.21),\ (x_2,\ 0.13),\ (x_3,\ 0.5),\ (x_4,\ 0.4)\}$ since,

$$\mu_{\tilde{A}\cap\tilde{B}^c}(x_1) = \min\ (\mu_{\tilde{A}}(x_1),\ \mu_{\tilde{B}^c}(x_1)) = \min\ (0.21,\ 0.3) = 0.21$$
$$\mu_{\tilde{A}\cap\tilde{B}^c}(x_2) = \min\ (\mu_{\tilde{A}}(x_2),\ \mu_{\tilde{B}^c}(x_2)) = \min\ (0.13,\ 0.84) = 0.13$$
$$\mu_{\tilde{A}\cap\tilde{B}^c}(x_3) = \min\ (\mu_{\tilde{A}}(x_3),\ \mu_{\tilde{B}^c}(x_3)) = \min\ (0.5,\ 0.77) = 0.5$$
$$\mu_{\tilde{A}\cap\tilde{B}^c}(x_4) = \min\ (\mu_{\tilde{A}}(x_4),\ \mu_{\tilde{B}^c}(x_4)) = \min\ (0.4,\ 0.65) = 0.4$$

Example 2 *If* $\tilde{A} = \{(x_1,\ 0.1),\ (x_2,\ 0.3),\ (x_3,\ 0.4),\ (x_4,\ 0.9),\ (x_5,\ 1)\}$ *and* $\tilde{B} = \{(x_1,\ 0.2),\ (x_2,\ 0.1),\ (x_3,\ 0.5),\ (x_4,\ 0.8),\ (x_5,\ 0.9)\}$ *are two fuzzy sets defined over the universe of discourse* $X = \{x_1,\ x_2,\ x_3,\ x_4,\ x_5\}$, *then compute the followings:*

(a) $(\tilde{A} \oplus \tilde{B})$ (b) $(\tilde{A} \cap \tilde{B})^c$ (c) $(\tilde{A}^c \cup \tilde{B}^c)$

Solution: Given that $\tilde{A} = \{(x_1, 0.1), (x_2, 0.3), (x_3, 0.4), (x_4, 0.9), (x_5, 1)\}$ and

$$\tilde{B} = \{(x_1, 0.2), (x_2, 0.1), (x_3, 0.5), (x_4, 0.8), (x_5, 0.9)\}.$$

Therefore, $\tilde{A}^c = \{(x_1, 0.9), (x_2, 0.7), (x_3, 0.6), (x_4, 0.1), (x_5, 0)\}$ and

$$\tilde{B}^c = \{(x_1, 0.8), (x_2, 0.9), (x_3, 0.5), (x_4, 0.2), (x_5, 0.1)\}$$

(*a*) $(\tilde{A} \oplus \tilde{B}) = (\tilde{A}^c \cap \tilde{B}) \cup (\tilde{A} \cap \tilde{B}^c)$

Again, $(\tilde{A}^c \cap \tilde{B}) = \{(x_1, 0.2), (x_2, 0.1), (x_3, 0.5), (x_4, 0.1), (x_5, 0)\}$ and

$$(\tilde{A} \cap \tilde{B}^c) = \{(x_1, 0.1), (x_2, 0.3), (x_3, 0.4), (x_4, 0.2), (x_5, 0.1)\}$$

Therefore, $(\tilde{A} \oplus \tilde{B}) = \{(x_1, 0.2), (x_2, 0.3), (x_3, 0.5), (x_4, 0.2), (x_5, 0.1)\}$

(*b*) $(\tilde{A} \cap \tilde{B}) = \{(x_1, 0.1), (x_2, 0.1), (x_3, 0.4), (x_4, 0.8), (x_5, 0.9)\}$

Therefore, $(\tilde{A} \cap \tilde{B})^c = \{(x_1, 0.9), (x_2, 0.9), (x_3, 0.6), (x_4, 0.2), (x_5, 0.1)\}$

(*c*) $(\tilde{A}^c \cup \tilde{B}^c) = \{(x_1, 0.9), (x_2, 0.9), (x_3, 0.6), (x_4, 0.2), (x_5, 0.1)\}$

Example 3 *If* $\tilde{A} = \{(x_1, 0.17), (x_2, 0.43), (x_3, 0.24), (x_4, 0.19)\}$ *be a fuzzy set defined over the universe* $X = \{x_1, x_2, x_3, x_4\}$, *then show that* $(\tilde{A}^c)^c = A$.

Solution: Given that $\tilde{A} = \{(x_1, 0.17), (x_2, 0.43), (x_3, 0.24), (x_4, 0.19)\}$.

Therefore, $\tilde{A}^c = \{(x_1, 0.83), (x_2, 0.57), (x_3, 0.76), (x_4, 0.81)\}$ since,

$$\mu_{\tilde{A}^c}(x_1) = 1 - \mu_{\tilde{A}}(x_1) = 1 - 0.17 = 0.83$$
$$\mu_{\tilde{A}^c}(x_2) = 1 - \mu_{\tilde{A}}(x_2) = 1 - 0.43 = 0.57$$
$$\mu_{\tilde{A}^c}(x_3) = 1 - \mu_{\tilde{A}}(x_3) = 1 - 0.24 = 0.76$$
$$\mu_{\tilde{A}^c}(x_4) = 1 - \mu_{\tilde{A}}(x_4) = 1 - 0.19 = 0.81$$

Similarly, $(\tilde{A}^c)^c = \{(x_1, 0.17), (x_2, 0.43), (x_3, 0.24), (x_4, 0.19)\} = \tilde{A}$

Example 4 *If* $\tilde{A} = \{(x_1, 0.1), (x_2, 0.4), (x_3, 0.9), (x_4, 0.7), (x_5, 0.6)\}$ *and* $\tilde{B} = \{(x_1, 0.9), (x_2, 0.3), (x_3, 0.1), (x_4, 0.5), (x_5, 0.2)\}$ *be two fuzzy sets defined on the universe of discourse* $X = \{x_1, x_2, x_3, x_4, x_5\}$, *then show that commutative laws holds good.*

Solution: Given $\tilde{A} = \{(x_1, 0.1), (x_2, 0.4), (x_3, 0.9), (x_4, 0.7), (x_5, 0.6)\}$ and $\tilde{B} = \{(x_1, 0.9), (x_2, 0.3), (x_3, 0.1), (x_4, 0.5), (x_5, 0.2)\}$ be two fuzzy sets defined on the universe $X = \{x_1, x_2, x_3, x_4, x_5\}$.

Therefore, $(\tilde{A} \cup \tilde{B}) = \{(x_1, 0.9), (x_2, 0.4), (x_3, 0.9), (x_4, 0.7), (x_5, 0.6)\} = (\tilde{B} \cup \tilde{A})$ since,

$$\mu_{\tilde{A} \cup \tilde{B}}(x_i) = \max\ (\mu_{\tilde{A}}(x_i), \mu_{\tilde{B}}(x_i))$$
$$= \max\ (\mu_{\tilde{B}}(x_i), \mu_{\tilde{A}}(x_i)) = \mu_{\tilde{B} \cup \tilde{A}}(x_i);\ i = 1, 2, 3, 4$$

Similarly, it can be shown that $(\tilde{A} \cap \tilde{B}) = (\tilde{B} \cap \tilde{A})$.

Example 5 *If* $\tilde{A} = \{(x_1, 0.15), (x_2, 0.71), (x_3, 0.56), (x_4, 0.25)\}$ *and* $\tilde{B} = \{(x_1, 0.1), (x_2, 0.25), (x_3, 0.02), (x_4, 0.01)\}$ *are two fuzzy sets defined on the universe* $X = \{x_1, x_2, x_3, x_4\}$ *, then compute the product* $\tilde{A}.\tilde{B}$*.*

Solution: Given that $\tilde{A} = \{(x_1, 0.15), (x_2, 0.71), (x_3, 0.56), (x_4, 0.25)\}$ and

$\tilde{B} = \{(x_1, 0.1), (x_2, 0.25), (x_3, 0.02), (x_4, 0.01)\}$.

Therefore, $\tilde{A}.\tilde{B} = \{(x_1, 0.015), (x_2, 0.1775), (x_3, 0.0112), (x_4, 0.0025)\}$

since, $\mu_{\tilde{A}.\tilde{B}}(x_1) = \mu_{\tilde{A}}(x_1).\mu_{\tilde{B}}(x_1) = (0.15)(0.1) = 0.015$

$\mu_{\tilde{A}.\tilde{B}}(x_2) = \mu_{\tilde{A}}(x_2).\mu_{\tilde{B}}(x_2) = (0.71)(0.25) = 0.1775$

$\mu_{\tilde{A}.\tilde{B}}(x_3) = \mu_{\tilde{A}}(x_3).\mu_{\tilde{B}}(x_3) = (0.56)(0.02) = 0.0112$

$\mu_{\tilde{A}.\tilde{B}}(x_4) = \mu_{\tilde{A}}(x_4).\mu_{\tilde{B}}(x_4) = (0.25)(0.01) = 0.0025$

Example 6 *If* $\tilde{A} = \{(x_1, 0.5), (x_2, 0.2), (x_3, 0.1), (x_4, 0.25), (x_5, 0.6)\}$ *and* $\tilde{B} = \{(x_1, 0.2), (x_2, 0.35), (x_3, 0.25), (x_4, 0.7), (x_5, 0.4)\}$ *are two fuzzy sets defined on the universe* $X = \{x_1, x_2, x_3, x_4, x_5\}$*, then show that the de Morgan's laws holds good.*

Solution: Given $\tilde{A} = \{(x_1, 0.5), (x_2, 0.2), (x_3, 0.1), (x_4, 0.25), (x_5, 0.6)\}$ and

$\tilde{B} = \{(x_1, 0.2), (x_2, 0.35), (x_3, 0.25), (x_4, 0.7), (x_5, 0.4)\}$.

Therefore, $\tilde{A}^c = \{(x_1, 0.5), (x_2, 0.8), (x_3, 0.9), (x_4, 0.75), (x_5, 0.4)\}$ and

$\tilde{B}^c = \{(x_1, 0.8), (x_2, 0.65), (x_3, 0.75), (x_4, 0.3), (x_5, 0.6)\}$

Now, $(\tilde{A} \cup \tilde{B}) = \{(x_1, 0.5), (x_2, 0.35), (x_3, 0.25), (x_4, 0.7), (x_5, 0.6)\}$

since, $\mu_{\tilde{A} \cup \tilde{B}}(x_1) = \max\ (\mu_{\tilde{A}}(x_1), \mu_{\tilde{B}}(x_1)) = \max\ (0.5, 0.2) = 0.5$

$\mu_{\tilde{A} \cup \tilde{B}}(x_2) = \max\ (\mu_{\tilde{A}}(x_2), \mu_{\tilde{B}}(x_2)) = \max\ (0.2, 0.35) = 0.35$

$\mu_{\tilde{A} \cup \tilde{B}}(x_3) = \max\ (\mu_{\tilde{A}}(x_3), \mu_{\tilde{B}}(x_3)) = \max\ (0.1, 0.25) = 0.25$

$\mu_{\tilde{A} \cup \tilde{B}}(x_4) = \max\ (\mu_{\tilde{A}}(x_4), \mu_{\tilde{B}}(x_4)) = \max\ (0.25, 0.7) = 0.7$

$\mu_{\tilde{A} \cup \tilde{B}}(x_5) = \max\ (\mu_{\tilde{A}}(x_5), \mu_{\tilde{B}}(x_5)) = \max\ (0.6, 0.4) = 0.6$

Thus, $(\tilde{A} \cup \tilde{B})^c = \{(x_1, 0.5), (x_2, 0.65), (x_3, 0.75), (x_4, 0.3), (x_5, 0.4)\}$

Again, $(\tilde{A}^c \cap \tilde{B}^c) = \{(x_1, 0.5), (x_2, 0.65), (x_3, 0.75), (x_4, 0.3), (x_5, 0.4)\}$

Therefore, we get $(\tilde{A} \cup \tilde{B})^c = \tilde{A}^c \cap \tilde{B}^c$

Similarly, it can be proved that $(\tilde{A} \cap \tilde{B})^c = \tilde{A}^c \cup \tilde{B}^c$

Example 7 *Let* $\tilde{A} = \{(x_1, 0.1), (x_2, 0.6), (x_3, 0.4), (x_4, 0.8), (x_5, 0.9)\}$ *and* $\tilde{B} = \{(y_1, 0.5), (y_2, 0.2), (y_3, 0.1)\}$ *be two fuzzy sets defined on the universes* $X = \{x_1, x_2, x_3, x_4, x_5\}$ *and* $Y = \{y_1, y_2, y_3\}$ *respectively. Then compute the fuzzy relation* $\tilde{R}$ *on* $\tilde{A} \times \tilde{B}$*.*

Solution: Let $\tilde{A} = \{(x_1,\ 0.1),\ (x_2,\ 0.6),\ (x_3,\ 0.4),\ (x_4,\ 0.8),\ (x_5,\ 0.9)\}$ and $\tilde{B} = \{(y_1,\ 0.5),\ (y_2,\ 0.2),\ (y_3,\ 0.1)\}$. Therefore,

$$\tilde{R} = \tilde{A} \times \tilde{B} = \begin{array}{c} \\ x_1 \\ x_2 \\ x_3 \\ x_4 \\ x_5 \end{array}\begin{array}{c} \begin{array}{ccc} y_1 & y_2 & y_3 \end{array} \\ \begin{bmatrix} 0.1 & 0.1 & 0.1 \\ 0.5 & 0.2 & 0.1 \\ 0.4 & 0.2 & 0.1 \\ 0.5 & 0.2 & 0.1 \\ 0.5 & 0.2 & 0.1 \end{bmatrix} \end{array} \text{ since,}$$

$$\tilde{R}(x_1,\ y_1) = \min\ (\mu_{\tilde{A}}(x_1),\ \mu_{\tilde{B}}(y_1)) = \min(0.1,\ 0.5) = 0.1$$
$$\tilde{R}(x_1,\ y_2) = \min\ (\mu_{\tilde{A}}(x_1),\ \mu_{\tilde{B}}(y_2)) = \min(0.1,\ 0.2) = 0.1$$
$$\tilde{R}(x_1,\ y_3) = \min\ (\mu_{\tilde{A}}(x_1),\ \mu_{\tilde{B}}(y_3)) = \min(0.1,\ 0.1) = 0.1$$
$$\tilde{R}(x_2,\ y_1) = \min\ (\mu_{\tilde{A}}(x_2),\ \mu_{\tilde{B}}(y_1)) = \min(0.6,\ 0.5) = 0.5$$
$$\tilde{R}(x_2,\ y_2) = \min\ (\mu_{\tilde{A}}(x_2),\ \mu_{\tilde{B}}(y_2)) = \min(0.6,\ 0.2) = 0.2$$
$$\tilde{R}(x_2,\ y_3) = \min\ (\mu_{\tilde{A}}(x_2),\ \mu_{\tilde{B}}(y_3)) = \min(0.6,\ 0.1) = 0.1$$

Similarly, the other elements of $\tilde{R}$ that are given above can be computed.

Example 8 *If $\tilde{A} = \{(x_1,\ 0.7),\ (x_2,\ 0.2)\}$ and $\tilde{B} = \{(x_1,\ 0.3),\ (x_2,\ 0.1)\}$ are two fuzzy sets defined on $X = \{x_1, x_2\}$, then compute $2\tilde{A} \cup 3\tilde{B}$.*

Solution: Let $\tilde{A} = \{(x_1,\ 0.4),\ (x_2,\ 0.2)\}$ and $\tilde{B} = \{(x_1,\ 0.3),\ (x_2,\ 0.1)\}$

Therefore, $2\tilde{A} = \{(x_1,\ 0.8),\ (x_2,\ 0.4)\}$ and $3\tilde{B} = \{(x_1,\ 0.9),\ (x_2,\ 0.3)\}$ since,

$$\mu_{2\tilde{A}}(x_1) = 2\mu_{\tilde{A}}(x_1) = 2(0.4) = 0.8;\ \mu_{2\tilde{A}}(x_2) = 2\mu_{\tilde{A}}(x_2) = 2(0.2) = 0.4$$
$$\mu_{3\tilde{B}}(x_1) = 3\mu_{\tilde{B}}(x_1) = 3(0.3) = 0.9;\ \mu_{3\tilde{B}}(x_2) = 3\mu_{\tilde{B}}(x_2) = 3(0.1) = 0.3$$

Therefore, $2\tilde{A} \cup 3\tilde{B} = \{(x_1,\ 0.9),\ (x_2,\ 0.4)\}$ since,

$$\mu_{2\tilde{A} \cup 3\tilde{B}}(x_1) = \max\ (\mu_{2\tilde{A}}(x_1), \mu_{3\tilde{B}}(x_1)) = \max\ (0.8,\ 0.9) = 0.9$$
$$\mu_{2\tilde{A} \cup 3\tilde{B}}(x_2) = \max\ (\mu_{2\tilde{A}}(x_2), \mu_{3\tilde{B}}(x_2)) = \max\ (0.4,\ 0.3) = 0.4$$

Example 9 *Compute the composition $\tilde{R} \circ \tilde{S}$ if $\tilde{R}$ and $\tilde{S}$ are two fuzzy relations defined on $X \times Y$ and $Y \times Z$ respectively such that*

$$\tilde{R} = \begin{array}{c} \\ x_1 \\ x_2 \\ x_3 \end{array}\begin{array}{c} \begin{array}{ccc} y_1 & y_2 & y_3 \end{array} \\ \begin{bmatrix} 0.5 & 0.3 & 0.2 \\ 0.3 & 0.7 & 0.9 \\ 0.4 & 0.8 & 1 \end{bmatrix} \end{array} \text{ and } \tilde{S} = \begin{array}{c} \\ y_1 \\ y_2 \\ y_3 \end{array}\begin{array}{c} \begin{array}{ccc} z_1 & z_2 & z_3 \end{array} \\ \begin{bmatrix} 0.2 & 0.9 & 0.1 \\ 0.5 & 0.1 & 0.6 \\ 1 & 0.3 & 0.8 \end{bmatrix} \end{array}$$

Solution: Consider $\tilde{R}$ and $\tilde{S}$ are two fuzzy relations as defined above. Therefore, the composition $(\tilde{R} \circ \tilde{S})$ is given as

$$\tilde{R} \circ \tilde{S} = \begin{array}{c} \\ x_1 \\ x_2 \\ x_3 \end{array} \begin{array}{c} \begin{matrix} z_1 & z_2 & z_3 \end{matrix} \\ \begin{bmatrix} 0.3 & 0.5 & 0.3 \\ 0.9 & 0.3 & 0.8 \\ 1 & 0.4 & 0.8 \end{bmatrix} \end{array} \text{ since,}$$

$$\mu_{\tilde{R} \circ \tilde{S}}(x_1, z_1) = \max_{y \in Y} (\min (\mu_{\tilde{R}}(x_1, y_1), \mu_{\tilde{S}}(y_1, z_1)), \min (\mu_{\tilde{R}}(x_1, y_2), \mu_{\tilde{S}}(y_2, z_1)), \min (\mu_{\tilde{R}}(x_1, y_3), \mu_{\tilde{S}}(y_3, z_1)))$$

i.e.,

$$\mu_{\tilde{R} \circ \tilde{S}}(x_1, z_1) = \max_{y \in Y} (\min (0.5, 0.2), \min(0.3, 0.5), \min(0.2, 1))$$
$$= \max (0.2, 0.3, 0.2) = 0.3$$

Similarly,

$$\mu_{\tilde{R} \circ \tilde{S}}(x_1, z_2) = \max_{y \in Y} (\min (0.5, 0.9), \min(0.3, 0.1), \min(0.2, 0.3))$$
$$= \max(0.5, 0.1, 0.2) = 0.5$$

$$\mu_{\tilde{R} \circ \tilde{S}}(x_1, z_3) = \max_{y \in Y} (\min (0.5, 0.1), \min(0.3, 0.6), \min(0.2, 0.8))$$
$$= \max (0.1, 0.3, 0.2) = 0.3$$

$$\mu_{\tilde{R} \circ \tilde{S}}(x_2, z_1) = \max_{y \in Y} (\min (0.3, 0.2), \min(0.7, 0.5), \min(0.9, 1))$$
$$= \max (0.2, 0.5, 0.9) = 0.9$$

$$\mu_{\tilde{R} \circ \tilde{S}}(x_2, z_2) = \max_{y \in Y} (\min (0.3, 0.9), \min(0.7, 0.1), \min(0.9, 0.3))$$
$$= \max (0.3, 0.1, 0.3) = 0.3$$

$$\mu_{\tilde{R} \circ \tilde{S}}(x_2, z_3) = \max_{y \in Y} (\min (0.3, 0.1), \min(0.7, 0.6), \min(0.9, 0.8))$$
$$= \max (0.1, 0.6, 0.8) = 0.8$$

$$\mu_{\tilde{R} \circ \tilde{S}}(x_3, z_1) = \max_{y \in Y} (\min (0.4, 0.2), \min(0.8, 0.5), \min(1, 1))$$
$$= \max (0.2, 0.5, 1) = 1$$

$$\mu_{\tilde{R} \circ \tilde{S}}(x_3, z_2) = \max_{y \in Y} (\min (0.4, 0.9), \min(0.8, 0.1), \min(1, 0.3))$$
$$= \max (0.4, 0.1, 0.3) = 0.4$$

$$\mu_{\tilde{R} \circ \tilde{S}}(x_3, z_3) = \max_{y \in Y} (\min (0.4, 0.1), \min(0.8, 0.6), \min(1, 0.8))$$
$$= \max (0.1, 0.6, 0.8) = 0.8$$

Example 10 *Consider the fuzzy sets* $\tilde{A}$ *and* $\tilde{B}$ *and defined on the interval* $X = [0, 7]$ *of real numbers, by the membership functions*

$$\mu_{\tilde{A}}(x) = \frac{1}{2x+1}; \qquad \mu_{\tilde{B}}(x) = \frac{1}{3^x}$$

Compute the mathematical formulae for the following membership functions of each of the following fuzzy sets.

(a) $\tilde{A}^c$ (b) $\tilde{B}^c$ (c) $(\tilde{A} \cup \tilde{B})$ (d) $\tilde{A} \cap \tilde{B}$

Solution: Given that $\mu_{\tilde{A}}(x) = \dfrac{1}{2x+1}$; $\mu_{\tilde{B}}(x) = \dfrac{1}{3^x}$

(*a*) $\mu_{\tilde{A}^c}(x) = 1 - \mu_{\tilde{A}}(x) = 1 - \dfrac{1}{2x+1} = \dfrac{2x}{2x+1}$

(*b*) $\mu_{\tilde{B}^c}(x) = 1 - \mu_{\tilde{B}}(x) = 1 - \dfrac{1}{3^x} = \dfrac{3^x - 1}{3^x}$

(*c*) $\mu_{\tilde{A} \cup \tilde{B}}(x) = \max\ (\mu_{\tilde{A}}(x), \mu_{\tilde{B}}(x)) = \max\ \left(\dfrac{1}{2x+1}, \dfrac{1}{3^x}\right)$

(*d*) $\mu_{\tilde{A} \cap \tilde{B}}(x) = \min\ (\mu_{\tilde{A}}(x), \mu_{\tilde{B}}(x)) = \min\ \left(\dfrac{1}{2x+1}, \dfrac{1}{3^x}\right)$

EXERCISES

1. Consider the set of people in the following groups.

0–10	10–20	20–30	30–40
40–50	50–60	60–70	70 and above

Represent a membership function graph for the fuzzy sets "young", "middle-aged" and "old".

2. If $\tilde{A} = \{(x_1, 0.23), (x_2, 0.73), (x_3, 0.3), (x_4, 0.45), (x_5, 0.63)\}$ and $\tilde{B} = \{(x_1, 0.27), (x_2, 0.65), (x_3, 0.13), (x_4, 0.55), (x_5, 0.43)\}$ are two fuzzy sets defined over the universe $X = \{x_1, x_2, x_3, x_4, x_5\}$, then compute the followings:

   (*a*) $(\tilde{A} \cup \tilde{B})$ (*b*) $(\tilde{A} \cap \tilde{B})$ (*c*) $(\tilde{A} - \tilde{B})$

   (*d*) $(\tilde{A}^c \cup \tilde{B}^c)$ (*e*) $(\tilde{A} \oplus \tilde{B})$ (*f*) $(\tilde{A} \cap \tilde{B})^c$

3. Let $\tilde{A} = \{(x_1, 0.35), (x_2, 0.74), (x_3, 0.13), (x_4, 0.45)\}$ be a fuzzy set defined on the universe $X = \{x_1, x_2, x_3, x_4\}$. Show that $(\tilde{A}^c)^c = \tilde{A}$.

4. If $\tilde{A} = \{(x_1, 0.25), (x_2, 0.7), (x_3, 0.6), (x_4, 0.35), (x_5, 0.55)\}$ and $\tilde{B} = \{(x_1, 0.2), (x_2, 0.6), (x_3, 0.15), (x_4, 0.35), (x_5, 0.03)\}$ are two fuzzy sets defined over the universe $X = \{x_1, x_2, x_3, x_4, x_5\}$, then compute $\tilde{A} \cdot \tilde{B}$.

5. Let $\tilde{B} = \{(y_1, 0.01), (y_2, 0.03)\}$ and $\tilde{A} = \{(y_1, 0.05), (y_2, 0.02)\}$ be two fuzzy sets defined on the universe $Y = \{y_1, y_2\}$. Show that the commutative laws holds good.

6. If $\tilde{A} = \{(x_1, 0.53), (x_2, 0.42), (x_3, 0.13), (x_4, 0.25), (x_5, 0.63)\}$ and $\tilde{B} = \{(x_1, 0.25), (x_2, 0.55), (x_3, 0.15), (x_4, 0.75), (x_5, 0.43)\}$ are two fuzzy sets defined over the universe $X = \{x_1, x_2, x_3, x_4, x_5\}$, then show that the De Morgan's laws holds good.

7. Consider the fuzzy sets $\tilde{A}$ and $\tilde{B}$ defined on the interval $X = [0, 5]$ of real numbers, by the membership functions

$$\mu_{\tilde{A}}(x) = \frac{x}{2x+1}; \qquad \mu_{\tilde{B}}(x) = \frac{1}{2^x}$$

Compute the mathematical formulae and draw the graphs for the following membership functions of each of the following sets.

(*a*) $\tilde{A}^c$ (*b*) $\tilde{B}^c$ (*c*) $(\tilde{A} \cup \tilde{B})$

(*d*) $(\tilde{A} \cap \tilde{B})$ (*e*) $\tilde{A} - \tilde{B}$ (*f*) $(\tilde{A} \cup \tilde{B})^c$

8. Let $\tilde{A} = \{(x_1, 0.2), (x_2, 0.7), (x_3, 0.9)\}$ and $\tilde{B} = \{(y_1, 0.1), (y_2, 0.3)\}$ be two fuzzy sets defined on the universes $X = \{x_1, x_2, x_3\}$ and $Y = \{y_1, y_2\}$ respectively. Then compute the fuzzy relation $\tilde{R}$ on $\tilde{A} \times \tilde{B}$.

9. If $\tilde{R}$ and $\tilde{S}$ are two fuzzy relations defined on $X \times Y$ and $Y \times Z$ respectively, then compute the composition $\tilde{R} \circ \tilde{S}$. It is given that

$$\tilde{R} = \begin{array}{c} \\ x_1 \\ x_2 \\ x_3 \end{array}\!\!\begin{array}{c} \begin{array}{ccc} y_1 & y_2 & y_3 \end{array} \\ \begin{bmatrix} 0.1 & 0.5 & 0.1 \\ 0.6 & 0.3 & 0.7 \\ 0.2 & 0.4 & 0.8 \end{bmatrix} \end{array} \text{ and } \tilde{S} = \begin{array}{c} \\ y_1 \\ y_2 \\ y_3 \end{array}\!\!\begin{array}{c} \begin{array}{cccc} z_1 & z_2 & z_3 & z_4 \end{array} \\ \begin{bmatrix} 0.8 & 0.3 & 0.5 & 0.2 \\ 0.2 & 0.4 & 0.1 & 0.7 \\ 0.1 & 0.5 & 0.3 & 0.8 \end{bmatrix} \end{array}$$

10. Consider a set $P = \{P_1, P_2, P_3\}$ of three persons, set $D = \{D_1, D_2, D_3\}$ of various diseases affecting the persons and $\tilde{S} = \{S_1, S_2, S_3, S_4\}$ be the common symptoms of the diseases. Let $\tilde{R}_1$ be a relation on P × D and $\tilde{R}_2$ be a relation on D × S. Obtain the association of the persons with the different symptoms of the diseases using max-min composition. It is given that

$$\tilde{R} = \begin{array}{c} \\ P_1 \\ P_2 \\ P_3 \end{array}\!\!\begin{array}{c} \begin{array}{ccc} D_1 & D_2 & D_3 \end{array} \\ \begin{bmatrix} 0.8 & 0.5 & 0.9 \\ 0.6 & 0.3 & 0.2 \\ 0.7 & 0.8 & 0.3 \end{bmatrix} \end{array} \text{ and } \tilde{S} = \begin{array}{c} \\ D_1 \\ D_2 \\ D_3 \end{array}\!\!\begin{array}{c} \begin{array}{cccc} S_1 & S_2 & S_3 & S_4 \end{array} \\ \begin{bmatrix} 1 & 0.3 & 0.8 & 0.5 \\ 0.1 & 0 & 0.3 & 0.9 \\ 0 & 1 & 0.7 & 0.1 \end{bmatrix} \end{array}$$

References

CHAPTER 1

1. B. Kolman, R.C. Busby, S.C. Ross: *Discrete Mathematical Structures*, 5th Edition; Prentice-Hall, Inc.; (2004).
2. C.L. Chang, R.C. Lee: *Symbolic Logic and Mechanical Theorem Proving*; Academic Press, New York; (1973).
3. C.L. Liu: *Elements of Discrete Mathematics*, 2nd Edition; McGraw-Hill Publishing Company Limited; (2002).
4. J.H. Gallier: *Logic for Computer Science*; Harper and Row; New York; (1986).
5. J.K. Truss: *Discrete Mathematics for Computer Scientists*, 2nd Edition; Addison Wesley; (2000).
6. J.P. Tremblay, R. Manohar: *Discrete Mathematical Structures with Applications to Computer Science*; McGraw-Hill Publishing Company Limited; (2003).
7. James L. Hein: *Discrete Structures, Logic, and Computability*, 2nd Edition; Jones and Bartlett Publishers, Inc.; (2002).
8. Martin D. Davis, Ron Sigal, Elaine J. Weyuker: *Computability, Complexity, and Languages – Fundamentals of Theoretical Computer Science*, 2nd Edition; Academic Press, Inc.; (1994).
9. P.T. Johnstone: *Notes on Logic and Set Theory*; Cambridge University Press; (1987).
10. R.R. Stoll: *Set Theory and Logic; Dover*, New York; (1979).
11. S.F. Barker: *The Elements of Logic*, 5th Edition; McGraw-Hill Company, New York; (1989).
12. W.J. Edgar: *The Elements of Logic*; SRA, Chicago; (1989).

CHAPTER 2

1. B. Kolman, R.C. Busby, S.C. Ross: *Discrete Mathematical Structures*, 5th Edition; Prentice-Hall, Inc.; (2004).

2. E.G. Goodaire, Michael M. Parmenter: *Discrete Mathematics with Graph Theory*, 2nd Edition; Prentice-Hall, Inc.; (2002).
3. H. Felix: *Set Theory*; Chelsia Publishing Co. New York; (1978).
4. J.K. Truss: *Discrete Mathematics for Computer Scientists*, 2nd Edition; Addison Wesley; (2000).
5. J.P. Tremblay, R. Manohar: *Discrete Mathematical Structures with Applications to Computer Science*; McGraw-Hill Publishing Company Limited; (2003).
6. Joe L. Mott, Abraham Kandel, Theodore P. Baker: *Discrete Mathematics for Computer Scientists and Mathematicians*, 2nd Edition; Prentice-Hall, Inc.; (2006).
7. Kenneth H. Rosen: *Discrete Mathematics and its Applications*, 5th Edition; McGraw-Hill Company; (2003).
8. P.R. Halmos: *Naïve Set Theory*; Springer Verlag; New York; (1974).
9. P.T. Johnstone: *Notes on Logic and Set Theory*; Cambridge University Press; (1987).
10. R.R. Stoll: *Set Theory and Logic*; Dover, New York; (1979).
11. S. Lipschutz, Marc Lipson: *Discrete Mathematics*, 2nd Edition; McGraw-Hill Book Company; (2001).

CHAPTER 3

1. B. Kolman, R.C. Busby, S.C. Ross: *Discrete Mathematical Structures*, 5th Edition; Prentice-Hall, Inc.; (2004).
2. C.L. Liu: *Elements of Discrete Mathematics*, 2nd Edition; McGraw-Hill Publishing Company Limited; (2002).
3. E.G. Goodaire, Michael M. Parmenter: *Discrete Mathematics with Graph Theory*, 2nd Edition; Prentice-Hall, Inc.; (2002).
4. G. Birkhoff, T.C. Bartee: *Modern Applied Algebra*; McGraw-Hill Company, Inc., (1970).
5. J.K. Truss: *Discrete Mathematics for Computer Scientists*, 2nd Edition; Addison Wesley; (2000).
6. J.P. Tremblay, R. Manohar: *Discrete Mathematical Structures with Applications to Computer Science*; McGraw-Hill Publishing Company Limited; (2003).
7. Joe L. Mott, Abraham Kandel, Theodore P. Baker: *Discrete Mathematics for Computer Scientists and Mathematicians*, 2nd Edition; Prentice-Hall, Inc.; (2006).
8. Kenneth H. Rosen: *Discrete Mathematics and its Applications*, 5th Edition; McGraw-Hill Company; (2003).
9. Norman L. Biggs: D*iscrete Mathematics*, 2nd Edition; Oxford University Press; (2003).
10. Rajendra Akerkar, Rupali Akerkar: *Discrete Mathematics*; Pearson Education (Singapore) Pte Ltd.; (2004).
11. Ralph P. Grimaldi: *Discrete and Combinatorial Mathematics*; Pearson Education, Inc.; (2003).

12. Richard Johnsonbaugh: *Discrete Mathematics*, 5th Edition; Pearson Education, Inc.; (2001).

13. Robert J. McEliece, Robert B. Ash, Carol Ash: *Introduction to Discrete Mathematics*; McGraw-Hill Book Company; (1989).

14. S. Lipschutz, Marc Lipson: *Discrete Mathematics*, 2nd Edition; McGraw-Hill Book Company; (2001).

CHAPTER 4

1. B. Kolman, R. C. Busby, S. C. Ross: *Discrete Mathematical Structures*, 5th Edition; Prentice-Hall, Inc.; (2004).

2. C.L. Liu: *Elements of Discrete Mathematics*, 2nd Edition; McGraw-Hill Publishing Company Limited; (2002).

3. E.G. Goodaire, Michael M. Parmenter: *Discrete Mathematics with Graph Theory*, 2nd Edition; Prentice-Hall, Inc.; (2002).

4. G. Birkhoff, T.C. Bartee: *Modern Applied Algebra*; McGraw-Hill Company, Inc., (1970).

5. J.K. Truss: *Discrete Mathematics for Computer Scientists*, 2nd Edition; Addison Wesley; (2000).

6. J.P. Tremblay, R. Manohar: *Discrete Mathematical Structures with Applications to Computer Science*; McGraw-Hill Publishing Company Limited; (2003).

7. Joe L. Mott, Abraham Kandel, Theodore P. Baker: *Discrete Mathematics for Computer Scientists and Mathematicians*, 2nd Edition; Prentice-Hall, Inc.; (2006).

8. Kenneth H. Rosen: *Discrete Mathematics and its Applications*, 5th Edition; McGraw-Hill Company; (2003).

9. Norman L. Biggs: *Discrete Mathematics*, 2nd Edition; Oxford University Press; (2003).

10. Rajendra Akerkar, Rupali Akerkar: *Discrete Mathematics*; Pearson Education (Singapore) Pvt. Ltd.; (2004).

11. Ralph P. Grimaldi: *Discrete and Combinatorial Mathematics*; Pearson Education, Inc.; (2003).

12. Richard Johnsonbaugh: *Discrete Mathematics*, 5th Edition; Pearson Education, Inc.; (2001).

13. Robert J. McEliece, Robert B. Ash, Carol Ash: *Introduction to Discrete Mathematics*; McGraw-Hill Book Company; (1989).

14. S. Lipschutz, Marc Lipson: *Discrete Mathematics*, 2nd Edition; McGraw-Hill Book Company; (2001).

CHAPTER 5

1. Alfred V. Aho, John E. Hopcroft, Jeffery D. Ullman: *The Design and Analysis of Computer Algorithms*; Pearson Education, Inc.; (2001).

2. Anany Levitin: *Introduction to The Design and Analysis of Algorithms*; Pearson Education, Inc.; (2003).
3. B. Kolman, R.C. Busby, S.C. Ross: *Discrete Mathematical Structures*, 5th Edition; Prentice-Hall, Inc.; (2004).
4. C.L. Liu: *Elements of Discrete Mathematics*, 2nd Edition; McGraw-Hill Publishing Company Limited; (2002).
5. E.G. Goodaire, Michael M. Parmenter: *Discrete Mathematics with Graph Theory*, 2nd Edition; Prentice-Hall, Inc.; (2002).
6. E. Horowitz, S. Sahni, S. Rajasekaran: *Fundamentals of Computer Algorithms*; Galgotia Publications Pvt. Ltd., New Delhi; (2000).
7. Michael O. Albertson, Joan P. Hutchinson: *Discrete Mathematics with Algorithms*; John Wiley and Sons; (2001).
8. P. Cull, E. F. Ecklund, Jr.: *Towers of Hanoi and Analysis of Algorithms*, Amer. Math. Monthly, 92 (1985), pp. 407–420.
9. R. Johnsonbaugh, M. Schaefer: *Algorithms*; Pearson Education, Inc.; (2004).
10. Richard Johnsonbaugh: *Discrete Mathematics*, 5th Edition; Pearson Education, Inc.; (2001).
11. S. Dasgupta, C. Papadimitriou, U. Vazirani: *Algorithms*; McGraw-Hill Company, Inc.; (2007).
12. Sara Baase, Allen Van Gelder: *Computer Algorithms–Introduction to Design and Analysis*, 3rd Edition; Pearson Education, Inc.; (2003).
13. Thomas H. Cormen, Charles E. Leiserson, Ronald L. Rivest: *Introduction to Algorithms*; MIT Press, Cambridge; (2000).

CHAPTER 6

1. Alan Tucker: *Applied Combinatorics*, 4th Edition; John Wiley and Sons, Inc.; (2003).
2. B. Kolman, R.C. Busby, S.C. Ross: *Discrete Mathematical Structures*, 5th Edition; Prentice-Hall, Inc.; (2004).
3. C.L. Liu: *Introduction to Combinatorial Mathematics*, McGraw-Hill Company, Inc., New York, (1968).
4. E.G. Goodaire, Michael M. Parmenter: *Discrete Mathematics with Graph Theory*, 2nd Edition; Prentice-Hall, Inc.; (2002).
5. F. Harary: *Graph Theory*; Addison-Wesley Publishing Company, Inc.; (2001).
6. F.P. Ramsey: *On a Problem of Formal Logic*, Proceedings London Math. Soc. Series 2, Vol. 30, pp. 264 – 286; (1930).
7. J.H. Van Lint, R.M. Wilson: *A Course in Combinatorics*; Cambridge University Press, New York; (1992).
8. J.K. Truss: *Discrete Mathematics for Computer Scientists*, 2nd Edition; Addison Wesley; (2000).

9. J. Riordan: *An Introduction to Combinatorial Analysis*; Wiley, New York; (1958).
10. N.Y. Vilenkin: *Combinatorics*; Academic Press; New York; (1971).
11. R.A. Brualdi: *Introductory Combinatorics*, 4th Edition; ELSEVIER, New York; (1997).
12. R. Graham, B. Rothschild, J. H. Spencer: *Ramsey Theory*; John Wiley and Sons, New York; (1980).
13. R.P. Stanley: *Enumerative Combinatorics*, Vol. 1; Cambridge University Press, England; (1999).
14. R.P. Stanley: *Enumerative Combinatorics*, Vol. 2; Cambridge University Press, England; (1999).
15. Ralph P. Grimaldi: *Discrete and Combinatorial Mathematics*; Pearson Education, Inc.; (2003).
15. S. Even: *Algorithmic Combinatorics*; MacMillan, New York; (1973).
16. V.K. Balakrishnan: *Combinatorics Including Concepts of Graph Theory*; McGraw-Hill Company, Inc.; (2005).

CHAPTER 7

1. B. Kolman, R.C. Busby, S.C. Ross: *Discrete Mathematical Structures*, 5th Edition; Prentice-Hall, Inc.; (2004).
2. G. Birkhoff, T. C. Bartee: *Modern Applied Algebra*; McGraw-Hill Company, Inc., (1970).
3. J.K. Truss: *Discrete Mathematics for Computer Scientists*, 2nd Edition; Addison Wesley; (2000).
4. J.P. Tremblay, R. Manohar: *Discrete Mathematical Structures with Applications to Computer Science*; McGraw-Hill Publishing Company Limited; (2003).
5. L.L. Dornhoff, F.E. Hohn: *Applied Modern Algebra*; MacMillan; (1978).
6. M.L. Lial, E.J. Hornsby, D.I. Schneider, C.D. Miller: *College Algebra*, 7th Edition; Addison Wesley, New York; (1997).
7. M. Sullivan: College Algebra, 5th Edition; Prentice-Hall, *Upper Saddle River*, N.J., (1999).
8. R. Lidl, G. Pilz: *Applied Abstract Algebra*; Springer-Verlag, New York, Inc.; (1984).
9. Rajendra Akerkar, Rupali Akerkar: *Discrete Mathematics*; Pearson Education (Singapore) Pte Ltd.; (2004).

CHAPTER 8

1. B. Kolman, R.C. Busby, S.C. Ross: *Discrete Mathematical Structures*, 5th Edition; Prentice-Hall, Inc.; (2004).
2. C.L. Liu: *Elements of Discrete Mathematics*, 2nd Edition; McGraw-Hill Publishing Company Limited; (2002).

3. G. Birkhoff, T. C. Bartee: *Modern Applied Algebra*; McGraw-Hill Company, Inc., (1970).
4. J. K. Truss: *Discrete Mathematics for Computer Scientists*, 2nd Edition; Addison Wesley; (2000).
5. J.P. Tremblay, R. Manohar: *Discrete Mathematical Structures with Applications to Computer Science*; McGraw-Hill Publishing Company Limited; (2003).
6. Rajendra Akerkar, Rupali Akerkar: *Discrete Mathematics*; Pearson Education (Singapore) Pte Ltd.; (2004).

CHAPTER 9

1. G. Birkhoff, T. C. Bartee: *Modern Applied Algebra*; McGraw-Hill Company, Inc., (1970).
2. J.K. Truss: *Discrete Mathematics for Computer Scientists*, 2nd Edition; Addison Wesley; (2000).
3. L.L. Dornhoff, F. E. Hohn: *Applied Modern Algebra*; MacMillan; (1978).
4. M.L. Lial, E. J. Hornsby, D.I. Schneider, C.D. Miller: *College Algebra*, 7th Edition; Addison Wesley, New York; (1997).
5. M. Sullivan: *College Algebra*, 5th Edition; Prentice-Hall, Upper Saddle River, N.J., (1999).
6. R. Lidl, G. Pilz: *Applied Abstract Algebra*; Springer- Verlag, New York, Inc.; (1984).
7. Rajendra Akerkar, Rupali Akerkar: *Discrete Mathematics*; Pearson Education (Singapore) Pte Ltd.; (2004).

CHAPTER 10

1. B. Kolman, R. C. Busby, S. C. Ross: *Discrete Mathematical Structures*, 5th Edition; Prentice-Hall, Inc.; (2004).
2. C.L. Liu: *Elements of Discrete Mathematics*, 2nd Edition; McGraw-Hill Publishing Company Limited; (2002).
3. G. Birkhoff, T. C. Bartee: *Modern Applied Algebra*; McGraw-Hill Company, Inc., (1970).
4. J.H. Gallier: *Logic for Computer Science*; Harper and Row; New York; (1986).
5. J.K. Truss: *Discrete Mathematics for Computer Scientists*, 2nd Edition; Addison Wesley; (2000).
6. J.P. Tremblay, R. Manohar: *Discrete Mathematical Structures with Applications to Computer Science*; McGraw-Hill Publishing Company Limited; (2003).
7. Richard Johnsonbaugh: *Discrete Mathematics*, 5th Edition; Pearson Education, Inc.; (2001).

CHAPTER 11

1. B. Kolman, R. C. Busby, S. C. Ross: *Discrete Mathematical Structures*, 5th Edition; Prentice-Hall, Inc.; (2004).

2. G. Birkhoff, T. C. Bartee: *Modern Applied Algebra*; McGraw-Hill Company, Inc., (1970).

3. J.K. Truss: *Discrete Mathematics for Computer Scientists*, 2nd Edition; Addison Wesley; (2000).

4. J.P. Tremblay, R. Manohar: *Discrete Mathematical Structures with Applications to Computer Science*; McGraw-Hill Publishing Company Limited; (2003).

5. S. Lipschutz, Marc Lipson: *Discrete Mathematics*, 2nd Edition; McGraw-Hill Book Company; (2001).

CHAPTER 12

1. A. Gibbons: *Algorithmic Graph Theory*; Cambridge University Press, Cambridge; (1985).

2. C. Berge: *Graphs and Hypergraphs*; North-Holland Publishing Company; Amsterdam; (1979).

3. C.L. Liu: *Elements of Discrete Mathematics*, 2nd Edition; McGraw-Hill Publishing Company Limited; (2002).

4. D.B. West: *Introduction to Graph Theory*, 2nd Edition; Pearson Education, Inc., (2001).

5. E.G. Goodaire, Michael M. Parmenter: *Discrete Mathematics with Graph Theory*, 2nd Edition; Prentice-Hall, Inc.; (2002).

6. F. Harary: *Graph Theory*; Addison-Wesley Publishing Company, Inc.; (2001).

7. Gary Chartrand, Ortrud R. Oellermann: *Applied and Algorithmic Graph Theory*; McGraw-Hill, Inc.; (1993).

8. John Clark, Derek Allan Holton: *A First Look at Graph Theory*; World Scientific Publishing Co. Pte. Ltd.; (1991).

9. L.R. Foulds: *Graph Theory Applications*; Springer-Verlag, New York, Inc.; (1992).

10. M.N.S. Swamy, K. Thulasiraman: *Graphs, Networks, and Algorithms*; John Wiley and Sons, New York; (2000).

11. N. Deo: *Graph Theory with Applications to Engineering and Computer Science*; Prentice-Hall, Inc.; (2005).

12. Robin J. Wilson: *Introduction to Graph Theory*, 4th Edition; Pearson Education, Inc.; (2005).

13. S. Even: *Graph Algorithms*, Computer Science Press, Rockville, Md., (1979).

14. V.K. Balakrishnan: *Combinatorics Including Concepts of Graph Theory*; McGraw-Hill Company, Inc.; (2005).

15. W.R. Scott: *Graph Theory*; Dover Publications, Inc.; (1987).

CHAPTER 13

1. A. Gibbons: *Algorithmic Graph Theory*; Cambridge University Press, Cambridge; (1985).

2. C. Berge: *Graphs and Hypergraphs*; North-Holland Publishing Company; Amsterdam; (1979).

3. C.L. Liu: *Elements of Discrete Mathematics*, 2nd Edition; McGraw-Hill Publishing Company Limited; (2002).
4. D.B. West: *Introduction to Graph Theory*, 2nd Edition; Pearson Education, Inc., (2001).
5. E.G. Goodaire, Michael M. Parmenter: *Discrete Mathematics with Graph Theory*, 2nd Edition; Prentice-Hall, Inc.; (2002).
6. F. Harary: *Graph Theory*; Addison-Wesley Publishing Company, Inc.; (2001).
7. Gary Chartrand, Ortrud R. Oellermann: *Applied and Algorithmic Graph Theory*; McGraw-Hill, Inc.; (1993).
8. John Clark, Derek Allan Holton: *A First Look at Graph Theory*; World Scientific Publishing Co. Pte. Ltd.; (1991).
9. L.R. Foulds: *Graph Theory Applications*; Springer-Verlag, New York, Inc.; (1992).
10. M.N.S. Swamy, K. Thulasiraman: *Graphs, Networks, and Algorithms*; John Wiley and Sons, New York; (2000).
11. N. Deo: *Graph Theory with Applications to Engineering and Computer Science*; Prentice-Hall, Inc.; (2005).
12. Robin J. Wilson: *Introduction to Graph Theory*, 4th Edition; Pearson Education, Inc.; (2005).
13. S. Even: *Graph Algorithms*, Computer Science Press, Rockville, Md., (1979).
14. V.K. Balakrishnan: *Combinatorics Including Concepts of Graph Theory*; McGraw-Hill Company, Inc.; (2005).
15. W.R. Scott: *Graph Theory*; Dover Publications, Inc.; (1987).

CHAPTER 14

1. George J. Klir, Bo Yuan: *Fuzzy Sets and Fuzzy Logic – Theory and Applications*; Prentice-Hall, Inc.; (2001).
2. George J. Klir, Tina A. Folger: *Fuzzy Sets Uncertainty and Information*; Prentice-Hall, Inc.; (2000).
3. L. A. Zadeh: *Fuzzy Sets*, Inf. Control, Vol. 8, pp. 338–353; (1965).

Index

A

Abelian group 174, 181, 183
Absolute function 78
Absorption law 260, 261
Absorption laws 26, 235
Acyclic graph 275, 309
Addition principle 141
Adjacency matrix 276, 277, 286, 323
Adjacent vertices 270, 319
Algebraic structure 173
And gate 228
Antecedent 3
Anti-reflexive 49, 51
    symmetric 50, 51
Arrow diagram 45
Associative law 26, 232, 234, 260, 261, 357
Asymmetric 49

B

Backtracking 101
    algorithm 319
Bertrand Russell 19
Bi-conditional 4
Bijective 72
Binary 173
    logic 364
    operation 178, 200, 283
    relation 41
    search 106, 110
    tree 311
Binomial 147
    coefficients 148
    theorem 147
Bipartite graph 280, 281
Bit 231
    matrix 276
Bi-valued logic 1, 365
Blaise Pascal 148
Block code 200, 201
Boolean algebra 227, 234
    expression 232
    function 238
    lattice 263
    matrix 276
Bound laws 26, 235
Bounded lattice 261
Breadth first search 317, 318
Binary operation 200, 283
Breadth first tree 318, 319
Bridge 312, 313

C

C.E. Shannon 227
Cantor 350
Cardinality 21

Cartesian product 41
Catalan numbers 151
Ceiling function 78
Central point 314
Centre 314
Characteristic equation 103
 function 79, 351
Child 310
Circuits 227
Circular permutation 143
Classical set 357
Closest insertion algorithm 327, 328
Closure of a graph 326
 relations 55
Code 200
 word 199, 200, 202
Coding problem 199
Co-domain 71, 74
Collision 80
Combination 141, 144
Combinatorial circuit 231, 233
Commutative group 174
 laws 26, 201, 232, 234 260, 357
Comparable 23
Complement 25, 284, 363
 laws 26, 232, 234
Complemented lattice 262
Complete binary tree 312
 bipartite graph 281
 graph 153, 274, 284
Composition 285
 of relations 46, 363
Concentration 354
Conclusion 3
Conditional 3
Conjunction 2
Conjunctive normal form 239, 240
Connected graph 279
Consequent 3
Constant function 77
Contra positive 5
Contradiction 7
Converse 5
Coordinate diagram 43
Coset leader 202
Crisp logic 365
 relation 360
 set theory 350
 set 357
Cube representation 238
Cut edge 312
Cut vertex 323
Cycle 274, 275
Cyclic group 181, 182

D

de Moivre 93
de Morgan 229
 laws 8, 27, 233, 235
Degree of vertex 273
Denes 269
Depth first search 317, 319
 first tree 320
 of tree 311
Diameter 314
Difference 23, 24, 357
 equation 97, 103
Digraph 272, 275
Dijkastra's algorithm 321
Dilation 354
Directed graph 272
Discrete numeric functions 93
Disjoint sets 28
Disjunction 3
Disjunctive counting 141
 normal form 239
 sum 357
Distance 200, 313
Distributed lattice 260
Distributive lattice 259
 law 26, 232, 234, 260, 261, 358

Divide-and-conquer 109
Domain 42, 71, 74
Double implication 4
Duality law 8
Dyck word 153

E

Eccentricity 313
Edge deleted subgraph 282
Elementary operation 283
Empty relation 45
    set 21, 22
Equal sets 22
Equality 352
Equivalence classes 56
    relation 52, 56, 180
Equivalent sets 21
Error correction 200
Eugene Charles Catalan 151
Euler 96
    graph 324, 325
    tour 324, 325, 326
    trail 325
Even-parity 230
Exclusive or 6
Exponential generating function 95
External node 310

F

F. P. Ramsey 153
Factorial 142
Fibonacci 99
    sequence 99, 103
Finite group 174
    set 20
Fleury's algorithm 324
Floor function 78
Floyd-Warshall algorithm 321, 323
Forest 313
Forward chaining 101, 102
Frege 19
Full binary tree 312
Function 69, 93
Fundamental product 30
Fusion algorithm 287
    of graphs 285
Fuzzy logic 364, 365
    proposition 365
    propositions 364
    relation 360, 362
Fuzzy set 350, 351, 357
    operations 355

G

Gate 227
Generating function 93, 96
Generator 181
George Boole 227
    cantor 19
Golden ratio 103
Graph 269, 270
    isomorphism 280
    of function 72
    relation 43
Greatest integer function 78
Group 200
    code 200, 173, 174

H

Hamiltonian cycle 326, 327
    graph 326
    path 326
Hash function 79
Hasse diagram 256
Height 311
Hierholzer's algorithm 324, 325
Homomorphism 183
Hypothesis 3

I

Idempotent laws 26, 235, 358
Identity element 175
    function 77
Identity laws 26, 232, 234, 358
    relation 45
Implication 4
Incidence matrix 277, 278
Indegree 273, 275
Infinite group 174
    order 175
Infinite set 20
Injective 71
Insertion sort 106, 107
Intersection 23, 24, 284, 355, 362
    of two I-V fuzzy sets 360
Interval 359
Into function 72
Inverse 5
    function 76
    relation 42
Involution laws 26, 235, 358
Irreflexive 49
Isolated vertex 273
Isomorphic lattices 263
i-v fuzzy sets 358, 359

K

Konig 269
Kruskal's algorithm 316

L

L. A. Zadeh 358
Lattices 255
Leaf 310
Left cancellation law 176
Left coset 183
Leibnitz 69
Leonardo di pisa 99
Level 311
Logical 227
Loop 271, 273, 316
Lotfi A. Zadeh 349
Lower bound 359

M

Many-many 45
    relations 45, 70
Many-one 45
Maximal spanning tree 347
Maximum likelihood criterion 201
Maxterm 240
Membership function 351
Method of contra positive 8
    contradiction 8
    extension 20
    induction 8
    intension 20
Minimum distance decoding 201, 202
Minimum spanning tree 315
Minterm 240
Monoid 174
Morse code 199
Multi graph 271, 276
Multiple roots 103
Multiplication principle 141

N

N space representation 238
Nand 6
    gate 229
Negation 2
Node 310
Nor 6
    gate 229
Not gate 227
Null set 21

O

Odd parity 230
Omar Khayyam 147
One-many 45, 70
One-one 45
    function 71
    onto function 72
Operation 173
Operations on graphs 283
Or gate 228
Order 270
    of a group 174
    of an element 174
    pair 29
Ordered pair 173
Outdegree 273, 275
Oystein ore 269

P

Pair set 21
Parallel edges 270, 271
Parent 310
Particular solution 104
Partitions of integers 95
Party problem 153
Pascal triangle 148
Pascal's identity 146, 147, 148
Path matrix 278
Path 273, 283
Pendant vertex 275
Permutation with repetition 143
Permutation 141, 142
Poset 255, 256
Power of a fuzzy set 354
Power set 23
Prim's algorithm 315
Product of graphs 284
    sets 29
Proper subset 22
Proposition 1
Pseudo graph 272

R

Radius 314
Ramsey number 153
Range 42, 71
Reachability matrix 278
Recurrence relation 97, 151
Reflexive 48, 50, 52
    closure 55
Regular graph 274
Relation matrix 46, 47, 50
Remainder function 79
Restricted combination 146
    permutations 144
Right cancellation law 176
Right coset 183
Root 309, 310
Roster method 20

S

Satisfiable 7
Selection sort 106
Semi group 174
Sequential counting 141
Set of sets 21
Set operations 23
Set-builder method 20
Sets 19
Sibling 311
Siblings 319
Signum function 79
Similar sets 21
Simple graph 271, 276
Single source shortest path 321
Singleton set 21
Sir Isaac Newton 147
Size 270
    of a tree 313

Spanning sub graph 315
Spanning tree 315
Specification method 20
Stanislaw Radziszowski 154
Statement 1
Strassen's matrix multiplication 106, 108
Strongly connected 279
Subgraph 281
Subgroup 178, 179, 200
Sublattice 263
Subset 22
Sum of products 253
Summation method 101, 102
Super graph 326, 327
Superset 22
Surjective 72
Symmetric 49, 50, 52
    closure 55
    difference 25

T

Tabular method 20
    representation 238
Tautology 7
Terminal node 310
Terminating node 321
Total solution 105
Tour 324
Towers of Hanoi 97
Trail 283, 324, 325
Transitive 49, 50, 52
    closure 55
    laws 358
Travelling salesman problem 327
Tree 309
Trial function 104
Triangle's inequality 201
Trivial walk 283
Two-optimal algorithm 327, 328
Two-valued logic 1

U

Unary operation 283
Union 23, 283, 355
Universal lower bound 261
    relation 46
    set 21
    upper bound 261
Unsatisfiable 7
Upper bound 359

V

Valued fuzzy set 359
Vander Monde's identity 149, 150
Venn diagram 24, 25, 28
Vertex deleted subgraph 282
Void relation 45
    set 21

W

Walk 282, 328
Weakly connected 279
Weighted graph 272, 276, 315, 316, 321, 323, 327
Well defined 19
Word 199

X

Xnor gate 230
Xor gate 229
Xor 6